Industrializing Iron Construction at the École Centrale Paris, 1829–1865

Cataloging-in-Publication Data is on file at the Library of Congress.

978-1-62671-193-8 (hardcover)
978-1-62671-194-5 (paperback)
978-1-62671-195-2 (epub)
978-1-62671-196-9 (epdf)

Industrializing Iron Construction at the École Centrale Paris, 1829–1865

Tom F. Peters

Purdue University Press
West Lafayette, Indiana

Table of contents

Motivation and acknowledgments

Culturally I cross back and forth between North America and Europe and constantly negotiate between the thinking patterns of architecture and structural engineering. So it is natural that I am concerned with developing methods for teaching the same material to people from differing intellectual and professional cultures. In 1982 I began to develop an entirely visual approach to statics and the strength of materials for architects at the University of California at Berkeley. My method used no mathematics, but only models and the analogy of the human body. I wanted to communicate my fascination with the technological approach to design to people who struggle to think in mathematical terms. My goal was to enable them to integrate a verifiable and fact-based, yet intuitive understanding of the 'flow' of forces and the principles of equilibrium qualified by material characteristics into design and construction. Unfortunately, when I left Berkeley the following year, I did not pursue this idea further, but the need to explore it never left me.

This urge had originated with my own teacher, the engineer Hans Hauri (1924–2018) who taught structural theory to architects at the ETH in Zurich. He had developed the idea to teach with models but without entirely abandoning mathematics. His approach was conceptually similar, as I discovered years later, to the method that Théodore Olivier had developed to teach descriptive geometry at the École Centrale 150 years earlier. Hauri's lively teaching was a revelation to me: It contrasted starkly with the bone-dry method of his eminent predecessor, who had insisted on mathematical calculations to the fifth decimal, and under whose teaching I had suffered for my entire first semester in 1962. Hauri combined visual and analytical thinking and thereby empowered his students to incorporate sensory perception into their understanding—at least it enabled me to do so, and his hybrid mode of reasoning resonated with my own developing way of thinking. This is the reason I was so intrigued by the pedagogical development at the École Centrale and its successful implementation in practical iron construction.

The questions that I ask in the following introductory chapter have bothered me for decades, and this book is the result of finally trying to answer them. Friends and acquaintances generously helped me by identifying incongruencies and pinpointing apparent contradictions that I had to resolve, and their questions and comments forced me to clarify my concepts. They also provided answers to my own questions and material for me to consider. My family not only tolerated my passion but encouraged and actually participated in it.

Acknowledgments

Two persons need to be especially emphasized: my wife Clotilde is by far my most intimate critic, and I treasure her perceptive questioning of the tentative ideas I presented at the dinner table, her analytical comments on their formulation, her overview from outside the field, and her insights and support. My friend Santiago Huerta's insightful opinions and support are of inestimable value to my development too. I bounce my ideas off him and his responses illuminate and clarify facts and issues in so many ways that I cannot detail all of them here. He discovered and introduced me to alternate sources of material and to information that I did not know, Rankine and Dorn among others. He proposed new viewpoints that I had not considered and enriched my argument with a depth that I had never suspected was there. Moreover, he kindly undertook the huge task of editor and publisher to help make what had been a passionate study into an attractive book. It is fair to say that this book would not have been published without his encouragement and hard work. His many suggestions, based on years of experience in publishing, were invariably excellent. I adopted them all. Justin Race and Katherine Purple of Purdue University Press have guided and accomplished this new edition with the same sensitivity and attention to detail. I could not have wished for three better editors.

I am further indebted to the long friendship and intellectual support of the eminent engineer Ernst Basler (1929–2024) who also generously provided financial backing for the publication of this book. His pioneering development of ecological engineering has for decades served as a stellar example of the possibilities inherent in engineering education. His border-crossing understanding of what constitutes professional knowledge inspired the development of the questions that I have attempted to elucidate here. Ernst Basler proved that the trans-disciplinary mindset established by the École Centrale can still serve as a template for modern education.

Many more people helped with the details of this study. Karl-Eugen Kurrer understands so much more structural theory than I ever will, and I am grateful to him especially for explaining Bélanger's significance to engineering theory and the relationship between his and Poncelet's contributions to the development of the field of structural statics. Discussions with Antonio Becchi provided valuable help in explaining the relationship between science and engineering practice in various cultures, especially with regard to material science. Waltraud Suwelack explained the distinction between didactics and pedagogy that helped me clarify the truly revolutionary aspects of the École Centrale's form of *vulgarisation.* Robert Carvais provided critique of the organization of the material, particularly regarding the bibliography, and brief discussions with Maria Teresa Como and Miguel Rotaeche at the XI Congreso Nacional de Historia de la Construcción in Soria, Spain, in 2019 propelled the importance of the distinction between 'looking at' or perceiving and 'seeing' to the fore. My daughter Clorinde provided thoughtful input on 'seeing' from the standpoint of her expertise in the visual aspect of cultural studies, and she also kindly undertook the task of reviewing and editing my writing. The distinction between perceiving and seeing had been an unconscious theme of this book, and the comments by these three scholars led me to rebalance my concept of the development and finally to comprehend fully and illuminate more clearly the import of what the École Centrale's teachers really did. Like my wife's, my daughter Alexandra's astute observations and critique stimulated me to clarify my thinking. Her Internet capabilities far outstrip my own, and she used them to establish the connection to the Roussel family, a contact that led to a more detailed examination of the role of the pre-industrial builders and that contributed important points to the complexity of this narrative.

Olivier Dibos of the Lavallée family was most forthcoming and helpful with information on his ancestor. Benoît and Christian Roussel provided access to all their documentation on the Entreprise Roussel. Both families graciously granted their permission to publish information and documents.

A special thanks is also due to Matteo Porrino, who after a three-month search in the collections of the École Centrale, now housed in the Archives nationales, finally identified and copied the all-important photograph illustrated as Fig. 79.

Many archivists need to be mentioned by name as they were most forthcoming and helpful with both information and advice above and beyond the call of duty. They are: Olivier Azzola, Responsable du Centre de ressources historiques-Mus'X, École Polytechnique; Ludovic Bouvier, Archiviste Mines Paris Tech; Amandine Gabriac and Steeve Gallizia, Service des archives at the Institut national de la propriété industrielle (INPI); Jean-François Hürst, Président du conseil de fondation du Centre de documentation et de recherche Pestalozzi; Anne Lacourt, Responsable du pôle patrimoine et archives, École nationale des ponts et chaussées (ENPC); Sarah Mottalini, Collections Manager/Registrar of the Permanent Collection, Union College; and Danièle Tosato-Rigo, Université de Lausanne and Présidente du conseil scientifique du Centre de documentation et de recherche Pestalozzi.

My sincere thanks to everyone mentioned above for their valued input, support, and advice, and to two anonymous reviewers of the manuscript, one in 2021 who made several suggestions that enhanced the integration of the material and called my writing 'punchy', which made me laugh, and the other in 2022 who suggested secondary, cultural sources that provided additional background to illuminate the argument.

Introduction

This is the story of the education that made Gustave Eiffel, William LeBaron Jenney, Henri Labrouste and so many other lesser-known engineer-designers who pioneered the use of iron so prominent and successful. Engineers belong to the professions that make things, and as Hélène Vérin pointed out (1984), one of the basic functions that distinguishes engineers from technologists is that they not only solve problems, but they innovate. But how do we educate engineers to become innovative? What pedagogy will guarantee success? The École Centrale des Arts et Manufactures in Paris provided an astonishingly successful answer in the first half of the nineteenth century. The school educated a multitude of civil and structural engineers, architects, contractors, and industrialists who became prominent players in the development of our modern form of iron construction. They led the development from the experimental into the industrial phase, defined its system characteristics, and self-assuredly proclaimed its aesthetic value.

Engineers are technological thinkers. Technology is an intellectual thought-form like science or art (Peters 2000). It is fundamentally a value-free human characteristic, although admittedly the ethical value of technology to society is a concern.[1] The aspect to be presented here is the development of the thought-form in the service of innovation: the creation of industrialized iron construction by engineers. How did that happen?

There are three levels of intellectual preoccupation with technology: technicians who can carry out solutions within an existing framework, technologists who can extrapolate from a framework to create new solutions to questions, and engineers who can invent new questions. The incentive for developing a successful pedagogy for the latter function was the profession's profound dissatisfaction with the abstraction of the scientific approach to knowledge as it had developed during the seventeenth and eighteenth centuries. Although it was clear to engineers that the advantages of a scientific understanding of structural and material behavior

1. Jacques Ellul, and many others before him, have perceived technology to be an independent, threatening social construct. This view is based on dialectical argument and is related to anarchist philosophy and Rousseau's understanding of a moral corruption of humanity through society. The issue is valid but does not feature here as it did not enter engineering and industrial discourse in the early nineteenth century.

were crucial to advancing the field of construction,[2] it was equally evident that an interface between understanding and application was lacking. Physics and mathematical analysis were too abstract and limiting to serve as tools for practical construction because the scientific method as used in the natural sciences and mathematical modeling excludes many minor or ostensibly extraneous factors that practitioners know are crucial to achieving a successful result.

Technology is not merely an application of science despite the commonly held belief that it is 'applied science'; rather, the relationship between the two fields is complex.[3] Scientists want to *understand* the world while technologists and engineers use science among other methods as a tool to *make* functioning objects. Or as William J. M. Rankine put it in 1856:

> In theoretical science, the question is—*What are we to think?...* But in practical science, the question is—*What are we to do?* (Rankine 1856, p. 20)[4]

Scientific thought in this sense is therefore quite distinct from technological thought. Not only are the goals of the two different, but so are their languages. For a scientist, like for the general public, the term 'detail' frequently—although not always—denotes a minor and therefore hierarchically less important part of a problem, while for a technologist the same term never carries the connotation of lesser importance: It only signifies the size of a part or the scale of a problem, the solution to which can be as critical to the functioning of an entity as the larger parts. For an engineer therefore, a detail is therefore small, but never minor.[5]

In reaction to what engineers experienced as the missing translation between these two very different but interdependent professional modes of thought, the École Centrale's founders determined to find and develop a fruitful middle ground between

---

2. An example was Nicolas de Chastillon's system at the École du génie de Mézières from 1748 that instituted a year of theory followed by a year of application—see: d'Orgeix 2011, esp. p. 60.

3. Equating applied science with technology is misleading. The term is meant to distinguish basic from applied research or to refer to the application of scientific methods to solve technological problems, and it appeared in the earliest school brochures of the École Centrale, although it may be even older. It has led to unfortunate misunderstandings as to what technology really is.

4. Thanks to Santiago Huerta for drawing my attention to Rankine's thinking.

5. This became very clear in the 1986 Challenger disaster, when the O-ring seals on the space shuttle Challenger proved to be insufficiently designed to withstand low temperatures at take-off. The engineers were concerned, but the administrators in charge of the mission authorized take-off because they erroneously misunderstood that the problem was merely a detail.

practical empiricism and theoretical science. As the founder Alphonse Lavallée (1791–1873) wrote in 1831:

> Today one can no longer doubt the usefulness of the natural sciences, but here, as in all the others, one must distinguish. For some, the search for truth is a passion to which they consecrate their lives and to which they sacrifice all their other interests. These few men study science for its own sake; they love to plumb its depths and conquer or at least fathom its complexities; and for them instruction can never be detailed or abstract enough. Most young men, on the other hand, regard science as a complement to their education, without which one is socially sidelined. They therefore seek quick, clear, and concise instruction. They require certainties, they fear ambiguities because they have neither the time nor the will to study and dispel doubts.
>
> Between these two distinct groups there is a third, far larger one, for whom instruction in science needs to be modified: it is this class of men destined by their profession to apply scientific concepts in their daily work.
>
> These men must be considered as having only their future position in view. Science is for them a means that must be sure and definite. In their view, everything in scientific instruction must be subordinated to their goal, which is that general concepts must find their intellectual place without speculative theories or complicated details. The phenomena attached to their use must on the contrary be studied deeply and in detail with a view to their practical application. (Lavallée/Dumas, 1831 p. 5)[6]

---

6. On ne peut plus mettre en doute aujourd'hui l'utilité de l'étude des sciences positives, mais dans celles-ci, comme dans toutes les autres, il est des distinctions à établir. Pour certains hommes, la recherches de la vérité est une passion, vive à laquelle toute leur existence est consacrée, à laquelle tous leurs intérêts sont sacrifiés. Ces hommes peu nombreux étudient les sciences pour elles-mêmes; ils aiment à sonder leurs profondeurs, à vaincre ou du moins à mesurer leurs difficultés; et pour eux l'enseignement ne saurait jamais être ni assez détaillé, ni assez abstrait. La masse des jeunes gens, au contraire, cherche dans l'étude des sciences un complément d'éducation, sans lequel on se trouve aujourd'hui déplacé dans la société. Elle veut en conséquence un enseignement rapide, clair et concis. Elle cherche des opinions toutes faites; elle craint des difficultés, parce qu'elle n'a ni le temps, ne la volonté de les étudier et de les vaincre.

Entre ces deux classes si distinctes s'en trouve une troisième, bien plus considérable que les deux premières, pour qui l'enseignement des sciences doit encore être modifié: c'est la classe des hommes destinés par leur état à faire une application journalière des conceptions scientifiques.

Ceux-ci doivent être considérés comme n'ayant en vue que l'intérêt de leur position future. Les sciences sont pour eux des instruments qui doivent devenir d'une application facile et certaine. A leur égard, tout doit être subordonné dans l'enseignement scientifique au but qu'ils ont en vue; les idées

The impetus for this need articulated by Lavallée was provided by the young railway industry that was the most important—and virtually the only—field in the 1830s and '40s that consistently used iron in an industrially standardized fashion rather than experimentally. From the outset, railways used iron for small-scale machinery and locomotive construction, for mid-scale track and infrastructure construction, and for large-scale station hall coverings. All the other early large-scale iron structures that architectural and construction historians have analyzed were experimental one-off structures, up to and even a little beyond the Crystal Palace of 1850.

My investigation of the middle ground between science and technology that the École Centrale's founders developed exposes what its salient elements were, how they and the school's teachers accomplished this translation, and how their educational principles differed quite remarkably from all previous forms. This study describes the educational and cultural background of the teachers who created the novel didactical model and developed the practical pedagogy that allowed science and technology to interact, how they went about harmonizing the two, and the synergetic effect that these teachers' interaction had with one another and with the professional world.

As far as the substance of the École Centrale's program is concerned, the courses that proved to be the most relevant to structural engineering and the development of iron construction were those taught by a group of teachers whose contributions have been mostly neglected in historical accounts: Jean-Baptiste Bélanger's[7] mechanics, Auguste Perdonnet's railway construction, Théodore Olivier's descriptive geometry, and Charles Walter de Saint-Ange, Auguste Faure, and Achille Ferry's metallurgy, machine theory, and machine construction. Louis-Charles Mary's architecture and construction course has received more attention from historians. Moreover, it was an integrated drawing course established by César Nicolas Louis Leblanc and then adopted by the school and taught by Nicolas-Auguste Thumeloup that helped fuse all these teachers' contributions into a coherent pedagogical system.

---

générales doivent se placer comme les faits dans leur esprit, sans entourage de théories incertaines ou de détails compliqués, les phénomènes qui se rattachent aux applications doivent au contraire être étudiés avec profondeur…, avec détail sous le point de vue pratique.

7. Often also spelled Belanger omitting the diacritic acute (for example in his former student and assistant Charles Bresse's *Cours de mécanique appliquée*).

Their efforts, as well as those of their assistants and colleagues, were reinforced by a solid interaction between the academic lecture-cum-exercise teaching format, of *amphithéâtre* and *atelier* as it was then formulated, and hands-on workshop apprenticeships in laboratories and industrial settings. The apprenticeships were novel in an academic setting, and they followed a model established by Jean-Daniel Colladon's steam engine course in the early years. The interaction between school and practice was guided by Alphonse Lavallée, the founder and spiritus rector of the school, and supported by his co-founders Théodore Olivier, Jean-Baptiste Dumas and Eugène Péclet. The last two founders' contributions were less directly focused on actual construction. Dumas and Péclet will therefore remain peripheral to this examination, although their basic courses in chemistry and general physics certainly contributed to the students' understanding of iron metallurgy and structural behavior.

Fig. 1. The *amphithéâtre* or lecture hall was where lectures were held and where the instructor and his assistants performed demonstrations. The students copied the diagrams from the blackboard and wrote down what the professor said into their notebooks. (Source: *La Science Illustrée*, 1888, vol. 1, p. 85)

Fig. 2. The *atelier* or *salle d'études* was the studio space up under the roof of the original school building where the students designed group or individual projects, completed their notebooks, and met with the assistants. (Source: Rousselet, *Nos grandes écoles,* 1888, p. 284)

The interaction between academics and practice occurred in both the mandatory apprenticeships and in post-graduate employment, especially in Eugène Flachat's firm—the first private French civil engineering company. Through this border-crossing interaction, the graduates developed a powerful method to translate *understanding* into *making*. The physical legacy of the school's successful pedagogy is still with us in our transportation and technical infrastructure.

The conceptual issues

History is always concerned with human thought—whether in an abstract form like philosophy or religion, in its multiple more tangible forms; in activities like writing, drawing, construction, war, business, and music; or in downright physical objects like iron structures. There is always a direct link among cerebral and sensory perception, the physical world, the actors who create and manipulate the environment, and how these actors think. This account relates how thinking develops in engineering, in particular the steps that lead to new thinking and thereby to novel concepts that condition professional behavior, and that ultimately lead to objects.

'Perceiving' versus 'seeing'

There is a basic distinction between looking at the world, which is our sensory perception of it filtered through our expectations, and 'seeing' it. Generally, when we 'perceive' something, we are bound by what we hope for, expect, or fear. This is why eye-witness accounts of events can vary so drastically from one witness to another. Seeing, on the other hand, is a creative mental activity that is triggered by critically analyzing what we perceive and the observations we draw from our perceptions. In other words, observation demands the questioning of perception and can lead to seeing—that is, to the discovery of the unexpected that lies beyond our cultural and personal filters. In contrast to perceiving, the ability to see can transcend artificial boundaries between categories of thought or professional fields. Seeing is thus freer of cultural restrictions than perceiving, and we can characterize the ability to see as a form of mental 'x-ray vision'. In religious terms seers are oracles; in science, art, technology, or other activities like business they are creators. The degree of freedom from cultural limitation that a seer is able to attain distinguishes the quality and the originality of that individual's ability to see. This mental capacity and the novel thinking and creative action it generates are the ultimate goals of education as opposed to training.

The creation of the physical environment that we know today, especially its transportation and technical infrastructure as well as our way of thinking technologically, originated to a large degree in the didactic model designed for the early École Centrale that enabled its students to see. Unfortunately, modern engineering education has long since abandoned much of the novel pedagogy developed in nineteenth-century Paris. The trend to narrow our focus on professional expertise as an accumulation of facts is training, not education. Coupled with a lack of historical awareness and the lack of a wider-ranging didactical vision, this trend has led to a noticeable slowing of innovation. Advances have certainly been made incrementally, and there have even been some genuine breakthroughs in engineering after the first decades of the twentieth century, but hardly to the revolutionary extent that the mid-nineteenth century experienced.

'Seeing' generates invention

This decline in inventiveness and the ability to see is not destiny. The proof was delivered both by the creativity of the École Centrale's graduates, which was demonstrably greater than that of their contemporaries who graduated from the other French engineering schools, the École Polytechnique, the École des Ponts et Chaussées, and the École des Mines. In another sphere, in the early twentieth century, the Bauhaus in Germany manifested a similar level of creativity in comparison to other industrial design schools, for which Walter Gropius had developed a comparable border-crossing and integrative didactic model. The industrial design prowess that the Bauhaus graduates exhibited throughout the twentieth century rivalled the novelty of the iron structures and infrastructure designed and built by the École Centrale's graduates in the nineteenth. Separated by almost a century, the graduates of these two schools learned to see in very similar ways and thereby determined the technology and impacted the physical appearance of our built and industrially designed world in a major way.

The background for these successful experiments lay in French engineering education in the eighteenth and nineteenth centuries, which is justly recognized as a determining element in the creation of modern civil engineering. The mature phase in this educational development coalesced largely at the École Centrale des Arts et Manufactures between 1830 and 1865, and it found expression in the intensely active fields of railway, canal, and iron construction. The ability to see established a common cultural baseline for the graduates of the École Centrale.

This view of the development is not meant to downplay the contributions of other cultures to modern technology, particularly the empirical development that had begun in the Ore Mountains ('Erzgebirge') of Bohemia and Saxony and in the Netherlands in the seventeenth century, and that then spread to Great Britain and Ireland in the late eighteenth, where it proliferated rapidly. Neither is it meant to downplay the role of energy in the development of industry. No industrial development used energy beyond natural animal, water, or wind power until the late eighteenth century. The first artificial energy creation using steam, its storage, and its transportation over distance was an English development that formed the basis for what we now call the Industrial Revolution.[8]

The great contribution that France made to the development of engineering was that it successfully incorporated the analytical aspect that science contributed to technology in the early nineteenth century with industry. Much of this was due to the didactic concept that the founders of the École Centrale created. The French development quickly permeated British and then North and South American engineering after 1830, just as the Anglo-Saxon empirical development had previously inspired progress in France, Sweden, Russia, Bavaria, the Netherlands, Belgium, Austria-Hungary, Prussia, Spain, and Italy. International and even intercontinental contacts in those times were certainly slower, but much closer than we generally realize.

The 'how' versus the 'what' of teaching

Several studies have looked at the curricula of the École Centrale's courses, at *what* the teachers taught their students; however, there are no in-depth examinations of *how* the instructors taught and how they interacted, although

---

8. Despite the fact that Hero of Alexandria's 'aeolipile' first used steam to rotate a sphere in the first century AD, that Jerónimo de Ayanz y Beaumont registered the first patent to use steam to expel water from a mine in 1606, and that Denis Papin invented a pressure cooker in 1679, it was Thomas Savery whose patent of 1698, coupled with Papin's ideas on cylinders and pistons from 1690 (*Acta Eruditorum*) turned it into a steam-powered 'atmospheric' machine by 1705 (Papin published a version in 1707). Thomas Newcomen improved it in 1711. James Watt improved it further with a condenser, and he and Matthew Boulton turned the result into an industry by 1776. See also Christian 1822–25, vol. 2, 1823, pp. 385–393 for more details.

Ulrich Pfammatter (1997) does discuss method.[9] Pfammatter recognized the essential elements of what was new in the École Centrale's pedagogy, especially in the teaching of Louis-Charles Mary, and this study amplifies what he discovered by describing how the teachers came to develop their ideas, how they interacted, and how their teaching was brought to bear on iron construction.

All three forms of knowledge—declared, procedural, and tacit—are involved in teaching. The first involves the *what*, the second is the *how* and the third is the experiential mode that constitutes the background on which both the what and the how are based. As any educator knows, the content of a course is only part of the process of learning. It is at least as important to understand the didactical and cultural background on which teaching is based and to recognize how the material is communicated, how the various subjects are linked, how the teachers interact with one another and with their socio-political environment, and how the students incorporate what they learn into their mode of thought that can then radiate into practice. This study therefore looks at the didactical system that determined the École Centrale's pedagogy. It looks at the methods that the teachers who concerned themselves with structural issues developed, how their courses meshed, and the cultural, political, and educational background on which their pedagogy developed.

In contrast to the École Polytechnique that—in its developed form, especially after 1816—relied almost exclusively on innovators who stressed science—that is, on teachers who developed new theoretical material and thereby advanced their respective fields—the École Centrale was geared toward industrial application· It therefore favored teachers who espoused an integrative view of the role of knowledge in society. Most of these teachers were border-crossers, or rather enablers who were able to transcend the distinction between innovation and application.

Many French engineers—from Bernard Forest de Bélidor in the eighteenth century to C.L.M. Henri Navier in the early nineteenth—had voiced their dissatisfaction with the existing scientific theories. These were so strict in their choice of limiting criteria that they neglected boundary conditions such as cost and feasibility, which are all-important to the practitioner. At the same time these theories were so generally

9. Weiss, on the other hand, concentrated on what the school was trying to develop as a *science industrielle* and how the curriculum conditioned the social standing of the engineers. Here I examine how the curriculum influenced the development of engineering inventiveness.

formulated that they lacked any practical applicability whatsoever.[10] Practitioners mistrusted theory with good reason. As Charles Eck wrote as late as 1841:

> Mathematical science has not yet attained that point of perfection that permits the formulation of the firm guidelines that *practice* requires. All *formulas* must needs be modified by what experience teaches us. (Eck 1841, vol. 2, unnumbered second page)[11]

A missing mutual understanding between the two ways of thinking and viewing the world was by no means exclusively a French problem. It was universal. In his 1835 treatise on the long and exasperating fight to tame a branch of the Danube in the Donaukanal at Vienna, the military engineer Ferdinand von Mitis described his frustration with the ignorance of practitioners and the lack of coordination between theory and the practice of hydraulic engineering:

> The failure to recognize the laws of Nature that often leads to the failure of plans that we thought were infallible does not confirm our failure to evaluate them correctly but rather highlights the diversity of opinion in those who should be capable of evaluating them correctly. (Mitis 1835, p. 42)[12]
>
> The empiricist is unconcerned with the root cause of a success and therefore thinks it to be the effect of an event that immediately preceded it.
>
> The abstract theoretician, on the other hand, presents existing conditions as axioms and deduces a number of conclusions from them that he often also considers to be as infallible as they might be if the natural conditions corresponded to the evaluated situation. (Mitis 1835, p. 43)[13]

---

10. Weiss, pp. 91–93, also Comberousse 1879.1, p. 29 ... *les arts industriels ne sont pas de simples annexes des sciences théoriques,* (the industrial arts are not simply addenda to theoretical science).

11. Les sciences mathématiques ne sont pas encore arrivées à ce point de perfection, qu'elles puissent nous donner à coup sûr les instructions que demande la *pratique;* il faut que toutes leurs *formules* soient modifiées par les enseignements que fournit l'expérience.

12. Dass man die Naturgesetze noch nicht kennt, welche das oftmalige Misslingen der für unfehlbar berechneten Plane [*sic*] veranlassten, beweiset nicht sowohl dieses Misslingen selbst, als vielmehr die Verschiedenartigkeit der Ansichten derer, welche am meisten in der Lage sein dürften, in dieser Sache richtig Urtheile fällen zu können.

13. Da der Empyriker sich wenig um die Grundursache eines Erfolges bekümmert, so hält er das für die Ursache einer Wirkung, was dieser letzteren zunächst unmittelbar vorangegangen ist.

Der abstrakte Theoretiker stellt Voraussetzungen als Axiome hin, stützet hieraus eine Reihe von Schlüssen, die er öfter für eben so unfehlbar hält, wie sie auch sein könnten, würden die Voraussetzungen der Natur der beurtheilten Sache entsprechen.

It was this incongruence in the relationship between theory and practice, in addition to the fact that the design implications of material characteristics, especially of the newly available wonder-material iron, were also neglected, that Alphonse Lavallée and his colleagues set out to address (Comberousse 1879.1, pp. 25–29). In 1831 Lavallée wrote:

> Theoretical studies differ essentially from industrial studies, both by nature and in their goals. In science the principal focus is on establishing general laws, and the elements that are of such importance to the various fields are neglected. In order to simplify calculations, one frequently assumes conditions that lead to results that differ from those made by observation. Thus, one does not take the number of operations, the time needed, or the amount of energy consumed into account in laboratory research, and these are indeed very important considerations in a field because they greatly influence cost. For example, several laws that concern hydraulic flow are inexact because the corresponding theory is based on hypotheses that are only valid under certain circumstances that are never present in practice. Added to this is the fact that up till now theoretical science is only made up of very incomplete generalities that do not allow one to deduce the exact laws that govern the plethora of complicated phenomena that even the simplest field manifests. (Lavallée/Dumas, 1831 p. 8)[14]

They resolved the dilemma by combining two approaches to the communication of knowledge: innovation and *vulgarisation.*

---

14. Les études théoriques diffèrent essentiellement des études industrielles et par leur nature et par leur but. En effet, dans les sciences théoriques l'on a principalement en vue d'établir des lois générales; presque toujours on néglige des éléments qui, dans les arts, sont d'une grande importance; souvent aussi, pour simplifier les calculs, on fait des suppositions qui conduisent à des résultats différens de ceux de l'observation. Ainsi, dans les recherches de laboratoire, on ne tient aucun compte, ni du nombre des opérations, ni du temps employé, ni de la quantité combustible consommée, et cependant ces divers éléments sont très-importans dans les arts, puisqu'ils ont une grande influence sur la dépense. Ainsi, plusieurs des lois sur le mouvement des fluides sont inexactes, parce que la théorie les a fait reposer sur des hypothèses qui ne sont vraies que dans certaines circonstances que la pratique ne présente jamais. Ajoutons que les sciences théoriques ne sont composées que de généralités trop incomplètes encore pour que l'on puisse en déduire les lois exactes de cette foule de phénomènes compliqués qui se rencontrent même dans les arts les plus simples.

Vulgarisation as a didactical concept

The French concept of *vulgarisation* when applied to technology is the basis for the pedagogical method of enabling, a form of translation between scientific thought and other thought modes. There is no English equivalent for the term. The common translations 'popularization' and 'simplification' miss the point. *Vulgarisation* fulfills the intellectual function of pedagogically incorporating scientific method into a field, not just diffusing scientific knowledge, and it stimulates inter- or transdisciplinary cross-fertilization. *Vulgarisation* primarily trains a user to dovetail scientific disciplines with other fields. It actively engages and enables the learner, whereas popularizing or diffusing is a passive form of transferring knowledge. It merely provides existing solutions that can then be applied to another field. The difference between *applying* and *enabling* is subtle but significant in this process: when a scientist or mathematician applies a process or a solution that was previously developed to solve a problem in construction or manufacture, it is the passive use of an existing tool. When an engineer is enabled to create an approach using mathematics or scientific method to define and then to solve a new technical problem, it is an active development or variation of a specific tool to do the job. The relationship between science and technology becomes creative: technology is active, not merely applied science.

In their discourse on historiography Langlois and Seignobos explained the term *vulgarisation* as a valid technique while cautioning that it be correctly applied:

> A *vulgarisateur* is exempted from original research but must be aware of everything that has previously been published on his topic. He must be conversant with it all and have worked through the conclusions of the specialists for himself. If he has not personally been engaged in such research, he must at least be well informed, and that takes work. (Langlois and Seignobos, 1898, p. 271. Consulted in the 2nd edition, 1899)[15]

The authors polemically, and perhaps a tad chauvinistically, thought it to be an essentially French technique. To give them due credit, French writers like Louis

---

15. Un vulgarisateur est dispensé de recherches originales; mais il doit connaître tout ce qui a été publié d'important sur son sujet, être, comme on dit, 'au courant', et avoir repensé par lui-même les conclusions des spécialistes. S'il n'a pas fait personnellement d'études spéciales, il faut donc qu'il s'informe, et c'est long.

Figuier and Camille Flammarion, as well as French technologists, as we shall see, were especially prominent in this endeavor:

> Thanks to their natural penchant for tact, skill, and analytical ability, the French generally excel in this area. (Langlois and Seignebos, 1898, p. 271)[16]

The importance of the French, and particularly Figuier's contribution to *vulgarisation* was indeed recognized. As the translator wrote in the preface to *Primitive Man*, the English version of *L'Homme Primitif*, in 1870:

> The important art of placing scientific knowledge, and especially new discoveries and topics of present controversy, within easy reach of educated readers not versed in their strictly technical details, is one which has for years been carried to remarkable perfection in France, in no small measure through the labours and example of M. Figuier himself. (Figuier, *Primitive Man* 1870, p. v)

But the French were not the only ones interested in *vulgarisation.* Johannes Kepler, the Swabian astronomer, mathematician, and astrologer, and the Hessian polymath Athanasius Kircher initiated popularized scientific writing in the German world in the seventeenth century, albeit in a fictionalized form and in Latin (Kepler, 1634, Kircher, 1656, 1657 and 1664/65). However, it only became widely popular in nineteenth-century France in the novels of Jules Verne (that contain more science in the original French than in most translations), and in strictly factual writing like that of Figuier's prolific production (Peters 2002).

Engineering *vulgarisation* did not simplify scientific information using analogy. French engineers gained scientific ability in their pedagogy by enabling students to develop solutions using theory and by integrating it into practice. They developed scientific theory in both algebraic and geometrical form using concrete examples.

This kind of multi-layered approach had been pioneered by Gaspard Monge, who developed descriptive geometry in the late 1760s and recast it in algebraic form shortly thereafter. The relationship between analytical theory and geometrical method had been picked up by his students and followers like Jean-Nicolas Pierre Hachette and then especially by Gaspard Riche de Prony, who was both a theoretician and a practitioner. They developed graphical means of

---

16. Les Français, grâce à des qualités naturelles de tact, de dextérité et de justesse d'esprit, excellent en général dans cet exercice.

calculating and designing machine parts and ways to solve structural problems as a pedagogical tool. The teachers at the École Centrale, many of whom had studied under one or both of these innovators[17] then expanded the idea of using various techniques to integrate theory and practice into a didactical principle.

Simply put, the founders of the École Centrale upgraded *vulgarisation* from a pedagogical method to a didactical principle. They did this by replacing the simplification of information, with a hierarchical prioritization and thoughtful choice of the scientific and other tools they used to communicate information, and by replacing analogy with real examples. The first prevented the bowdlerization of information and the second rendered each theoretical postulate tangible. How they did this will become clear in the example of Bélanger's organization and development of modern analytical statics and mechanics, and in Colladon's and Olivier's use of visual and haptic tools. The founders were convinced that they would serve French industry best by enabling engineers to function both in theory and in practice, rather than simply learning how to apply predeveloped knowledge to practice.

It was an experiment, and the four founders hoped that the shift that they proposed through the conceptualization of *vulgarisation* would render the science-based process of analyzing that deconstructs the world compatible with the creative process of design that synthesizes it. They knew that what they proposed would have to be defended within the traditional educational system in which these two modes of thought were clearly segregated. It was both an intellectual issue and an emotional one, and Théodore Olivier took it very personally in a socio-political way when he polemically wrote:

> Theoreticians who style themselves *pure scientists* think that they constitute an aristocratic body with the right to command and to dominate practitioners whom they treat as *serfs* and *peasants*. They therefore forget, these *pure scientists*, just like the *nobles* of all types and places that it is *work* that all humans are intended for, and that this *work* that is useful to humanity must be recognized and recompensed before all other *amusements, mind games*, or *circus tricks*. (Olivier 1851, p. iv)[18]

---

17. Both taught at the École Polytechnique, and Prony was also director of the École des Ponts et Chaussées.

18. Les théoriciens qui prennent le nom de *savants purs* se regardent comme devant former un corps aristocratique, ayant le droit de commander et de dominer les praticiens, qui sont traités par eux

Oliver's diatribe needs qualification right at the beginning. It must be said that the École Polytechnique's theoreticians were (and still are) very capable and useful to the development of engineering. In the nineteenth century they were largely responsible for the development of the elastic theory of structural behavior. This theory enabled many technological innovations and developments that engineering practitioners participated in and that fostered industrial development. Both the theoretical and the practical approaches contribute to that development. The dichotomy between theoreticians and practitioners that existed still exists today. It needed and still needs to be bridged to enable a fruitful technological development. Many of the most profound thinkers in both categories recognized the value of the other approach. A prominent example of a border-crossing thinker was the abstract theoretician and practical *vulgarisateur* Jean-Victor Poncelet, who began his career as a theoretician and then taught mechanics to military engineers and to workmen and the general public. He later became director (then *commandant*) of the École Polytechnique 1848–50. In the 1820s, the École Centrale's founders had already felt that bridging this divide had become urgent, especially in the service of industry, and this bridging is what they set out to provide.

The mix of theoretical innovation via analysis and making that the École Centrale achieved through *vulgarisation,* as well as the interaction between the teachers—most of whom functioned to varying degrees and in different ways as both innovators and *vulgarisateurs*—determined the novelty and the success of the program and the influence that the École Centrale exerted worldwide. The ability to see helped the students to cross back and forth freely between theory and practice, and it constituted an important component of the *vulgarisation* approach developed by the school.

The focus of this study lies on construction, principally on the work and thought of two chief innovator-*vulgarisateurs*, Poncelet's former classmate at the École Polytechnique and friend Jean-Baptiste Bélanger (1790–1874)[19] and

---

comme des *ilotes* ou des *manants*. Ils oublient donc, les *savants purs*, comme les *nobles* de toutes les espèces et de tous les pays, que la condition imposée à l'homme est le *travail*, et que le *travail* utile à l'humanité doit être honoré et récompensé avant les *amusements* que procurent à l'homme et les *jeux d'esprit* et les *jeux de cirque*.

19. Bélanger's contribution to structural statics was chiefly but not exclusively *vulgarisation.* He also innovated in the field of hydraulics.

Auguste Perdonnet (1801–67). They were seconded by Théodore Olivier (1793–1853) and the physicist Jean-Daniel Colladon (1802–93) whose pedagogical contribution during the earliest formative phase was determinative as he was the first to teach and *vulgarise* theory through objects. These teachers were further supported by the contributions of those inclined more exclusively toward application: Charles Walter de Saint-Ange (1793–1851), Achille Ferry (1798–1882), Auguste Faure (1807–63), Nicolas-Auguste Thumeloup (1804–54), and Louis-Charles Mary (1791–1870). Mary's contribution was a special case. He coordinated the whole program through adaptation rather than *vulgarisation* or application.

In 1970 Harold Dorn wrote about the complex relationship between the fields of science and industry and speculated on the reason why the two fields only 'married', as he put it, in the nineteenth century (preface, especially pp. iv-vi).[20] He examined applied science rather than *vulgarisation.*

Several engineers in England did propose the use of mathematics to determine the best form for arch bridges early in the eighteenth century (Dorn 1970, pp. 126–137). Leonhard Euler and Daniel Bernoulli quantified how beams bend in 1750, and Euler also provided a formula to determine how much load a column can bear without buckling in 1757. However, virtually no practitioner knew how to use these results, a fact that one contemporary English researcher complained of bitterly (Dorn 1970, pp. 135–136).

There were very few who did try to bridge the gap between science and engineering. Philippe La Hire had attempted to understand how the two were connected in the late seventeenth century,[21] and a generation later, a remarkable attempt was made by Leonhard Christoph Sturm. In 1713, thanks to an attack on his practical expertise by a military engineer who considered him a mere theoretician, Sturm published a polemical treatise, *Gründlicher Unterricht... Von Häng-oder Sprengwercken,* in which he described in great detail the practical use of what he called *mathesis* and *statica* to build a functioning truss (Peters 2015). The reach of Sturm's treatise was limited by its form: he struggled with language, incorporating many Latin words and writing incomprehensibly contorted sentences in an attempt to express concepts that had never before been

20. I am grateful to Santiago Huerta for drawing my attention to Dorn's dissertation.

21. Thanks to Antonio Becchi for the information on the border-crossing efforts of de La Hire and his circle, which also included Giovanni Domenico Cassini and de La Hire's son Gabriel Philippe.

expressed in German. He tried, but for most builders who were usually not well read and whose knowledge had been gained through visual and experiential means, the abstraction that was mathematics had no point of contact with physical building. This disconnect was the primary complaint of practicing engineers. The didactically reinterpreted *vulgarisation* of science, that turned it from an application into the action of enabling, was missing.

The theoretician, pedagogue and practical engineer Rankine was one of those who, a century and a half after Sturm, was able to cross the boundary between the two worlds successfully. He had studied carefully why it was that practitioners could not understand theoreticians and what was needed:

> …the symbols of algebra, when employed in abstruse and complex theoretical investigations, constitute a sort of thought-saving machine, by whose aid a person skilled in its use can solve problems respecting quantities, and dispense with the mental labour of thinking of the quantities denoted by the symbols, except at the beginning and the end of the operation. In treating of the practical application of scientific principles, an algebraical formula should only be employed when its shortness and simplicity are such as to render it a clearer expression of a proposition or rule than common language would be, and when there is no difficulty in keeping the thing represented by each symbol constantly before the mind. (Rankine 1856, p. 20)

But even those rare engineers who did cross between physical practice and intellectual abstraction, and who did understand mathematics and tried to use it, often made errors:

> These errors do not reflect any misunderstanding of the theoretical principles involved; rather they suggest a lack of experience with the application of the principles to engineering problems. (Dorn 1970, pp. 146–147 footnote 1)

Translation between science and practice was what was required, and as will be shown, the missing link was this new conceptual form of *vulgarisation.* Although many engineers, mathematicians, and physicists attempted to bridge the divide between the two modes of thought, some with individual success like Rankine, it was the École Centrale that successfully forged this link pedagogically, influencing the development of engineering worldwide. The medium of this success was the curriculum, and the means were its faculty. One of my goals is therefore to demonstrate the value of *vulgarisation* (as distinct from the diffusion of knowledge) to the history of science and technology generally and to the history of construction specifically.

Historical study has traditionally focused predominantly on the importance of innovation—that is, on the discovery of new knowledge and its spread. Innovation is of signal importance to history because it supports the concept of the advancement of human knowledge, but it needs to be supplemented and balanced by an explanation of how knowledge is made useful. Usage is the main criterion for the creation of technical knowledge in the first place. Louis-Charles Mary's work, discussed in Part II, demonstrates how this was accomplished.

By examining the characters, interests, built works, and writings of all these teachers and the educational backgrounds that they brought to bear on their pedagogy, I attempt to clarify the trends that formed a distinctive mid-nineteenth-century French school of thought in civil engineering education, however without postulating simplistic causal relationships.

There are two further aspects that will be touched on repeatedly throughout this account but not dealt with separately in detail. These are the influence of the Protestant attitude toward education and its influence on *vulgarisation* and French engineering, and playfulness and enthusiasm as constituent elements of innovation and education.

Research method, basic sources, and questions

My research focuses on ephemera and rara as well as on manuscript and visual material from private sources. The Dibos (Lavallée), Peters, and Roussel Collections contain documents that have not previously been accessible to researchers. These are supplemented with information from the archives of the École Centrale as investigated and published by Dominique Barjot and Jacques Dureuil, Bruno Belhoste, Jean-François Belhoste, Jean-Yves Dupont, Daniel Gourisse, André Grelon, Valerie Nègre, Ulrich Pfammatter, and John Hubbel Weiss, and with information on the intellectual background of the development discussed in essays and articles by many scholars, among them a theoretical study by William Addis in 1986. The École Centrale's archive—as I discovered and the sociologist Grelon had previously noted (2000, p. 6, note 1)—remains, despite subsequent improvement, incomplete for the early period of the school.

The unfortunate 'triage' that was made before, and again when the École Centrale's voluminous archives were transferred to the Archives nationales discarded much that would have been of value to this study (see Billaud et al. 2018, esp. p. 8). I was able to use my sources to fill some of the lacunae in what my predecessors had discovered. The goal was to gain a more holistic picture of the genesis of our modern form of engineering education. I adopted a point of view that differed from these scholars by analyzing the didactic concept and the resultant pedagogy that the École Centrale's founders and early teachers developed. The synergy of their interactions was an aspect that drew my attention. My conclusions were contextualized with additional information and source material from many French and some foreign archives. They, and the material I used are listed at the beginning of the bibliography and credited at the appropriate places. I also used many of the school's lithographed, internally circulated course transcripts, some of which are available in libraries, and others that I found on the antiquarian book market. I consulted many of the teachers' and graduates' published treatises as well, which are also documented in the bibliography.

My method is discursive as well as analytical because the cultural, religious, political, and personal elements are rarely expressed in the writings of engineers and can often only be inferred from statistically relevant concatenations of facts and from statements about the teachers by their students and others. Among these, the most informative are the writings of Francis Pothier (1887), a graduate of 1842 and Lavallée's son-in-law, Charles de Comberousse (1879), a graduate of 1850 and from 1875 professor at the school, whose collected obituaries at the end of his book illuminate the characters of many teachers, and Colladon (1893) who taught at the École Centrale in the earliest years 1830–35 and who had been involved in the preliminary discussions that led to its foundation.

I extract the pedagogy that formed the professional logic and self-image of the École Centrale's graduates from the content and form of the teachers' and graduates' texts. Their introductions often provide insights into their intentions or motives for publishing, as Théodore Olivier's diatribe does in the just-cited introduction to his 1851 treatise *Mémoires de Géométrie Descriptive*.

My own collection of fourteen previously unknown student notebooks from the period are another source that show how the information that the teachers provided in their courses was filtered into the students' understanding. These supplement those notebooks in the École Centrale's archive that previous scholars have drawn from. The students' understanding and, on occasion, their 'creative misunderstanding' of the instruction they received provided them with the raw material for innovation. The pedagogy that these transcripts, treatises, and notebooks reveal and the innovation that they engendered were affected by cultural input from various sources: a century and a half of influences by educators, but also by builders, politicians, and theoreticians who formed a specific intellectual system shared both by those who taught and those who learned. What were these sources and how did they differ from those in other educational systems? How did the École Centrale's system develop from earlier French didactical concepts and yet distinguish itself from them and from other forms prevalent in German-speaking lands or in Britain? Who were the creators of these concepts and how did they think? How did the didactical concept that the school's faculty developed inform the pedagogy or method they used? The overarching intention in pursuing these questions is to learn how the school formed its students' approach to innovation: what is similar to the way other schools of thought functioned in this regard, what is different, and why?

As far as the quotations in the text are concerned, I have rendered their meaning, not their precise wording or formulation in my translations. That is why I have included all the original quotations for those who wish to experience the mindset of the authors, correct any misinterpretations that I may have made, or find more appropriate formulations and extract further nuances.

Websites like 'geneanet' or 'Graces Guide' were helpful in identifying innovators who are otherwise only known by their surnames, after which they could be researched and traced more easily. The online 'MacTutor History of Mathematics Archive' hosted by the venerable University of St. Andrews in Scotland, a pioneer in science teaching in the English-speaking world, is especially valuable when checking facts on mathematicians, physicists and related scholars as it quotes authored sources. 'Wikipedia' is of course invaluable. Like most Internet resources it provides unsubstantiated, anonymous *information* that must be checked in treatises, manuscripts, or recordings, in other words substantiated through repositories of *knowledge* that are authored. These are where one can authenticate facts and opinions and where their originators' credibility can be checked for veracity or at least plausibility. Both information and knowledge are important. Each has a role to play in research.

Many of the manuscripts, transcripts, and texts are now available online, so much so that travel to libraries or archives (whose guardians are usually very accommodating in providing scans) is often no longer necessary. Online scanned texts, particularly those available on *Gallica*, the French national library's (BnF) digital archive, or on that of the Ponts et Chaussées, make it easy to consult almost whatever one wishes and even search a text for individual words. I often used both the scan and the original document. The first makes specific searches for information easy, and the second invites discovery through the pleasure of physically leafing through a treatise or a manuscript. I have found that chance juxtaposition through comparing several open texts on one's table may spark knowledge in a way that the linearity of computer reading cannot. Even the physicality of a document, its size, volume, and sequence within a treatise can communicate information. A well-worn book tells one that it has been much used. If it tends to fall open to a particular section or certain pages are uncut, that too may prove meaningful.

As a construction historian I expanded the Braudelian interest in multiple standpoints, time scales, and minutiae from the human to the artifact and followed the third generation of the *Annales* school of history by expanding their *mentalité* approach that I think of as context, to include physical objects. This is a technologist's approach to details, those small-scale issues that can be as telling for the whole as the larger issues. The intention was to 'weave a fabric' in Braudel's tradition and to view the development from outside the culture. I therefore did not follow the threads of the historical development by cherry-picking and prioritizing certain trains of thought over others as historical studies usually do, but integrated as many relevant minor players, facts, and longer-term connections as I could, in order to manifest the overall pattern. The reader will judge whether this has been successful.

Nevertheless, there are limitations to what any historical analysis using both virtual and physical objects can discover. It should be evident that a good education as I have tried to define it here is no guarantee of inventiveness in problem solving. The elusive issues of individual character, gifts, proclivities, or even happenstance are invariably present. Education is discovery, and its value lies in seeing rather than in simple perception or the accumulation of material information. The École Centrale was, in its early years, committed to this form of innovative discovery at a time when industry and capitalism were liberating people from the elitist hierarchies, of science and aristocracy (of course these then eventually created their own hierarchies). The foe of education is the blind support of such hierarchies and that is why education must always redefine itself. However, even under ideal conditions, education can only trigger or support tendencies or gifts in responsive individuals. It can influence or guide them toward seeing, but nothing more.

The goal here is to identify such ideal conditions. Nevertheless, it will always be impossible to discover how innovative thought really occurs; one can only approach the protagonists and their truths obliquely. Even if the innovators themselves wrote on the topic, and very few did (Louis Bruyère was a very rare exception, as we shall see), one must take what they communicate with a very large grain of salt. A source's self-delusion, misunderstanding, and intentional obfuscation fog an elusive and basically illusory reality. The attempt is nonetheless worthwhile if we wish to uncover at least something of the mysterious relationship between education and invention.

Part I: The Development of the Pedagogy

A group of students

A host of eminent civil engineers and architects graduated from the École Centrale (ECP)[22] around the mid-nineteenth century and pursued influential careers in many fields and on all continents. One could take any number of them and trace how their education impacted their professional development and work.[23] Most of these graduates remain virtually unknown, but some contributed interesting individual projects or structures that merit attention, like Théophile Seyrig, who collaborated with the far more famous Gustave Eiffel and with Henri de Dion, and who contributed substantially to their successes. Others were Louis Gellusseau, who helped design the reinforced concrete structures that made the Perret Brothers famous (see chapter 'Connaissance des matériaux'); Émile Lavezzari, who proposed the first version of the 1889 Galerie des Machines (see the discussion of Contamin's career in the last chapter), and Sylvain Périssé, who built the iron- and ceramic-clad house that Viollet-le-Duc had famously proposed in 1872 (see the end of

22. The usual abbreviation for the École Centrale des Arts et Manufactures (ECAM) is now ECP, which stands for École Centrale Paris. Although the only École Centrale that existed at the time, and the only one that I refer to, was in Paris, I will still use this common acronym. The acronym for the École des Ponts et Chaussées is now ENPC (École Nationale des Ponts et Chaussées). It will stand here to represent all historical forms of the name: École royale…, École impériale…, École nationale… as well as its predecessor, Perronet's Bureau des géographes et dessinateurs du roi. The École Polytechnique is normally abbreviated in France as 'X', the origin of which is obscure, possibly deriving as an intellectualist abstraction from the algebraic unknown 'x'. Because this is confusing to a non-French reader, I will use the less common 'EP' for clarity. Moreover, the French usually state the year a candidate entered a school because the entrance requirements are so strict that they constitute the main barrier to success. Here I will follow the Anglo-Saxon practice and indicate the graduation year whenever I can. The full official name of the École des Mines is École Nationale Supérieure des Mines de Paris or ENSMP. However, this cumbersome acronym is now rare, and the current usual form is simply École des Mines. I also use the English form of capitalization of the school names, titles and names of institutions that is not usual in French.

23. Barjot and Dureuil 2008 name many of these.

Fig. 3. The six students who became the most prominent developers of iron construction: Camille Polonceau, Henri de Dion, Gustave Eiffel, William L. Jenney, Armand Moisant, and Victor Contamin. (Source: Polonceau: public domain; de Dion: Wikimedia Commons; Eiffel: Félix Nadar 1888 public domain; Jenney: Industrial Chicago, 1894, vol. 1, Goodspeed Publishers; Moisant: Guillet 1929, pl. 64; Contamin: *L'Exposition de Paris* 1889, p. 8)

the chapter 'ECP graduates join the entrepreneurs de serrurerie'). However, there were famous ones too. This study looks closest at the school's best-known graduates, who predominantly influenced the development of modern iron construction in the nineteenth century. They were: Camille Polonceau (1813–59), Henri de Dion (1828–78), Gustave Eiffel (1832–1923), William Jenney (1832–1907), Armand Moisant (1838–1906), and Victor Contamin (1840–93). Two of the less prominent engineers, Alexis Barrault (1812–65) and Émile Baudet (c.1830–94), made relevant contributions that will serve to document the importance of the school's pedagogy in the transition from experimental, pre-industrial iron construction to the fully developed industrial form and its architectural expression after 1850. Including everybody who contributed to the development would become repetitive and tedious. The primary innovators were the six: Polonceau, de Dion, Eiffel, Jenney, Moisant and Contamin, while Barrault and Baudet serve as prominent examples of those who enabled the change. These eight engineers, seven French and one foreign, became leaders of the development despite their cultural and social differences.

The transition in France from aristocratic privilege to meritocracy by way of the Revolution was a factor that enabled the development of modern industry. The six students I look at in detail who became professionally eminent came from very different backgrounds: one each from the state bureaucracy, the aristocracy, the professions, commerce, a farming background, and the working class (the last described only as 'modest background').

Most of the ECP graduates who remained in France formed a close-knit social and professional group that transcended former social divides. Their cohesion was defined by their common ability to use observation to see problems through the same lens, their unique ability to pass easily from an analytical scientific mode of thought to a creative design mode and to combine the two fruitfully, and their status as alumni of the school. Their abilities were such an important part of their success and way of working that Jenney's example evidently communicated them to the group of architects and engineers with whom he pioneered the skyscraper in Chicago at the end of the century.

How did the pedagogy that the school developed enable these men to become so innovative in the development of modern iron construction? All of them experienced a unique form of scientific-cum-engineering-cum-architectural education under the same teachers: Barrault, Polonceau and de Dion in the formative phases of the school's didactical model, and Eiffel, Jenney, Baudet, Moisant, and Contamin in the fully developed phase. All of them followed a virtually identical course sequence that made their educational paths similar to one another's but distinctly different from the educational culture of other schools.

The other French engineering traditions, like those of the École des Ponts et Chaussées (ENPC see note 22), the Écoles d'Arts et Métiers, or the École des Mines also contributed to the development of construction systems, but the ECP was unique in the breadth of its purview, in its industrial rather than governmental constituency, and in its concentration on the incorporation of theory and practice that inspired intellectual border-crossing.

In contrast to the École Polytechnique (EP or X in France, see note 22) and the ENPC, which habitually classed their graduates by the year they entered for reasons that will be discussed later, the ECP always classed them by the year they graduated. Camille Polonceau and Alexis Barrault graduated in 1836, Henri de Dion in 1851, Gustave Eiffel and William Jenney one year apart in 1855 and 1856, Émile Baudet in 1858, Armand Moisant the following year in 1859, and Victor Contamin a year after him in 1860. Others who graduated in that quarter century displayed similar professional traits, albeit with less incisive implications for iron construction.

The ECP's concept also differed from the two older models. The most important difference was that the school was conceived as a private institution designed to serve the burgeoning and initially less socially prestigious but increasingly powerful bourgeois industrial base. Conceptually the founders based their curriculum on incorporation of theory and practice through their expanded version of *vulgarisation.*

The cultural background into which Jenney returned led to his very different career in the United States from Polonceau, Barrault, Baudet and Contamin's in France, or de Dion, Eiffel, and Moisant's all over Europe, Latin America and the French colonies in Asia. But other influences and interests notwithstanding, there are parallels in the way all of them attacked novel problems that appear to stem from their common education rather than from their individual backgrounds. All eight, and especially Eiffel, Jenney, Moisant, and Contamin, who survived longer than the others and were able to develop more mature careers, profoundly influenced the course of iron construction worldwide.

The earlier graduates Polonceau and Barrault (who had been classmates), as well as de Dion, were fully involved in their professional careers by the time that the most famous of the graduates, Eiffel and Jenney, entered the school. There is no indication that the younger students were more than possibly fleetingly acquainted with any of them, or that Eiffel and Jenney were even acquainted with each other during their studies. As they came from very different backgrounds this is not surprising. The two young men also differed in their goals, Eiffel opting in his second year for a major in chemistry—obviously intending to prepare for entering his uncle's factory—and Jenney for a major in construction. Baudet entered while

Jenney was an upperclassman, and Moisant the year Jenney left and, although they all opted for the same major, there cannot have been any meaningful contact between them at school. It is likely that some of those who studied just after mid-century were passingly acquainted because certain classes were taught in common to all second- and third-year students irrespective of their intended majors, but the student body was growing ever larger in the years 1852–60. In the earliest years the number of students in each year had been small; Barrault and Polonceau's year had 22 students, and de Dion's 58. In Eiffel and Jenney's case each year was around 70

Fig. 4. All schools had uniforms that the students were required to wear, but the École Centrale's were intentionally casual. In school they wore a smock, known as the *tenue d'étude* (analogous to the *biaude* of the students in the Écoles d'Arts et Métiers) that they were allowed to personalize with decoration, and in public and on formal occasions they wore the *tenue de ville* that was simply a young gentleman's standard jacket decorated with the school's insignia, and trousers with a special cap, leather gloves, and a cane. (Source: Rousselet, *Nos grandes écoles*, 1888, pp. 277 and 276)

strong (71 and 69 by graduation),[24] and the total student body in all three years of the program stood around 200. Baudet's had 77, while Moisant and Contamin's years were considerably larger (112 and 116 by graduation). Nevertheless, an *esprit de corps* and a common professional culture were encouraged in many ways by the founders. This expressed itself in many forms, from the collaborative courses to the later temporary introduction of uniforms similar to those in the military academies. In general, the students wore a so-called *tenue d'étude* in the lecture hall, the laboratories, and the studios, and a type of anti-uniform, the *tenue de ville*, which they were required to wear outside the school and on formal occasions.

A further encouragement to the formation of an *esprit de corps* took the form of a common canteen where it was indeed possible that students from different years could meet and socialize.

So it is conceivable that some of the eight students may have met at school or at a school-related function, or even perhaps when Jenney returned to Paris for several years in 1858, but they could hardly have influenced one another professionally at that point. However, all of them—except Jenney who immediately left France after graduation—were subsequently connected through their work experience. They also associated professionally and socially with one another as members of the Société des ingénieurs civils and the Association amicale des anciens élèves. Although both societies were only officially founded at mid-century, in the early years an efficient informal network had already formed that the school supported (Lavallée/Dumas 1834, p. 12). As early as 1831, the school brochure had noted:

> Upon graduating, all those who receive the diploma remain in regular contact [with the institution]; they receive all information that may be of interest to them regarding industries they may be associated with and about

24. Various sources cite sometimes widely differing numbers for the size of each class. In part this must have been due to the large initial attrition rate of 60% that decreased to around 30% at mid-century (Comberousse 1879.1, pp. 41/173–174). I have used the school's publication, *Les Anciens Élèves de l'École Centrale 1832–1888*, (*École centrale 1889*), which lists what it claims are all graduates by later activity and year of graduation in the chapter *Liste par profession de tous les anciens élèves de l'École centrale*. However, many graduates are listed several times if they were active in different fields, and it is quite a chore to sort out these multiple listings to arrive at what here appear to be more or less correct total numbers.

Fig. 5. From mid-century on the classes were so large that the students from different studios and years might only meet at mealtimes in the canteen, *le réfectoire*. (Source: Rousselet, *Nos grandes écoles*, 1888, p. 295)

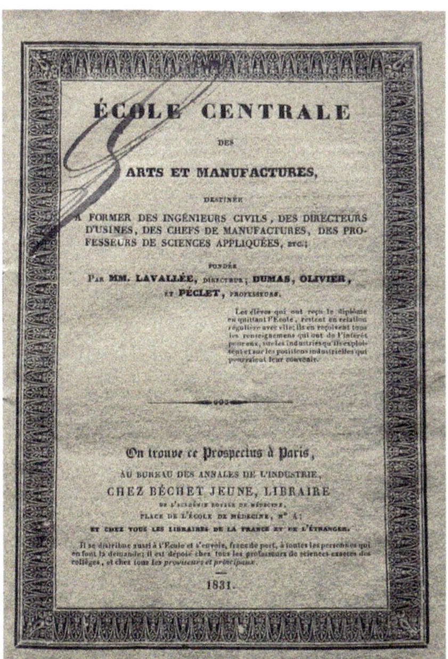

ÉCOLE CENTRALE

DES

ARTS ET MANUFACTURES,

DESTINÉE

A FORMER DES INGÉNIEURS CIVILS, DES DIRECTEURS D'USINES, DES CHEFS DE MANUFACTURES, DES PROFESSEURS DE SCIENCES APPLIQUÉES, ETC.;

FONDÉE

PAR MM. LAVALLÉE, DIRECTEUR; DUMAS, OLIVIER, ET PÉCLET, PROFESSEURS.

Les élèves qui ont reçu le diplôme en quittant l'École, restent en relation régulière avec elle; ils en reçoivent tous les renseignemens qui ont de l'intérêt pour eux, sur les industries qu'ils exploitent et sur les positions industrielles qui pourraient leur convenir.

On trouve ce Prospectus à Paris,

AU BUREAU DES ANNALES DE L'INDUSTRIE,

CHEZ BÉCHET JEUNE, LIBRAIRE

DE L'ACADÉMIE ROYALE DE MÉDECINE,

PLACE DE L'ÉCOLE DE MÉDECINE, N° 4;

ET CHEZ TOUS LES LIBRAIRES DE LA FRANCE ET DE L'ÉTRANGER.

Il se distribue aussi à l'École et s'envoie, franc de port, à toutes les personnes qui en font la demande; il est déposé chez tous les professeurs de sciences exactes des colléges, et chez tous les *proviseurs et principaux*.

1831.

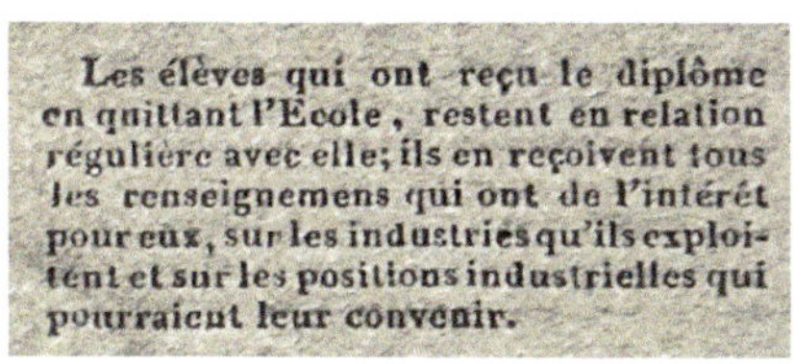

Les élèves qui ont reçu le diplôme en quittant l'École, restent en relation régulière avec elle; ils en reçoivent tous les renseignemens qui ont de l'intérêt pour eux, sur les industries qu'ils exploitent et sur les positions industrielles qui pourraient leur convenir.

Fig. 6. Cover of the 1831 ECP brochure with the text announcing continued support of the students (Source: Peters Collection)

> positions in industry that may be relevant to their goals. (Lavallée/Dumas 1831, front cover)[25]

Polonceau and Barrault worked for a time together in the railway pioneer Eugène Flachat's office after graduating and were certainly in professional and personal contact. Polonceau stayed exclusively with railways while Barrault gravitated quickly into iron metallurgy and the construction of iron mills and structure, only later returning again to railways. Eiffel and de Dion worked together several times in the late 1850s and early 1860s, and Eiffel and Moisant were in professional contact with one another as well because they were direct competitors and occasional collaborators, both having founded their own contracting firms in 1866–67. Eiffel possibly knew Polonceau, too, in the formative phase of his career. However, most earlier contacts tended to be more social than professional, so their similarity in mindset and in their approaches to design and construction stemmed in the main from their education: they were all taught by the same men in consecutive periods. The fact that some of them followed different majors at school played no role, as we shall see further on in the discussion of the school's organization.

Through changes in his family situation and the need for an income, Eiffel did not stay with chemistry after graduation but immediately moved into railway construction, and then via railway bridges into component manufacture and contracting. While Eiffel's trajectory was dictated in part by chance, Barrault and Polonceau entered Flachat's office directly through their teacher Perdonnet's patronage. There Polonceau worked on railways while Barrault initially concentrated on industrial buildings. De Dion, who had graduated in mechanical engineering, also initially worked for Flachat and spent years with railways, working sporadically in that field, first in France and then in Spain for two decades before returning home. Baudet first worked for the contractor Joly together with Moisant, who had first briefly worked at the Chemin de Fer du Nord under Eugène Flachat's erstwhile partner Jules Petiet. Then Baudet joined the contractor Leturc as partner. Eiffel and Moisant followed as independent contractors a few years later. Eiffel developed into an industrialist as well, while Moisant also focused extensively and in multiple capacities with

25. Les élèves qui ont reçu le diplôme en quittant l'École, restent en relation régulière avec elle; ils en reçoivent tous les renseignemens qui ont de l'intérêt pour eux, sur les industries qu'ils exploitent et sur les positions industrielles qui pourraient leur convenir.

the series of World Exhibitions (*École centrale 1889*, p. 19). Contamin, who graduated from the mechanical division like de Dion, worked first for two years in Spain and then under Petiet at the Chemin de Fer du Nord where he remained based for his entire career. Jenney, on the other hand, left the country and gained part of his practical experience for a year in Mexico on Judah Benjamin and John Slidell's ill-fated Louisiana Tehuantepec Company's project for an Atlantic-Pacific railroad connection. Then he worked for the military engineer Hiram Berdan, and later on fortifications during the American Civil War. So while their professional experience in railways occurred in various ways from happenstance to mentoring, the fact remains that Barrault, who launched the transitional phase, and Polonceau, de Dion, Eiffel, Jenney, Moisant and Contamin, who represent the century's most influential iron constructors, all began there. Their education at the ECP prepared them well for this first step.

Their background in railway construction with Perdonnet and Flachat

What was therefore common to all but Baudet, and common to many others among the school's civil engineering graduates in those decades, was that they cut their teeth professionally in railway construction, almost half of them in Eugène Flachat's firm.[26] This experience proved relevant to their education as it placed the self-made engineer Flachat and the railway pioneer Perdonnet's influence squarely in the focus of their career development.

As previously explained, this initial professional step was into the only established field that used iron consistently as an industrially produced material rather than in an experimental fashion, and it was the young and dynamic railway industry in France that had defined the need for the school in the first place. In fact, railways were so ingrained in the self-image of the school and its graduates that even as late as the 1880s, the majority of the articles in the *Mémoires et compte rendu* of

---

26. Aside from Polonceau, Barrault, and de Dion, among many others who were later involved in iron construction who began their careers in Flachat's office, were, according to J.-F. Belhoste 2004, p. 70, and 2008, p. 63: Ernest Deligny (grad. 1842), Ivan Flachat (grad. 1850), Ferdinand and Henri Mathieu (grad. 1838 and 1841), Léon Molinos (grad. 1851), Jules Petiet (grad. 1832 who, after briefly working for Paulin Talabot, became a partner in the firm from 1835–42), and Charles Pronnier (grad. 1851). (Moisant and Contamin both worked for Petiet, who had left Flachat to become director of the Chemins de fer du Nord in 1848). And then there were others, like Émile Muller (grad. 1844), whom Polonceau had recommended to a position at the chemin de fer de l'Est (Cauvet in Birlë et al. 1889, p. 2).

the Société des ingénieurs civils was devoted to their construction, organization, maintenance, running, and funding. The symbiotic relationship between the school, the iron industry, and the railway industry was one of the principal contributions that nineteenth-century France made to modern engineering education.

Railway development in Great Britain, by contrast, did not directly lead to an educational initiative despite the fact that from the very beginning the practical use of iron in structures had required a corresponding analytical structural understanding there too.[27] The difference between the two countries can be attributed in part to the contrast between the high social standing of academics and engineers in post-Revolutionary France where crossover between the two was fluid, and aristocratic Britain where the two categories were distinct and the engineer was classed among the craftsmen and definitely 'not a gentleman'. Sir Lyon Playfair discussed the problem in 1852 (p. 7), and Rankine also described the persistent derogatory evaluation of engineers by British academics at length in the introductory lecture to his class in civil engineering and mechanics at Glasgow University in 1856 (see also Channell 1982, p. 44).

The close connection between the ECP and the railways was the result of French industrialists' perception that British industry was in the lead, and their ensuing call for a new educational approach beyond the purely scientific that the EP provided. Figuier was the first to refer to Flachat as *le créateur du génie civil en France* (the founder of civil engineering in France) (Figuier, *L'Année scientifique et industrielle*, vol. 17, 1873, p. 545),[28] and Jean-François Belhoste confirms the role of the practitioner in the creation of the academic engineer. Especially Flachat's office, the first private French engineering consulting firm, was a major factor in the development of the ECP and the key impetus for modern French engineering:

> It is he who was the founder of French engineering. In 1834 he effectively created a consulting firm that immediately recruited an important number of young École Centrale graduates whose first projects chiefly involved the iron industry. (J.-F. Belhoste 2008, p. 60, see also Malo 1873, p. 52)[29]

---

27. Dorn, pp. 163–168, discusses the earliest known example of this dependency in Skempton (1956, pp. 1034–1037) and in Johnson and Skempton regarding William Strutt's use of calculation to design structural iron beams at the very end of the eighteenth century.

28. This was hyperbole, as the profession of 'civil' engineer was known, but Flachat certainly established the profession firmly as a private, industrial enterprise.

29. C'est lui qui fut le fondateur de l'ingénierie française. Il créa en effet dès 1834 un cabinet

Fig. 7. Eugène Flachat, pioneering railway engineer and founder of the first private engineering firm in France. (Source: albumen print c. 1865–67, Peters Collection)

Flachat was from the outset a firm supporter of the ECP and of its graduates. He was in reality an unacknowledged co-founder of the program and a major influence in the development of its concept. In the school's first four decades, he served as an unofficial post-graduate mentor of students, many of whom found their first employment in the office that he had founded only two years after the first ECP class graduated:

> He was a mentor especially to the École Centrale's students. He loved to surround himself with them, coached them with his vast experience, employed them early on in his projects and liked to encourage them to become agents of his thinking. The reason was that the École Centrale corresponded to one of his oldest and most enduring aspirations: it had developed a program that was at once both theoretical and practical, a combination that he had

---

de conseil, dont les premières interventions portèrent principalement sur l'industrie sidérurgique, recrutant d'emblée bon nombre de jeunes centraliens.

> missed in his own education, a lack that that his powerful intellect and his remarkable mathematical instinct had succeeded in filling, but at the price of what relentless striving and intellectual exhaustion! One could infer it from his happiness at knowing this lack filled in those whom he referred to as his collaborators, and they who proudly called themselves his students.
>
> ....
>
> This great and lively affection that M. Flachat lavished on those he surrounded himself with was by no means fruitless. He did not limit himself to educating them: he worried about their physical well-being, their advancement in the profession, and he did not neglect any opportunity to highlight their personal worth and to assure that they were accorded due professional credit. (Malo 1873, pp. 48–49)[30]

This last comment is borne out for example in de Dion and Lasvignes 1861 (p. 5 footnote 1), in which Flachat carefully detailed his collaborators' contributions. Flachat's intense interest in mentoring young professionals was one that Perdonnet shared. In Perdonnet's case his interest had probably resulted from his experience as a youngster in Johann Heinrich Pestalozzi's institute in Yverdon (see later under 'The details of an unusual education'). In Flachat's case the source is unknown, but as Perdonnet was one of his close collaborators, this characteristic may have been due to him. Flachat's own education does not provide a clue.

In 1873 Léon Malo (pp. 7–14) informs us that the entrepreneur and businessman Eugène Flachat had, after studying mathematics, begun his engineering education

---

30. Il fut particulièrement le protecteur des élèves de l'École centrale; il s'entourait d'eux volontiers, les instruisait de sa grande expérience, les associait de bonne heure à ses travaux et se plaisait à en faire les instruments de sa pensée. C'est que l'École centrale répondait à l'une de ses plus anciennes et de ses plus persistantes aspirations; elle réalisait cet enseignement théorique et pratique à la fois dont les leçons avaient manqué à son éducation première; lacune que sa puissante intelligence et son remarquable instinct mathématique étaient parvenus, il est vrai, à combler; mais au prix de quels labeurs intimes et de quelles fatigues d'esprit ! On le devinait en voyant combien il était heureux de sentir cette lacune remplie chez ceux qu'il appelait ses collaborateurs et qui, eux, se disaient fièrement ses élèves.

....

Cette grande et vivace affection que M. Flachat vouait à ceux dont il s'entourait n'était à aucun point de vue stérile. Il ne se bornait pas à les enseigner, il s'inquiétait de leur situation matérielle, de leur avancement; il ne négligeait aucune occasion de mettre en lumière leur valeur personnelle, et d'attirer sur eux les récompenses honorifiques.

by working on a project for a canal from Paris to Le Havre (1827–29) with his half-brother Stéphane Mony, who had attended but not graduated from the École des Mines in 1821. He then worked on a drilling enterprise for artesian wells with their other brother Adolphe Flachat, probably around Tartarbunary in the Odessa Oblast (1831–32), and finally he learned more through his contacts in England, particularly his friendship with Marc and Isambard Brunel and George and Robert Stephenson. No school existed yet that could have provided such a varied education. Upon returning to France in 1833, he founded his own firm and worked with Mony, and then from 1835 with Mony's childhood friend Émile Clapeyron and Clapeyron's classmate Gabriel Lamé, who had both graduated from the École des Mines in 1820 and had recently returned from Russia. Flachat was associated with them all on the construction of France's first passenger railway, the Paris–Saint-Germain Line.[31] The young ECP graduates who later worked for Flachat on his railway enterprises in the early years were in contact with these men and with British developments as well. Flachat himself had studied British structures closely and published reports internationally, for instance (in German translation) on three major London docks in the *Allgemeine Bauzeitung* (1837, vol. 2, pp. 129–133) as a partial translation of a longer report (Flachat 1836).

British engineers and contractors had aggressively participated in the French and French Belgian (Walloon) development, too, establishing industries, building bridges, and laying entire railway lines in the earlier phase. The Cockerills, William and his sons John and Charles James, moved to Liège (then French) in 1799 and built up a successful iron industry. Thomas Bonehill established himself in Charleroi (then Netherlands) a little later in 1824. Aaron Manby set up a machine works in Charenton-le-Pont just southeast of Paris at the confluence of the Seine and the Marne in 1822 to avoid import duties on British machinery, and later owned and modernized the ironworks at Le Creusot between 1826 and 1836 when it was sold to the Schneider brothers. Thomas Brassey worked as contactor on multiple railways and bridges between 1840 and 1860, and Oliver York came to France to work for him. The Fox and Henderson company (Charles Fox and John Henderson) contracted for the Paris-Lyon Line and later the Lyon to Geneva

31. Although it was the first French line designed specifically for passenger traffic, it was actually the second to carry passengers. The first had been the Saint-Étienne to Andrézieux line, founded for coal transport in 1827, which had also carried passengers from 1832.

railway, and Joseph Locke was engineer on the Paris-Rouen-Le Havre line (the first in France to be built with British capital and engineers) that superseded the canal proposal that Flachat had previously worked on. Locke associated with William Mackenzie and Brassey as contractors for the French-English banking firm of Laffitte, Blount et Cie. (1841–43) on the Paris-Rouen segment and then for its successor Charles Laffitte et Cie. from 1843. He also built the Mantes-Cherbourg segment that opened in 1858. Mackenzie and Brassey built a total of four lines in France, among them the segments between Orléans and Bordeaux of the Paris-Bordeaux railway (1843–53). Robert Stephenson consulted for Paulin Talabot on the Beauvoir-Alès segment of the Alès-Beaucaire Line (1837–40). The high point and end of this cross-channel collaboration fell in the years 1857–58, just after the worldwide commercial and financial crisis of 1857. Thereafter, French firms, many of which employed ECP graduates in key positions, took over and fewer British firms were involved.

The iron industry and railway construction were certainly not the only avenues open to ECP graduates in the 1830–60s, although the most innovative constructors in the early period took that route. They became true *ingénieurs industriels* in the sense that the founders defined—not in our modern sense of the term as engineers who organize industrial processes, although they did that too—but in a more general sense as engineers who worked for industry and used industrial means to solve their problems. They became a very different breed of engineer from the military engineers that the EP and Metz educated, or the government-employed civil engineers who graduated from the ENPC.

Although two bridges built by ENPC engineers in Paris, the Passerelle des Arts (1804) and the Pont d'Austerlitz (1806) did use iron inventively in the early period, they were experimental, unique structures that were produced manually in small foundries. They were icons, but had no direct influence on industrial development. The engineers of these structures were Louis-Alexandre de Cessart and Jacques Dillon for the Passerelle des Arts and Joseph Becquey de Beaupré and Corneille Lamandé for the Austerlitz. The engineer of the 1811 iron dome of the Halle aux Blés, François Brunet, who had entered the ENPC as a student in 1767–68 (Lacourt), but is apparently otherwise not traceable,[32] acted more as a consultant

32. Porrino 2021.2, p. 42 footnote 65 considers Brunet more a technician than an engineer and credits Becquey de Beaupré (whose 1807 memorandum on the structure indicates his intensive concern with the technical detailing) and others with the technical design of the structure, as well as the

for the calculation of the structure than as the builder. There was one more, Louis Bruyère, who built another experimental structure in 1808, and he became a model for the development of the ECP's education, as we shall see.

Flachat's men, on the other hand, were industrially oriented. They used industrial products to design and build their bridges, virtually all of them in iron. They built many, and other structures as well, and each influenced the design of the next in rapid succession. They had, in fact, participated in the development of structural sections like the I-beam (see chapter 'Flachat and the I-beam'). This conditioned the ECP engineers' approach to design and iron construction in a completely different way. Since the railway industry was the only field that used iron in an industrialized form at the time, their use of iron would reflect the industrial setting they were trained in.

Among the early graduates who became these *ingénieurs industriels*, Jules Petiet (grad. 1832), Alexis Barrault and Polonceau (grad. 1836), and later Contamin (grad. 1860) were among the most active in railways: they committed a large part of their careers to the development. Railways were certainly the largest construction sector at that time in France, but other paths led elsewhere. Barrault, Eiffel, Jenney, Baudet, and Moisant built their reputations on many types of iron structures. Others gravitated all over the world into very diverse careers, some of them in multiple areas. The broad spread of professional expertise had not previously existed in engineering. Industry and infrastructure development opened the field up, and the flexibly educated ECP graduates rose to the occasion. Each segment of the field subsequently expanded to demand its own educational curriculum, some even several, so no single professional program today can even attempt to prepare its graduates in the same way. A modern curriculum would have to try to achieve a corresponding degree of flexibility in a different manner.

The professional fields that the ECP graduates chose

The curricular breadth that enabled the graduates to be so flexible was one of the features that no other engineering program has ever encompassed, previously or since. For example, there were students like Émile Trélat (grad. 1840), Émile Muller (grad. 1844), Émile Lavezzari (grad. 1853), René Demimuid (grad.

---

architect Bélanger, who had visited England several times and had even proposed using iron for a bridge design over the Seine in 1787 (Porrino 2021.2, pp. 58–59, also J.-F. Belhoste, pp. 113–200).

1858), or Jules Denfer (grad. 1861), who became architects and prominent border-crossing members of the Société des ingénieurs civils. Muller became Mary's successor as professor for *architecture civile* in 1864 and president of the Société in 1872, Lavezzari was a builder of hospitals and the developer of the tuberculosis spa Berck-sur-Mer, as well as the director of a gas factory and the designer of municipal lighting. Demimuid was the designer of the ECP's new building in the 1880s and a professor at Trélat's École Centrale d'Architecture, and from 1872 on at his alma mater. Denfer wrote books on several technical aspects of architecture, including iron construction, and carried out the new school building after Demimuid's death in 1881, succeeding him as professor at the ECP. Muller, Lavezzari, Demimuid, and Denfer were honored members of the Société and members of its steering committee. This professional flexibility among the graduates was astonishing: ECP graduates worked in many fields and also fulfilled many roles simultaneously. The multidisciplinary character strengthened as the program matured, and was certainly a factor that contributed to the adaptation of iron to construction. It also led to the international spread of the ECP's influence, not only among foreign students but also among the French engineers who worked all over the world.

As Charles de Comberousse, professor at the ECP and himself an 1850 graduate, told the audience of 4000 that filled the Salle des Fêtes of the new Palais du Trocadéro for the 50th anniversary celebration of the school on June 21, 1879, the key to the school's success lay in its generalist breadth, and that the graduate was fully aware that in order to succeed in industry it was

> ...not a single key that he required, but a whole regalia of keys.[33]

A listing compiled by the school (*École Centrale 1889*) documents the professional spread. However, it only provides a broad indication of trends over the years and is far from complete. De Dion's case shows its limitations: the list only shows him in two roles, railway engineer and teacher, although he actually worked as a designer, a theoretician, a teacher, an administrator, and in industrial management at different times during his life. His career was more varied than most and is detailed in the last chapter.

---

33. ...non pas une clef, qu'il lui faut, mais un trousseau de clefs. *La Semaine des Constructeurs*, nr. 52, Sat. 28 June 1879, p. 1.

Fig. 8. The Salle des Fêtes in the Palais du Trocadéro. (Source: Simon de Vandières, *L'Exposition universelle de 1878 illustrée* 1879, pl. 68, p. 129)

I have not attempted to list all the countries and regions in which the graduates worked, but Spain and the Middle East were among the most prominent, and later eastern Europe, Russia, Vietnam (then known as Indochina), and China. The Suez and Panama Canals and their related machine industries and projects were the major engineering fields that employed ECP graduates overseas from mid-century on, both in planning offices and in contracting firms. Many other countries in northern Africa, Latin America, and the French island colonies drew them abroad too. The tendency toward internationalism was already discernible as early as 1834–35, when the school brochure for that year listed the professions and domiciles of the 57 graduates that the school could positively confirm (Lavallée/Dumas 1834, pp. 22–24).

While, as we saw, British engineers built a great many railways in the initial period in France and influenced the early French development, much of the subsequent spread of engineering expertise worldwide was due to the activity of French financiers like the Pereire brothers, James Mayer de Rothschild, Charles Laffitte, or Paulin Talabot, who was both an engineer and a banker. Family connections were relevant and often close: the railway and bridge engineer Paul Régnauld was Émile Pereire's nephew. The contractor Ernest Goüin (whose firm later became Société de construction des Batignolles) had married a cousin of Émile Pereire's wife, and his firm had been financed through Rothschild. He later also became a regent of the Banque de France. Nicolas-Didier Riant, the entrepreneur who headed the Mignon iron foundry, one of the early entreprises de serrurerie, later became a partner of the Laffitte railway enterprise. The French banking world was closely allied with engineers and entrepreneurs, and together they funded and built many of the largest international projects.[34] Paris functioned as a powerful center of international finance in the early to mid-nineteenth century. Eiffel documented the worldwide spread of French engineering expertise in great detail in his maiden speech as president of the Société des ingénieurs civils on January 4, 1889 (*Mémoires et compte rendu* 1889.1, pp. 26–44). He cited the French funding of great projects in Russia, Italy, Spain, and Austria that were being carried out under French engineers and with French firms supplying materials and machinery. He listed the names of many of his colleagues working on projects in Argentina, Belgium, Brazil, China, Denmark, Egypt, England, Greece, Holland, Japan, Peru, Portugal, Serbia, Romania, Tunisia, and Turkey, and explained the advantages of the French practice of riveting trusses over the pinning practiced in the United States.

Among the students in Barrault and Polonceau's small graduating class of 22 in 1836, most seemed to have pursued single fields, and mainly in continental France. Seven worked as general civil engineers, four worked in railway construction (among these Barrault, who would also build iron mills and the celebrated 1855 Paris Exhibition building), one in the machine industry, three in metallurgy, one in industrial production, and two primarily as professors.

After mid-century and the Revolution of 1848, the school grew quickly, and the border-crossing structure of professional involvement expanded along with it as the participation of ECP engineers in sites abroad gradually grew more common. De Dion's graduating class of 58 in 1851 was more than double the size of Barrault and Polonceau's fifteen years before. In his class, eleven worked in public works and construction, seventeen as general civil engineers, and twelve in railways (the most prominent among whom were de Dion himself, Léon Molinos, and Charles Pronnier). Two more worked in the machine industry, one in metallurgy (this was Molinos who was already counted under railways), one in

34. Steiner (1984, pp. 35–36) discusses the importance of financiers and bankers to the development of iron in France.

mining, and fifteen in industrial production, chemistry, or agriculture (whereby several are double entries like Molinos). Five, including de Dion himself and two other double entries, became professors.

Eiffel's 1855 graduating class had 71 students, among whom he ranked thirteenth. From mid-century on, double or even multiple professional fields became common.[35] Fifteen of the students, including Eiffel himself, went into public works and construction, eighteen into private practice as civil engineers, ten into railways (including one double entry), four into the machine industry (including Eiffel as a double entry), fifteen into metallurgy or mining (four multiple entries), fifteen into various industries (two double entries), five into teaching (three double entries) and three into business or the military. A number were also prominent in politics.The eleven most prominent graduates were mentioned in Léon Guillet's 1929 publication celebrating the school's first century, among the most noteworthy being Tommaso Agudio, the Italian railway engineer and inventor of a stationary traction system for the Genoa-Milan Railway, who also entered politics.

In Jenney's class of 1856, seventeen of 69 graduates entered the field of public works and construction, twelve went into private practice as civil engineers, nineteen chose railways (four double entries), nine chose the machine industry (four double entries), thirteen chose metallurgy or mining (five double entries), and thirteen chose various industries (four double entries). One became a professor, and ten were especially noted by Guillet: four in industry, three in engineering, two in education and two in politics.

Moisant graduated in 1859, twenty-eighth out of a class of 112, of whom eighteen entered into public works and construction, thirty-two into private practice as civil engineers, twenty-two into railways (five double entries), eight into the machine industry (three double entries including Moisant), twenty-three into metallurgy or mining (eight double entries, among these André Wurgler, who later became professor of mining at the ECP), seven became professors (four double entries, including Wurgler), twenty-seven entered into various industries (six double entries), one into law, and one who is laconically merely listed as 'proprietor'. Seventeen are especially mentioned by Guillet, most of whom became industrialists, with several in the iron industry and at least two of them in construction.

---

35. This is traceable in the *Liste par profession* (*École Centrale 1889*).

Contamin, whose academic record was consistently excellent, graduated in 1869, second in his class of 116, among whom eighteen chose public works and construction, many of these on the Suez Canal or the abortive French effort in Panama. Thirty-four went into private engineering practice, thirty into railways, which was Contamin's own primary field, (two double entries), ten into the machine industry, eight into metallurgy or mining (two double entries), twenty-four ended up in various industries (five double entries), and one in agriculture. Eight became professors, including Contamin himself, as well as four other double entries.

The listing is surely incomplete—Jenney's brief stint in railway construction is not mentioned, for example, but it serves to hint at the increasingly border-crossing occupations of the ECP's graduates over time, their growing industrial involvement, and the growth of private civil engineering firms.

The political and educational background

The first third to mid-nineteenth century was a time of political and social upheaval all over Europe and especially in France. The 1830 revolt *Les Trois Glorieuses* initiated a development that the 1848 anti-monarchical Revolution then spread from Paris to rock the already shaky socio-political structure of the whole continent. The first French president, Charles-Louis Napoléon Bonaparte, elected in 1848, staged a coup d'état in 1851 and declared himself Emperor Napoleon III in 1852.[36]

The situation was politically and socially turbulent, and it is impossible to characterize it in terms of a positivistic interpretation of development. When the ECP opened in 1829–30, the ultra-conservative King Charles X was trying hard to dismantle all the democratic gains the French Revolution had made and to suppress the industrial grande-bourgeoisie that was then a progressive segment of society. At the same time, the industrialized capitalism that this bourgeoisie promoted exploited the working class, which in a social sense concurred with what the reactionaries under the king were trying to achieve by returning to a quasi-feudal model. On the other hand, many supporters of the egalitarianism that the Revolution had engendered harbored what we would today consider reactionary views and were against the concept of compulsory education for everyone because it did not guarantee the freedom of choice to refuse education. Across the Channel, British politicians were quickly and brutally solidifying their worldwide colonial regime, as were the French in northern Africa and southeast Asia. At the same time, first Napoleon and then the British, although representing aristocratic societal structures, imported the concept of democracy into the repressive autocratic social structures they

---

36. He was counted as the third because his uncle, Napoleon I's four-year old son, had officially reigned for 16 days in 1815.

suppressed. There were negative and surprisingly even some positive aspects to the development everywhere; the times were truly chaotic.

Conquest and the resultant internationalization of contacts and political awareness led to the first genuinely international exhibition that merited the name, the Great Exhibition of the Works of All Nations in London, which closed in October 1851. The London Exhibition is particularly remembered for its astonishing Crystal Palace as a showcase for the industrialization of building materials and processes. Dublin and New York held copycat exhibitions in iron-and-glass structures in 1853, and Munich built an innovative Glaspalast for its own exhibition in 1854.

With the exception of the development of iron production and construction in Russia, which appears to have been driven by iron-founders but was completely unknown in the West (Kosykh 2021), most of the interesting structures were built by men who came from fields other than iron founding, architecture, or structural engineering. They were builders who had railway or mechanical experience and they led the way in this early phase of structural experimentation with the new material, iron. The London Crystal Palace was built by the gardener, railway investor, and ducal estate manager Joseph Paxton after successful experiments with glass hothouses, albeit with critical help from the railway engineer Charles Fox. Munich's Glaspalast was built in in 1854 by the locksmith and what we would now term mechanical engineer Johann Ludwig Werder, based on both the London structure and his successful market hall, the Schrannenhalle of 1851–52. These and all similar structures—many of which were built in France by *serruriers* (locksmiths)—will be discussed later in the chapter 'Implementation: the architect Labrouste adopts iron structure'. In general, locksmiths and a few ironmasters led the early development. Notable among their structures were the Sayn Foundry in Bendorf on the Rhine by Carl Ludwig Althans in 1830 and the influential Palm House at Kew by Richard Turner in 1848. These were innovative machine-buildings that simultaneously housed and performed mechanical functions. All of them were interesting one-off experiments using cast or wrought iron. In contrast to the many early Russian iron structures, they were all published and influenced the development as models, but it would be up to the graduates of the ECP to create modern iron construction by directing the development toward standardized structural systems and their industrialized production. ECP engineers began exerting their influence on industry and architecture in the transitional decade 1850–60, which is the period during which Barrault and Baudet contributed.

Before the early experimental period, construction had always meant either wood or the socially more prestigious masonry. Education in their use in design and construction was either practical through on-the-job apprenticeships or academically formalized at the ENPC and the Académie Royale d'Architecture (the predecessor of the École des Beaux-Arts). After the Revolution it was also taught at the new undergraduate school, the EP, which had been founded in 1794 as the École Centrale des Travaux Publics. The EP graduates supplemented their knowledge further with study at either the ENPC or the École des Mines. Pothier traces the development of scientific and technological education in France in detail (1887, pp. 371–416. See also Gourisse and Picon 2009).

After its first decade, however, the EP mutated into a military academy under Napoleon to produce military engineers to replace the professionals who had fled or been executed. Thus, it was not surprising that the school had gravitated away from its original purpose and had grown into a more theoretically weighted institution that educated scientists, military engineers, and engineer-administrators for government service, and not practical builders. Part of the problem was that the EP began to hire its own graduates as professors before they even had a chance to gain practical experience in the field, and also that it used their theoretical texts combining abstract mathematical analysis and geometry with calculus rather than with practical construction. While it is true that the ECP later also preferred to hire its own graduates as *répétiteurs* (teaching assistants) and then, if they proved worthy, as professors, they did this because of their ability to function equally well in practice and in theory, and the school only elevated them to full professorial level when they had proved themselves in the field.

As far as the building arts were concerned, Joseph-Mathieu Sganzin taught construction at the EP from 1796 to 1812, and Jean-Nicolas Louis Durand taught architecture from 1795 until his death in 1834. While Sganzin was an accomplished practitioner with vast experience, he left early, and both he and the theoretician Durand, who developed a celebrated design method, provided little more than overview courses. Durand's method, while entirely abstract, did influence the ECP's curriculum. It will be discussed in detail under 'Part II: Consolidation and Implementation'.

The several graduate écoles d'application that the Revolutionary government developed or reorganized on Jacques-Élie Lamblardie's suggestion had as their goal the professionalization of the basic knowledge taught at the EP. The application of this basic knowledge to engineering practice was the role that the ENPC adopted. It provided an advanced professional training, but the courses simply applied existing scientific knowledge to practical problems. The students were not enabled to create and develop theoretical approaches to the practical problems they encountered themselves, and instruction was biased toward construction for the state, such as roads, bridges, official public buildings, and fortifications. The two practical courses that did exist were provided by Charles-François Mandar (architecture as construction) and Louis Bruyère (materials and technology as construction). Their influence on the ECP will also be discussed in Part II. These courses did provide a professional basis, but without the crucial crossover to theory or a connection to industry. An educational gap had grown at the beginning of the century that was especially apparent to the country's fledgling industrial base. Industry needed professionals who could develop theoretical as well as practical approaches to new problems.

This gap came into being as soon as the government's undergraduate EP abandoned the educational concept that Gaspard Monge had originally devised. The school was militarized in 1804 under pressure from the emperor that was supported by a group of economists and the military school at Metz (formerly Mézières).[37] Napoleon also revoked the financial support system at the EP for less-affluent students and introduced a Latin requirement for admission with the argument that young men from 'better' families and with prior academic education lessened the danger of revolutionary tendencies.

The increasing concentration on science after Napoleon's downfall eventually led to the high intellectual level of scientific education that the undergraduate EP still enjoys today. However, the socio-intellectual shift toward the military combined with the reorganization of the school in 1816 destroyed what remained of practical engineering education as one of the ECP's founders Théodore Olivier wrote in his diatribe in 1851 (pp. xi-xii, cited by Sakarovitch 1988, p. 322) and Francis Pothier later repeated (1887, p. 26). The EP's theoretical bias inspired only about 20–25% of the EP's graduates to opt in their graduate education for the civil side of engineering, and virtually all of these

37. The school had moved to Metz from Mézières in 1794. It had lost a significant part of its aristocratic graduate body through flight or execution during the Revolution. The school later merged with the École d'Artillerie de Châlons-en-Champagne in 1807 to form the new École d'Application de l'Artillerie et du Génie as the principal graduate school for the EP.

then worked in government. The rest entered the military branches (Lavallée/ Dumas 1834, p. 6). None even remotely considered entering the socially less prestigious industrial sector.

Lavallée's founding group reverted to the socio-economic roots and practical focus of academic engineering education by modelling the ECP on Gaspard Monge and his colleagues' original 1794 concept for the École Centrale des Travaux Publics (Pothier 1887, p. 26 footnote 1), and on a modified version of the combined theoretical and practical system that Jean-Rodolphe Perronet had originally developed for the ENPC. Perronet and Monge had been the two towering minds in French technical education in the late eighteenth century, and their pedagogical models provided powerful prototypes for the new school's pedagogy.

Perronet's development

Perronet's influence on engineering education was earlier and of a different kind than Monge's. Until the ECP opened its doors in 1829, the ENPC had been the only remaining academic school in France in the early nineteenth century that offered a technically based civil engineering program. Its origins in 1747 were based on practice. The founder, Jean-Rodolphe Perronet, was of Swiss Protestant extraction. His father David was a French-Swiss soldier from what was considered a 'good family'. He was therefore accepted as a *cadet* and later probably a non-commissioned officer in French royal service. Perronet's mother was a young Catholic girl. Perronet was born out of wedlock (Vacant 2006, p. 11), and although David lived with the mother of his son, they may never have married. The issue of religion in pre-Revolutionary France was important. Protestants stressed a completely different attitude toward secular education than the Catholic church, and this impacted engineering as well.[38]

According to a later, unsubstantiated account, Perronet was influenced by a great-grand-uncle in his youth, the mathematician and Protestant philosopher Jean-Pierre de Crousaz in Lausanne.[39] As Perronet later personally mentored

38. See Peters 2015. Like Prince Consort Albert, who stressed the importance of technical education two generations later in England, Perronet had evidently been raised in the Protestant confidence that secular education and work of any type were pleasing to God.

39. Although Vacant disputes this, erroneously stating that de Crousaz was in Sweden during the

Fig. 9. Jean-Rodolphe Perronet, portrait by Louis Michel van Loo in 1764. Perronet was the founder of the first academic civil engineering school. (Source: © École nationale des Ponts et Chaussées Collection)

the mathematics course at the ENPC (Vacant 2006, p. 89), this influence sounds plausible. At any rate, because of this early education, and after his apprenticeship in an architect's office in Paris, he was far better educated than most other architect-engineers of his day. His later work in Alençon testifies to the quality of his education (see Vacant 2006, p. 81). Perronet became a prominent *Encyclopédiste* and among the most eminent eighteenth-century engineering practitioners. He was convinced that a solid scientific education was mandatory for success in any modern technical field.

boy's youth (Vacant 2006, pp. 30–31), he was a professor of philosophy in Lausanne until 1724. Vacant also cites the fact that Perronet's father had converted to Catholicism and was therefore likely to have been estranged from his family—although he continued to practice as a Protestant (Vacant 2006, p. 29). This is a contradiction that does not exclude the possibility that his well-to-do paternal grandmother's family concerned itself at least to some extent with the boy and his education.

Protestantism and education

Protestants were a small minority in France, but they contributed out of all proportion to their number to the development of universal secular education and to the professional education of engineers as well. Before the Enlightenment, in which French philosophers played a major role, there were two main opposing attitudes toward education in Western culture. The first and older was that of the Catholic Church, for which education was an elite affair. The Bible and other religious texts were only available in Latin and the reasoning was that they needed expert interpretation in order to be correctly understood. Without education—and that was religious instruction reserved almost exclusively for the clergy—supposedly 'false' interpretations endangered salvation, and therefore the laity was required to trust the clergy and to believe, not to reason. Through belief, their salvation was guaranteed.

This almost hermetic educational principle was first countered by John Wycliffe and his group in Oxford, who translated the fourth-century Vulgate (Bible) into the vernacular in 1382–84, and by the Czech Jan Hus, who wrote several religious tracts a generation later in his vernacular as well. What drove their actions was their conviction that each individual was personally responsible for salvation, and that was only possible through direct knowledge of the texts. The individual needed therefore to be able to read, reason, and act. Supported by the political backing that Martin Luther and his followers received from German princes a century later, it became a credo that education and individual action were God-given commands in both religious and in secular matters. The belief was that action—both religious and secular—in this world would accrue reward in the next.

Luther therefore required compulsory primary education for both boys and girls. The doctrine was, as Luther wrote in 1520 in *An den christlichen Adel deutscher Nation*, a universal priesthood which has since been termed a priesthood of all believers. The Catholic Church recognized this as an attack on its authority. In response, the leaders of the Counter-Reformation, especially the Jesuits, reacted to what they could not stop by providing lay education as well, albeit ecclesiastically controlled, and also by being active in secular fields of knowledge, by promoting science, often even publishing in the vernacular themselves.[40] Lay education was then espoused by the French *Enclopédistes* in the mid-eighteenth century.

Over and over again in eighteenth-century treatises on secular topics written by Protestant scholars—especially in the German-speaking lands that spearheaded this development—one reads that the deity had blessed the writer with a discovery or invention, and that communicating this new knowledge in a publication was a duty that was pleasing to God.[41] The drive to understand and communicate is where fields like construction profited, and this was the mindset in which Perronet had been educated.

The distinction between the two opposing educational ideas eventually grew less pronounced in the French Revolution under the influence of the Encyclopédistes. Even before then, in predominantly Catholic France, Protestant French writers avoided invoking the deity in their forewords in order not to raise theological issues. But the Protestant educational concept was strong, fed by developments in Alsace and the French-speaking Swiss Cantons and dependencies, especially in Calvinist Geneva. Two Swiss pedagogues, Jean-Jacques Rousseau and Johann Heinrich Pestalozzi were particularly influential. Many studies have examined the relationship between universal secular education and Protestantism, but none so far have looked at its impact on engineering and construction.[42]

The Protestant Swiss were open to educating all social classes, including those involved in what were called the trades. By the 1830s it had become clear in the German-speaking countries that industry needed scientific knowledge. Another Swiss pedagogue Christoph Bernoulli, who had studied under Johann Beckmann in Göttingen, published his *Elementares Handbuch der industriellen Physik* in 1834–35,[43] and wrote in the introduction:

---

40. For example, see the Jesuit lay brother Andrea Pozzo's pioneering treatise *Perspective pictorum atque Architectorum* published in a dual Latin-Italian edition in 1693/1700 with a partial edition in French in 1700 Rome (only part 2 traceable), Latin-German 1707/09 Augsburg, Latin-English (only vol. 1) 1707 London, and Dutch-French 1710 Brussels.

41. An example is Leonhard Christoph Sturm's innovative *Gründlicher Unterricht... Von Häng- oder Sprengwercken* of 1713, to date the earliest known book on bridge construction.

42. See Peters 2015 for a partial discussion of this influence in the development of wooden truss construction in the eighteenth century.

43. He had previously published a *Vademecum des Mechanikers* in 1829.

> There is no doubt that we Germans have no lack of books, and excellent ones, too, in all branches of the physical and mechanical sciences. But most of these are not written with an eye to the true needs of the industrial classes, and do not contain that material which, under the given circumstances, the vast majority of such readers require to obtain and learn. (Bernoulli, p. viii)[44]

As a result of the stress that Protestant thinkers laid on education, many innovative French engineers and engineering educators of the eighteenth and early nineteenth centuries—a far larger percentage than the tiny French Protestant population warranted—were either of Protestant background like the concrete pioneer Louis Joseph Vicat, or had been educated in Protestant-run institutions or by Protestant scholars, among them Perronet. The reason is evident: by taking active responsibility for education instead of passively relying on belief, and by taking the initiative in the development of our world instead of focusing on the hereafter, engineering and technology generally became socially acceptable and even desirable human preoccupations.

Perronet's career

Thanks to this educational culture and the fact that he had already founded a drafting office with an educational bias in Alençon (Vacant 2006, p. 77), Perronet was designated in 1747 by Louis XV's finance minister Daniel-Charles Trudaine[45] to be director of an enlarged Bureau des géographes et des dessinateurs du roi[46] that was designed to coordinate the cartographers' work in the then 32 *généralités*.[47] On the basis of a memorandum he then wrote to Trudaine (Vacant 2006, p. 80), he was empowered to organize the education of the civil engineers of the Corps des ponts et chaussées for the state as well. Recognized for his

44. Unstreitig fehlt es uns Deutschen nicht an Büchern, und an trefflichen, über alle Zweige der physikalischen und mechanischen Wissenschaften. Die allermeisten scheinen aber wenig die eigentlichen Bedürfnisse der industriellen Stände im Auge zu haben und nicht vorzugsweise eben das zu enthalten, was die grosse Mehrzahl solcher Leser bei den einmal vorhandenen Verhältnissen zu erfahren und zu lernen verlangt.

45. Trudaine had been appointed finance minister in 1743 and founded the 'school' three years later.

46. The Bureau des dessinateurs du roi had originally been established in 1744 to map France.

47. The term *généralité* corresponded roughly to the more modern administrative unit of the *préfecture*. By the time of the Revolution in 1789 there were 36 such *généralités*.

innovative work as an engineer and educator by the royal government, Perronet was named *Premier ingénieur du roi* in 1763 and was ennobled that year as one of the very few technologists to be so honored under the *ancien régime*.

The system that Perronet introduced in this first 'civil' as opposed to military engineering establishment was experimental. Students were accepted from all social strata, which was a novelty at the time in an official school (which the ENPC was not yet), but acceptable because the *aspirants* (candidates) were usually recruited from among the draftsmen from the provinces. Science and mathematics were important basic skills,[48] but the *bureau* had no permanently employed teachers. Perronet based his didactical concept on a combination of peer teaching, by which the older students, most of whom originally came from military schools with some knowledge of their field, taught the younger,[49] and on the traditional apprenticeship system that had been usual in the building trades up until his time. Perronet did some of the teaching himself and also occasionally called upon colleagues and friends to teach courses, among others the eminent Bernard Forest de Bélidor (Vacant 2006, pp. 86–87 and footnote 17). This was not all he did, however, and he supplemented instruction with organized site visits that aimed to link theory and practice.

Initially the in-house subjects were geometry, algebra, mechanics, and hydraulics, with the rest delegated to occasional outside instructors. It was only Perronet's say-so as both director of the school and head of the *corps* that determined whether a student was considered advanced enough to graduate into the Corps des ponts et chaussées (Vacant 2006, p. 106). The conflation was unique, as Hélène Vérin pointed out (1993, p. 203), and potentially negative because it could easily lead to arbitrary decisions. Gradually, from the mid-1750s (Vacant 2006, pp. 95/334), the bureau transformed itself into a school. The arbitrariness ended in 1775 under the reorganization that the *Encyclopédiste* Anne Robert Turgot and the educational reformer Nicolas de Condorcet instituted. Perronet established a regime of exams and lectures when the bureau formally became a school the following year as the École Royale des Ponts et Chaussées (ENPC),

48. Antoine Picon traces the development of the mathematization of scientific and especially engineering knowledge in France and its relationship to educational policy and development in the eighteenth century (Picon 2009).

49. This was a very early monitorial system as was later propagated for British colonial primary education in the early nineteenth century.

and he supplemented the original core curriculum with more mathematics, architecture, stone cutting, and timber framing. Despite his own keen interest in science, being the practitioner that he was, Perronet resisted going overboard in the direction of theory, in contrast to those two generations later who espoused the transformation of the EP in 1816.

Although technological subject matter had also been taught at the Académie royale d'Architecture by François Blondel from 1671, Perronet's hybrid form of education was unique, and so was the fact that he was simultaneously in charge of both the *corps* and the school. The only other French academic engineering schools were the more formalized army establishments with militarily focused curricula and technical input that was restricted to the army's needs.[50] These were the several Écoles de l'Artillerie founded in various French cities from the end of the seventeenth century,[51] especially the one at La Fère from 1720 under Bélidor and the École du Génie de Mézières where Monge later began his career, and that had been founded as a military counterpart to Perronet's *bureau* in 1748 under Nicolas de Chastillon.[52] Thanks to the fact that both Charles Bossut and Monge taught at Mézières, Picon considered the school to be the most modern school of the eighteenth century and one of the templates for the development of the EP (1988, p. 117, note 11).

It is probable that the model on which the French military schools had been established had originated in Spain. The Spanish crown had founded its first military academies that taught scientific subjects in the sixteenth and seventeenth centuries. They were all what we might call proto-engineering schools. A forerunner, the Academia Real Matemática, had been founded in

50. The only other 'civil' technical school around that time, the École d'Arts et Métiers, originated in 1780 as an informal private school for training military orphans and foundlings in trades. By 1788 it had become the École des Enfants de l'Armée, under royal patronage. It was of far lesser importance and had no academic pretensions. It was only at the beginning of the nineteenth century that it became what today would be termed a trade school, and formal engineering diplomas were only instituted in 1907.

51. The first at Douai in 1679.

52. The École des Mines, the only other French civil (as opposed to military) engineering school of the eighteenth century, was only founded in 1783, modelled on the Bergakademie Freiberg of 1765. A knowledge of German was compulsory at the school at the beginning (Kosykh, p. 49). The school disappeared during the Revolution but was reestablished in 1794. For the reorganization of French engineering education at the beginning of the French Revolution, see B. Belhoste 1989.

1582 in Madrid under Juan de Herrera, who was a pioneering proponent of mathematical instruction for all the technical arts (Portuondo 2009, pp. 82–86). The Academia apparently expanded its offerings to include engineering, but only for a brief interlude in 1607–11. This was followed by what may have been the first true military academy,[53] the Academia Militar del Ejército de los Países Bajos in Brussels founded by Sebastián Fernández de Medrano in 1674 (or possibly even earlier, in 1671 by Francisco Pavan de Cecati).[54] However, not only were the curricula of all the military academies more technologically restricted than Perronet's but, as far as France is concerned and probably the others, too, they were reserved exclusively for a select group of aristocratic cadets.

In contrast to the institutions that trained officers whose objectives were exclusively military, Perronet's school trained civil servants who built infrastructure for the state. Like the aristocratic military hierarchy, the ENPC graduates also formed a socially elite group, but as the school was open to all social strata, its graduates were by definition upwardly mobile, and this, too, was uncommon in eighteenth-century France. However, as Perronet's students also originally self-monitored the moral and social behavior of their fellows (Vacant 2006, pp. 107–108), another somewhat suspect and arbitrary practice, they did guarantee a certain level of intellectual and social conformity and cohesion.

Initially those who did graduate into the *corps* worked in the public sector as was originally intended. Over time however, especially after the Revolution with the burgeoning of private or semi-public projects for industry, more and more chose to go into private practice instead. Among those in private practice by the mid-nineteenth century we count many engineers like Jean-Baptiste Krantz, (grad. 1840?), who is said—albeit somewhat doubtfully—to have engineered the novel cable-suspended roof of Jacques Ignace Hittorff's

---

53. This is contested by the Military Academy of Savoy founded in 1669 in Turin, today the Military Academy of Modena.

54. These were followed by the Compañia y Academia de Guardias Marinas in Cádiz (1717) and finally the Real y Militar Academia de Matemáticas in Barcelona (1720) established by Jorge Próspero de Verboom who had studied under Medrano in Brussels. Three other early academies with relevant curricula were the Søe Cadet Compagniet in Copenhagen, now the Søværnets Officersskole (Royal Danish Naval Academy) that provided some basic training in mathematics in 1701, the k.u.k. Technische Militärakademie (Royal and Imperial Technical Military Academy, 1717) in Vienna, and the Royal Military Academy in Woolwich (1741) in England that all trained military engineers.

1839 Rotonde des Champs-Élysées while still a student, and who later built the 1867 Paris Exhibition building; Stanislas de La Roche-Tolay, a graduate of 1841 who, together with Paul Régnauld (grad. 1846) developed a series of iron truss bridges including the 1858–60 Bordeaux railway bridge on which Eiffel gained his initial practical experience; and Wilhelm Nordling, who also built iron bridges.[55]

Time had changed the original form of the ENPC, which had adjusted its pedagogy and its organization in reaction to the political pressure that regularized graduation procedures and forced the introduction of more science when it became an official school after 1775 (Michel 1981). Another factor that forced changes was the organizational difficulties that arose when new laws were enacted in 1791 after the Revolution (B. Belhoste 1989, pp. 26–29).[56] At the same time, the new laws also revitalized the whole higher education system of the military schools, as these had degenerated during the last pre-revolutionary phase (Comberousse 1879.1, pp. 3–4). The result was the creation of the organized form of education with formal lectures and exams that we are familiar with today and that have been attributed to Monge's innovations.[57] The foundation of the EP and its designation as an undergraduate preparatory school with a scientific bias was a hiatus in practical engineering education. It undermined the integrative format that Perronet had originally established, and the ENPC became an école d'application that applied theoretical knowledge to practice. But even before this, changes occurred at the ENPC under successive directors: the EP co-founder Lamblardie who followed Perronet (1795–97), then Antoine Chézy for a single year (1797–98).[58] Perronet's erstwhile protégé, the theoretician and practitioner Gaspard Riche de Prony followed Chézy and exerted a major influence on the development of the school. He influenced a number of students who became ECP teachers and was still its director when Lavallée founded the ECP in 1829.

---

55. Wilhelm von Nördlinger from Württemberg, later naturalized in France as Wilhelm Nordling, was still a foreigner when he was permitted to attend the ENPC, but only as an auditor (1843–44).

56. Bruno Belhoste details these difficulties and the changes citing Antoine Picon's work on the topic.

57. Bruter 2008 suggests a more complex origin in the public lectures of the eighteenth century and its transformation into the educational establishments of the time (citing B. Belhoste 2008).

58. Chézy had already unofficially served as co-director from 1763, when Perronet was appointed Premier ingénieur du roi, and his appointment became official from 1782 (Vacant 2006, p. 109).

Fig. 10. Gaspard Monge, the main force behind the founding of the École Polytechnique and the development of engineering pedagogy. (Source: open source, Wikipedia)

Monge's background

Gaspard Monge was forty years younger than Perronet and had been recognized for his mathematical ability as a schoolboy. Two decades after Perronet had founded his school, Monge began his professional career as a draftsman at the École Royale du Génie in Mézières in 1765 based on a plan of Beaune that he had drawn. It was there that he developed the new field of descriptive geometry. At first his system was a graphical method for designing fortifications that derived from a long history of geometry, but it quickly proved to have major implications in other fields as well, for instance in ballistics for the calculation of trajectories and later also for technical design generally. The mathematics professor at the school, Charles Bossut, recognized Monge's gifts and the potential importance of his work.

As a non-aristocrat, Monge could not be admitted to the academy as a student, so Bossut hired him as his *répétiteur* for mathematics in 1766 in order to be able to foster his development. Bossut left the academy for Paris to teach hydraulics at the Académie royale des sciences two years later,[59] but he was able to designate Monge as his successor at the school, a position that he took on in 1769. He had made such

59. The Académie was housed at the Louvre and served as a proto-university for science instruction.

an impression that the physicist Jean-Antoine Nollet apparently also appointed him *répétiteur* in physics, and Monge succeeded him as well the year following Nollet's death in 1770. As he was broadly interested in mathematics, science, and nature, he took his students on excursions into the countryside around Mézières and combined theoretical and practical teaching in geology and physics. They were captivated by his border-crossing intellectuality and his enthusiasm (Dupin 1819, p. 10).

The school at Mézières had classified descriptive geometry as a military secret. It was not even to be made known to the other rival military academies, and this prevented Monge from becoming known by publishing his method. As his disciple Olivier wrote in 1844,

> …since he was forbidden as professor at the *Mézières* military academy to publicize his geometrical methods, he was obliged to reinterpret the results *analytically* that he had so straightforwardly developed by geometrical projection. (Olivier 1843/1844, vol. 1, pp. vi–vii)[60]

By avoiding sanction in this way, he contributed innovative developments to algebra too. Monge had also encountered stiff resistance from traditionalists at the school in his attempts to broaden the applications of his technique, and he may have encountered class discrimination as well. In compensation, perhaps also in frustration, he began researching several other branches of mathematics and the sciences. His inadvertant border-crossing, through which Monge contributed to the development of algebraic theory and other fields, was to prove beneficial for the inclusive concept that he later developed for the foundation of the EP.

The social climate was changing in France even before the Revolution, and Bossut continued to foster Monge's polymath development from Paris. He and Condorcet helped the young mathematician to publish his first papers at the Académie des sciences, upon which he was subsequently elected as an associate in 1780. On the basis of his publications, Monge began teaching hydrodynamics under Bossut at the Académie that year and, because of his teaching experience there and at Mézières, he was offered a position as Examinateur des gardes du pavillon et de la marine three years later, in 1783. This was a position that Étienne Bézout's death had opened up, and it involved developing urgently

60. …comme il lui était defendu de faire connaître ses méthodes géométriques, attendu qu'il était attaché comme professeur à l'école du génie de *Mézières*, il fut obligé de traduire en *analyse*, les résultats auxquels il était parvenu directement par la méthode des projections.

needed curricular reforms. The educational situation at all the military academies had slowly eroded over time even though teachers like Bézout had written important and popular textbooks. The new position provided Monge with further experience in *vulgarisation.* He initially commuted the 250 kilometers at six-month intervals between Mézières in the Ardennes and Paris. For a time his younger brother Louis helped him to fulfill the role of examiner (and continued after Monge left). In December 1784 Monge could no longer cope with the overload and he resigned from the army's academy. This allowed him to devote more energy to reforming the navy curriculum, and he developed a statics course for the cadets in 1786 (*Traité élémentaire de statique*, published in 1788) while continuing to write papers on many widely ranging topics, from combinatorial analysis in *Réflexions sur un tour de cartes* in 1773,[61] the movement of knights on a chessboard in 1780 that apparently remained unpublished,[62] optics, tides, and geometry, to metallurgy, chemistry, and physics. The two studies on a card trick and chess moves testify to Monge's specific qualities as *vulgarisateur*, not so much in the sense of applying science to practical problems, but rather as a method of using visualization to study even the most abstract issues by means of a concrete example. They also illuminate what might be a sense of playfulness in his character, and perhaps they also indicate an acceptance of playfulness as a component of innovation in science. De Launay reported that the editor of the *Mémoires de mathématique et de physique* wrote:

> Discussions of this type are of interest as they represent a totally new category, and it would therefore be hoped that M. Monge would examine other less simple laws of permutation in order, if possible, to derive a general theory from them. We judged this memorandum worthy of inclusion in the recueil des Savants. [63] (de Launay 1933, chapter 1)

61. Monge presented a total of three mathematical *mémoires* in this volume of the *Mémoires de mathématique et de physique*. His paper on the probability of the card trick had been preceded by Montmort, *Essay d'analyse sur les jeux de hazard* in 1708, de Moivre, *The doctrine of chance* in 1718, and de Mairan, Sur le jeu de pair ou non, in: *Histoire de l'Académie royale des sciences, année 1728*, Paris 1853, pp. 53–57.

62. Reported, for example, by Jelliss, citing Bastian.

63. Ces discussions sont intéressantes parce qu'elles sont d'un genre tout à fait neuf, et il serait à désirer que M. Monge examinât de même d'autres lois moins simples de permutation, afin de parvenir, s'il est possible, à une théorie générale. Nous croyons ce Mémoire digne d'être inséré au Recueil des Savants étrangers.

Monge was quickly recognized as one of the foremost scientists in France, and he collaborated with eminent colleagues of his time in several disciplines.

The Revolution intervened in 1789. Monge enthusiastically supported the change but remained politically on the sidelines and focused on his educational duties with the navy. During this time also he became one of the founders of the *Annales de chimie* and a member of the Académie des sciences's committee on weights and measures. These activities guaranteed him a certain degree of professional stability in the troubled times. Even after the Academy itself was temporarily abolished in 1793, the committee that included Laplace, Lagrange, and several others, continued under an altered name and eventually led to the adoption of the metric system in 1799.

When France declared war on Austria and Prussia and the monarchy was abolished in 1792, Monge was appointed Minister of the Navy (Ministre de la marine et des colonies). This proved too much even for the workaholic that he was, and he resigned in April of the following year to return to research and teaching. Repeatedly, Monge seems to have preferred work to position and to deemphasize the honors and high offices that the government accorded him.

After the reorganization of military and civil education during the Revolution, Monge began to teach descriptive geometry openly in 1793, assisted by his former student Jean Nicolas Pierre Hachette, the mathematician Joseph Eisenmann, and Sylvestre-François Lacroix (who had followed Monge's course in hydrodynamics). They first taught it briefly at the short-lived teacher college, the École Normale de l'An III[64] in 1793, and then in 1795 when teaching began at the École Centrale des Travaux Publics that he had helped to establish the year before. The lectures were partly recorded and then published in the *Journal des séances des écoles normales, recueillies par les sténographes et revues par les professeurs* (1795, vol. 2, pp. 338–368). Hachette, who took charge of teaching and developing the subject further, subsequently published the first nine of Monge's original thirteen lectures as a treatise in 'An VII' (1798–99) without Monge's knowledge, as he was in Egypt at the time, but with his wife's permission. He had developed them from the lectures that had been recorded for the *Journal*.[65]

---

64. The school was subsequently recreated by Napoleon in 1808 and again as the 'École préparatoire' in 1826, adopting the name 'École normale' in 1830 and adding 'supérieure' in 1845.

65. Hachette added the rest of the lectures and his own notes in 1811, which makes this edition more

Monge explained his pedagogical intentions in the forward to this seminal treatise. We cannot be sure how much Hachette contributed to this statement, as it was he who edited and published it while Monge was abroad, but the two men collaborated so closely in this period that it can be attributed to both of them.

> In order to lift France out of its current dependence on foreign industry one must first orient national education policy toward the knowledge of objects that demand accuracy, a need that has been totally neglected up till now, and to habituate our artisans' hands to the manipulation of all kinds of instruments that serve to render and calibrate work precisely. Thereafter, consumers, having become accustomed to precision, will demand it and pay the necessary price for various goods, and our artisans, familiarized with it from a young age, will be able to deliver it.
>
> Secondly, one must popularize the knowledge of a great number of natural phenomena that are indispensable to the development of industry, and then profit from the happy circumstance of the availability of the chief necessary resources to advance the popular education of the nation.
>
> Finally, one must spread the knowledge of creative processes in the crafts and in the building of machinery that have as their goal either the reduction of manual labor or the ability to render the results of labor more uniform and precise, and in this regard, one must admit that we have a lot to learn from foreign nations.
>
> One cannot satisfy all these requirements without reorienting national education. (Monge 1799, p. 1)[66]

---

complete than the first of 1799. 'An VII' (1798–99) had also been the year Monge himself published the augmented third edition of his *Traité élémentaire de statique*.

66. Pour en tirer la nation française de la dépendance où elle a été jusqu'à présent de l'industrie étrangère, il faut, premièrement, diriger l'éducation national vers la connoissance des objets qui exigent de l''exactitude ce qui a été totalement négligé jusqu'à ce jour et accoutumer les mains de nos artistes au maniement des instrumens de tous les genres, qui servent à porter la précision dans les travaux et à mesurer ses différens degrés : alors les consommateurs, devenus sensibles à l'exactitude, pourront l'exiger dans les divers ouvrages, y mettre le prix nécessaire ; et nos artistes, familiarisés avec elle dès l'âge le plus tendre, seront en état de l'atteindre.

Il faut, en second lieu, rendre populaire la connoissance d'un grand nombre de phénomènes naturels, indispensable aux progrès de l'industrie, et profiter, pour l'avancement de l'instruction générale de la nation, de cette circonstance heureuse dans laquelle elle se trouve, d'avoir à sa disposition les principales ressources qui lui sont nécessaires.

Il faut enfin répandre parmi nos artistes la connoissance des procédés des arts, et celle des machines qui ont pour objet, ou de diminuer la main-d'œuvre, ou de donner aux résultats des travaux

Monge was certainly in agreement with this statement even if he had not formulated it himself because it was repeated verbatim in the 1811 complete edition of the treatise, with which he certainly had been involved (pp. vii-viii). It was a clarion manifesto of the intentions that had led to the foundation of the EP program, and it influenced both the founders of the ECP profoundly and the future of technical education worldwide. Moreover, it appears to be the first expression of the idea of quality in design and of the concept behind design in manufacture, earlier than the beginning that Guillerme posits with Gérard-Joseph Christian in 1819.[67] Far from being a scientist who only considered intellectual abstractions, Monge had gained personal experience in industry. His wife Marie-Catherine, whom he had married two decades earlier in 1777, had inherited a foundry from her first husband at Forge du Prince, Brûly-de-Pesche, then the Austrian Netherlands (now in Belgium), just over the French border, close to Mézières. Monge, who was then still based at the military academy, concerned himself with running the forge, and it inspired him as part of the expansion of his interests to research metallurgical questions from the late 1770s. In 1793 he co-published a memorandum on the manufacture of steel and another as sole author on manufacturing cannon in 1794, and this brought his work to the attention of the government. He befriended Napoleon while on a government mission looting art works in defeated Italy for shipment to Paris in 1796. Throughout the following period and until Napoleon's final defeat in 1815, Monge was fully supported by the First Consul and then Emperor in his scientific work, involved in cultural matters, and highly honored by the empire.

Monge as a teacher

In addition to his diverse interests in mathematical, technological, and industrial research, Monge was involved in multiple public duties that appear to have been less to his liking despite his adulation of Napoleon. Monge was first and foremost a passionate researcher and teacher. His educational impulse and experience in incorporating science into technical instruction became the driving force behind the development of the École Normale de l'An III and the École Centrale des

plus d'uniformité et plus de précision ; et à cet égard, il faut l'avouer, nous avons beaucoup à puiser chez les nations étrangères.

On ne peut remplir toutes ces vues qu'en donnant à l'éducation nationale une direction nouvelle.

67. *Prodromes et conditions de l'essor du 'Design'*, 1993, reprinted in Guillerme 2008, pp. 309–316.

Travaux Publics (EP). He had founded the latter in 1794 with the civil engineer Jacques-Élie Lamblardie, a former Perronet student, and two of his own former Mézières graduates, the mathematician and politician Lazare Carnot and the military engineer Claude-Antoine Prieur-Duvernois. They conceived the new school as a replacement for both the ENPC and Mézières, which the government had closed down (Dupont 2000, p. 1). This idea was quickly abandoned as unrealistic, however, and the government reestablished the two older institutions as graduate schools, while the EP became their preparatory undergraduate institution. Lamblardie was appointed as the EP's first director, and Monge, who had formulated its original didactical goals in two documents (Dupont 2000, p. 1)[68] and then explained them in the manifesto quoted above, followed as director in 1798. However, he only filled the position for a short period that was interrupted by Napoleon's scientific mission in Egypt. When he returned to Paris late in 1799, he taught several mathematical subjects at the EP and published widely. His student and collaborator Hachette became responsible for teaching descriptive geometry, the field that came to be of prime importance to mechanical and industrial engineers. The relevance of this field to the development of an 'engineering language' is described by Karl-Eugen Kurrer (2002, pp. 41–42), and Monge himself explained it in the forward to the treatise on descriptive geometry. He considered drawing a language, one that an engineer must master:

> This skill [descriptive geometry] serves two primary objectives. The first is to reproduce objects precisely that are three-dimensional and that can be defined clearly in drawings that have only two.
>
> From this standpoint it represents a language that is necessary for an engineer to know in order to conceive a project, for those who must carry it out, and finally for those artisans who must manufacture the various parts themselves. (Monge 1799, p. 2)[69]

68. The respective roles of Monge and Lamblardie in the definition and foundation of the new school is a complicated question; see Bruno Belhoste 1994. Thanks to Olivier Azzola, Archiviste, Responsable du Centre de Ressources Historiques-MusX, École Polytechnique for this reference.

69. Cet art [la géométrie descriptive] a deux objets principaux. Le premier est de représenter avec exactitude, sur les dessins qui n'ont que deux dimensions, les objets qui en ont trois, et qui sont susceptibles de définition rigoureuse.

Sous ce point de vue, c'est une langue nécessaire à l'homme de génie qui conçoit un projet, à ceux qui doivent en diriger l'exécution, et enfin aux artistes qui doivent eux-mêmes en exécuter les différentes parties.

Monge's lecturing and mentoring style fascinated and captivated his students. As Charles Dupin reported:

> The most remarkable thing about Monge's lessons was that his gaze took in the whole vast auditorium despite his intense concentration on his topic. He was able to perceive whether his explanations were easy to follow or whether they confused his listeners, even those [seated] farthest away from him. As soon as he saw confusion cloud a face, that expression on someone who can no longer follow a train of thought, he immediately stopped and considered what steps his thoughts had taken that had been too quick, what more elementary path he ought to follow. And then, with charming affability, he apologized for his failure to be clear and began explaining again, presenting a more cautiously and more simply developed method that was often even more elegant than the first.
>
> This is how Monge taught in our lecture halls. Before and after the lectures he received us with a fatherly kindness and his students presented their difficulties and their attempts at developing geometrical research. He judged these attempts less on the merits of their absolute value and more on indications of their logic and the imagination they inspired in him. He always tended toward indulgence in his evaluations. If he pointed out logical mistakes in the path one had followed it was only to help one to avoid them in the future. He showed one how to rectify the path and suggested ways that seemed more promising to avoid intellectual roadblocks. This is how Monge received my first attempts. It was he who taught me to appreciate how inspiring and sublime the simplest praise could be when spoken by an eminent mind. My heart still beats higher at the thought of Monge, seventeen years ago, encouraging the feeble efforts of a young man with no support, no advocate, and no recommendation to his name. (Dupin 1819, pp. 80–81)[70]

---

70. Ce que Monge avait de plus remarquable encore dans les leçons, c'est que, malgré la forte attention qu'il donnait à son sujet, son regard embrassait tout un vaste auditoire, et démêlait sur les figures mêmes les plus éloignées, si ses préceptes s'offraient sous une forme facile, ou n'étaient plus que difficilement entendus. A peine voyait-il se peindre sur les physionomies, cette contention pénible de l'homme qui cesse de suivre le fil de la vérité, il s'arrêtait brusquement: il examinait dans sa pensée quel pas trop rapide il venait de faire, quelle route plus élémentaire il convenait de suivre; et alors, avec une charmante bonhomie, il s'accusait de n'être plus assez intelligible; il recommençait son explication, et présentait une méthode plus développée, plus facile et souvent plus élégante que la première.

Tel était Monge professant dans nos amphithéâtres. Avant et après les leçons, il accueillait avec une bonté paternelle, et les élèves qui lui soumettaient leurs difficultés, et ceux qui lui présentaient leur

The three basic skills that Monge promoted at the EP were mathematics, graphics, and experimentation (Dupont 2000, pp. 1–2). It was apparently he who established the lecture format as a central teaching method, and it was he who first blurred the borders between science and technology, especially through descriptive geometry. He communicated his enthusiasm for both fields to his colleagues and students. Monge was particularly interested in the practical incorporation of science in industrial processes that he had developed in his wife's foundry since the late 1770s, as well as descriptive and differential geometry and the theory of machines. His conviction of the importance of visual education through geometry and drawing underpinned his urge to teach. Monge's insistence on the importance of visual learning was a novel aspect in higher education. It was his most significant contribution to the future of engineering and a concern that would later form a major theme in the ECP curriculum.

As Dupin had written, he identified with his students and their problems and he even supported some of them financially (Dupin 1819, p. 41). Many of these, like Dupin, Barnabé Brisson, or Gaspard Riche de Prony, whom he taught privately (Dupin 1819, p. 14), became leaders in education. Other former students were active in the development of the use of mathematics as a new language in engineering, like Jean-Victor Poncelet or Bélanger. As far as the ECP was concerned, it was Gaspard-Gustave de Coriolis, Bélanger, and Mary who followed his lead in teaching. Olivier was another Monge disciple. Like the others, he also studied under Hachette who had established a course on the theory of machines in 1806 based on Monge (Dupont 2000, pp. 2–3) and who championed the same views and enthusiasms.

Both Monge and Hachette lost their positions after Napoleon was defeated in 1815 and the political winds shifted. The theorist Pierre-Simon Laplace, a friend and colleague of theirs, gained precedence in the EP reform of 1816. Laplace and the newly appointed Augustin-Louis Cauchy had entirely different views

---

essais dans la carrière des recherches géométriques. Il jugeait ces essais moins sur leur valeur absolue que sur les indices de raison et d'imagination qu'ils pouvaient lui fournir. Il penchait toujours du côté de l'indulgence; s'il montrait les erreurs de la route qu'on avait suivie, c'était uniquement pour qu'on les évitât à l'avenir. Il indiquait, avec les moyens de rectifier cette route, ceux qui lui semblaient les plus propres à en reculer les bornes. C'est ainsi que Monge accueillit mes premiers travaux, c'est lui qui m'apprit à connaitre combien est inspirant et sublime le plus simple éloge décerné par une illustre voix. Mon cœur palpite encore à la pensée de Monge encourageant, il y a dix-sept ans, avec sa bonté généreuse, les faibles efforts d'un jeune homme sans appui, sans prôneurs et sans recommandation.

on education from Monge and Hachette and espoused a far narrower and more stringent view of the curriculum as a strictly mathematically driven scientific program. Dupin railed against this didactical reform in his 1819 obituary of Monge (pp. 81–84; see also Pothier, p. 26), as did Olivier years later, sneering about those he called *algébristes* (Olivier 1851).[71] It was left to Monge's disciples to maintain and develop his unique mentoring attitude and his border-crossing interests that bridged science, technology, and industry at the ECP in the following generation.

Both Perronet and Monge exerted great influence on engineering education in France and abroad, especially through their disciples' work on the development of the ECP's curriculum. Monge's concept of the original EP as well as the experience of the ENPC's practicing engineers informed the professional knowledge base on which the ECP's program was founded. ENPC engineers also supplemented the students' subsequent development in the field. Both de Dion and later Eiffel were later employed by ENPC engineers and gained much of their initial practical experience from them. They also profited equally from many of their fellow *centraliens*.[72]

There was much in common between the two schools, but there were also significant differences. One of the most telling was the self-image of their graduates. ECP engineers, following Monge, moved naturally and quite unselfconsciously back and forth between theoretical abstraction and practical construction, while ENPC engineers were often wary of what they felt was an overeager incorporation of new theoretical ideas into practice. Nordling showed himself to be skeptical of Contamin's 1866 presentation of a novel assumption in the calculation of truss behavior. In a feisty dispute in the Société des ingénieurs civils on January 5, 1866 the *Mémoires et compte rendu* reported:

> Analytical research manifests a charm to which M. Nordling is far from insensitive, and this kind of research is not so strongly in vogue that one would have to discourage it. However, if science has its uses, so does practice. This holds as true in mechanics as it does in philosophy: every new theory that arises and that seems to reject what went before has the effect

71. Olivier polemicized in several of his writings against the changes that Laplace had made (see Sakarovitch 1988, pp. 321ff).

72. These contacts to other *centraliens* like Émile Trélat (grad. 1840) the founder of the first modern architecture school, Léon Molinos (grad. 1851), the industrialist Ernest Chabrier (grad. 1847), Alexandre Gottschalk (grad. 1853), and Jules Charton (grad. 1862), are detailed in Seitz, chapter 2.

> of creating skepticism in the mind of the practitioner. (*Mémoires et compte rendu* 1866, p. 43, quoted in part by J.-F. Belhoste 2008 in Barjot and Dureuil, p. 75)[73]

The discussion carried over to the next meeting on January 19, where Contamin made it clear that his calculations were quite orthodox and based on Bélanger:

> M. Contamin declared that the principles on which his theory were based were those developed by Bélanger, and, if he had not stated this from the outset, it was only because the development was his own and that he had to assume responsibility for his own work: he wanted to raise this discussion to clarify the question, and Bélanger's respected name and authority might possibly have prevented useful criticism. (*Mémoires et compte rendu* 1866, p. 55)[74]

Nordling's objection is certainly in part attributable to generational differences—Contamin at twenty-six was almost a generation younger than Nordling at forty-five—but the distinction between the two modes of thought was also founded on the clear demarcation that Nordling drew between theory and practice and his thinking in distinct categories. Nordling's professional worldview as it had been formed through his education at the ENPC was diametrically opposed to the blurring of the boundary between the two by means of *vulgarisation.* This was the Monge approach that the ECP founders didactically enhanced and on which they based their curriculum. We shall see how the ECP taught theory through objects, and how that was conceptually very different from all other engineering programs.

Non-elitist education in the service of industry was very much at the forefront of political concern by the first third of the century in France. After its founding in 1829, the ECP rapidly became the country's most forward-looking and

73. Les recherches analytiques ont un charme auquel M. Nordling est loin d'être insensible, et ces sortes de recherches ne sont pas tellement dans le goût du jour qu'il faille les décourager. Mais si la science a ses droits, la pratique a les siens. Or il arrive en mécanique ce qui arrive en philosophie: chaque nouvelle théorie qui surgit et qui semble renverser ses devancières a pour effet de répandre le scepticisme dans l'esprit des praticiens.

74. M. Contamin déclare que les principes de sa théorie appartiennent à M. Bélanger, et, s'il n'a pas fait cette déclaration tout d'abord, c'est que, livré à lui-même pour leur développement, il devait assumer la responsabilité de son travail; il voulait d'ailleurs soulever la discussion pour éclairer la question et le nom respecté et autorisé de M. Bélanger aurait peut-être empêché des critiques utiles de se produire.

Fig. 11. The logo of the École Centrale was the worker bee as a symbol of industriousness. The connection to Napoleon's emblem is possible but unknown. The ECP logo appears differently, with wings spread.* (Source: Pothier, *Histoire de l'École Centrale*, 1887, cover vignette)

practical engineering school. A decade and a half later it would be reorganized after the 1848 Revolution during which its students (like those of the EP) manned the barricades and organized *ateliers nationaux* to educate workers (Pfammatter 1997, p. 121), much to the chagrin of director Lavallée (Pothier 1887, pp. 205–207). When the school reopened in late 1848 or early 1849, it reaffirmed its non-elitist but nevertheless rigorously organized commitment to the industrial, laboring world. The school's logo, the 'worker bee' had also been one of Napoleon I's personal emblems. Who knows whether this may have helped curry imperial favor from his educationally concerned nephew, Napoleon III?[75]

This was the turbulent world in which aspiring industrial engineers in France began their studies, well aware of the revolutionary issues and the newest developments in industrial and construction thinking.

* The bee had been the symbol of Saint Ambrose and of the Barbarini family, but Napoleon is said to have chosen Childeric I's bee for its similarity to an inverted fleur-de-lys, relating his symbol simultaneously to the founder of the Frankish kingdom and to the royal symbol of France.

75. Napoleon III's interest in education may have been nurtured both by the empress and by his mentor at the military academy in Thun, Switzerland, the engineer Guillaume-Henri Dufour, a former student of the École des Mines.

Developments in the rest of Europe

Teaching technological subject matter in Europe has a prehistory that is longer than that of the French schools. Many eminent educators in other cultures had prepared the way for the new thinking that Perronet, Monge, and their colleagues introduced, and the French innovators were certainly aware of some of this. Although the development that they initiated proved to be so powerful that it dominated the evolution of engineering from the late eighteenth century on, it was not autonomous. French engineers were influenced by what had preceded them, and what they accomplished needs to be seen in context. Their novel intellectual thrust with its unique stress on science made the educational system that Lavallée and his co-founders later conceived radically different from what had gone before. This is particularly true regarding both the breadth of the program that they offered and its flexible applicability to multiple professions. A comparison with Great Britain, France's chief technological 'competitor' at the time, and a brief overview of the general situation of engineering throughout Europe highlight the importance of what occurred in France.

Traditional engineering instruction in Britain was a focused training for a trade rather than a broadly-based intellectual education for a profession. Throughout the first half of the nineteenth century, it still derived its form from the earlier, guild-dominated era. Engineering training was primarily through apprenticeships in offices and was entirely geared to what an engineer needed to know in a particular field. It neglected any organized form of scientific ability or technical knowledge. An academic educational system that was adapted to modern industrial needs only began to be implemented in Britain around mid-century upon the urging of Prince Consort Albert and Sir Lyon Playfair, among other influential people. The Royal School of Mines opened in 1853, in part under their influence, although the subjects of engineering and mining had previously been briefly taught at Durham University from 1838.[76] Prince Albert had received a Lutheran education in Coburg and had studied science and philosophy in Brussels and Bonn, and Playfair had crossed the border between science and industry when he had studied agricultural chemistry under Liebig at the University of

76. The Durham program was established by the astronomer Temple Chevallier and the chemist James Finley Weir Johnston (Chevallier and Johnston 1838). Despite support from industrialists, it was abandoned after a few years as too expensive to maintain.

Giessen. Both were thoroughly convinced of the importance of a scientifically based academic education for industrial development.[77]

Most of the early initiatives in Britain were private. Various workmen's education societies sprang up in the early to mid-nineteenth century, in which the dilettante architectural critic John Ruskin, the engineering and architectural theoretician Robert Willis, and the practicing engineer William Fairbairn[78] were active. They attempted to rectify the lack of technical education among craftsmen, but in an ad hoc and less organized manner than pedagogues like Charles Dupin in France. The earliest workman societies were established in Edinburgh in 1821; Glasgow, Liverpool, and London in 1823;[79] and Ipswich and Manchester in 1824. These were not academic institutions, and in a parallel development, under the influence of Playfair and Prince Albert, who had arrived in England in 1840, civil engineering chairs were established at a few public and private universities and colleges to educate engineers too.

The first professor of engineering in Britain was Lewis Gordon, who was named Regius Professor of Civil Engineering and Mechanics in Glasgow in 1840. Regius professorships are a uniquely British institution of personal royal appointments. Gordon's nomination to a chair established by Queen Victoria was certainly due to the importance that her husband and advisor Prince Consort Albert attached to engineering education. In most cases, schools founded by aristocratic patrons were sponsored by rulers who however did not personally involve themselves in their organization or in nominations. This case was different, as Prince Albert directly concerned himself with professional education.

Gordon was an *innovateur-vulgarisateur* who had worked under Marc Brunel on the celebrated Thames Tunnel and then studied at the Bergakademie Freiberg in Saxony and at the EP. His signal professorship, the only direct royal appointment in engineering at the time,[80] was followed by other, less presitigious professorships in Ireland, first at Trinity College Dublin with the nomination of the railway engineer

---

77. Smith (1992) documents the subsequent spread of a theoretical approach to engineering in British education 1850–90.

78. See introduction to Fairbairn (1856).

79. The Liverpool Mechanics Library was formed in 1823 and the Institute followed with lectures from 1825. Manchester and Liverpool became the two most influential institutes.

80. Queen Victoria created a total of fourteen Regius professorships during her reign. Only one was another engineering professorship, at the University of Edinburgh in 1868, seven years after Prince Albert's death.

and former Telford assistant, John Benjamin Macneill in 1842 and then at two of the three new University Colleges in Galway and Cork with Thomas Drane in Galway in 1849–1850, followed by William Bindon Blood 1850–60, and finally by James Thomson, one of Gordon's protégés,[81] at the Queen's University that united the three Irish colleges in Belfast in 1855. England followed slightly later than Scotland and Ireland when William Pole, who had previously taught engineering between 1844 and 1848 at Elphinstone College in Bombay (now Mumbai), was appointed professor at University College London in 1859.[82] These men were all required to 'promote the application of scientific principles to practical purposes' as Rankine said in his inaugural lecture at the University of Glasgow (Rankine 1856, p. 16).

Rankine, who had been Gordon's assistant, succeeded him in 1856 and continued working to establish engineering as an academic subject. Despite the fact that they both enjoyed the prestige of a Regius chair, they experienced resistance from the scientific establishment, which considered what Rankine termed 'engineering science' as a devaluation and pollution of pure science (Channell 1982, p. 44). Rankine saw the problem of the variant definition of the term 'detail' in science and in technology:

> ...no material object, however small—no force, however feeble,—no phenomenon, however familiar, is insignificant, or beneath the attention of the philosopher; that the processes of the workshop, the labours of the artizan, are full of instruction to the man of science; that the scientific study of practical mechanics is well worthy of the attention of the most accomplished mathematician. (Rankine 1856, p. 8)

Gordon had maintained contacts in France and was aware of French developments. It is therefore likely that Rankine, as his assistant and later a practicing

81. Thomson, who left Belfast for Glasgow as Rankine's successor in 1873, was the famous physicist Lord Kelvin's older brother. There had been earlier academic engineering appointments at University College, but all of them temporary: John Millington who had previously taught mechanics at the Royal Institution 1817–29, was originally appointed when the college was founded in 1827, but he never taught there because of a pay dispute. Charles Blacker Vignoles taught from 1841 until he left for Kiev in 1847, when Eaton Hodgkinson replaced him for the specialty of Mechanical Principles of Engineering, but Pole was the first permanent full appointment in 1859.

82. University College London was England's third university (after Oxford and Cambridge) and its first public and secular one. It was founded in 1826.

civil engineer and scientist working in the field of hydraulics (among others), would have known of Bélanger's contemporary professional work and of his participation in the pedagogical experiment at the ECP.

The early professorships were all individual appointments. Fully fledged academic engineering departments appeared later. The University of Manchester traces its engineering department back to the 1824 Manchester Mechanics Institute. However, it only became a reality when Osborne Reynolds, who had been educated both in the shipbuilding industry and in mathematics at Cambridge, was appointed to the new Chair of Civil and Mechanical Engineering in 1868 at what was then Owens College. Fairbairn, who had already supported the 1824 institution, supported this new development too. Cambridge followed Manchester's example after Willis's death in 1875, when James Stuart was appointed to a newly created professorship of Mechanism and Applied Mechanics. Although Cambridge had not had a previously designated engineering professorship, Stuart's predecessor Willis, the Jacksonian Professor of Natural Philosophy, had concerned himself with engineering.

Despite a growing recognition of the need for a broad, modern engineering education, the traditional form of contractual or articled apprenticeships in the offices of established civil engineers remained the most common form of engineering instruction in Britain for many years. The very title 'engineer' was classed socially with 'mechanic' in England as a lower-class craft rather than as a profession. This was a negative connotation that it carried well into the twentieth century.

On the upper-class side of society, the British Association for the Advancement of Science was founded in 1831 in reaction to the growing elitism and conservatism in the Royal Society.[83] One of its four sections, Physics, included mathematics and the so-called 'mechanical arts'.

Historically, Scotland and Ireland had closer contacts to France and the European continent generally than England, and were therefore active earlier in scientific and engineering education. As early as in 1839, for instance, the Scottish

83. This was a new development. In the eighteenth and early years of the nineteenth century, the Royal Society had been open and welcoming to technology. Smeaton had been a member. He had presented many papers and been awarded the Society's Copley medal in 1759 'On account of his curious Experiments concerning Water-wheels and Wind-mill Sails, communicated to the Society. For his experimental enquiry concerning the powers of water and wind in the moving of Mills.' He also prepared the 1794 first republishing of this and two of his later papers (1776 and 1782) in a volume by the Society before he died in 1792 (note in Smeaton 1794, p. 93). In 1814, the Society had also republished all 18 papers that he had presented there over the years in book form.

Duke of Buccleuch, Walter Francis Montagu Scott, a prominent promoter and funder of major harbor works and railways, had established the short-lived private College for Civil Engineers and of General, Practical and Scientific Education, where Playfair (who was also a Scot), taught.[84]

Parallel to this development, theoretical works by non-academic engineers gradually began to appear in England after mid-century. These were treatises that applied mechanics to iron construction for special cases and were inspired by their authors' construction experience in the field. Francis Sheilds's *The Strains on Structures of Ironwork* (1861) was one that had resulted from his work on the rebuilding of the Crystal Palace in 1854. Another was William Humber's *A Complete Treatise of Cast and Wrought Iron Bridge Construction, Including Iron Foundations,* also in 1861. It had grown out of his previous *A Practical Treatise of Cast and Wrought Iron Bridges and Girders* of 1857, and this in turn had been written in reaction to the official inquiry into the collapse of Robert Stephenson's Dee Bridge in 1848. They were all specialized volumes that addressed practitioners, not academic treatises, and they could not replace the academic educational programs that slowly began to form in the latter half of the century. Similar treatises appeared also in France, of course. Examples were the bridge treatises by Léon Molinos and Charles Pronnier, *Traité théorique et pratique de la Construction des Ponts Métalliques* in 1857, or Paul Régnauld's later *Traité pratique de la construction des ponts et viaducs métalliques* in 1870. These also contained some analysis, but they were backed by theoretical studies that had been produced at the EP, to which they referred.

Apart from Britain, the early engineering schools were almost always sponsored by interested rulers who appointed a host of gifted technologists of various descriptions and scientists known for their expertise in multiple disciplines to organize them.

Sweden as a mining nation and major producer of iron had the longest history of formal engineering education. What may have been the first engineering school anywhere was founded by the mechanic, clockmaker, and later mining engineer and industrialist Christopher Polhem. The Laboratorium Mechanicum that he set up under royal patronage on the island of Kungsholmen in Stockholm in 1697 included a workshop that was also a laboratory and exhibition space for

---

84. Buccleuch was elected an honorary member of the Institution of Civil Engineers in 1842 and Playfair became president of the British Assocation for the Advancement of Science in 1885.

his 'mechanical alphabet', a collection of eighty wooden models that visually and haptically demonstrated mechanical principles. Over the following century the laboratory gradually evolved and became the Mekaniska Skola in 1798 and then the Teknologiska Institut in 1827. Despite its new form being modelled on the EP, it remained similar to a British workman's institute until it was reformed twenty years later by the mining school director and physicist Joachim Åkerman into a modern school in 1846.[85] A second, unrelated institution was the Chalmersska Slöjdskola (craft school). It was founded in Gothenburg in 1829 by a legacy left by William Chalmers, a director of the Swedish East India Company to train craftsmen, and it, too, grew into an engineering school around 1850. The Chalmers Tekniska Högskola remained a private institution until 1937 when it was incorporated as the second national technical university. Agrarian Denmark followed Sweden in 1829, when the chemist and physicist Hans Christian Ørsted established the Danske Polytekniske Læreanstalt (now Danske Tekniske Universitet DTU) as a more science-focused institution on the model of the EP, which he had had visited as a young man. He headed it until his death in 1851.

Further south, the Netherlands that had regained its independence from France in 1814 founded its first engineering school as the Royal Academy in Delft in 1842. It presumably functioned as a combination of an academy and a tradeschool because the founding document stated that its purpose was 'for the education of civilian engineers, for serving both nation and industry, and of apprentices for trade'. It was therefore more comprehensive than the original Chalmers institute in Gothenburg, and it primarily graduated engineers for the Dutch East India Company and for government. The first director was Antoine Lipkins, who had studied as an external student at the EP and worked first as a topographical engineer in France, and then after liberation in Holland, where he was responsible for draining the Haarlemmermeer in North Holland, the first polder to be drained using steam engines. In stages, the academy eventually became today's Technische Universiteit Delft.

The highly industrialized Belgium that had split off from the Netherlands in 1830 opened its first engineering program in Mons as the École des Mines du Hainaut[86] in 1837. The program was broader than that of a simple mining school,

85. It became the Kungliga Tekniska Högskolan in 1877.

86. It eventually became the Faculté Polytechnique of the Université de Mons.

however, as it was the first technical institution to be based directly on the ECP. Its two founders were early *centraliens*: Adolphe Devillez (grad. 1835), who taught mechanics and *constructions civiles*, and Théophile Guibal (grad. 1836), who was responsible for machine construction and mining (Guillet 1929, p. 282).

In the many German counties, duchies, principalities, and kingdoms, schools were government foundations rather than private enterprises. As Prussia led twenty-five of these states into a new German empire between 1866 and 1871, the pre-existing schools did not coordinate into a unified national organization. In contrast to centralized France, the German development retained its diversity. The partially German-speaking Austro-Hungarian empire[87] and Switzerland remained separate political entities. Each retained its individual characteristics, but with a common basic cultural approach to education in which mechanics and builders were generally more broadly educated than in Britain but also initially less scientifically than in France.

Playfair (1852) describes the relationship between the two German technically based school systems, the *Realschulen* ('real' as in 'practical' schools) and the *Gewerbeschulen* (trade schools). Builders followed a program in these schools with a professional orientation that was based in part on the ENPC's original curriculum or on those of the military academies and a traditional apprenticeship. An alternate Prussian system devised by Wilhelm von Humboldt in the first years of the nineteenth century foresaw the *Gymnasium* (humanistic high school) as leading to the university. It was also a system of general education, but with another pedagogical goal. It postulated the integration of the arts and sciences with research leading to *Bildung* (general character development) as a holistic formation of individuality rather than as a professionally focused education. Students were to participate in their teachers' research activities to enable them to develop their own approach to knowledge in a 'seminar' format. Although there was clearly a social hierarchy between the two types of education, they were seen as complementary. As we will see further on, the Hungarian example of 1782 and Karlsruhe (1825) presaged a partial integration of the two systems.

Humboldt's system was never fully implemented in the German or any other university system, but his idea did serve as a conceptual basis for an upgrade

87. The empire included what are today Austria, Hungary, the Czech Republic, Slovakia, Slovenia, Bosnia, Croatia and parts of Poland, Romania, Italy, Ukraine, Moldova, Serbia, and Montenegro.

of the *Realschulen*, first into the 'upper practical school' (*Oberrealschule*) and then into the full technical or scientific *Gymnasium*, and from there to the establishment of the modern scientific and technical research university, first in Prussia and then in the united German Empire. The partial amalgamation of the technological and humanistic approaches to education explains the strength of German technical and scientific education through the nineteenth and into the early twentieth century.

The earliest German institution to introduce a hybrid, albeit rudimentarily scientific-practical education, the Collegium Carolinum (1745) in Braunschweig, was established half a century before this. It was an Enlightenment project proposed by Johann Jerusalem, a Protestant theologian and advisor to the Duke of Braunschweig-Wolfenbüttel. The school began small, with two teachers, and soon had ten. However, only one of them, probably the physicist and mathematician Johann Ludwig Oeder, taught any scientific or technological material. Jerusalem conceived his school on a broad scale and advertised his program internationally in English and French. It formed the nucleus of today's Technische Universität Braunschweig.

The foundations that followed Braunschweig were more narrowly focused on practical instruction. They were mining schools that supported industrial development or hydraulic engineering to harness waterpower or facilitate river transportation. The first was a small mining school, one of the few private institutions in the German-speaking world. It was founded in 1735 by the royal engineer Samuel Mikviny. It was named the Bergakademie Schemnitz in today's Banska Stiavnica, in the metal-mining area of Slovakia. Mikviny, who was a member of the Prussian Academy of Sciences, was considered a polymath and taught all the courses in mathematics, mechanics, hydraulics, and surveying himself. Slovakia was then then part of the Austro-Hungarian empire, and the region was of economic interest to the empress Maria Theresia, who raised it to the status of a state school in the early 1760s. By 1762 imperial patronage had developed it into a mining college with several renowned teachers, and although it remained small, the technological universities of Hungary, the Czech Republic, and Slovakia all trace their ancestry to it.

The most influential mining college of the eighteenth century was the Kurfürstlich-Sächsische Bergakademie zu Freiberg in the Electorate (Principality) of Saxony. It was founded in an attempt to improve the royal budget after the devastation of the Seven Years' War, and it was organized as the first fully equipped mining academy in 1765 by two mining administrators, the former Braunschweig mining director Friedrich Anton von Heynitz[88] and Friedrich Wilhelm von Oppel, the local mining supervisor who became its first director. It immediately hired four teachers who between them taught mathematics, drawing, mechanics, physics, metallurgical chemistry, assaying, mineralogy, and mining. The foundation was a success, and it remains the oldest continuously existing technical university with an unbroken lineage.[89]

The Freiberg program taught both scientific and practical subjects, and it directly influenced the foundation of the Saint Petersburg Mining School in 1773 under the empress Catherine II by the head of the state mining board (Collegium of Mines), Mikhail Fedorovich Soimonov, who simultaneously reorganized mining in Russia.[90] The Bergakademie influenced the founding of many further schools, among them the Academia de Minería y Geografia Subterránea de Almadén in Spain 1777,[91] and even one as far afield as the Real Seminario de Minería in Mexico City in 1792, at which Alexander von Humboldt taught. The École Royale des Mines in Paris was founded by the Royal Council in 1783, also on the model of the Freiberg Academy, to supplement Peronnet's ENPC, which had been established in the previous reign. It opened under the directorship of the chemist and mineralogist Balthazar Georges Sage, but initially only with Sage and the mining engineer Guillot-Duhamel as teachers. A sign of Freiberg's influence on the École des Mines and mining education generally was that German remained a required course right up to the period when Perdonnet studied there in the late 1820s. These foundations were all specialized mining schools with a narrowly defined mission.[92]

In 1782 a more generally focused technical institution appeared in Hungary

---

88. The Braunschweig ruling families, the Duchy of Braunschweig-Wolfenbüttel and the Electorate of Braunschweig-Lünenburg, and their mining directors were signally active in initiating technical schools.

89. The school became Königlich-Sächische (Royal) from 1806, and today the Technische Universität Bergakademie Freiberg.

90. Soimonov organized the school based on the ideas of Mikhail Vasilevich Lemonosov, the Russian polymath who had studied in Freiberg, among other German universities, and who had died shortly before.

91. The Almadén academy relocated to Madrid in 1835 as part of the Madrid Polytechnic University.

92. Freiberg also influenced the formation of other schools, like the trade school in Clausthal (Electorate Braunschweig-Lünenburg) in 1775 founded by yet another former Braunschweig mining administrator, Nikolaus Friedrich von Reden. In 1810 Clausthal became a college, in 1864 a fully academic Bergakademie, and today a technical university.

that signaled a first attempt to integrate technological teaching (practice) into a general (humanistic) system. The Institutum Geometricum Hydrotechnicum had a broader focus than the earlier schools because it was not founded as an independent institution, but as part of the Faculty of Liberal Arts of the 150-year-old Buda University. Its primary focus was on surveying, hydraulic engineering, and river management, and initial lectures on mechanics were held by physicists. Since academic instruction in the Austro-Hungarian Empire was in Latin and German at the time, it may have served as example for the subsequent development in Germany. After the Institutum separated from the university and merged with the newly founded 1844 Joseph College of Technology in 1850, an independent department of mechanics and machinery was established in 1857 with instruction initially in German and from 1860 in Hungarian. This new entity was recognized as a full university on an equal standing with its former host institution. It gradually evolved into today's Budapest University of Technology and Economy.

The concept of technology had appeared on the German academic radar for the first time when the economist Johann Beckmann defined the term in its modern sense.[93] He introduced it, albeit as a uniquely economic factor, at the University of Göttingen in the early 1770s and published it in 1777. Inspired by the *Encyclopédie* that had appeared in France (1751–66), Beckmann then published a five-volume historical account of all technical fields, *Beyträge zur Geschichte der Erfindungen* (1780–1805). As in France, technology was valued as a social phenomenon, but not yet recognized as a legitimate academic preoccupation. Beckmann followed his historical study with another more theoretical work, *Entwurf einer algemeinen Technologie* in 1806, in which he proposed a way that technical ability and knowledge could be categorized by function. This study influenced the curricula at several universities through pedagogues like Christoph Bernoulli in Switzerland or Gérard-Joseph Christian in France, until Beckmann's approach was replaced by the French science-based program of the ECP at mid-century. Under the influence of Beckmann's publications, the Carolinum modified its focus around 1833, and two years later August Uhde, a physicist and

93. Guillerme cataloged the history and shifting meanings of the term in 1984 without mentioning Beckmann's crucial innovative use of the term. See Guillerme *Les liens du sens dans l'histoire de la technologie*, republished in Guillerme 2008. Picon 1994 records Beckmann's influence on Gérard-Joseph Christian in the first decades of the nineteenth century.

mathematician who had studied in Göttingen shortly after Beckmann died, joined the school. Uhde influenced the curriculum and brought in several more similarly interested teachers (Krafczyk 2016, pp. 27–29).

The further German development can be seen on the background of Wilhelm von Humboldt's concept, and the schools that were not devoted exclusively to mining were all founded by men who were professionally active in multiple fields. Karlsruhe in the Grand Duchy of Baden founded its Polytechnische Schule in 1825. It was the first German school to be entirely devoted to technical subjects through the merger of three private schools: Friedrich Weinbrenner's architecture school (1800), Johann Gottfried Tulla's engineering school (1807), and the Karlsruhe Lyceum, a high school.[94] In contrast to the Braunschweig faculty, which was oriented to Britain through the reigning Duke, Tulla had studied briefly in France and Freiberg. He was particularly interested in the new French decimal system, which he introduced into the Grand Duchy, and his specialty was river correction, which he carried out for the Rhine and the Linth in Switzerland. Weinbrenner had worked in Switzerland and studied in Vienna and Dresden. Tulla had adopted the original EP model for his school, but he resisted the change that had occurred in France and remained practically oriented. This resistance was in large part due to Weinbrenner's empirical formation as a carpenter and his work as a builder in Lausanne and Zurich, which then founded its own Industrieschule in 1832. Unlike Karlsruhe and most of the early high schools in Germany, however, Zurich's Industrieschule did not develop into a technical university, but remained part of the high school system.[95]

The German development was characterized by a variety of establishments: Prussia's Königliche Gewerbeakademie was founded by the statesman and promoter of industrial development Wilhelm Beuth in Berlin in 1821. In 1829 the Württemberg Königliche vereinigte Real- und Gewerbeschule opened in Stuttgart and was renamed the Polytechnische Schule in 1840. The Höhere Gewerbeschule in Hannover (Polytechnische Schule from 1847) followed in 1831. It was founded by Karl Karmarsch, who had studied in Vienna and who combined chemistry, mechanics, and construction in his teaching. Both the

94. The Lyceum had been founded in 1586 as the 'Gymnasium illustre' and adopted the name 'Lyceum' in 1806.

95. The Industrieschule, known by its students as the 'Indili'. became the Kantonale Oberrealschule in 1928, and finally the Mathematische-Naturwissenschaftliche Gymnasium MNG in 1974.

northern Guelph-ruled states (the Duchy of Braunschweig and the Kingdom of Hannover) had contacts with Britain that fostered an early interest in technology. The British royal family was Hanovarian, and there was a close Scottish family tie to the key pedagogical innovator at the Carolinum, August Uhde (Krafczyk 2016, p. 26). The Carolinum had also catered to international students and even taught English speakers in a separate class.

Although other professional programs of a rudimentary type existed in many German states from the early 1800s on (some even earlier), most of these were ad hoc training situations. One might term them proto-schools that did not really even merit the designation. These 'schools' had a single teacher, no formal curriculum or oversight, and very few students. It was Beckmann's influence in the background and the French models of the EP and the ENPC in the foreground that influenced the establishment of professional schools in most states. And it was in large part the influence of the ECP's success that helped them transform into institutes of technology after mid-century, based on the model of Auguste Perdonnet's dream of a *Sorbonne industrielle* (Comberousse 1879.1, p. xiii).

The German schools that focused on building originally combined architectural and engineering programs that provided even more architectural material than the ECP, and the graduates often worked interchangeably in both fields. Although Timoshenko credits the German schools with being the first to establish a connection between engineering and industry (Timoshenko, p. 102), the German system had become to a great extent a reflection of the ECP's pedagogy. The shift to the ECP-influenced model, both there and elsewhere, was often accompanied by a name change to suit their altered status.

The ENPC in France kept its name but began to broaden its purview after 1851, almost certainly due to several ECP teachers like Bélanger and Mary who had held dual appointments there. Several new schools followed. The École Polytechique Fédérale de Lausanne (EPFL) began in 1853 as the École Spéciale de Lausanne, founded, like the École des Mines in Mons fifteen years before, by a group that included the ECP graduates Jules Marguet (grad. 1840) and Louis Rivier (grad. 1843). It followed Mons as the second new foundation based directly on the ECP. Zurich's Eidgenössische Polytechnikum (today's ETHZ) then followed Lausanne as a federal (*eidgenössisch*) foundation. It did not evolve from the Industrieschule of 1832, but became the first German-speaking school on the new ECP system in 1854. One of the Polytechnikum's founders, Johann Konrad Kern, who later became ambassador to France, knew the ECP's Perdonnet personally.

The Eidgenössische Polytechnikum enjoyed a special relationship with the ECP. Paul Buquet (grad. 1853), the president of the Association amicale des élèves de l'École Centrale, thanked the Polytechnikum's graduates at a banquet during the festivities at the 1889 Paris Exhibition for not having neglected *la vieille mais toujours jeune École Centrale, qui est l'ainée de l'École de Zurich* (the ancient but ever young École Centrale, the older sibling of the Zurich school—*Mémoires et compte rendu* 1889.1, p. 903).

The parallel Czech development in what was then the Austro-Hungarian Empire was a special case. What is today the Czech National University in Prague was founded even earlier than any school in Germany, but as one of the proto-schools: a private, so-called 'fortification school' with a single teacher and twelve students. It was set up in 1717 by the military engineer Christian Josef Willenberg[96], who had learned engineering on the job in the German Régiment d'Alsace in France. It was eventually taken over by the government and chartered in 1787 as a non-military engineering school, probably under the influence of the ENPC. A comprehensive program of engineering education in the empire began in earnest with the foundation of the Böhmische Ständische Polytechnische Institut in Prague by the scientist and railway pioneer Franz Josef von Gerstner in 1803. It opened its doors in 1806 along the lines of the original, more practical version of the EP. This was followed by the Polytechnische Institut in Vienna,[97] founded by the educator Johann Joseph von Prechtl in 1810–15.

At first, the same dispersed pattern held true in Russia, with short-lived mining schools in 1713 in Olonets near the Finnish border, in the Ural region from 1723, and the school in Yekaterinburg in 1733 (Kosykh 2021). The first national school was the militarily organized Saint Petersburg Mining Institute in 1773,[98] founded with imperial permission and patronage by Soimonov on the initiative of an industrialist from Perm and partially funded by a voluntary contribution from all the iron mining firms of the empire.

Modern engineering education in Russia was a foreign import, however. After the 1807 treaty of Tilsit, a short-lived agreement between Napoleon and Czar Alexander

96. The first Austro-Hungarian k.u.k. ("königliche und kaiserliche") Technische Militärakademie was founded the same year in Vienna.

97. Today's Technische Universität Wien.

98. Now the G. V. Plekhanov Saint Petersburg State Mining Institute.

I led to the hiring of several French-trained teachers. Agustín de Betancourt y Molina was the first in 1808, and he recruited four more: Pierre-Dominique Bazaine, Jean-Antoine Destrem, Charles-Michel Potier, and Jacques-Alexandre Fabre. They entered Russian service in 1810 and taught under Betancourt at the new 1809 Institute of Transportation Engineers[99], which Betancourt modelled on the ENPC that he had attended, and on the Escuela Técnica Superior de Ingenieros de Caminos, Canales y Puertos de Madrid that he had founded in 1802. Together these engineers established Russia's modern technical education on the lines of the ENPC and the newly founded EP. Of this group only Betancourt, Destrem, and Potier settled permanently in Russia. They contributed longer to the development of the country's transportation infrastructure than the others, in collaboration with mainly Scottish iron-founders like Matthew Clark and Charles Baird.[100] Alexander I had also called other Western engineers to Russia in this period, among them the southern German Wilhelm von Traitteur. He was an autodidact who did not teach, but worked from 1814 to 1832 in St. Petersburg, first under Betancourt, and most notably built several pioneering chain suspension bridges between 1823 and 1826 (Fedorov 2000). After the end of the Napoleonic War, Betancourt brought Émile Clapeyron and Gabriel Lamé to continue the French academic tradition. They came to St. Petersburg to teach in 1820, and Antoine Raucourt joined them in 1821.[101] Most foreign engineers eventually left again after the Czar's death in 1825: Raucourt in 1827, Bazaine in 1828, Clapeyron in 1830, Lamé and von Traitteur in 1832, and Fabre in 1833, upon which their former student Pavel Petrovich Melnikov (grad. 1825) became the first Russian professor of applied mechanics (Rieber 1990, p. 558). It was he who initiated the independent development of engineering and engineering education in Russia.

Napoleon's invasion and conquest of Italy in 1796–99 and Spain in 1808 imposed the introduction of French systems of government and education in both countries in the early years of the nineteenth century. Earlier autochthonous developments in Spain, where Juan de Herrera had been the first builder to introduce the study of mathematics in 1582, with the first military academies in the seventeenth century that taught technological subjects, the 1777 Almadén

---

99. Today's Emperor Alexander I Saint Petersburg State Transport University.

100. This is considered the beginning of modern industrial iron construction in Russia even though craftsmen and iron-founders—especially the Demidov family—had been producing iron and building iron structures since Nikita Demidov founded the manufacturing dynasty in 1702 (Kosykh 2021, p. 64).

101. For the full complexity and details of the development see Rieber 1990.

mining academy, and Betancourt's 1802 school in Madrid, were overlaid with the French model and this persisted after the fall of Napoleon with the entire national system concentrated in Madrid.

The situation in Italy, with its unified culture and fragmented political organization, was more complex. Eminent theoreticians from Alberti, Leonardo, and Galileo in the fifteenth and sixteenth centuries, to Cassini, Boscovich, and Lagrange in the eighteenth and nineteenth, to name only a very few of those who were most significant for construction, had long influenced the international development of science and its introduction into engineering. The peninsula's engineering tradition and the activism of the Ducal House of Savoy in particular, as well as the participation of Italian-French Savoyard scientists like Cassini and Lagrange in developments in France prepared an intellectual base for new academic engineering programs (see Bocquet 2007). The first civil engineering department based on the EP had been established at the university of Naples in 1811. It was abolished after the defeat of the French in 1815 but reinstituted four years later as the Scuola di Applicazione di Ponti e Strade. The Napolitian school was followed by the first separate foundation, the Scuola di Applicazione per gli Ingegneri in the Savoyard capital Turin in 1859, and then the Istituto Tecnico Superiore in Milan, today's Politecnico Milano in 1863, shortly after Italian unification.[102]

While the development in Russia and Spain continued to follow the centralized, state-driven EP-ENPC tradition, Italy only unified as a nation-state in 1861 and Germany in 1866–71, and both remained organizationally diverse enough to pursue multiple paths. The development was particularly notable in Germany, where a transformation of the larger regional schools began after mid-century, influenced by the ECP's growing success. The name changes of many of these reflected this development.[103] However, until the transformation of all

102. Naples became a separate Scuola Superiore Politecnica in 1905, and Turin became the Regio Politecnico di Torino in 1906. The Milan school was founded by the civil engineer and mathematician Francesco Brioschi.

103. In 1862 Braunschweig's Collegium Carolinum became the Technische Universität, Karlsruhe's Polytechnische Schule morphed into the Polytechnische Hochschule in 1865. Munich's Polytechnische Schule was a new foundation that replaced older and simpler versions in Munich, Augsburg, and Nürnberg in 1868. Two of these smaller trade schools had previously existed in Munich itself: the Polytechnische Centralschule of 1827 and the Polytechnische Schule of 1833. Dresden's 1828 Technische Bildungsanstalt was transformed into the Königlich-Sächsisches Polytechnikum in 1871. Vienna's Polytechnisches Institut became the Technische Hochschule Wien in

these schools took place in the second half of the nineteenth century, German technical education remained geared towards government employment. It was not designed to cross borders into industry.

There was academic resistance to or at least ambivalence about this transformation too. The influential developer of graphic statics Carl Culmann, who had been partly educated in France, expressed the opinion that only his teaching at the Eidgenössische Polytechnikum together with mathematics—as taught at the University of Zurich and the French system of the EP with its graduate continuation in the écoles d'application—combined the necessary theoretical and practical skills that engineers really needed (Culmann 1866, pp. viii-x). His statement in 1866 reads like a self-conscious statement of educational ranking, in which the ECP did not deserve to be included. Johann Konrad Kern, Culmann's immediate supervisor and president of the Polytechnikum's board of directors (Schweizerischer Schulrat) that had appointed him, had discussed pedagogy with Perdonnet in 1857. Additionally, the Genevan engineer Guillaume-Henri Dufour who had been a member of the original planning commission had specifically referenced the ECP, so it was unlikely that Culmann was unaware of the school and its influence on the development of his own institution. Nevertheless, he ignored it, even though his own work owed much to the concept of *vulgarisation.* One may speculate whether his antipathy softened after his contact with Eiffel through his student Maurice Koechlin in 1879.

This was the complex intellectual environment of engineering education on the European continent in the early nineteenth century to which the founders of the ECP reacted and from which they imported those elements that they found useful to mold their new, industrially focused approach. As a result, the young engineers at the ECP were educated very differently from their predecessors as well as their contemporaries in France and elsewhere. Only after 1850, with very few exceptions, among them Mons, Lausanne, and Zurich, which were based on the ECP from their foundation, did the success of the ECP model gradually began to

---

1872, and in 1879, the Berlin Bauakademie absorbed Wilhelm Beuth's Gewerbeinstitut to become the Technische Hochschule Charlottenburg, followed by Hannover's Polytechnische Schule the same year as the Königliche Technische Hochschule. The same development can be followed in Sweden, Denmark, Belgium, and the Netherlands. Switzerland's Polytechnikum, which was already based on the ECP's system, only changed its name to 'Technische Hochschule' in 1911, and even today it is still colloquially known as the 'Poly'.

insinuate itself into the pedagogy of the many other pre-existing schools. Playfair wrote in 1852 that the ECP was the most important industrial school with the best teachers in France, and presciently predicted that the men it was educating were destined to become preeminent in their field:

> It possesses the most eminent men of France as its professors, and it has reared those who promise to be her future brightest ornaments. As a commercial speculation it has been singularly successful, and it still remains under the business direction of the original enterprising capitalist M. Lavallée. The government now gives to it a certain number of exhibitions [i.e., stipends] to educate poor students of extraordinary talents, and the Councils General of twenty-nine departments of France also do the same. (Playfair 1852, p. 26)

Thanks to the ECP, the predominance of French iron construction persisted right up until the end of the century, as the *centralien* Eiffel pointed out in his maiden speech as president of the Société des ingénieurs civils in 1889:

> As far as iron construction is concerned, the French school [of construction] continues to be one of the most influential, and our builders still have little to fear from international competition. Even the Americans tend to abandon their use of pins and adopt our French method of rigid assembly by riveting. (*Mémoires et compte rendu*, 1889.1, p. 45)[104]

The seminal role that the ECP played in engineering education and the evolutionary influence it exerted worldwide has not been fully acknowledged. Its broad, general, science-practice-based engineering tradition was specifically designed to cross borders between fields, and especially between the public sector and industry.

---

104. Pour les construction métalliques, l'École Française continue à tenir l'un des premiers rangs et nos constructeurs redoutent peu de courir les chances des concours internationaux. Les Américains eux-mêmes ont une tendance à revenir à nos habitudes françaises en abandonnant leur système d'assemblage par axes, pour adopter nos assemblages rigides par rivure.

The founders define the ECP

The initiator of the ECP in 1829 was Alphonse Lavallée. He was a lawyer who had gained his industry and business experience in partnership with his brother-in-law, the arms manufacturer Chrétien-Charles Haëntjens in Nantes (Gourisse n.d., p. 52; Luquet 2015; Dibos personal communication).[105] In 1827 Lavallée moved to Paris, where he became financially involved in the railway industry.

As one of the directors of the Compagnie du chemin de fer de Paris à Orléans, France's second railway company, Lavallée was in close contact with practitioners like Émile Clapeyron and Perdonnet, a young convert to the cause, as well as with the financiers James Mayer de Rothschild and Émile and Isaac Pereire. Lavallée was therefore aware of the urgent need for practically oriented, industrial rather than theoretically oriented, academic engineers. One of his closest contacts was the enterprising self-made engineer Eugène Flachat, the chief engineer of the other early railway company, Paris–Saint-Germain, who as we have seen was a strong supporter of the concept of technological *vulgarisation* (Malo 1873, p. 12).

Like most of his peers, Lavallée embraced capitalist doctrine and Liberal ideology that believed in *individual* sovereignty rather than in the Republican ideology that believed in *majority* sovereignty. Through his friendship with Paul-François Dubois, one of the founders of the short-lived liberalist journal *Le Globe*, Lavallée came into contact with other thinkers who were interested in pedagogy. The Liberals were a heterogeneous group with various political concerns. Among them was a fellow investor in the journal, the Protestant François Guizot, one of the leaders of the *Doctrinaires*, and like Dubois a member of the anti-Bourbon Aide-Toi Society.[106] Later, as minister under King Louis-

---

105. Personal communication, Dec. 2019, and supplementary source information from the Lavallée family archive kindly provided by Olivier Dibos, Lavallée's great-great-grandson.

106. The *Doctrinaires*, led by Duc Élie Descaze, were a group that supported a tightly controlled constitutional monarchy. The Aide-Toi Society was a political group formed in 1824 that opposed the Bourbon government by supporting Liberal candidates in the 1827 elections. Its motto was *aide-toi*

Fig. 12. Alphonse Lavallée, the initiator, principal founder, and driving force behind the École Centrale. Sketch dated around 1840, ten years after founding the school. This sketch served as model for the image in Guillet, *Cent ans de la vie de l'École centrale*, 1929, p. 1. (Source: Olivier Dibos Collection)

Philippe and with the philosopher Victor Cousin's support, Guizot introduced the law establishing compulsory primary education in France. From 1828 until *Le Globe* was hijacked by radical Saint-Simonians in 1830 (and then suppressed in 1832), the circle around the journal was active in promoting *vulgarisation* through publishing and education.[107] This group of men already recognized in the mid-1820s that industry was in urgent need of flexibly educated engineers who understood scientific theory and were at the same time able to adapt to a rapidly changing industrial world (Grelon, 2000).

As a former export-oriented industrialist, Lavallée had a broad outlook through his contacts in the Anglo-Saxon world. He was not alone in this view in France. Economists like the Protestant Jean-Baptiste Say, who had lived and worked in England before the Revolution, the Dauphin Louis-Antoine, Duc d'Angoulême[108],

et le ciel t'aidera (God helps him who helps himself), a Protestant sentiment. Guizot, whose speech later helped spark the 1830 revolt, the so-called *Trois Glorieuses*, was president of the society for a time. The society was disbanded in 1832.

107. Among Lavallée's many supporters at *Le Globe* was the Saint-Simonian publisher and supporter of *vulgarisation* Édouard Charton, the father of the ECP engineer Jules Charton (grad. 1862) who became Contamin's second-in-command at the Exhibition of 1889.

108. The dauphin had sponsored the second edition of Dupin's *Voyages dans la Grande-Bretagne*,

who had been in exile there and had even fought (together with his father) with Wellington's army against Napoleon, liberal thinkers like Joseph Cordier, a supporter of private enterprise, and many others shared his internationalist views. Lavallée's conviction of the importance of industry in the development of France was therefore very much in the air at the time, made all the more more urgent as a result of the recent military defeat and the awareness of British superiority in manufacture. While Britain's success in industry rested on private initiative and had contributed to the British victory in the recent war, it was clear that France enjoyed an advantage in the well-organized state support of science. The question was therefore how the French educational advantages in scientific research and education could be applied to a liberal development. The answer seemed to be a private institution based on *vulgarisation.* Private schools were unheard of in French higher education, but the concept of *vulgarisation* had been prepared by Gaspard Monge's approach to teaching engineering.

Fig. 13. Pierre Laurans, Lavallée's father-in-law and the initial funder of the École Centrale. (Source: Olivier Dibos Collection)

Lavallée's conviction that his goal could only be achieved through privatizing technical education was most unusual in France and somewhat of a mystery. It

1825–26, see vol.1, pp. ix–xiv, thereby protecting Dupin's work against critics who accused it of 'anti-French' sentiment.

can perhaps be explained by Liberal theory, or could have been a result of the general fascination with all things Anglo-Saxon. It may also have been due to Flachat's influence, or that of Lavallée's father-in-law Pierre Laurens, a United States citizen of French birth who had married well in Louisiana and had become a wealthy plantation owner and businessman.[109] Laurens had been a client of Lavallée's brother-in-law Haëntjens in Nantes (Gourisse n.d., p. 52; Luquet 2015; Dibos personal communication),[110] and Lavallée married Laurens's daughter Adèle in 1825. Following Adèle's plea for help in a letter that she wrote to her father three days after her husband and the other ECP founders had signed their agreement of association on January 20, 1829, he agreed to back Lavallée and his plan financially (Dibos personal communication, Pothier 1887, p. 86). One can hardly envision how Laurens dared risk supporting his 33-year-old son-in-law to the tune of 100,000 francs in an uncharted adventure from abroad unless there had been some sort of previous discussion of the issue between them. It was only a year later, in 1830, the year the school began its first semester, that Laurans returned to settle in Paris and could follow the development firsthand.

Lavallée's own father Robert had categorically refused to support his son's venture (Pothier 1887, p. 86; Dibos personal communication). Since Robert only died in 1858 after the ECP had been transferred to the state, it is not clear that he ever permitted any part of the family fortune to be used for his son's project. Laurens's possible intellectual influence on the ideology behind the founding of the ECP has not been explored, but it was clear that his funding was what made it possible, especially his support during the crisis of 1832 when Lavallée almost succumbed to cholera.[111] According to family tradition, Lavallée is said to have stated: *sans les dollars de la Louisiane, il n'y aurait pas eu d'École Centrale* (without the dollars from Louisiana there would have been no École Centrale), and as he wrote to his father-in-law,

---

109. Laurans's plantation was situated in the parish of Pointe Coupée, about 250 kilometers north of New Orleans on the Mississippi. He lived in New Orleans and also had properties there. Sources for this information are multiple sites on Internet listing census and court cases, in which Laurans was a defendant. In some transactions he appears as Jean-Pierre Laurans (researched under 'Pierre Laurans Louisiana'. Dates and personal information were found under Geneanet, 'Olivier Dibos family'. (See also Luquet 2015).

110. The purchase of arms by planters in Louisiana was probably the reaction to the revolt of slaves in Pointe Coupée in 1795 and several further insurrections that continued until 1811.

111. Dibos, personal communication 31.12.2019, comment on Pothier, p. 86.

> ...you show trust in me to a degree that a son does not even expect from his father. I am anxious to hear your opinion on a matter that will determine my very existence. (Jacques Lavallée 1983)[112]

Lavallée was obviously a gifted negotiator, and not only with his father-in-law. He was able to reconcile conflicting interests and counter the entrenched political attitudes of government educational policy. He also coordinated and managed a strong-minded faculty, represented in this account by the examples of Perdonnet and Olivier, whose characters will be more closely described. We do not know, however, how Lavallée interacted with the students. The lack of information is likely due to the awe that everyone in his environment held for the school's founder, who was known in the Lavallée family reverentially as *Monsieur Père* (Pierre Lavallée, cited in Dibos, Wikisource). Outwardly reserved and sober in appearance, he was inwardly passionate about the idea of education and had been interested in founding a school as early as 1826 (Pothier 1887, p. 22). He may have been influenced in that direction by his brother-in-law's model farm and proposed agricultural school in Nantes (Pothier 1887, p. 22). In this case, however, his energy was to be directed toward the needs of industry (Charles Callon in Comberousse 1879.1, p. A83). When he moved to Paris the following year, Lavallée found support from the group surrounding *Le Globe* and inspiration for action from hearing the young Jean-Baptiste Dumas (1800–84) lecture at the scholarly society L'Athénée[113], where he was impressed by Dumas's gift for *vulgarisation* (Pothier 1887, p. 23). After discussing the idea with Dumas, Lavallée decided that the time was ripe to establish a novel form of institution.

Dumas was evidently, and despite his youth, a brilliant and lucid lecturer and a pedagogue with a broad vision (Pothier 1887, pp. 55–56, Comberousse 1879.1, p. 51). According to his famous student Louis Pasteur, 'he could set fire to the soul' (Debré, p. 23, English edition 1998). He had a broad vision that covered all aspects of chemistry, which at that stage in the development of the field was still possible, and he had the ability to explain the most complex issues elegantly and lucidly (Comberousse 1879.1, p. 51).

---

112. ...vous me montrez une confiance qu'aucun fils n'attendrait de son père. Je suis bien impatient de savoir votre opinion sur une affaire qui décide de mon existence.

113. Colladon, pp. 187–191, condolence letter to Hervé Mangon, Dumas's son-in-law 1884.

Fig. 14. Jean-Baptiste Dumas was the youngest of the four founders of the school, only twenty-nine when the school opened, but he was the most active in developing the pedagogy. (Source: Muspratt, *Chemistry, theoretical, practical and analytical*, 1860, vol. 1, 2nd frontispiece)

Dumas had been educated in Geneva. He was a pioneering chemist and pharmacist and had very decided ideas about the organization and pedagogy of secondary education and the new market for industrial engineers. The idea for a school to educate them was in the air at the time, and others had developed similar thoughts. The influential mathematician François de Sainte-Preuve was willing to support an engineering school, but a small one with a narrowly defined curriculum, and definitely only under the university system (Pothier 1887, p. 23). Dupin, who had established the first chair of *mécanique appliquée aux arts* at the Conservatoire National des Arts et Métiers a decade previously, may also originally have been of the same opinion. That was, of course, not at all what Lavallée or Dumas had in mind. In the succinct definition that Comberousse attributed to Perdonnet, what Lavallée and his co-founders were after was what we now understand as a private Institute of Technology: *l'École Centrale, c'est la Sorbonne industrielle* (Comberousse 1879.1, p. xiii). In this definition, industry was the envisaged client rather than an abstract governmental-scientific-technological constituency.

Philosophical concepts and their influence on the school: public versus private

Higher education was a state monopoly in France, and the revolutionary issue of independence from governmental control that the Liberal Lavallée envisaged as necessary in order to address such a client was a delicate political problem. So when serious discussions began in 1828, Dumas consulted Jean-Daniel Colladon, a Swiss friend from his student days in Geneva 1817–22 (Colladon 1893, p. 187, and Pothier 1887, p. 32), with the not-so-hidden agenda of using his contacts.

With the help of the Aide-Toi Society, a new and somewhat more liberal government than its predecessor had just been elected under the repressive Charles X at the end of 1827. Lavallée and Dumas were quick to take advantage of the change. Colladon had moved to Paris three years before to collaborate with Joseph Fourier on scientific experiments, and was quite willing to be of service. Colladon contacted Fourier and convinced him of the viability and value of the idea, and through him, as was the intention, he caught the ear of the new Minister of Education, Antoine de Vatimesnil. Vatimesnil, head of the new ministry that had just separated from the Ministry of Religion, was conservative and religious but he actively supported the split between education and religion, possibly in agreement with Guizot.[114] Like Dumas, Vatimesnil was interested in educational policy, and according to the ECP brochure of 1831 (Lavallée/Dumas, 1831, p. 4), was enthusiastic about the idea. Supported by a committee, he agreed to the foundation of a new school that was to be independent of the university system (Pothier 1887, pp. 11, 33 and 59–60 and Comberousse 1879.1, pp. 33–34). This was a bold idea in France at the time and it ran strictly counter to tradition (Colladon 1893, p. 177).

The issue was theoretically one of private sponsorship—Liberal in concept—rather than state (at that time Monarchist), or perhaps unvoiced in the back of their minds, Republican sponsorship. Hélène Vérin (1993, p. 37) speculates that there may also have been a political motive behind the formation of the school, namely an attempt to balance the monopoly of the state engineers with a corresponding group of civilian professionals. The thought may not have presented itself to the lawyer Vatimesnil, but it may have occured to the industrialists and financiers in his advisory committee, which included the scientists, mathematicians, and industrialists Jean-Antoine Chaptal, François Arago, Louis Jacques Thénard, and the bankers and politicians Jacques Laffitte and Casimir-Pierre Périer, among

114. For the history of the power of religious and political use of education, see: Bruter 2011.

others (Foulques 1829, p. 82). It was certainly not one of the founders' concerns. Vatimesnil's decision apparently did refer to the lobbying that Charles Dupin had previously undertaken in technical education (Foulques 1829, p. 84). Whatever the motives of those involved may have been, the need was imperative for the development of French industry.

One of the arguments that Lavallée and Dumas put forward was that the existing specialist schools that educated technologists in France—the two Écoles d'Arts et Métiers in Chalon-sur-Saône and Angers, as well as the many workmen's colleges established on Dupin's suggestion all over the country—were focused on educating workmen and supervisors. Other institutions, like the École des Mineurs in Saint-Étienne, founded in 1816, were more narrowly devoted to a single profession. Progressive private groups had begun to initiate courses as well. In 1826, for instance, the newly founded Société industrielle de Mulhouse, a group of twenty-two Alsatian Protestant industrialists, most of them members of the Masonic lodge La Parfaite Harmonie, had instituted evening classes. One of their articles of association stipulated

> the propagation and consolidation of the following bourgeois, liberal values among the working classes: love of work, economy, and education. (Wikipedia)[115]

The relationship between technological education, Protestant educational philosophy, and the non-religious French form of Freemasonry was close.[116]

All the technical schools and groups taught a great deal of practice but no theory, and what was needed in industry was a combination of both. The elite Conservatoire National des Arts et Métiers, the third educational entity founded in 1794 (along with the EP and the École Normale de l'An III), also provided lectures in technological and industrial fields. But the Conservatoire served primarily as a continuing education resource for established professionals in government or industry. Although César Leblanc educated technical draftsmen there when the institution became more oriented toward technical education

---

115. la propagation et la consolidation dans la classe ouvrière des valeurs libérales bourgeoises suivantes: l'amour pour le travail, l'économie et l'instruction.

116. It is worthy of note that another Alsatian Protestant association devoted to education, the Société pour l'encouragement de l'instruction primaire parmi les protestants de France, was founded three years later in 1829, the same year as the ECP.

in 1819, it mostly held lectures (Lavallée/Dumas 1832.2, p. 2) and was not a degree-granting school with an organized curriculum, exercises, examinations, or internships that would have been capable of educating young civil and industrial engineers professionally.[117] Most of its courses were general-interest offerings and not specifically directed toward industry, as those at the ECP would be (Lavallée/Dumas 1834, p. 7). The traditional écoles d'application, the ENPC, the École du Génie in Metz, and the École des Mines did not admit practitioners. In fact, they only accepted scientifically trained entrants from the EP.

Although it was never mentioned by the founders, the participatory practical-theoretical program as Lavallée and Dumas defined it bears a conceptual resemblance to the Liberal approach to education that Wilhelm von Humboldt had devised in Prussia the previous decade.

Presumably with the help of Dumas, Lavallée enlisted three more like-minded scientists from the group surrounding *Le Globe* to help him: Eugène Péclet, a physicist, mathematician, and chemist; Théodore Olivier, a mathematician recently returned from Sweden who specialized in descriptive geometry and who researched its industrial application; and the military topographer and mechanical engineer Philippe Benoit. According to Comberousse (1879.1, chapter 1), it was Péclet, Olivier, and Dumas who first had the idea. After their disappointment with the mathematician Sainte-Preuve, who had advocated a small, government sponsored school, they had been approached by Lavallée. Pothier and Colladon who had been part of the founding group (the former citing documentary evidence), downplay and seemingly contradict Comberousse's statement. Whatever the sequence of events really was, all of them were thoroughly convinced of the need for a new and independent school and were fully on board with the idea of *vulgarisation* in their respective fields in service of industrially focused engineering education (Pothier 1887, pp. 26–30).

Benoit withdrew from the project before teaching began for reasons that remain somewhat muddy, although he remained on the books as a member

117. *Rapport fait à la société d'encouragement, sur l'école centrale des arts et manufactures, le 12 juin 1833, par MM. le Baron Thénard, le Comte de Lasteyrie, le Baron Ch. Dupin, Francoeur, Mérimée et Payen, Rapporteur*, p. 12; *Notice sur l'École centrale des arts et manufactures, adressée, par une commission représentant les professeurs de cet établissement et par le directeur, au Conseil Général des Manufactures, le 27 février 1833*, p. 4 and *Rapport fait au Conseil Général des Manufactures, le 30 mai 1833, Par une Commission composée de MM. Bérard, Roman, Desrousseaux, Saulnier, et D'Arcet, rapporteur*, p. 8 (Lavallée/Dumas, 1832.2).

of the *conseil de perfectionnement* (Lavallée/Dumas, 1831, p. 1) at least for a year. The others remained life-long contributors. Eugène Flachat, Colladon, and Pierre-Charles Gourlier, the technologically oriented architect and secretary of the state approval board (the *Conseil des bâtiments civils*), were also involved in the preliminary discussions of the group that met at Olivier's apartment in 1828 (Colladon 1893, pp. 176–177), and the latter two in the initial teaching program as well. However, they were not counted as official founders because neither the practitioner Flachat nor the two others who did teach in the first years of the school were willing to commit to teach for twenty years as stipulated in the articles of association that the founders signed in January 1829 (Colladon 1893, p. 193). Nevertheless, these three men influenced the curriculum. Colladon taught for six years and wrote his memoirs that recorded his contributions, and Flachat was involved professionally with many of the graduates and became the first president of the alumni association, so they are recorded in the official histories of the school. Gourlier only taught for a single year.[118] He also successfully adapted the building that the school occupied and apparently continued to give individual lectures after resigning. César Leblanc, the pioneer of technical drafting and the first drawing teacher, also taught for a single year, but neither he nor Gourlier received the attention that they would merit in histories of the development of the school.[119] Both were a decade older than the founders, and Leblanc had years of pedagogical experience at the Conservatoire. Judging from his published thoughts (see chapter 'The importance of drawing'), he may have influenced the didactical concept of the program.

Dumas took the lead in developing the concept despite, or perhaps because of the fact that he was the youngest member of the founding group (Colladon 1893, pp. 187–191).[120] It was also he who edited the first curriculum (Pothier, p. 368) and possibly also co-authored the admissions brochures that described the program. In fact, Colladon, who was later uninvolved in the school,

118. Colladon (1893, p. 193) writes two years, but probably meant the academic year 1829–30.

119. Gourlier is mentioned occasionally. Comberousse mentions Leblanc only once (1879.1 p. 50), and Pothier too (1887, p. 63), but Guillet and Colladon do not mention Leblanc.

120. Condolence letter from Colladon to the engineer and meteorologist Hervé Mangon, Dumas's son-in-law, 1884. According to Colladon, Olivier, who also claimed a leading role for himself, was a vain man who slightly slanted his account of the founding to give himself more credit than he really deserved (Colladon 1893, p. 190).

unequivocally stated that Lavallée and Dumas alone were the mainstays of the whole project (Colladon 1893, p. 190). This was an opinion that the directly implicated and emotionally involved Comberousse and Pothier could not endorse.

The state-run schools were all Republican in philosophy, and most of them were organized militarily. In contrast, the individualist, Liberal ideals on which Lavallée and his associates based the ECP adhered firmly to civilian, bourgeois, and industrially defined qualities (C. Lavollée 1872, p. 419). The connection between the early development of the school that was supported by Protestant educational philosophy lies principally in the role of the Liberal doctrine of personal independence and individual action.

Politically programmatic symbolism

An indication of the political difference between the Liberal and Republican philosophies and a sign of the attention that Lavallée and his team paid to the importance of visual communication, was the '*centralien*' *tenue de ville*. This was a dress code and not a uniform in a military sense. It varied only slightly from the standardized school uniforms of the time, for instance those worn by students at the preparatory Collège Sainte Barbe, a secondary or middle school that many students at the ECP had previously attended.[121]

The ECP's *tenue de ville* included the bourgeois attributes of gloves and a fashionable cane in contrast to the collège uniform, and it differed decidedly from the elaborate military uniforms, the *tenue d'intérieur* (BD) or even the everyday, so-called *petite tenue*, worn by the *cadets* of the EP that included white gloves and ceremonial swords, the so-called *tangentes*. The name given to this *épée* or sword that the EP students in the *Ingénieur Polytechnicien* program still sport as part of their dress uniform (*grande tenue*) describes its position at a tangent to the line of the leg, in the same way as 'X', the name the EP gives itself, probably relates to the algebraic unknown. The abstraction of these mathematical terms underscores the self-consciously elitist and somewhat affected quasi-scientific self-image of the school in contrast to that of the ECP.

---

121. A French *collège* corresponds to a secondary or middle school. It educates 11- to 15-year-olds, while a *lycée* is a high school in the American sense and provides education between the ages of 15 and 18.

Fig. 15. The École Centrale's *tenue de ville*, anonymous student c. 1890. (Source: photographer Mayer, 67 rue des Batignoles, Paris. Wikimedia Commons) It was not much more elaborate than the standard preparatory-school uniform of the day, for example of the Collège Sainte-Barbe c. 1884 on the right side. (Source: Internet site *Les petites mains, histoire de la mode enfantine*, 03.09.2018. Detail of *Histoire de l'uniforme scolaire en France. classe de 4e, Collège Sainte-Barbe, 1884–1885*, Archives Collège Sainte-Barbe)

This difference demonstrated the social gradient between the two schools. It was intentional and an almost English form of understatement on Lavallée's part and certainly politically programmatic. It served to stress the contrast between the two: the EP as elitist and the ECP as industrial and down-to-earth. In the years following the 1830 Revolution, the ECP did briefly adopt a simple uniform and a sword (Guillet 1929, pls. 40 and 41), and during the Revolution of 1848, the student body voted to revive military dress (Guillet 1929, pp. 419–420 and pl. 41). However, these incidents were short-lived aberrations, and the ECP military uniform quickly disappeared again (Mattatia 2018, pp. 38–39).[122] ECP students were and still are *étudiants*, while those at the EP were enrolled as military *cadets*. Although the EP has been a civilian institution since 1970 (albeit still under the Ministry of Defense), it has never relinquished the *cadet* (now known as *élève-officier*) status of the French students in its *ingénieur polytechnicien* program or its formal *grande tenue* uniform and *tangente*. As reflected by the dress code, Lavallée recognized that something non-nationalistic, non-military, and socially broadly based was needed in industry.

122. At the very end of the nineteenth century, military training once again made a brief reappearance (Gourisse n.d., pp. 64–65).

Fig. 16. The École Polytechnique's military everyday *petite tenue* with sword and white gloves and the formal *grande tenue* uniform that is still worn by the *élèves-officiers* on official occasions today. (Source: anonymous nineteenth-century engravings)

The name the founders chose for the school was politically programmatic, too, both reminiscent of and in conscious contrast to the original name for the EP: the École Centrale des Travaux Publics. The name blatantly asserted a national, centralistic role for the institution in contrast to its avowed private charter. Lavallée and Dumas wanted to link to Monge's original concept for the EP thirty-five years earlier, which he had intended as the premier national and practical engineering school. In fact, the brochure of 1831 specifically stated that this was their intention (Lavallée/Dumas, 1831, p. 7), but theirs was to focus on private industry rather than on public works. It would, of course, concentrate more on practical issues than on higher mathematics (Pothier, p. 42) in contrast to the direction in which Laplace and Cauchy had since led the EP. Some claimed the term *centrale* had a further meaning: it was the hope that such schools would be established centrally all over France to aid the development of industry in each region.[123] However, the primary idea was to supplement the national schools of higher education that served the sciences (EP), the liberal arts (Sorbonne), and the visual arts (École des Beaux-Arts), by crossing the boundaries between science and practice to serve a new category of industry, the 'industrial arts'. The self-understood intention was to establish industry as a cultural force on an equal social footing with the others:

> The École Centrale raises industry to the level of the liberal arts. It lends it focus, a faculty, and a Sorbonne from whence emerge new doctors of industry in the guise of civil engineers. (Lavallée/Dumas 1834, p. 3)[124]

Saint-Simonianism and Freemasonry

Lavallée, Dumas, Péclet, and Olivier were formed by the French Revolution's social mores and were egalitarian in their attitudes. They were all self-made men and embraced intellectual and manual intelligence with equal fervor. Moreover, they were influenced by Adam Smith's and David Ricardo's late eighteenth-century value theory of labor.[125]

There were conceptual differences between the ECP founders' interest in manual labor in Paris and Beckmann's at Göttingen. First of all, Beckmann had only been concerned with economics, and the ECP system focused on industry. While the two are closely linked, they are not the same. The German approach to education and to knowledge was predicated on Protestant theology, which encouraged questioning authority and declared any kind of work to be pleasing in the eyes of God (Peters 2015). Industrialists, especially those influenced by Calvinist Geneva agreed, but

---

123. The schools that followed the ECP model in the course of the nineteenth century were Lille 1854, followed by Lyon (1857), and Marseilles (1890). Today there is one in Nantes that was founded in (1919), and more recently three abroad: Beijing (2005), Casablanca (2013), and Hyderabad (2014).

124. L'École Centrale élève l'industrie au rang des arts libéraux, lui donne un centre, une faculté, une Sorbonne, d'où sortent, sous le nom d'ingénieurs civils, les nouveaux docteurs industriels.

125. The appreciation of Smith's manufacturing theories in the *Wealth of Nations* was propagated among engineers through Prony's use of his division of labor, transposing it from pin manufacture to mathematical calculations in the compilation of his logarithmic and trigonometrical tables for the French *Cadastre* 1793–96 using unemployed wigmakers as human calculators. Prony is said to have boasted 'I'll do my calculations like manufacturing pins' (*Je ferai mes calculs comme on fait des épingles*) (Roegel).

while Protestant theology preached equality before God while still accepting the social distinctions between organizational and manual labor on earth, the founders of the ECP were democratically egalitarian in their political views. As a result, the founders were strongly impressed by Saint-Simonianism. Olivier was the only truly convinced Saint-Simonian among them, while the others were essentially Liberals, certainly influenced by the doctrine but not as orthodox as he. As Weiss points out, it was a watered-down Saint-Simonian doctrine in an economic sense that these men subscribed to (Weiss 1982, pp. 52–53).

In its original form, Saint-Simonianism was an idealist doctrine that fostered technocratic socialism combined with state capitalism as the basis for social equality and industrial development. It had a pragmatic and secular focus that made it popular among French industrialists, engineers, railway financiers, and contractors in the early to mid-nineteenth century. The attitude they took from this doctrine was positivist and geared to industrial development, and engineers and entrepreneurs who remained based in France were virtually all Saint-Simonians in this sense. Among these were contractors like Ernest Goüin, iron-founders like Émile Martin, railway engineers like Gabriel Lamé and Émile Clapéron or Eugène Flachat, his relative Ivan Flachat, and many of his erstwhile employees like de Dion, Barrault, and Molinos, Flachat's brother Adolphe Flachat and his stepbrother Stéphane Mony, Paulin Talabot, who was both railway engineer and banker, or railway financiers like the brothers Émile and Isaac Pereire and Émile's nephew, the engineer Paul Régnauld. While their focus was primarily proto-socialist and positivist, there were many Liberal or Protestant-educated engineers in their group as well. The doctrinal boundaries were malleable.

Another influence common among engineers in France was Freemasonry. Many eighteenth- and early nineteenth-century French financiers, engineers, and even contractors belonged to this group. James Mayer de Rothschild the banker, Perronet of the ENPC, or Monge and Carnot of the EP were among the most prominent, but there were others of less exalted background, like the *entrepreneur de serrurerie* (locksmith-contractor) Jean-François Roussel, who reached the degree of Master Mason in the Grand-Orient de France lodge (Roussel 2015). Eiffel, who some claim was also a Freemason, later collaborated with the Freemason sculptor Bartholdi to create a flaming-torch symbol of the belief in progress in the 1883 Statue of Liberty. Its original name was La Liberté Éclairant le Monde (Liberty Illuminating the World).

Although there were many men who were both Saint-Simonians and the particular French brand of secular Freemasons that the Revolution had brought forth, the latter group was definitely more intellectual and elitist, and this led to conceptual differences between the French schools. In particular, the divide was evident between the Saint-Simonian-influenced, private ECP and the state-run, Freemason-influenced, undergraduate EP and other state-run graduate schools like the ENPC or the École des Beaux-Arts. Partly as a result of this, but especially by virtue of their almost exclusively theoretical bent, the students at the state schools held themselves to be higher-ranking 'intellectuals'. They mocked the didactic models that the ECP was developing as intellectually inferior to engineering science and pejoratively dubbed its students 'pistons'. In this, the ECP approach was closer to the hybrid educational models developed in the eighteenth-century German and Austrian mining schools, or the concepts on which the Institutum had been founded in Hungary, albeit with the advantage of the French excellence in theory. The new school combined the best of both worlds.

> …the École Polytechnique's students, wont as they are to hover in the higher regions of science, often display a singular aversion to stoop to the tedious details of practice, details that one nevertheless cannot neglect without seriously imperiling the success of an undertaking. It has been recognized, for example, that they cannot fulfill the role that civil engineers occupy in England who, originating from the working classes, are certainly a great deal less educated but who, on the other hand, are capable of industrial ability and a most dogged perseverance. (Lavallée/Dumas 1832.1, *Notice sur l'école centrale* pp. 4–5)[126]

The 'pistons' reacted with pride at their working-class, Saint-Simonian image. They had their worker-bee logo and their civilian dress code. It was a point of honor that they got their hands and shoes dirty on sites, in workshops, and industry. The 'pistons' experimented in smelly laboratories, and they studied materials, machines, and practical construction in far more technical detail than did the more theoretically minded, elegantly clad, and white-gloved intellectuals of the EP and the ENPC. In addition to their background in theory, the ECP graduates were concerned with men, mud, and machines.

---

126. …les élèves de l'École Polytechnique, habitués à planer dans les hautes régions de la science, ont souvent montré de la répugnance à s'abaisser aux détails fastidieux de la pratique; détails que cependant on ne peut négliger sans compromettre gravement le succès des entreprises. On a reconnu, par exemple, qu'ils ne pouvaient remplir exactement la place qu'occupent en Angleterre les Ingénieurs civils, qui, sortis de la classe des ouvriers, certainement sont beaucoup moins instruits, mais qui, en revanche, sont doués d'un tact industriel et d'une persévérance toute particulière.

The first years

The program as devised by Lavallée and Dumas evidently touched a nerve at the time. Even before the school year began, various provincial governments began offering stipends to prospective students. As early as May 1829 (teaching only began in November of that year), the Département Haute-Vienne offered 300 francs or half the tuition costs to successful candidates in the entrance exam that the local government had organized (Foulques 1829, p. 85).

And yet, success came slowly despite the support for the idea from those engineers involved in the early railways (Barbin 2021), who mostly would have preferred a school under state supervision. Setbacks occurred immediately in the regressive, brief reign of Charles X that sought to reverse the rise of bourgeois power and capitalism, through two revolutions in 1830 and 1848 and two cholera epidemics in 1832 and 1848. The founding group persevered through all these difficulties, and for decades thereafter Lavallée, who remained the director as well as the actual owner of the institution, devoted the family's fortune (Lavallée's father-in-law who had originally funded the enterprise died in 1848) and all his time to running his private school. His goal was to educate, as he said, *médecins des usines et des fabriques* (factory and workshop physicians). This curious expression, which sounds as strange in French as it does in translation, was in all probability modelled on the nickname 'engine-doctor' that had been given to the young George Stephenson around 1806 (Smiles 1857, p. 43, Perdonnet 1859, p. 304). Flachat and Perdonnet may both have learned of it from their contact with the Stephensons in England.

The school was off to a rough start. Even before opening, the Liberal government under Martignac fell, and the supportive education minister Vatimesnil left office in August 1829. Lavallée had to renegotiate for patronage from the newly established Ministry of Agriculture, as the Ministry of Education had grown unsympathetic to its goals.[127] The constellation was propitious as

127. The new minister of religion and education, Montbel, was extremely conservative, but nevertheless refused to abandon the educational reforms that Guizot and Cousin had set in motion. He was rabidly anti-Liberal, however, and this may have been the reason that he did not support Lavallée's school.

the interim Agriculture Minister was Adrien de Gasparin, a Protestant and a close friend of Lavallée's aquaintance Guizot, who would soon briefly become Minister of the Interior in 1830 and then Minister of Education from 1832. Both were interested in education. Lavallée proved so successful in negotiating that, despite royal antipathy to innovation, the new ministry even helped furnish stipends for students (Comberousse 1879.1, pp. 110/121/158). The funding mechanism evidently also functioned through the industry group, the Société d'encouragement pour l'industrie nationale that, like the Département Haute-Vienne, offered four half-stipends (Lavallée/Dumas 1834, p. 5). The dynamism of the founders, the devotion of its young faculty and students, and the support of regional governments, professional organizations, and industry, forced it to succeed:

> Despite repeated political shocks and the effects of a disastrous epidemic, the success of all these projects was due to a vitality in this so young school that exists only thanks to those it has formed. Motivated by the same ideal and united as one man, the professors and students all march together with a force that nothing can disrupt. They have rivaled one another in their devotion and self-sacrifice whenever the nascent institution appeared to be menaced by public events. (Lavallée/Dumas 1834, p. 3)[128]

Still, the first years were a constant battle for fee-paying students. Lavallée advertised the school through all the media at his disposal. In 1836 the Bureau des Longitudes published a seven-page addendum to its *Annuaire pour l'an 1837* as an appeal to the readers. It took the form of a pamphlet urging industrialists to give their sons an adequate education that would prepare them to inherit their factories. Lavallée's tone was upbeat, but one may detect a slightly frantic note:

> In approaching once again fathers and young men who are unsure of their future, the founders of the École Centrale now feel comfortable in judging confidently the results of six years of solid educational performance. It is no longer a question of projects that might fail: the École Centrale is in full

128. Si tous ces projets se sont accomplis, malgré les secousses réitérées de la politique et d'une épidémie désastreuse, c'est qu'il existe une force de vie dans cette École si jeune encore, qui n'est comprise que de ceux qui y ont passé. Professeurs et élèves, tous animés d'une même pensée, unis comme un seul homme, marchent avec un ensemble que rien n'a pu troubler; tous ont rivalisé de dévouement et d'abnégation quand l'Institution naissante a paru menacée par les événemens publics.

> swing and for the past three years has provided industry with the young civil engineers it undertook to prepare. (Lavallée/Dumas 1836, p. 1)[129]

With combative rhetoric, Lavallée went on to stress the need for a scientifically based instruction to prepare the transfer of factories to a new generation of leaders, and to explain that the empirical preparation that the pioneering first generation had experienced no longer sufficed:

> The father has accomplished all that a high-school education can provide, but in order to go farther one needs a scientific education.... With its aid, one becomes capable of fine-tuning one's industry; one's vision becomes more able to seize the opportunity and to recognize the source of the obstacles still to be conquered.... In a word, let us provide these young men who are destined to combat the elements and the forces of nature with a scientific education....
>
> Sixty years ago, industry outpaced the sciences; today the sciences dominate and control industry. Those who refuse to accept this fact are wrong and are caught in the past without recognizing what is happening today. (Lavallée/Dumas 1836, pp. 3/6)[130]

The founders' anxiety was understandable. The group of young idealists had set themselves up against the vested interests of French higher education, and they also shouldered the added task of convincing the first generation of bourgeois industrialists that their acknowledged successes were not an adequate basis for projecting future developments. Lavallée was supported by his co-founders and a

129. En s'adressant de nouveau aux pères de famille, aux jeunes gens incertains de leur avenir, les fondateurs de l'École Centrale se croient autorisés à leur parler avec assurance des résultats d'un enseignement de six années permet de juger maintenant avec certitude. Il ne s'agit plus de projets dont l'accomplissement pourrait être entravé; l'École Centrale en pleine activité livre à l'industrie depuis trois ans les jeunes ingénieurs civils qu'elle s'est chargée de préparer.

130. Le père a fait tout ce que permettait l'éducation de collège; pour aller plus loin, il faut une éducation scientifique.... Par elle, il se verrait capable de perfectionner son industrie; son œil plus clairvoyant saurait saisir le fil qui lui permettrait de remonter à la cause des difficultés qui restent à vaincre....En un mot, donnons une éducation scientifique à ces jeunes gens destinés à lutter avec les éléments et les forces de la nature....

Il y a soixante ans, l'industrie avait devancé les sciences, aujourd'hui les sciences dominent et maîtrisent l'industrie. Les personnes qui refusent à admettre ce point de vue sont dans l'erreur et se laissent préoccuper par les faits du passé sans tenir compte des événements de l'époque actuelle.

tenacious faculty. Influential politicians and members of the government supported him too, but the government was unstable, and the obstacles were immense.

The argument he presented underscored the founders' goal of combining empiricism with science. Lavallée, Dumas and their team had recognized that there was an urgent need for a new breed of engineer for industry. They aimed to combine the best of the two systems—Britain's private initiative and France's state-supported scientific education—and they modelled their program on the post-Revolutionary development of Perronet's school, on Monge's original EP, and on the new technical knowledge and entrepreneurial approach that Flachat, Perdonnet, and others had brought back to France. They based their didactical concept on the definition of the engineer that Henry Robinson Palmer[131] had formulated in a paper he had read before the future founders of the Institution of Civil Engineers in 1818:

> An Engineer is a mediator between the Philosopher and the working Mechanic; and like an interpreter between two foreigners must understand the language of both. (*A brief History* 1928, p. 11)[132]

Lavallée had to convince French industrialists that this stance really did reflect their interests. After the Napoleonic War had ended, the widespread awareness that the victorious Great Britain, which both the Revolutionaries and then many of the Imperialists had pejoratively looked down upon as 'a nation of shop-keepers',[133] had completely outstripped France in industry and engineering. The Société d'encouragement pour l'industrie nationale that had been established during Napoleon's consular period in 1801 with the aim of rectifying what had already then been perceived as an imbalance in industry, had modelled itself on

131. Palmer, who shortly after this paper became Telford's assistant, is best known as the inventor of corrugated iron (Herbert 1978, pp. 34–35).

132. The author of *A Brief History of the Institution of Civil Engineers* was probably Henry Homan Jeffcott (1877–1937), then Secretary of the Institution, and his cited sources were the Council Report of 1886 and an article by John Henry Tudsbery Tudsbery (1859–1939) in *Min. Proc. ICE*, vol. 205, 1917/18, pt. 1, p. 216, who, however, had only paraphrased the quotation. Guedes (2010, p. 235, footnote 196) found the full quotation in the manuscript copy of Institution's Minutes of Proceedings dated Fleet St. 2 January 1818.

133. This expression was originally attributed (somewhat doubtfully) to Napoleon, or more probably to the opportunistic revolutionary Bertrand Barère de Vieuzac. See also Perdonnet's dislike of the British focus on business and the accumulation of wealth further on.

the British Society for the Encouragement of Arts, Commerce, and Manufactures that had existed since 1754. Its members and French engineers knew that Britain had long possessed a mutually supportive body of civil engineers in the Society of Civil Engineers that John Smeaton had founded in 1771.[134] They were also aware that a professional Institution of Civil Engineers had been established on Palmer's principles, that the celebrated Thomas Telford had been prevailed upon to become its president in 1820, and that it had been officially recognized by Royal Charter in 1828.

The French attitude toward Great Britain had certainly changed after the war when many Liberals and exiled aristocrats returned. Government agencies sent engineers to England as soon as the borders were open again to learn what they could and publish their findings. Among the best known engineering reports are Dutens's *Mémoire sur les travaux publics de l'Angleterre* 1819 on canals and other hydraulic works, bridges, roads, and even a note on the earliest (horse-drawn) 'iron roads' with a concentration on the economic aspects and private versus government funding,[135] Navier's *Rapport à Monsieur Becquey* of 1823 on suspension bridges, and Charles Dupin's magisterial multiple-volume and multiple-edition *Voyages dans la Grande-Bretagne* 1820–24 on the entire breadth of British infrastructure.[136]

These three men were not the only engineers that the government sent abroad. Around the time he sent Navier, Directeur général des ponts et chaussées Louis Becquey also sent Polonceau's father, Antoine-Rémi Polonceau to England to study roads and the locks on the Thames (A.R. Polonceau, 1829, p. 19 footnote). In his report of 1823 and later in his treatise of 1829 he especially recommended the new roadbuilding technique that John Loudon McAdam had pioneered. There were many other reports, some even critically comparing French and British engineering and defending French professional pride and

134. Now a private association called the Smeatonian Society of Civil Engineers

135. Page 176 cites a letter between John Rennie and Prony, proving that professional contact between French and British engineers had been quickly reestablished after the war.

136. Dupin's study, part of which had already appeared in 1818, was so novel and comprehensive that partial English translations and adaptations appeared in 1818, 1822, 1825, and 1829. The tradition of state-sponsored reporting on foreign infrastructure development continued right up to Émile Malézieux's two-volume *Travaux Publics des États-Unis d'Amérique en 1870*, published in 1873. (Antoine Comolli's later *Les Ponts de l'Amérique du Nord* 1879 appears to have been undertaken entirely on his own initiative.)

national competence. One of these, by the canal and harbor engineer Joseph Cachin, *Mémoire sur la Digue de Cherbourg, comparée au Breakwater ou Jetée de Plymouth* of 1820 criticized the execution of John Rennie's project and corrected a statement published in an addendum to the *Encyclopedia Britannica* of 1817. Cachin maintained that his breakwater (originally proposed together with Lamblardie in 1792) was less expensive to construct per meter and more stable than the other.

However, the enthusiasts dominated. Dupin was far more admiring of Rennie's feat at Plymouth than Cachin (Dupin, *Voyages,* vol. 2, 1825 ed., tome II, Études et travaux, pp. 280–290). Flachat and Perdonnet also visited England to learn about British technology and railways from Marc and Isambard Brunel, and George and Robert Stephenson. Perdonnet formed a friendship with George Stephenson, and Flachat, who stayed longer, became friends with all four (Malo 1873, p. 10). Flachat also charged the young Ernest Goüin (EP 1834), who had studied for a year as an external student at the ENPC in 1837 (Lacourt) and had worked in England from 1838, to supervise the building of locomotives that he had ordered there (Montel 1995, p. 218, footnote 12).

There were other engineers besides Goüin who had worked in England and then brought their experience and British technology back to France, for instance Paulin Talabot, who had met and had possibly even briefly worked for the Stephensons. Alexandre Lavalley was another prominent example. After graduating from the EP in 1842, he had worked his way up from simple laborer to iron constructor in the English iron industry and then returned to France to collaborate, after a short stint at the chemins de fer du Nord (*Mémoires et compte rendu*, 1892.2 p. 1003), as technical director with Goüin at Batignolles from 1846.[137]

---

137. Ernest Goüin et Cie. imported British construction techniques, developed novel construction machinery, and erected many bridges with Lavalley. Among these last were the bridges Clichy-Asnières and Langon that de Dion worked on. See Molinos and Pronnier 1857 for multiple references to these bridges and to Lavalley and Goüin's expertise in riveting and construction (eg., pp. 122–125). When Lavalley left Goüin he was replaced in 1865 by Émile Nouguier, Eiffel's later employee. He then entered a partnership with Paul Borel in the firm Borel, Lavalley et Cie. to become the contractor for the dredges on the Suez Canal.

The original concept in detail

In addition to basing their approach on the two available pedagogical models—Monge's original EP and the modified Perronet ENPC—the concept that the founders and their group developed for the new school responded to impulses from outside France and particularly to the industrial development in Great Britain. They based it on the principle of self-help as it had been defined in Protestant religious and educational theory and as propagated in the pioneering work of Pestalozzi and by liberal groups like the Aide-Toi Society.[138] The founders chose teachers who were all young and enthusiastic about the possibility of developing novel courses in their fields. These men saw the opportunity to use what they taught to further their own research agendas (Comberousse 1879.1 p. 58), and this, too, contributed to the academic excitement that the experiment generated.

The didactic means that the ECP employed in teaching were three-fold: intellectual (both through the spoken word and written communication), visual, and haptic. All three are cognitive because they increase knowledge. Furthermore, instead of Perronet's original combination of peer teaching and apprenticeship that had proved slow—the education of an engineer could last anywhere between four and twelve years—the ECP followed the then current form of teaching and employed professors to give ex-cathedra lectures that the students were required to record in notebooks. These were regularly inspected for correctness and completeness and certified by assistants called *répétiteurs*.

Lectures were supplemented with mechanically reproduced transcripts, exercises led by the assistants, a series of exams, and practical internships in workshops.

The founders laid great emphasis on a close working relationship between the school and the professional world. Neither the military establishments nor the EP had done this before, and the original ENPC only marginally addressed the professional sphere through the recruitment of its student body from practice. The internships may have owed their origin in part to the educational system prevalent in Germany at the time.[139] They aimed at inculcating the idea of collaboration on

138. In 1859 the idea was enshrined in the 'bible' of mid-Victorian morality in Britain by Samuel Smiles as *Self-Help: With illustrations of character and conduct.*

139. Perdonnet had visited Germany as a student in 1826, and others had probably done so too.

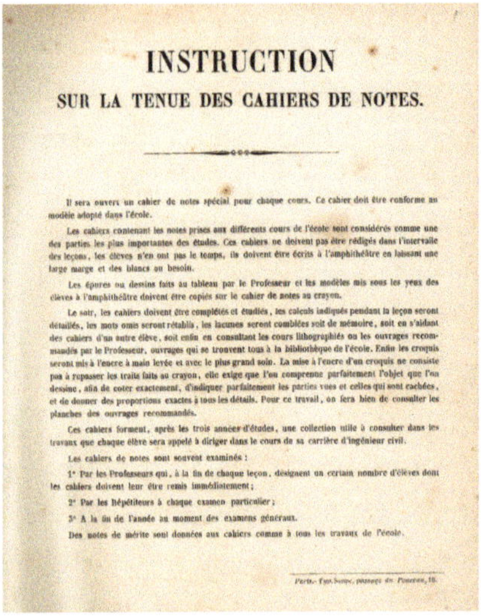

INSTRUCTION

SUR LA TENUE DES CAHIERS DE NOTES.

Il sera ouvert un cahier de notes spécial pour chaque cours. Ce cahier doit être conforme au modèle adopté dans l'école.

Les cahiers contenant les notes prises aux différents cours de l'école sont considérés comme une des parties les plus importantes des études. Ces cahiers ne doivent pas être rédigés dans l'intervalle des leçons, les élèves n'en ont pas le temps, ils doivent être écrits à l'amphithéâtre en laissant une large marge et des blancs au besoin.

Les épures ou dessins faits au tableau par le Professeur et les modèles mis sous les yeux des élèves à l'amphithéâtre doivent être copiés sur le cahier de notes au crayon.

Le soir, les cahiers doivent être complétés et étudiés, les calculs indiqués pendant la leçon seront détaillés, les mots omis seront rétablis, les lacunes seront comblées soit de mémoire, soit en s'aidant des cahiers d'un autre élève, soit enfin en consultant les cours lithographiés ou les ouvrages recommandés par le Professeur, ouvrages qui se trouvent tous à la bibliothèque de l'école. Enfin les croquis seront mis à l'encre à main levée et avec le plus grand soin. La mise à l'encre d'un croquis ne consiste pas à repasser les traits faits au crayon, elle exige que l'on comprenne parfaitement l'objet que l'on dessine, afin de coter exactement, d'indiquer parfaitement les parties vues et celles qui sont cachées, et de donner des proportions exactes à tous les détails. Pour ce travail, on fera bien de consulter les planches des ouvrages recommandés.

Ces cahiers forment, après les trois années d'études, une collection utile à consulter dans les travaux que chaque élève sera appelé à diriger dans le cours de sa carrière d'ingénieur civil.

Les cahiers de notes sont souvent examinés :

1° Par les Professeurs qui, à la fin de chaque leçon, désignent un certain nombre d'élèves dont les cahiers doivent leur être remis immédiatement ;

2° Par les Répétiteurs à chaque examen particulier ;

3° A la fin de l'année au moment des examens généraux.

Des notes de mérite sont données aux cahiers comme à tous les travaux de l'école.

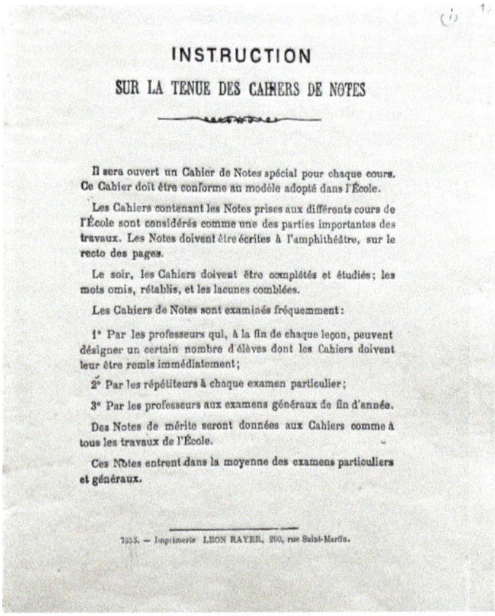

INSTRUCTION

SUR LA TENUE DES CAHIERS DE NOTES

Il sera ouvert un Cahier de Notes spécial pour chaque cours. Ce Cahier doit être conforme au modèle adopté dans l'École.

Les Cahiers contenant les Notes prises aux différents cours de l'École sont considérés comme une des parties importantes des travaux. Les Notes doivent être écrites à l'amphithéâtre, sur le recto des pages.

Le soir, les Cahiers doivent être complétés et étudiés; les mots omis, rétablis, et les lacunes comblées.

Les Cahiers de Notes sont examinés fréquemment :

1° Par les professeurs qui, à la fin de chaque leçon, peuvent désigner un certain nombre d'élèves dont les Cahiers doivent leur être remis immédiatement;

2° Par les répétiteurs à chaque examen particulier;

3° Par les professeurs aux examens généraux de fin d'année.

Des Notes de mérite seront données aux Cahiers comme à tous les travaux de l'École.

Ces Notes entrent dans la moyenne des examens particuliers et généraux.

1555. — Imprimerie LEON RAYER, 200, rue Saint-Martin.

Fig. 17. Two printed leaflets of instructions pasted into Jules Carimantrand's student notebook from Achille Ferry's *Cours de métallurgie* (1856–57) and in C. Lafont's student notebook from Maurice Lévy's *Cours de mécanique* (1888) detailing the rules for recording, checking, and grading lecture notebooks. These rules remained in place for the better part of a century. (Source: Peters Collection)

projects and a non-elitist respect for the knowledge that could be gained from manual labor. This was, of course, a very different intention from Beckmann's far simpler observation of labor. The philosophical background for this high regard for manual labor may lie in an estimation of the economists Adam Smith's and David Ricardo's theories.

The insistence on the workshop entering the university environment was a logical extension of the ECP's integrative principle. Workshops were unknown in French schools,[140] but common in Germany, especially in the mining schools that were based on Freiberg's curriculum, and the first were installed at the ECP for Colladon's steam-engine course when the school opened in 1829–30 (see later). Steam engines naturally involved the study of iron, and so the new material was present in the curriculum from the very beginning. Colladon's course was reminiscent of Beckmann's site visits to workshops and manufactories that he had explored in his courses on economy and manufacture at Göttingen in the 1770s.

140. Laboratories for analytical experiments did exist, for example in the École des Mines, but not workshops where the students built objects.

Fig. 18. In contrast to the École Polytechnique and the École des Ponts et Chaussées, education at the École Centrale was hands-on and down-to-earth. The students experimented in smelly laboratories and dirty workshops like this first-year chemistry laboratory (Source: Rousselet, *Nos grandes écoles*, 1888, p. 301)

But where Beckmann's students merely observed manual labor with the intention of understanding the involved economic principles, the ECP's curriculum went further and demanded active participation. The first drawing teacher, César Leblanc, had previously also taken his students at the Conservatoire National des Arts et Métiers into workshops where they drew and recorded machinery. Colladon and the other ECP professors knew this, of course, and they were also familiar with the Perronet system that had similarly involved site visits, and they expanded the excursions into actual hands-on traineeships designed to serve the needs of industry.

During the early years, hardly any hierarchical distance existed between the professors and their students (Weiss 1982, pp. 193–194). This was unique in higher education but understandable in this case because most of the professors were themselves still young and relatively inexperienced; some of their more mature students were almost the same age as they. Lavallée himself had barely turned thirty-three when the school opened in 1829. In fact, an alumnus of the first graduating class, Gustave Loustau, a classmate of the more famous

Jules Petiet, termed his fellow students *collaborateurs et en quelque sorte des associés* (collaborators and in some sense, associates) of their teachers (Loustau in Comberousse 1879.1, pp. A86–87). This is why the graduates of the first three years—who had so devotedly contributed to the school's survival during the difficult years of Charles X's reign as well as the revolt and the epidemic of 1830–32—were later officially designated *élèves-fondateurs* (founder-students), also in recognition of their contribution to the development of the school's didactical mission. The early years were an exhilarating departure from the norms of European education.

After this startup period, as the age differences between professors and students gradually grew greater (Bélanger was forty-eight when he began teaching in 1838), the relationship slowly reverted to the more distanced and standard European form of hierarchical collaboration. However, for the whole early period until major curriculum changes were instituted between 1862 and 1865, contacts between those who taught and those who learned still remained stronger than was usual in other French schools of higher education. The mechanism for this continued closeness was two-fold. The teachers who had studied with Monge at the EP included an informal participatory seminar format in their teaching that they themselves had experienced (Dupin 1819).[141] This was supported in the curriculum by the courses that combined theory and practice and that were taught to the combined second- and third-year students until 1864 (Pothier 1887, pp. 340–341). They brought successive years into daily contact with one another and with their instructors as a group over a two-year period. It was a system that enhanced the esprit de corps and that bound the students and their teachers together.

The programmatic concept was that the first year would concentrate on the basic theory of each field. These were courses that all students were obliged to take, after which the students were required in the middle of their second year to choose one of five majors that they wished to pursue in more detail. The five majors originally offered by the school were machine construction and mechanics, construction and what were called the physical arts, inorganic chemistry and its industrial applications, organic chemistry and agriculture, and

141. These were Bélanger, Mary, and Olivier. Bruyère had also used ad hoc seminars in his teaching at the ENPC (Jullien 1832), and Wilhelm von Humboldt had stressed this type of collaboration in a more formalized fashion in learning and research too.

mining and metallurgy. Inorganic and organic chemistry were later combined into a single major (Comberousse 1879.1, p. 63).

The second (and the quickly added) third years were then devoted to the incorporation of theory in practice in each field. In 1839–40 the balance was adjusted, and three semesters were devoted to the basic courses (Pfammatter 1997, p. 119). Despite their choice of major, all students still had to participate in all upper-level courses, but with the difference that the exercises were divided into general and specialized categories. The students had additional, more detailed exercises to complete in the courses that corresponded to their chosen major, while the others were only required to do the general exercises (Comberousse 1879.1, pp. 62–63). This ingenious scheme was unique to the ECP. It resolved the apparent contradiction of providing both comprehensive general knowledge to all students and professional specialization to the majors in a subject in the same course. To my knowledge, this has never been implemented anywhere else.

The workload was demanding. The school day lasted from 8:30 to 16:30 six days a week, with lecture notes to be revised and completed in the evenings and on Sundays. There appears to have been no break at the end of the year, with the two semesters running together consecutively.[142] Discipline was draconian by today's standards. The students had to appear for roll call at 8:00 every day, and they were only allowed reading material that was related to their work:

> Literary works and political or literary journals will be removed from those students who bring them. (Lavallée/Dumas, 1831, p. 14)[143]

In 1840 the school increased the workload even further by demanding a summer project from the students. This was to be measured drawings of a machine or a building from the first year and a comprehensively illustrated and detailed analysis of a factory organization for the second (Comberousse 1879.1, pp. 97–98). At the end of the three-year program, and after an adjustment of the requirements at the end of the experimental period in 1832 (Comberousse 1879.1 p. 78), a final project led either to a diploma or a certificate of competence.

---

142. Antonio Montero's notebook of Olivier's course in my collection dates lessons on December 20, 23, 27, 30 and January 3 and 7.

143. Les ouvrages de littérature et les journaux politiques ou littéraires seraient enlevés aux élèves qui les apporteraient.

From the outset, between 25% and 30% of the ECP student body was foreign. In fact the first brochure was also published in a German translation in 1830 (Lavallée/Dumas 1830). This was unusual, and it stood in contrast to the EP and the ENPC's policy that only admitted very few foreigners, and those that it did were almost exclusively auditors, not regular students. Perronet had been politically prevailed upon to admit individual students sponsored by foreign governments from 1786 (Vacant, p. 109), but this was the exception, not the rule. Although no information appears to be readily available on early regularly admitted foreign students at the ENPC, the first appeared only after changes were made in the organization of the school in 1852 (Montel 1995, p. 217 footnote 7) under the decree of 1851 that opened the school to a larger pool of applicants.

Lavallée's school also placed no upper age limit on entrants as the EP did, where the cutoff age was twenty. The ECP's lower age limit was fifteen (Pothier 1887, p. 46). This increased to sixteen in 1835, corresponding to that at the EP (Comberousse 1879.1, p. 39), and to eighteen in 1856 or 1858 (Comberousse 1879.1. pp. 99/161).

The ECP was organized as a day school—that is, the students boarded externally. This was another feature that was unusual in France. The EP had also originally been conceived as a day school until Napoleon militarized it as an academy in 1804, where the *cadets* were of course barracked from 1805 on (Mary 1862, *Notice sur la vie...,* p. 184). In the French public engineering schools, particularly the graduate écoles d'application, students were officially enrolled as cadets with uniforms and drill, and many of the *lycées* (high schools) of the time were modelled on military academies too. Lavallée may have based his model on the original form of the EP, perhaps also with an eye to how the Americans, the English, or other countries had organized their new civilian engineering schools. Flachat may also have influenced the unusual decision to organize the school in a non-military fashion rather than as an academy. He was himself a graduate of the Lycée Condorcet (then named the Lycée Bonaparte), that was one of the rare French high schools at the time that was not organized as a boarding school.

The ages of the entering students were diverse in the early years, ranging between fifteen and the late twenties, so that a boarding school would have been impossible anyway. Lavallée therefore entered into a number of contracts with surrounding boarding houses to alleviate the misgivings that parents of the youngest applicants had voiced (Lavallée/Dumas 1834, p. 11). But despite concerns that the youth of the students and the lack of close monitoring of their personal lives would present problems, there were few disciplinary issues because the severely overworked youngsters simply had no leisure and little energy for extra-curricular activities of any kind, whether socially acceptable or unacceptable.

The students and their teachers were exceedingly hardworking, far more so than we are accustomed to today. Lectures lasted 1½ - 2 hours each and the students were required to take approximately between 957 (until 1837) and 1049 (until 1857) lessons during their three years at the school (Pothier 1887, p. 357). If Pothier's calculations are correct, this workload is inconceivable from a modern standpoint.

> Workshops, drawings, sketches, weekly exams, general exams, painstakingly executed notebooks, mandatory rough drafts, detail projects, comprehensive projects, vacation homework, factory and diverse site visits, supporting memoranda, everything is calculated and organized so as not to leave the students a single moment of idleness to interrupt their work, so as to strengthen their moral fiber, to prepare them for both practical problems and life's vicissitudes. (Comberousse 1879.1, p. 124)[144]

Aside from sons of industrialists who were eager to gain more useful knowledge—almost 35% of who (48 out of 140) were over twenty-one in the school's first year (Comberousse 1879.1, pp. 34–35)—there was also a younger group that had passed the EP's entrance exams, the *concours*, and had been ranked *admissible* but had not gained entry for lack of space. Many of these then opted for the ECP, where they were admitted on the basis of an interview and without further examination. This, too, was an unconventional approach to recruitment in French higher education, as was the unheard-of novelty that the interviewers looked more for aptitude than for previously acquired knowledge (Comberousse 1879.1, pp. 40–41). As a fledgling institution, the ECP intentionally instituted flexible entrance requirements in order to generate a viable revenue and a sufficiently large student body. This contrasted with the entrance requirements at both the EP and the

144. Manipulations, travaux graphiques, épures, examens de semaine, examens généraux, cahiers de notes soigneusement tenus, croquis exigés, éléments de projets, projets d'ensemble, travaux de vacances, visites d'usines et d'exploitations diverses, mémoires à l'appui, tout est calculé et réuni pour ne pas laisser une minute de désœuvrement aux élèves pour les rompre au travail, pour affermir leur moral, pour les préparer à la fois aux problèmes de la pratique et aux difficultés de la vie.

ENPC that were (and still are) exceedingly stringent. They were intended as a filter and the most important hurdle to higher education at these institutions. Once entrance was secured, students were virtually guaranteed success no matter when they finished the course. The ECP students, on the other hand, originally had an easy access to the program, after which they continually had to prove themselves. This is why, although the system has now changed, alumni of the EP and ENPC are often still classed by their year of entry as a 'class', not their year of individual graduation like the alumni of the ECP (see Guillet 1929, pp. 511–524). Lax entry requirements were unheard of in French higher education, and only a private institution could go against traditional admissions policy.

This unusual policy proved critical for the successful development of the school. It had the effect in the early years of guaranteeing a rapid acceptance of Lavallée and Dumas's experiment, but with the serious drawback of a low initial graduation rate of roughly 40% (Comberousse 1879.1, pp. 173–174).

It was not easy to filter candidates for admission. In 1834, Olivier, as the first directeur des études received a letter from a 24-year-old prospective student, Charles Frénet, in Lyon. He was studying on his own, hoping to be accepted into the second year, as he felt himself too old to burden his parents with more than two years' worth of tuition fees. He had written to Lavallée, who had referred him to Olivier.[145] Frénet, whose eagerness came across in the letter as a little immature, was evidently refused, or at least never graduated, as he cannot be traced in the list of graduates of the school and there is no record of him online.

The administration made a change in 1837 that improved the situation: prospective students were required to write essays in geometry and literature and to present material from the preparatory school that they had attended (Pfammatter 1997, p. 119). After the school was taken over by the state in 1857 this system was abandoned in favor of the traditional French *concours*, with the intention of eliminating the possibility of favoritism (Comberousse 1879.1, p. 41). This change increased the graduation rate to around 70%. A few places were also reserved for what we would now call continuing education and for auditors who wished to take courses that were not offered elsewhere (Comberousse 1879.1, pp. 64–65). In 1833, three years after opening, the school had an equal

145. Charles Frénet to Théodore Olivier, Sept. 8, 1834. Date stamped the same date, Peters Collection.

number of students who were sons of industrialists, and students who had no previous connection to industry.[146]

Perhaps the earliest school to adopt the ECP's approach to education for industry—aside from Mons in 1837—was the second civil engineering school in the United States, Union College in Schenectady in upstate New York (Comberousse 1879.1, pp. 135–136). Union College is said to have graduated its first engineering students in 1835. However, a formal program was only established by William Mitchell Gillespie in 1845.[147] Gillespie had graduated from Columbia College in 1834,[148] and had visited France where he attended some engineering classes, most likely at the ECP, during the decade he spent in Europe (1835–45).[149]

Decades later Flachat recalled the need for a comprehensive, border-crossing industrial engineering education that he and the founders had identified in the 1820s:

> ...a large factory could not be created comprehensively with a design like it can today. Designing, calculating, and drafting were relatively simple affairs. Establishing it took the most effort. Today building an iron mill, however important it may be, is a question of planning and general organization for the engineer, and when done well, the contractor does the rest, even saving the engineer a considerable amount of detail work. Nothing comparable existed then.

---

146. *Notice sur l'École centrale des arts et manufactures, adressée, par une commission représentant les professeurs de cet établissement et par le directeur, au Conseil Général des Manufactures, le 27 février 1833* in: Lavallée/Dumas 1832.1, p. 6

147. The first American civil engineering school was Rensselaer Polytechnic Institute in neighboring Troy, NY. It was established in 1824 and modeled after the military engineering program at West Point, which had been initiated in 1802 but only academically established after changes were made in the school's organization in 1817. Both of these had been based on the EP model, while the only other analogous establishment, the Liverpool Mechanics Library and Institute in England, founded in 1823–25 (although there had been earlier, less influential ones 1821–24 in Edinburgh, Glasgow, London, and Ipswich), was a workmen's education society similar in type to the Association Polytechnique pour le développement de l'instruction populaire of 1830 in Paris.

148. The undergraduate Columbia College, now part of Columbia University, did not offer engineering.

149. Certain reports intimate that he was an external student at the ENPC, but foreign students were rarely accepted into that school before 1852 and his name does not appear in the online listing (1744–1920) of the ENPC students (Lacourt, 2020).

> On the site chosen for a factory, one had to model, mold, cast, and adjust the cast-iron members, forge and manufacture the wrought iron, level the site, lay out the works, the masonry, the timbering, the shelters—in other words, all the installations needed for a large factory. The engineer had only local workers and materials at his disposal. And these resources had to be created, too, in the sense that he needed to study the local rocks that would provide the lime, manufacture the hydraulic mortar—virtually unknown at the time in civil construction—choose the timber in the forest cuttings, build the carpentry shop for making the models, the cupolas [for the iron furnaces], the towers, etc.: indeed, he had to train the local workmen, all the more difficult a task as all the methods of manufacture were novel. The engineer took lodgings on site and shared the life of the workmen. (Flachat 1866, p. 535)[150]

The experimental idea at the ECP was that the school would establish an integrated 'industrial science' following Dumas's dictum *la science de l'ingénieur est une* (the science of the engineer is an indivisible entity, Guillet 1929, p. 87), or as Lavallée wrote in the school brochure (1834, p. 8) and Pothier reiterated in 1887 (p. 43), *la science industrielle est une*. Aside from science and technology, the school was to concentrate on the economy of industry and manufacturing processes as necessary components of a rounded education. The founders recognized the need to convince potential employers that industrial engineers were capable of organizing and directing viable enterprises (Lavallée/Dumas 1834, pp. 7–9).

---

150. …une grande usine ne pouvait pas, comme aujourd'hui, être créée de toute pièces avec le dessin. Projeter, calculer, dessiner était relativement facile. Exécuter était l'œuvre principale. Aujourd'hui la construction d'une usine à fer, quelque importante qu'elle soit, est, pour l'ingénieur, une affaire d'étude et de dispositions générales, et quand l'étude est bien achevée, les ateliers de construction se chargent du reste. Ils épargnent même à l'ingénieur une multitude de détails. Alors rien de pareil n'existait.

Sur le lieu où l'usine devait être établie, il fallait modeler, mouler, fondre et ajuster les pièces de fonte; forger et façonner les pièces de fer; faire les nivellements, les tracés des ouvrages, les maçonneries, les charpentes, les couvertures, toutes les installations, en un mot, d'une grande fabrique. L'ingénieur ne disposait pour cela que des ressources locales en matériaux et en ouvriers. Ces ressources étaient même à créer en ce sens qu'il fallait étudier la qualités des roches voisines, de la pierre à chaux, faire des mortiers hydrauliques, chose presque inconnue dans les constructions civiles; choisir le bois dans les coupes, établir des atelier de menuiserie pour les modèles, des cubilots, des tours, etc.; enfin former les ouvriers de la localité; c'était là une difficulté d'autant plus grave que toutes les méthodes de travail étaient nouvelles. L'ingénieur prenait alors domicile sur place et y menait la vie d'ouvrier.

A collapse illustrates the need for an integrated program

There were several events at the time that underscored the need for an engineering education that combined scientific understanding with industrial practice. The collapse of the Anthæum, an iron conservatory at Hove near Brighton, in 1833 was an extreme case that perfectly illustrated the pitfalls of a purely empirical, non-scientific approach and a lack of understanding of an organized and industrialized building process. The collapse shocked both the public and the profession, and although Lavallée made no reference to it or to any other similar event, occurrences of this type were on all engineers' minds at the time.

Fig. 19. The Anthæum was to have been the world's largest dome. Its collapse in 1833 demonstrated the dangers of segregated professional fields that did not communicate. (Source: section of: Amon Henry Wilds, Panorama of Brighton, 1833)

The Anthæum was to have been the world's largest dome. Only the vaguest information has survived on the structure that took the form of a single cast-iron dome of 50 meters in span, 20 meters high, and topped by a cupola.[151] During its construction, disagreements had grown between the three principals. The contractor had hired an architect and an engineering consultant who apparently flattened the dome, upon which the architect resigned. Then the contractor altered the design further, weakening the upper level observation gallery under the cupola that should have functioned as a compression ring at the top, omitting the original central pillar and struts for reasons of expediency, upon which the engineer left. Although details are no longer available, it must have been a totally dysfunctional

151. Most of this information is presented by Dale 1947 (pp. 153–155), however without citing the sources he used, many of which appear to have been private (p. vi). Berry 2012 also provides details, but also without sources.

building process from start to finish. The cast-iron arcs were buried three meters deep in the ground and not precisely symmetrically aligned, while the upper compression ring was faulty and poorly connected—anecdotal reports imply an unprofessional process and a fatal lack of structural understanding. Erection began in 1832, and the removal of the scaffolding caused it to fail the day before it was to open in 1833 (Pappworth, 1872).[152] The failure was all the more remarkable as

> ...it is almost as difficult to build a dome that will fall as it is to build a vault that will stand. (Fergusson 1855, vol.1, p. 441)[153]

Of course Fergusson's comment implies not only knowledge of the flow of forces involved, but also an understanding of the physical characteristics of the material used and of the detailing that guarantees structural integrity and force transference. Whatever the actual details of this case may have been, this understanding was evidently missing. The ruin remained a tourist attraction for over twenty years and was only cleared around 1855.

This lack of understanding of structural and construction problems and the missing communication and coordination between disciplines was something that an ECP education was designed to avoid. From the very beginning the curriculum was to function as a whole, in which all courses from structural theory and mechanical design to metallurgy and industrial organization crossed the boundaries between theory and practice and were geared toward this common concept, each course supporting and supplementing the others.[154]

> All the school's courses form one single entity, and they complete or illuminate one another. It is the unity of these courses that imparts industrial science to the students. But by focusing specially on one part of the studies,

---

152. There is no reliable source of information on this structure. The conservatory was apparently contracted for by a builder named Henry English who associated with the Griffin Foundry. He hired the architect Amon Henry Wilds for the design and an engineer named Peter Hollis as structural consultant. Only the architect is traceable. The Griffin Foundry was in Chesterfield, not Clerkenwell as Pappworth claimed. There was, however, a Phoenix Ironworks in Clerkenwell. Neither of them has been associated with any builder named English and no engineer named Hollis (either Peter or 'C' as another source claims) can be identified in that period.

153. Thanks to Santiago Huerta for this quotation.

154. Weiss 1982, p. 97 citing Guillet 1929, as well as Dumas's unpublished founding document *Rapport à présenter à M. le Ministre*.

> each student acquires an in-depth knowledge of the particular field that he has chosen. (Lavallée/Dumas, 1831 p. 8)[155]

The idea was not to educate specialists, but knowledgeable and adaptable generalists who would then intelligently be able to direct their interests into special areas while retaining an informed overview of the whole (Lavallée/Dumas 1834, p. 10). This comes very close to what I have elsewhere defined as the concept of technological thought (Peters 2010, pp. 127–152 / 2003, vol. 3, pp. 1629–1637 / 2000, pp. 118–129).

The elements of an industrial education consisted of

> first, a general scientific education suited to its purpose; second, a general industrial education; third, a form of apprenticeship or a specialized, practical education. (Lavallée/Dumas 1832.1, *Notice sur l'École centrale*, p. 5)[156]

The border-crossing concept of the school explains how de Dion's and Contamin's choice of mechanical engineering or Eiffel's choice of industrial chemistry, as their majors did not disadvantage them in their career shifts to construction. As the prospectus published by the school for the year 1847–48 declares,

> The chemist who graduates from this school is not only adept in the theory and practice of laboratory techniques, but he has studied mechanics, without which there can be no profitable business enterprise. He has learned how to use fuel economically. He can draft plans for buildings for the industry he will run and judge the best ways to build them. The student who plans to be an architect or build infrastructure is not only limited to the specialized knowledge of his craft, he does not only know how to calculate the stability of structures, but chemistry and mineralogy teach him how correctly to judge the quality of the materials he uses, applied physics serves to guide him in understanding heating, lighting, and the ventilation of workspaces

---

155. Tous les cours de l'École s'en forment qu'un, et se complètent ou s'éclairent l'un par l'autre. C'est l'ensemble de ces cours qui donnent aux élèves la science industrielle. Mais chaque élève, par l'attention spéciale qu'il est obligé d'apporter à une partie des études, acquiert la connaissance approfondie de l'art auquel il se destine.

156. 1° D'une éducation scientifique générale appropriée à sa direction; 2° D'une éducation industrielle générale; 3° D'une sorte d'apprentissage ou d'une éducation spéciale et pratique.

> and dwellings. When the mechanical engineer graduates from the school, he carries a useful knowledge of chemistry and physics with him, and the metallurgist is at the same time both mechanical engineer and chemist. This general education is common to all students. While satisfying the need to present an overview of science that informed industrialists need and that modern life demands, it has the advantage of making these young people able to develop into the very diverse careers that fate may decree for them, and one can also cite the cases of students who, after having graduated in one specialty, have successfully worked in another branch of industry. (Lavallée/Dumas 1847, pp. 3–4)[157]

This prospectus shows how the founders integrated the courses in basic and industrial chemistry, physics, material technology, and practical machine construction. As a result, this integrative approach to technical education inspired many graduates to develop their careers in related fields and even to cross boundaries to explore new combinations of fields or to define whole new directions in engineering. The resulting professional flexibility and inventiveness were new, and they became the hallmarks of nineteenth-century engineering.

---

157. Le chimiste sorti de cette École n'est pas seulement exercé à la théorie et à la pratique des opérations de laboratoire; il a étudié la mécanique, sans laquelle il n'est point d'exploitation profitable; il connaît les règles de l'emploi économique des combustibles, il peut dresser le plan et indiquer les meilleurs moyens d'exécution des édifices consacrés à l'industrie qu'il doit diriger. L'élève qui se destine à l'architecture ou aux constructions publiques ne s'est pas borné aux connaissances spéciales à cet art, il ne sait pas seulement soumettre aux calculs les questions relatives à la stabilité des édifices; la chimie, la minéralogie lui enseignent à apprécier sûrement les qualités des matériaux qu'il emploie; la physique appliquée lui sert de guide en ce qui concerne le chauffage, l'éclairage, la ventilation des ateliers et des habitations. En sortant de l'École le mécanicien emporte des connaissances positives en chimie, en physique, et le métallurgiste est à la fois mécanicien et chimiste. Cette éducation générale et commune à tous les élèves, en même temps qu'elle satisfait à la condition aujourd'hui nécessaire de présenter la science au point de vue d'ensemble qui fait les habiles industriels, a l'avantage de rendre ces jeunes gens aptes aux carrières les plus diverses que des circonstances souvent imprévues peuvent les conduire à embrasser: aussi peut-on citer de nombreux exemples d'élèves qui, après avoir obtenus le diplôme d'une spécialité, ont exercé avec succès une autre branche de l'industrie savante.

The teachers and their courses

In the spirit of Palmer's 1818 definition of civil engineering, the school deliberately recruited multifaceted practitioners who were willing to collaborate both internally and with the profession rather than hiring pure theoreticians as instructors. Enthusiasm for a novel approach to education was certainly a criterion, and although it was not stated, the related issue of playfulness in learning may have resonated in the hiring process. Playfulness underlies the concept of self-directed study in Pestalozzi's pedagogy, which certainly influenced education at the time, and it also underlaid the highly esteemed Monge's unorthodox approach to serious research through the apparent whimsy of chess moves and card tricks.

Aside from descriptive geometry, mechanics, metallurgy, machinery, public works (infrastructure), and building construction, the fields included mineralogy, mining, drawing, physics, chemistry, *hygiène industrielle* (actually physiology), economics, natural history, geology, and English. Common to all these courses was that they were organized with industry in mind according to the ECP's declared concept, and they included explanations of their relevance to various industrial fields (Comberousse 1879.1, pp. 54–56). While some courses like English soon disappeared, others like drawing were modified, and some split to spawn further subjects like railway construction and prime movers, which at the time meant steam engines.

Among these courses, physics taught by Péclet and chemistry taught by Dumas were among the mainstays of the school. The physics of material behavior and the chemical theory behind metallurgy were certainly important for the development of iron, but my focus here lies on the part of the curriculum that would have the most direct relevance for innovation in iron construction and its involvement with industry.

From this point of view, the most important courses and the relevant professors who developed and taught them were Jean-Baptiste Bélanger, who taught that part of physics that became mechanics, Auguste Perdonnet, the railway pioneer who began with mining and quickly shifted his focus to railway construction, Théodore

Olivier, with his passion for descriptive geometry, its application to industry, and visualization through drawing, and Louis-Charles Mary for public works, architecture, and building construction. It appears that these men were the most active and effective pedagogues in the development of this part of the curriculum.

They were seconded by Charles Walter de Saint-Ange, Achille Ferry, and Auguste Faure, who together and individually taught the metallurgy of ferrous metals, machine theory, and machine construction, as well as César Leblanc and his successor Nicolas-Auguste Thumeloup, who implemented Olivier's, Bélanger's, and Péclet's emphasis on drawing. There were several others who contributed important pedagogical ideas to the school during the early period, like the mathematician Gaspard-Gustave de Coriolis and the physicist Colladon, and although Coriolis did not teach any of the eight innovators featured in this study, and although Colladon only taught Polonceau and Barrault, they influenced the curriculum for all. There were many other teachers who were relevant to industrial and chemical engineering, agriculture, or mining, but this was the core group as far as construction was concerned.

Most historical accounts focus on Mary's construction courses as the key to the education of Eiffel and Jenney, particularly (see Pfammatter 1997). These courses were certainly one of the keys, but not in the sense that is generally thought. Mary's influence was more complex than the straightforward teaching of architectural and construction content. As Pfammatter first explained, Mary's contribution lay in his ability to modify theory to practice. His job was to connect the various parts of the curriculum and to develop a design method that became central to the pedagogy of the program.

From his first year at the school in 1833, Mary was intimately involved in the definition and organization of the curriculum and its methodology. Aside from linking material characteristics to architectural form, most of the content of his courses in construction (excepting the brief segment on design method—see chapter 'Mary's method') was traditional and unrelated to iron construction. Nevertheless, Mary was central to the program, which was why—together with Perdonnet and the two founders still active in 1857, Lavallée and Péclet[158]—

158. Olivier had died in 1853 and Dumas had left active teaching when he was appointed Minister of Agriculture and Commerce for a year in 1850 (Pothier 1887, p. 201), although he remained very active in matters of technical education and a member and then president of the *conseil d'études* until it was disbanded in 1862 (Guillet 1929, p. 44).

Mary was co-signatory as *délégué du conseil des études* of the letter to Emperor Napoleon III offering the transfer of the school to the state (Guillet 1929, p. 15).

The reason for this offer was the shock caused by Olivier's death in 1853 (Comberousse 1879.1, pp. 127ff) and likely also the cumulative effect of Walter's death two years before and Thumeloup's death the year after that, all of them at quite an early age (60, 57½, and 50).[159] The surviving founders, especially Lavallée, realized that the school was entirely dependent on them and that they, too, were mortal. In offering his school to the nation, Lavallée underscored its importance to the emperor, citing Playfair's opinion (Grelon 2000, p. 5), who had, as we saw, spoken so highly of the school in 1852. Coming from a source that the emperor knew to be a close advisor to the English crown, Playfair's opinion of course carried weight. The transition was finalized in 1857.

After the transition, Péclet died at the age of 64 in late 1857. Lavallée continued as director for another half-decade until 1862, but these critical losses, compounded by the shift in ownership, had immediate repercussions in the teaching of construction, which both Perdonnet and Mary were involved with. The changes—especially those instituted in a second shift in the program after 1862 when Lavallée relinquished the directorship, as well as the rise of other civil engineering programs that had learned from the founders' pedagogical models—initiated a gradual fading of the pioneering importance of the ECP in engineering education. The ENPC had opened its doors to foreign students in 1852 and had begun accepting applicants from institutions other than the EP. At the same time, it widened its scope to encompass broader goals than the education of French state employees. It became a direct competitor to the ECP (Comberousse 1879.1, p. 104, footnote), albeit with the difference that it aimed more at educating civil engineers than *ingénieurs industriels*.

Bélanger bridges between science and technology

Jean-Baptiste Bélanger was the son of a locksmith, and became a canal and railway engineer, mathematician, and physicist. He had graduated from the EP in 1810 and from the ENPC in 1815, where the interface between mathematics and

159. In 1850 an average Englishman could expect to reach 63, and an upper-class Frenchman a few years more, perhaps closer to the Swedish average of almost 69. Lavallée died at 76, Bélanger at 84, Perdonnet at 66, Mary five days shy of 79 and Dumas at 83 and 9 months.

Fig. 20. Jean-Baptiste Bélanger, one of the founders of modern statics, as a young man. He was both a theoretician and a practical engineer in equal measure and thus able to amalgamate theory and practice in a way that no one else could. (Source: nineteenth-century engraving, published on wonders-of-the-world.net - Eiffel Tower, consulted 03.08.2022)

construction had evidently attracted him. The theoretician of mechanics Prony was director of the ENPC, but he was preoccupied with his research, and his teacher there was the mathematician Joseph Eisenmann.[160]

EP graduates in those years like Olivier, Coriolis, Mary, and Bélanger had been inspired by Monge with a drive—or as he put it, a *devoir d'instruire*—to teach young professionals (Pothier 1887, p. 156). However, Bélanger only began his teaching career after having practiced as a hydraulic engineer for twenty-six years, during which time he worked with Poncelet and Antoine-Rémi Polonceau (Camille Polonceau's father) and made significant theoretical contributions to the field. These twenty-six years, as well as his background as the son of a craftsman, conditioned him to appreciate both theory and practice in equal measure.

160. Eisenmann, from Württemberg, had studied mathematics from 1773 at the university of Freiburg in Breisgau and lived as a priest in the Prémontré monastery near Laon and at the Collège des Prémontrés in Paris. (Scheurer 1991). He had worked as a cartographer for the monastery and continued to study as an autodidact. After the dissolution of his order in 1790, he applied for acceptance into the ENPC in 1794, naming many mathematical works that he had studied, but none in engineering. He was admitted in 1794–95 (An III), curiously as *élève et professeur* (Scheurer 1991, p. 457), and became professor for stereotomy in 1796. He had also served as Monge's assistant for descriptive geometry together with Hachette, and then for stereotomy at the EP in 1795 (Dupont 2000, pp. 3–4/9). When Bruyère joined the ENPC faculty in 1799, Eisenmann shifted to structural mechanics.

Initially he taught one course at the ECP, *analyse géométrique et mécanique générale* from 1838 to 1840. He then split it into two courses: a new *mécanique générale* from 1841 to 1846 that he later relinquished to another teacher, and *théorie des machines*, which included the stability of structures and dynamics of machines that he taught for his entire ECP career until 1864.[161] He also taught his specialty, hydraulic engineering, as part of the third-year mechanics course and in separate lectures that he collected together and published in transcript form (along with series on construction and mechanics) as yearly *Notes* for both the ECP and the ENPC (1842–51). Charles de Comberousse, who graduated from the ECP in 1850 and later came back to teach, considered Bélanger's and Dumas's contributions to be the two anchors of the institution (Comberousse 1879.2, cited in Weiss 1982, pp. 148–149).

Bélanger's interest in *vulgarisation* of theory for practitioners may have been kindled by Prony, the director of ENPC when he studied there. Although his mechanics teacher had been Eisenmann, Prony had established the curriculum in 1799 and published a seminal paper that dealt with vulgarisation of theory for engineers and laborers using graphical instead of algebraic means, his 1802 *Instruction-pratique sur une méthode pour déterminer les dimensions des murs de revêtement, en se servant de la formule graphique* (Practical instruction in a method to determine the dimensions of retaining walls using graphical means). This short, twelve-page paper translated the design of retaining walls using analytical theory into a simple graphical method for innumerate builders. The paper was first issued for practitioners and reissued as a longer 44-page treatise the following month for both EP and ENPC students.[162] It was possibly the earliest example of conscious *vulgarisation* in engineering (see chapter "Graphic statics" and Fig. 48).

Even though Prony was not one of the program's teachers, he was the director, and his border-crossing approach to engineering education was very much present at the school. Bélanger was to follow Prony's path in systematizing mechanics and in formulating it for practitioners. He was familiar with Prony's ideas in the *Mécanique philosophique* of 1800, and adopted the concept of

---

161. As a joint appointee he then also taught at the at the ENPC (1841–55) and also at the EP (1851–60) after that school's curricular changes.

162. The title page of the first paper is dated as Germinal An X (22.03–20.04.1802), and the contemporary sticker on a copy of the second scanned by Google states 'Floréal an 10' (21.04.–20.05.1802). The title page of the reissue states that it was intended for both the EP and the ENPC.

vulgarisation but did not use graphics. Instead, he developed his own approach within the bounds of mathematical analysis using language as his vehicle.

Bélanger was a slow and deliberate lecturer, who often hesitated while searching for the precise term he needed. Nevertheless, he was much appreciated by the students for the organizational clarity of his thinking. One example was his redefinition of the term 'force', which became universal (Pothier 1887, pp. 222 and 322–323). Bélanger had discussed theoretical issues with Coriolis since 1819 (Coriolis 1829, p. ii), and Coriolis had probably been the first to use the term in 1829.[163] As Santiago Huerta writes, 'any deep change requires a substantial change in thinking and this requires a new language with new words and new meanings' (personal communication 01.11.2020), and such shifts were occurring in structural theory at the time.

Words and concepts are inextricably connected like the proverbial chicken and the egg; one cannot conceive of something one cannot name, just as one cannot name something one cannot conceive of. An original thinker may be able to envision the need for a new concept and work slowly toward a concrete form, but until it has been formulated, it remains nebulous. This was the problem Bélanger, Coriolis, and many others were struggling with in the creation of structural theory.

Whereas empirical practice in building often used (and still uses) imprecise terms that reflect local usages or the influence of various languages and fuzzy creative thinking, it was important to Bélanger that the messy business of construction be subjected to meticulous analysis. This was a difficult task, and it would help bridge the divide between science and technology. As Pothier, an early student and friend wrote:

> M. Bélanger attached great importance to method, to precise language, and to correct style. (Pothier 1887, p. 321)[164]

Bélanger strove to reconcile the ad hoc world of practice—where he had spent a quarter of a century—with the clear analysis of science. This was a difficult

---

163. According to Bélanger (1866, *préface*, p. xx), Coriolis was also the first to formally define *travail* (work) in 1826. Dugas (1988, p. 128) traces the concept back to Salomon of Caus in the sixteenth century.

164. M. Bélanger attachait une grande importance à la méthode, à la précision du langage, à la correction du style.

endeavor in which many before him had failed. In choosing language—both written and algebraic rather than graphical—Bélanger insisted on precise definitions, both in words and in mathematics, and through them he communicated the need for mental precision to his students. This emphasis appears to have been a near obsession for him. The exact meaning of a word permits a clear delimitation of a problem. It defines the parameters of an examination, which is what Bélanger needed to determine what he should define as belonging within the field of mechanics and what he should relegate to another subject. Step by step he created order in the impenetrable thicket that had developed in structural knowledge,[165] which also helped him realize and define statics as a special case of dynamics within the overall field of mechanics.[166] This was a variant didactical form of *vulgarisation*, and it required an overview and a mastery of the subject from two very different points of view: theory and practice.

Before developing his own course, Bélanger had served two years as *directeur des études* at the ECP 1836–38 (Guillet 1929, p. 44). In that capacity, pedagogy was his responsibility. He concerned himself with Perdonnet's and Mary's earliest endeavors in curriculum development. This involved the coordination of student assignments, the yearly *concours*, the meetings of the *conseil des études*,[167] and the development of industrial projects across all the majors. Each group of students—chemists, mechanics, constructors and so on—would contribute different aspects of this coordination from their respective fields (Pothier 1887, pp. 154–155).[168] While Perdonnet was the promotor and organizer who was comfortable with the give-and-take of political and social thinking, and Mary was the designer who encouraged multidirectional experimentation in construction, Bélanger must have realized that it would fall to him to provide the sharpness of analysis for the pedagogical balance.

When Bélanger relinquished his appointment as *directeur des études* in 1838 and began to develop his own course, his aim was to define the relationship between the mathematical theory of mechanics as Coriolis had conceived of it

---

165. Theoretical structural knowledge was contained in an undifferentiated form within classical mechanics at the time. Bélanger's own EP professor Joseph-Louis Lagrange had begun to reformulate the field analytically in 1788–89, and William Rowan Hamilton began to do so in 1833.

166. Monge had previously distinguished between them as two entirely separate fields in the introduction to his *Traité élémentaire de statique* of 1788 (p. 3 in the 3rd ed. of An VII = 1798–99).

167. Almost all the faculty were members of this council. It was finally disbanded in 1862.

168. This was extremely time-consuming, and in later years the projects were not often so linked.

and the technical detailing of machine construction. He set himself the goal of *vulgarising* classical mechanics, the study of bodies in movement. Benoit had recognized the need to translate theory into practice from the outset, but neither he nor his replacement Narcisse Didiez[169] had dealt with it (Pothier 1887, p. 63).

Engineering mechanics was an uncharted field when Bélanger began teaching. It included aspects of physics, metallurgy, and construction. Benoit had left the founding group before he could address the problem. Péclet, Coriolis, and Colladon succeeded him. Péclet took care of the basic physics course and gave it an industrial bias with his research on lighting and gas combustion, while Coriolis took another part in the direction of theory. Colladon, who was both an experimental and a theoretical physicist, served as Péclet's co-teacher and dealt with steam in the third year. He also briefly stepped in for Coriolis, who was only able to start teaching the mechanics course in January 1830 (Colladon 1893, p. 194). It was not at all easy to develop an appropriate concept for the subject, nor was deciding what would be better treated as parts of other courses (general physics, metallurgy, and machine construction) and which parts should constitute the core of mechanics. Bélanger also had to decide how to choose the mathematics he needed appropriately.

He had evidently already concerned himself with the issue of standardizing the field before 1829 (Coriolis 1829, p. iii), but it took him several more years of hard work to accomplish it. His lithographed transcript *Cours de mécanique générale et de théorie dynamique des machines* that appeared from 1840 to 1842 was an intermediate stage, and he published the final result[170] in its refined form in 1847. The title of his *Cours de mécanique ou résumé de leçons sur la dynamique, la statique et leurs applications à l'art de l'ingénieur* reflected the course's clarification, as it distinguished between dynamics (that he linked with kinematics) and statics.

The course was a masterfully crafted achievement. In it he presented his ideas as they had evolved from the general mix of geometry, mechanics, and mathematics taught up to that time by his own teacher Eisenmann at the ENPC. Eisenmann had divided the course *mécanique appliquée* into three parts: 1. *Résistance*

---

169. Didiez is always named as N. J. Didiez, but according to obscure genealogical information on the Internet he was almost certainly Narcisse Didiez (1796–1861). See also Lorenat, pp. 85–86.

170. This was only final insofar as the pedagogy of the school was concerned. After he retired in 1864, Bélanger reached a final formulation of his thinking in three volumes (Bélanger 1864/1864/1866).

des matériaux. Établissement des constructions en terre, en maçonnerie et en charpente, 2. *Mouvement et résistance des fluides—Conduite et distribution des eaux*, and 3. *Établissement des machines* (Eisenmann 1801–02, ENPC ms. 2002–3). However, Bélanger was an engineer, where Eisenmann had been a mathematician. As the very title of Eisenmann's course stated, he applied mathematics to a host of different problems in practice. So while Bélanger gradually expanded this basic material in the direction of analysis, he always kept in mind that his task was to unify abstract science and practical engineering rather than to apply the one to a hodge-podge of issues in the other. At the same time, he sought to reconcile scientific theory and its concomitant search for understanding with technological theory and its goal of making.

As Bélanger was both a builder and a scientist, he could not disregard the one aspect or the other. Like so many before him, he realized that the conundrum of the relationship between practice and theory had to be resolved. His solution was to split the original machine course into two parts. One he devoted to machine construction, which Walter de Saint-Ange ('Walter' as the school referred to him) was charged with organizing, and the other was devoted to mechanics, which he took on himself. Despite his analytical bent, and in contrast to all others who were working on the development of mechanics at the time, Bélanger insisted on conceiving the course according to industrially useful rather than theoretical criteria. The difficulty he faced was to do this without compromising on clarity or completeness and without bowdlerizing any aspect of scientific accuracy. As Rankine—who certainly knew of French developments and might even have met Bélanger—later explained in his inaugural lecture in 1856:

> There are certain characteristics in the mode of treating the subjects, by which this practical-scientific instruction ought to be distinguished from instruction for purely scientific purposes.
>
> In the first place, it will be universally admitted, that as far as is possible, all mathematical intricacy ought to be avoided.
>
> In the original discovery of a proposition of practical utility, by deduction from general principles and from experimental data, a complex algebraical investigation is often not merely useful, but indispensable; but in expounding such a proposition as part of practical science, and applying it to practical purposes, simplicity is of the first importance:—and, in fact, the more thoroughly a scientific man has studied the higher mathematics, the more fully does he become aware of this truth,—and I may add, the better qualified does

> he become to free the exposition and application of scientific principles from mathematical intricacy. (Rankine 1856, p. 19)

Once he had established a theoretical order that satisfied him, Bélanger could attack the second part of the problem and shift his focus from that of a scientist to that of a *vulgarisateur*. He did this by linking every aspect of his theory to practical issues in the Monge-Hachette tradition, and he used this link to supplement the approximation and ad hoc reaction that professional practice requires. His evenhanded focus on theory and practice and his ability to encompass both the didactical (systematic) and pedagogical (communicative) aspects of *vulgarisation* is what distinguished Bélanger's mind from the purely scientific bias of his own professors Lagrange and Eisenmann, and of his contemporaries at the EP. Clarity and analytical thinking in industrial and construction processes became characteristics that his students subsequently carried with them into the profession with great success. Unambiguous formulation dominated Bélanger's whole career. In his very last treatise in 1866 he wrote:

> An essential condition of success in this subject [the study of mechanics] is that one must attach the utmost importance to the scientific significance of certain terms taken from everyday language, but that gain a special and unique connotation in mechanics. (Bélanger 1866, p. vi)[171]

Coordinating diverse approaches to engineering demanded flexibility, and Bélanger was able to provide that in his role as mediator between theory and practice (see Dupont 2009, p. 61). Flexibility encouraged adaptation and collaborative understanding among students and teachers. It also served to enhance the integration of different aspects of industrial engineering and it supported the border-crossing goals of the school. Bélanger must have been something of an organizational control freak.[172] He was obsessed not only with a precise use of language, but also with the curricular minutiae that helped carry that precision into practice.

Guillet gave a clear example of this precision in Bélanger's excruciatingly detailed definition of the drawing teacher's role. The sub-committee that

---

171. Une condition essentielle du succès de cette étude est d'attacher la plus grande attention à la signification scientifique de certains mots empruntés au langage commun et recevant en mécanique un sens spécial qui doit seul leur être attribué.

172. See the introduction to his treatise of 1842 later in this chapter.

Bélanger chaired, consisting of Péclet, Olivier, and himself defined the drawing course right down to the number of students permitted in each class section, the process the students were to use in drafting, when to use pencil or ink, the quality, thickness and type of lines to draw, how much pressure to apply in drawing these lines, and the color of the ink they were to use (Guillet 1929, p. 94 and Pothier 1887, p. 168). One may speculate on how much the initial drawing teacher César Leblanc, who was a pioneer of technical drafting, may have contributed to the discussion, but the concentration on drawing that Bélanger helped to promote proved to be one of the critical elements of the program and among the school's several important contributions to innovative teaching. This focus introduced visual learning into the curriculum. It followed Monge, who had considered descriptive geometry a visual language (Picon 2000, p. 27 and Monge 1799, p. 2).

Because he was also an experienced practitioner—in contrast to most of his own teachers and colleagues at the EP—he never lost sight of the fact that his students needed to be enabled to create objects using what he taught. His pedagogical intentions were therefore different and broader than those of his fellow physicists at the EP, and his hard-won border-crossing approach completely fulfilled Dumas's ideal of a *science industrielle*.

Bélanger was neither an intellectual snob nor a pedant. He did not regard science as an aristocratic attribute as some others did, for example Alexis de Tocqueville (Weiss 1982, p. 94), nor did he treat practice as a lesser field like his contemporary academics at the EP and in Britain. He was a true disciple of Monge and Hachette, and he took great pains to develop a clarity in his teaching that left nothing of importance out while gearing his theories to the very specific needs and the mindset of practitioners. His effectiveness as a teacher and as a model for engineers to emulate rested on the success he achieved in developing this clarity.

> It is the union of the various branches of the theory of mechanics and practical knowledge gained through the workshop that alone can educate the mechanical engineer. (Bélanger, *Traité de la cinématique* 1864, pp. xxi–xxii, cited in Pothier 1887, p. 322)[173]

Understanding the limitations and accepting approximation as a valid structural concept was a very big step in the development of our modern technological

173. C'est l'union des diverses branches de la mécanique théorique aux connaissances pratiques acquises par la fréquentation des ateliers, qui peut seule former un ingénieur constructeur des machines.

mode of thought. It emancipated technology from being merely regarded as an inexact stepchild of science.[174] Approximation is a central tenet of engineering thought as opposed to physics. It is one of the characteristics of the distinction between science and technology. As Bélanger's students Léon Molinos and Théophile Seyrig wrote in 1879,

> The engineer's point of view is not the same as the scholar's, who is concerned with pure science. He has no need of an exact mathematical solution; he needs adequate and above all safe approximations within certain limits. (in: *Bulletin de l'Association amicale des anciens élèves de l'École Centrale*, 1879, cit by Jean-François Belhoste in Barjot and Dureuil 2008, p. 74)[175]

Both Molinos and Charles Pronnier had directly experienced Bélanger's teaching during this process of intellectualization, and they expressed their esteem of his methods in the introduction to their 1857 book on bridges:

> ...in the first part of the treatise we presented mathematical calculations for various bridge systems, indicating simplifications that allow them to be abbreviated. We reproduced the method that the learned professor M. Bélanger, who was the first to treat this question in a general way, has taught since 1842 at the École des ponts et chaussées, and more fully since then at the École centrale. (Molinos and Pronnier 1857, p. vi)[176]

In 1879 the ECP mathematics professor Comberousse seconded this opinion when he wrote: "As far as we are concerned, he created the true field of engineering mechanics" (*pour nous, il a créé la vraie Mécanique de l'ingénieur*. Comberousse 1879.1, p. 95).

---

174. See Peters 2000. This form of approximation had apparently first been suggested by Leonhard Euler's amanuensis Nikolaus Fuss when he commented that the parabola rather than the catenary is sufficiently precise for the design of a suspension bridge (Peters 1987, p. 75).

175. Le point de vue de l'ingénieur, n'est pas le même que celui du savant qui s'occupe de science pure. Il n'a pas besoin d'une solution mathématiquement exacte, il lui faut des approximations suffisantes, et surtout sûres dans certaines limites.

176. ...nous avons donné dans la première partie de l'ouvrage des méthodes de calcul pour les différents systèmes de ponts, en indiquant les simplifications qui permettent de les abréger; nous avons reproduit la méthode de M. Bélanger, enseignée depuis 1842 à l'École des ponts et chaussées, et plus complétement [*sic*] ensuite à l'École centrale, par ce savant professeur, qui, le premier, a traité cette question d'une manière générale.

Bélanger was not working toward this goal in a vacuum; there were others who were striving in the same direction based on what Monge had pioneered and the systematization that Prony had attempted but not completed in 1800, albeit with different foci. Bélanger's friend Poncelet, another of Prony and Monge's former students, had previously published two transcripts that he had used at the École d'Application de l'Artillerie in Metz: *Cours de mécanique applique aux machines* in 1826, and the more general *Introduction à la mécanique industrielle* in 1829. Comberousse later claimed that these courses were more of a popular character and directed toward workmen rather than to professional engineers (Comberousse 1879.1, p. 27). He obviously meant 'workmen' pejoratively referring only to the public lectures that Poncelet had held in Metz (1827–32) using the text *Introduction*, because in truth Poncelet's main course at Metz (1825–34) was geared to the needs of artillery officers. In contrast to Comberousse, Henri de Dion especially treasured Poncelet's *vulgarisation* (*Mémoires et compte rendu*, 1877.1, pp. 75–76).

Navier had also collected and organized all the research that had previously been accomplished by others in his lectures, contributed essential new material[177] in the first part of his course published in 1826, and again in a second volume published posthumously in 1838, *Résumé des leçons données a l'École des ponts et chaussées, sur l'application de la mécanique à l'établissement des constructions et des machines*. However, Poncelet and Navier leaned more toward theory than practice in their treatises. Bélanger's contribution lay primarily in his balance. He systematized statics for technical purposes (Pothier 1887, p. 323) as an expansion of what Poncelet had previously accomplished, and developed a theory of torsion as part of the strength of materials that was subsumed under technical mechanics at the time.[178] Poncelet, who as *commandant* of the EP (1848–50) had become the advocate of the scientific engineer, was one of those profound thinkers who could appreciate the need for bridging the gap between the scientific and the practical approach to engineering, and he valued Bélanger's efforts. He was his friend and colleague and regretted that Bélanger did not publicize the innovative character of his work (Pothier 1887, p. 324).

177. Navier's published class notes: 'contained seminal contributions in beam theory, truss theory, arch theory, and general analysis and design approaches that transformed structural engineering'. Gasparini and Buonopane 2019, p. 12.

178. I am grateful for this clarification in a personal communication from Karl-Eugen Kurrer, 28.01.2019.

We can appreciate the intellectual effort it took and how complicated a task it was to accomplish the delicate reconciliation between theory and practice, this *vulgarisation scientifico-technique*. He had explained what the ECP's curriculum required of him in the introduction to the first edition of his earlier mathematical work in 1842:

> The extent of mathematical knowledge that was hitherto required for admission to this school was restricted to arithmetic, elementary geometry and some branches of algebra that include the resolution of first- and second-degree equations, logarithmic theory, and Newton's binomial theorem. The summary of the lessons published here aims at supplementing the students' instruction in pure mathematics up to the point necessary for them to be able to usefully follow the courses in rational [i.e., classical] mechanics, hydraulics, and the dynamic theory of machines. Together with the other sciences applicable to industry, these come within the purview of instruction at the École Centrale.
>
> Some of the topics that this preparatory course in higher mathematics discusses go beyond what is currently taught in university colleges. However, its scope is limited so that the students only need four or five months to master it. This is what the École Centrale des Arts et Manufactures' curriculum requires, and I attempt here to satisfy this condition without sacrificing the rigorous logic of the argument without which mathematics can degenerate into a misleading demi-science. I do this not by compromising clarity through excessive brevity, but by choosing those parts of analytical geometry and infinitesimal calculus that every educated engineer needs to acquire, and particularly those parts which are necessary for the study of mechanics viewed from the standpoint of its practical application to industry. (Bélanger 1842, pp. v-vi)[179]

---

179. Les connaissances mathématiques exigées jusqu'à présent pour l'admission à cette École se bornent à l'Arithmétique, à la Géométrie élémentaire et à la partie de l'Algèbre qui comprend la résolution des équations des deux premiers degrés, la théorie des logarithmes et la formule du binome [*sic*] de Newton, les leçons dont nous publions le résumé ont pour objet de compléter l'instruction des élèves sur les mathématiques pures, jusqu'au point nécessaire pour les mettre en état de suivre utilement les cours de Mécanique rationnelle, d'Hydraulique, et de Théorie dynamique des Machines, qui entrent, avec les autres sciences applicables à l'industrie, dans le cadre de l'enseignement de l'École centrale.

Ce cours préparatoire de mathématiques supérieures s'élève, par quelques-uns des sujets qu'il traite, au-delà de l'enseignement actuel des collèges universitaires; cependant son peu d'étendue est tel, qu'il n'exige des élèves auxquels il est destiné que quatre ou cinq mois d'études. Ainsi le voulait le plan d'enseignement de l'École centrale des Arts et Manufactures; et cette condition, nous avons tâché de la remplir, non en sacrifiant dans les démonstrations la logique sévère sans laquelle les

Five years after writing this, Bélanger successfully completed the *vulgarisation* of his fully developed mechanics course in 1847, where he innovatively treated statics as a limited case of dynamics. This distinction enabled him to link the fields of structural statics and machine mechanics, thereby forming one of the connections that the curriculum offered between machine construction in iron, and iron in building and bridge construction.

The system that Bélanger taught from 1842 and published five years later possibly contributed, in conjunction with Mary's design method, to the difference between the individual problem-solving approach of the three earlier graduates Polonceau, Barrault, and de Dion and the more general system-oriented concerns of Eiffel, Jenney, Baudet, Moisant, and Contamin. Polonceau and Barrault had not experienced Bélanger as a teacher at all—although Polonceau probably knew him personally as a former colleague of his father's—and all three had graduated before Mary's curriculum had fully matured from *Architecture* to *Construction*.

Bélanger's border-crossing dovetailed perfectly with the *vulgarisation* of knowledge that the founders, together with Perdonnet and Mary, were trying to achieve. Bélanger's ability to establish the field of structural mechanics is one example of the influence that he had on the overall development of the curriculum and on his students and their future professional abilities too.

Bélanger's pedagogical goal in this course, as well in everything else that he taught the students, was nothing less than the coordination of all relevant aspects of the field while reducing their complicatedness to a simple form, but without compromising the complexity of the subject. Only when he had fully understood the problem himself could he render and communicate it lucidly. This highest level of intellectualization that he attained using precise language over and beyond his definition of structural statics represented an important step in the development of industrialized iron construction, its conceptualization as a system, and especially its detailing. Even more importantly, Bélanger's ability to intellectually separate the needs and criteria

---

mathématiques deviennent une demi-science souvent trompeuse, non en compromettant la clarté par une excessive concision, mais en choisissant dans la Géométrie analytique et dans le Calcul infinitésimal les parties que tout ingénieur instruit doit posséder, et notamment celles qui sont nécessaires à l'étude de la Mécanique considérée au point de vue de son utilité pratique dans la direction des travaux d'industrie.

of science from the needs and criteria of practice was a rare intellectual gift. He valued them both equally and could therefore insist unreservedly on using approximation as an adequate technique for construction purposes. Despite his insistence on precision in language, he knew that there was no need to cling to scientific exactitude like his colleagues at the EP did (and unfortunately many modern engineering theoreticians too).

His analytical clarity not only enabled him to translate[180] theory into practice and to unify scientific and technological theory, but it also served his students as a model of how to cross the boundaries between the two in order to solve a network of practical problems in the real world. This can be seen in several aspects of Henri de Dion's work, for example in his solution of the impending collapse of the tower at the crossing in Bayeux cathedral (1855–59) or in his complex coordination of the design, calculation, manufacture, and erection of the iron structures for the Paris Exhibition of 1878. De Dion's career will be discussed in the final chapters. Bélanger's protégé Contamin was another masterful intellectual border-crosser. He was the chief engineer of a railway company, a researcher and teacher in applied mechanics, and a government inspector and expert for the review and verification of the stability of all the iron structures at the 1889 and the 1900 Paris Exhibitions. The structures he reviewed included Eiffel's tower. Contamin also designed, detailed, and dimensioned the structure of the record-spanning Galerie des Machines in 1889 (see Fig. 137, Contamin letter 1889). This, and several other structures that he designed for Moisant, demonstrate his complete command of material technology and statics on the one hand, and structural design, detailing, and industrial manufacture on the other. Bélanger's former students designed and carried out many more structures at the state-supported International Exhibitions in Paris between 1855 and 1900. They contributed to the development of French engineering expertise in iron construction and to its international influence.

Quite apart from the theoretical aspect of his work, Bélanger concerned himself with the relationship between the machine and construction industries and the chemical and metallurgical fields. He followed the development of the newly established industries in the *banlieues* outside metropolitan Paris in the early years of the Second Empire. Metallurgical and chemical industries were established at La Villette and Grenelle, and in structural work, Pierre-François

---

180. For the definition of translation in this sense see Peters 2000.

Fig. 21. Bélanger photographed toward the end of his career. (Source: Photograph taken around 1860, open source)

Joly's firm, Constructions Métalliques d'Argenteuil, had around 350 men, which was the largest number at Argenteuil. Jean-François Cail had far more: 1500 in Grenelle and Chaillot, and Ernest Goüin employed 1000 at Batignolles. Smaller firms also moved to the *banlieues* in this period; the Entreprise Roussel established itself with around 50 employees at Batignolles in 1852–53 (Roussel 2015), and after 1868, Eiffel and Théophile Seyrig at Levallois-Perret and Moisant at Grenelle would further swell these numbers.

Bélanger continued to serve practice and industry after his retirement in 1864, when he reworked and published all that he had developed and taught in three volumes. He issued the first two parts of his mechanics that year as *Cinématique* and *Traité de la Dynamique d'un Point Matériel*, and two years later the third part as *Traité de la Dynamique des Systèmes Matériels*. In all three volumes he divided the material strictly into two sections for the benefit of practitioners in the field: pure theory and the practical application to the problems of machinery and structures.

Perdonnet's contribution

The premier engineering school, the academic ENPC, had existed for just over half a century. While it had already developed a dominant tradition in the engineering profession, the ECP was an experiment in a different direction. Everything was fresh and untried in the new industrial environment. The changing profession had to be explored and tested by the teachers as well as the young graduates as they entered the working world.

As we have seen, many ECP graduates, and particularly the constructors among them, began their careers in railway construction. This was in part because all the industries connected with the railways in France were dynamically expanding, and the railways themselves represented the largest single construction field of the period 1830–60. Its popularity among the young *centraliens* was also because Perdonnet stood squarely astride this development. Perdonnet was a determined advocate for the new means of transportation, and his core course in the subject influenced his students, as did his professional collaboration with Flachat.

Fig. 22. Auguste Perdonnet, pioneer and promoter of railways in France, who mentored many of the early iron constructors. (Source: *Carte de visite* c. 1862–64 by Franck, official photographer of all the *grandes écoles.* Albumen print from the Peters Collection, also reproduced in Guillet, *Cent ans de la vie de l'École centrale*, 1929, pl. 15, opposite p. 38)

Perdonnet's influence, as determining as it was, does not explain the whole story. According to J.-F. Belhoste (2004, p. 67), by 1850 40% of all ECP graduates had begun their careers somehow connected with railway construction. It is not possible to verify this number, but according to *École centrale 1889,* an approximation of those who worked longer in the railway industry during the course of their careers was four in Polonceau and Barrault's graduating class of 22 (18%), ten in de Dion, Pronnier, and Molinos's class of 58 (17%), also ten in Eiffel's class of 71 (14.1%),[181] twenty in Jenney's class of 69 (28.9%) (although Jenney, who did work for a year in railways, is missing from the number), sixteen in Baudet's class of 77 (20.8%), twenty-two in Moisant's class of 112 (19.6%), and thirty in Contamin's class of 116 (25.9%). So in fact the actual percentage may have been lower. Among the students who did, however, we find seven of the eight who contributed most to the development of iron construction. This is not a coincidence. The railways constituted an entirely new engineering field, and it was dominated by the new material.

Although many students did not work in railways, the use of iron and its propagation in the course, was an important element of the school's development. Perdonnet's gift for mentoring played a decisive role in the students' awareness of the new material. The clearest case is that of Camille Polonceau, the developer of many locomotives and the successful composite truss that bears his name. Perdonnet considered him more a colleague and a close friend than as a student even before he had graduated in 1836 (Guillet 1929, p. 322, and Perdonnet in Comberousse 1879.1, p. A118). He placed Polonceau with several of his earliest railway projects and made him co-author of the multi-volume *Portefeuille de l'ingénieur des Chemins de Fer* too.

Perdonnet was generous to his friends and students with all his resources, his contacts, his knowledge, and even his finances (Dumas in Comberousse 1879.1, p. A90. See also Martelet 1867). He divided his time between his many concerns: the education of youth, the mentoring of his former students' careers, the betterment of workmen's situations, general adult education through the Association Polytechnique pour le développement de l'instruction populaire that he had co-founded, and the administrative roles that he fulfilled both in professional societies and in railway enterprises.

His character was not all as positive as was reported, however, and he did have weaknesses, as we shall soon see. His private life also shed an interesting light on his character in another way. He lived with a working-class woman, Jeanne Émilie Mary, with whom he had a daughter in 1851, although they only married in 1861.[182] This was most unusual at the time. Although a wealthy member of the grande bourgeoisie himself, he was clearly not a snob, and he was open and

181. Eiffel is not mentioned by that source, but his contributions were railway bridges.

182. Upon his marriage, he adopted his daughter, Claire Augustine, later Lefèvre (Sandras 2020).

fostering in his approach to people of all social classes. His character was not a simple one, possibly the most complex and contradictory of any of the ECP's teachers, and he could sometimes be severe in his judgments. It seems that even his harshest critiques were invariably accepted, however; the recipients of his criticism could sense his devotion to the issue at hand, his sense of human dignity irrespective of social standing (Flachat in Comberousse 1879.1, p. A103), and the underlying benevolence of his intentions (Dumas in Comberousse 1879.1, p. A93). I question this overly generous characterization in the discussion of Fig. 79 further on.

Both Jules Petiet and Eugène Flachat referred to Perdonnet as a friend and a benefactor of youth, of the workingman, and of the profession (Petiet and Flachat in Comberousse 1879.1, pp. A94–97). Through his several research trips to Britain, Perdonnet became convinced, even before his involvement in the school and his later voyages with Polonceau (Perdonnet in Comberousse 1879.1, p. A120), that the involvement of private industry rather than state control was what was required to foster the development of iron and the modernization of infrastructure in France (Flachat in Comberousse 1879.1, p. A99). He shared this conviction with Flachat and Lavallée. His egalitarian views and his ability to disagree without alienating an adversary (Flachat in Comberousse 1879.1, pp. A99–100) may in part have been the result of his educational background with Henrich Pestalozzi,[183] as well as the Swiss cultural penchant for political compromise. This non-threatening approach to argument contributed to his success in the integrative educational program at the ECP.

The details of his most unusual education

Perdonnet was possibly the most pedagogically experienced and flexible of all the teachers at the new school, and his heterogeneous, chaotic, and varied educational development subsequently influenced the ECP's pedagogy. The didactical principles that he had experienced as a youth, especially Pestalozzi's precept of the involvement of all the senses in learning, affected the teaching at the ECP. Of course, Perdonnet was not the only later ECP teacher to be influenced by Rousseau's thinking and by Pestalozzi. The other two Geneva-

183. Another railway engineer of similar egalitarian conviction was Adolphe Jullien, who had also been educated by Pestalozzi.

educated faculty members, Dumas and Colladon, supported the same attitudes to learning, but Perdonnet appears to have been the only faculty member educated by Pestalozzi himself. Aside from this formative influence, the ordeals he had repeatedly suffered at all the schools he had attended, and that he had learned to overcome, also help us to understand his mature and benevolent manner toward his advisees and the ability to mitigate potential conflicts (despite his own forceful character) that his contemporaries so admired. These characteristics had an impact on Perdonnet's influence in the school.

As a child Perdonnet lived with his banker, businessman, and diplomat father Vincent, with his stepmother, and with a growing number of half-siblings in Paris.[184] He had lost his mother at the age of one. At ten the boy began his formal education at the Collège Sainte-Barbe, where he studied from 1811 to 1814.[185] This famous institution based its pedagogy on Masonic principles as propagated by the French Revolution.[186] He was not an easy student, and the comments of his teachers show how very difficult he was. On April 17, 1812, one of them, clearly a perceptive man, wrote:

> Solid ideas, but poorly directed mentally; hence, success one day and failure the next. If this chaotic intellect can be tamed, M. Perdonnet will be good at his studies. We already note fewer of those grand explosions with which he used to threaten us, and which exhausted us; he suffers from … tantrums that are stronger than he. He disappoints those who are appalled by his behavior. In his own eyes he does not consider himself at fault. (Archive of the Collège Sainte-Barbe in the Archive de Paris, D50Z_275)[187]

---

184. There were two in the household by the time he was sent away to school (one more had died at birth) and later a total of eight, all except three of whom died young.

185. 01.10.1811–01.1814, information from the Archives de Paris, 23.07.2020.

186. The Collège Sainte-Barbe was, until 1999 when it disbanded, the oldest surviving middle school in France. Founded around 1430, it had revoked its monastic background in 1798 and mutated into a novel institution, closely allied with the secular Masonic Lodge Grand Orient de France, that is with liberal Continental Freemasonry under the pedagogical leadership of Victor de Lanneau, an aristocrat and defrocked priest who had repudiated Christian education and embraced the Revolution. By 1835 it became the preferred preparatory school for entrance into the *grandes écoles*.

187. Des moyens réels, mais mal dirigés par sa tête; de là, succès un jour, défaite un autre. Si cette tête bouillante peut se calmer, M. Perdonnet fera de bonnes études. Déjà nous avons moins de ces grands coups de tonnerre dont souvent il nous menaçait et nous fatiguait; il a … de tonner plus fort que lui. Malheureux avec lui celui qui s'en épouvanterait. Il n'y a pas de manquemens dont il ne se rendît coupable à son égard.

And later:

> ...he moves from melancholy and gloom to joy, to dissipation with incredible rapidity; but is this last condition any better, because his moods and liveliness are not positive. (Archive of the Collège Sainte Barbe in the Archive de Paris, D50Z_280)[188]

He was therefore taken out of school and sent to the pioneering pedagogue Pestalozzi at Yverdon-les-Bains in Switzerland, possibly through the intercession of a maternal uncle only known as 'Dutoit-Bridel' in Vevey (1996, *Johann H. Pestalozzi: Sämtliche Werke und Briefe, Registerband 1*, p. 114), where he studied for two years (1816–17).[189]

Normally Pestalozzi admitted students as young as possible,[190] but he evidently made an exception in Perdonnet's case. Peer teaching had previously been one of Pestalozzi's educational experiments at Stans in 1798, and Pestalozzi may have regarded the 15-year-old Perdonnet as a possible recruit for a student teacher role, at least at first, as teacher education was also a focus at Yverdon. But this was not to be.

The good in human nature, compassionate sympathy, social justice, and equality were hallmarks of the school. Children were to learn through action and work. Action resulted from sense perception coupled with educated observation (*Anschauung*)—in other words, seeing. The goal was to experience the essence of a thing (*das Wesen einer Sache*), and the experimentation that resulted from seeing led to reflection and then gradually to evaluation. Heart, hand, and mind were all involved in the learning process. Pestalozzi's pedagogy stood under the complex influence of the contradictory thoughts of Jean-Jacques Rousseau but focused on group education rather than individual tutoring. Civilization was not an enemy of Nature as it had been for Rousseau, but a complement to it. All students, the younger

188. ...passe de la tristesse, de la morosité, à la joie, à la dissipation avec une rapidité incalculable; mais encore ce dernier état vaut-il mieux, car ses humeurs et ses vivacités ne sont pas bonnes.

189. Information from Jean-François Hürst, President of the Pestalozzi Foundation and Professor Danièle Tosato-Rigo, Lausanne University, President of the Scientific Council of the Centre de documentation et de recherche Pestalozzi, 24.07.2020.

190. *Johann H. Pestalozzi: Sämtliche Werke und Briefe,* vol. 10. *Briefe aus den Jahren 1816 und 1817*, 1968, letter 4276 to an Englishman, p. 109: *J'aime beaucoup pour ce motif à recevoir des enfans très jeunes, dont rien n'a pu altérer les dispositions primitives.* (I prefer for these reasons to admit very young children whose original tendencies nothing has yet altered.)

and the older at Pestalozzi's institution were encouraged to follow their own pace of development and spontaneous interests and to engage in self-motivated study. This demands and encourages self-discipline, which is what Perdonnet lacked.

He was thrown out of Yverdon in 1817. His two years there had not been without conflict. A letter that Pestalozzi wrote to his uncle describes a fight with another student, his lie to attempt to absent himself from school, and his destruction of a desk in rage.[191] Ten days later, on May 23 of that year, Pestalozzi wrote to Perdonnet's uncle that he was going to expel the young man:

> As a result, I have fixed Good Friday as the date that he will leave our institute. Moreover, under the pretext of the space he demanded in preparing for his confirmation, he has alienated himself to a point at which we can no longer recall him to his senses [even if we were to use] severely repressive means, which we loathe to use and which we do not think could profit him in any way.

Pestalozzi suggested that

> a stay in a house with few people and where he would be subjected to strict surveillance would be better for him than to remain in the environment he would be in here. And as we are convinced that it would be good for him to be provisionally housed in the home of one of his relatives until his father has made a definite decision as to his future, we will undertake to send him to Vevey and we recommend him to your care as a young man who could develop into a good person if he could be placed into a situation that would force him to reflect [on his behavior] and where he could not avoid the means used for the completion of his education. (*Johann H. Pestalozzi: Sämtliche Werke und Briefe,* vol. 10. *Briefe aus den Jahren 1816 und 1817*, 1968, letter 4667 to Dutoit-Bridel, pp. 296–297)[192]

---

191. Letter from Pestalozzi to Dutoit-Bridel 13.05.1816 (*Johann H. Pestalozzi: Sämtliche Werke und Briefe,* vol. 10. *Briefe aus den Jahren 1816 und 1817*, 1968, letter 4293, p. 117).

192. Nous avons en conséquence fixé l'époque où il sera admis à la sainte Cène, comme celle, où il devient nécessaire qu'il quitte notre institut. De plus sous le prétexte d'une plus grande liberté, qu'il revendiquait pour se préparer à la confirmation, il s'est émancipé à un tel point, que nous ne pouvons plus le rappeler à l'ordre par des moyens de répression sévères, qu'il nous répugne d'employer et que nous ne croyons pas devoir lui profiter.

...le séjour dans une maison composée de peu de personnes et où il y aurait une surveillance sévère, lui serait plus convenable que de rester dans les relations où il se trouve parmi nous. Et com-

A week later, on May 30, Pestalozzi described details of further recalcitrant behavior, but agreed to keep Perdonnet at school until his father could make a final decision,

> provided that I have no more major problems with him. (1968 *Sämtliche Werke*, vol. 10, letter 4671, pp. 299–300).[193]

But Perdonnet's case obviously continued to intrigue Pestalozzi who, despite the fact that he had expelled him, was not one to give up easily on a difficult student. On June 24 he wrote the Genevan publisher and bookseller Jean-Jacques Paschoud (*Sämtliche Werke* 1968, vol 10, letter 4687, p. 310) to send Perdonnet (and several others) a copy of Antoine-Gervais Girard's *Préceptes de rhétorique*. He may have hoped that Perdonnet might profit by reading the sections of the first chapter titled *preuves, mœurs, passions, imagination, sensibilité,* and *discernement* (5th edition 1815). In Girard's definition these translate as reasoning, morality, passions and the conditions necessary for exciting these passions, imagination, emotional sensitivity, and intellectual discernment.

There is no indication whether Perdonnet's father followed any of Pestalozzi's suggestions or what course he chose to adopt for his son's further education. In 1817 Vincent Perdonnet purchased an estate as a future retirement home in Lausanne, so it is not known whether the troubled youngster ended up there (the rebuilding of the estate lasted several years) or at home in Paris. At any rate, in the three years that followed, he evidently learned to tame his rebellious nature somewhat, and he may have begun either there or in Paris to prepare himself for the EP entry *concours*.

The competition for the elite school was fierce. It required concentrated application and perseverance. Perdonnet had obviously acquired this ability in the interim, and he succeeded in gaining one of the coveted places by 1820. However, his time there was also brief. He spent only a few months at the EP in 1821 before being expelled once again the following year along with his whole studio under suspicion of subversive *carbonariste*—that is, anti-royalist—

---

me nous sommes persuadés que ce serait un grand bien pour lui de pouvoir rester provisoirement dans la maison d'un de ces parens, jusqu'à ce que son père aye pris un parti définitif à cet égard, nous l'engagerons à se rendre à Vevey, et nous vous le recommandons comme un jeune homme, qui pourra devenir un bon sujet, s'il vient à être place dans des relations qui le forceront à réfléchir et où il pourra ne point éluder des moyens employés pour le perfectionnement de son éducation.

193. ...pourvu que je n'aie point de trop grands excès de sa part.

political activities.[194] According to one of the other expelled students, the future EP architecture professor and later director of the ENPC Léonce Reynaud,[195] it was later discovered that only two students in the studio, to which all the more undisciplined among the students had been assigned, had actually been *carbonariste* members (Dartein, p. 17). Who these were, Reynaud never revealed. In hindsight, the many troubles that Perdonnet had suffered as a rebellious youth seem to have encouraged a sense of balance and perhaps also a tolerance toward others later in life.

In 1823, a year after his expulsion from the EP, he embarked on the first of his many voyages to England (Pothier 1887, p. 74 footnote). It was a research expedition with the geologists Armand Dufrénoy, who was then substitute professor of mineralogy at the École des Mines, and Jean-Baptiste Élie de Beaumont.[196] This trip, probably arranged for him through one of his father's connections, turned out to be a key event in Perdonnet's development. The two eminent scientists sparked his interest in geology and metallurgy, and as soon as he returned to Paris, he began attending courses at the École des Mines.[197] He was officially admitted to the school the following year on the basis of having been a student at the EP,[198] (although his expulsion was only to be officially rescinded after King Charles X abdicated in 1830).

---

194. The clandestine *carbonariste* movement in France, among whose sympathizers was Lafayette (as an organizer among many), was a protest movement of Republicans, Bonapartists and others against the Bourbon Restoration. The future emperor Napoleon III had also been a sympathizer, but with the *carbonari* in Italy in their fight to establish Italian unification under the House of Savoy.

195. Reynaud had been a classmate in the same studio as Perdonnet at the EP, and he was expelled at the same time and for the same political reasons, but he went on later to study at the ENPC and eventually become its director, while Perdonnet studied at the École des Mines. According to Reynaud 1870, it was Mary who proposed him for the directorship of the ENPC in 1869.

196. Although their former professor André Brochant de Villiers had headed the mission, it appears that he did not take part in the various detailed visits that the other two did.

197. Minutes of the meeting of the Conseil de l'École Royale des Mines November 9, 1824, (p. 373): *Mr. le D.teur G.al fait connaître au Conseil qu'il a autorisé Mr. Perdonnet qui depuis un an suit les cours de l'École à assister jusqu'à la reprise des cours, aux exercices du laboratoire.* (The Director General informed the council that he has authorized Mr. Perdonnet, who has followed courses at the school for the past year, to participate in the laboratory exercises until the school year resumes.) Ludovic Bouvier, archivist at Mines-ParisTech, kindly provided the information on the organization of the École des Mines and Perdonnet's studies there (30.06.–03.07.2020).

198. Minutes of the meeting of the Conseil October 7–9, 1824, (p. 374) *M. le D.teur G.al prévient le Conseil qu'il a déclaré admissible et autorisé à se présenter au Concours général pour*

There were three categories of students at the École des Mines: élèves ingénieurs, élèves externes, and élèves étrangers. As originally Swiss, Perdonnet would have been classed in the third category, but as a former EP student and son of a recently naturalized French citizen, he was initially classed as élève externe[199]—an exception that he quickly declined as soon as the group was scheduled for an exam[200] that he might not have felt ready for, and perhaps also because he rejected the straightjacket and wanted to set his own curriculum.

His three official years at the school were characterized by many exceptions. As a student in the élève étranger category it seems that he was permitted to pick and choose the courses he wanted to take. He also took the opportunity to travel in Germany in 1826 (Sandras 2020, endnote 20), and that is when he may have learned about mining schools that had integrated workshops into their programs. With all his extra-curricular activities and exceptions one can only speculate that Dufrénoy,[201] his mineralogy professor and the associate director of studies at the school, probably advocated for him as his sponsor and mentor.

When he finished his studies in 1827, he was ranked among the best students in the upper level courses that he did take in his final year: *exploitation* (mine management), *docimasie* (assaying), *minéralurgie, minéralogie* (two separate

---

l'admission aux places d'Élève externes les Candidats suivants: ... 2° Mr. Perdonnet, ancien élève de l'École Polytechnique. (The Director General informed the council that he has declared [the following candidates] admissible and authorized them to take part in the general competition as external students...second, Mr. Perdonnet, former EP student).

199. His diplomat, businessman, and banker father had taken French citizenship after 1801.

200. Minutes of the meeting of the Conseil 16 November 1824 (p. 377): *Mr. le Dteur. G.al transmet au Conseil la liste des Élèves que le Jury a reconnu admissibles dans les services publics et qui ont été classés dans celui des mines... Ces élèves au nombre de cinq sont M. M. ... Perdonnet.. L'ordre du jour appelle l'examen des candidats aux places d'Élèves externes. Ces candidats sont: M. M. Perdonnet .... L'Inspecteur des études informe le Conseil ... 2°. que Mr. Perdonnet a déclaré renoncer à la place d'Élève externe*. (The Director General submits to the council the list of the students that the jury recognized as admissible to public service and who have been classed in the mining service... These five students are ... Mr. Perdonnet... The agenda calls the [following] candidates to apply for a place as external student. These candidates are: Mr. Perdonnet.... The Director of studies informs the council ...2nd that Mr. Perdonnet has declared his intention to renounce his slot as external student).

201. Dufrénoy had been associate professor (suppléant) and inspecteur adjoint des études at the École des Mines since 1825 and held the chair from 1835.

subjects), and geology.[202] He had previously taken chemistry and laboratory courses in the earlier years, but evidently did not take all the other upper-level courses that the regular students did, like drawing, drafting, analysis, German, and English.[203] He was also not evaluated on attendance or diligence (*assiduité en application*).

Nevertheless, on May 15, 1827 the *conseil de l'école*, (presumably supported by Dufrénoy although the minutes of the *conseil* make no mention of it), judged that he had completed his studies and awarded him the *certificat* (certificate of competence), which the school would issue if he so requested.[204] It seems that he never requested his certificate from the administration, and that appears to be the reason that his name does not figure on any of the official lists of the regular and external students or of prominent alumni of the school.

At the École des Mines he had pursued a clear idea of what he wanted from the school, and ever since his successful application to the EP in 1820, he demonstrated that he possessed the determination and ability to apply himself to achieving any goal he might choose to pursue. He was clearly confident (surely also due to his family's wealth) that he would make his way even without an official certificate. This tortuous educational background, coupled with his determination and independence, would serve him well in his advocacy of the railway against all opposition, in the development of his novel railway course, and in his empathic mentoring of students.

From the facts of his record at Sainte-Barbe, his expulsion from Yverdon and from the EP, and the unusual trajectory of his time at the École des Mines, it seems that Perdonnet had, to put it mildly, a complex and difficult character. He could be quite opinionated. This manifested for the most part positively in his introduction of passionate arguments for railways in every possible presentation.

---

202. Minutes of the meetings of the Conseil 02.05.1826 (p. 12), 09.05.1826 (p. 15), 08.05.1827 (p. 119), and 15.05.1827 (pp. 121, 124–125).

203. Minutes of the meetings of the Conseil 01.05.1827 (pp. 116–117), and 15.05.1827 (pp. 122–123). Perdonnet, who had spent some time in Germany by then, presumably already spoke both languages.

204. Minutes of the meeting of the Conseil May 15, 1827 (p. 128): Le Conseil de l'École Royale des Mines considère que, 1° M.M. … Perdonnet, Élève Étranger … ont terminé leurs études théoriques à l'École Royale des Mines, 2° Il sera statué sur le Diplôme ou Certificat à leur délivrer lorsque ces Élèves en feront la demande. (The Council of the Royal School of Mines considers that, first, Mr. Perdonnet, foreign student… [has] finished [his] studies at the Royal School of Mines, second, [he is] to be awarded either the diploma or the certificate that will be delivered to [him] upon request.)

Especially when talking to young EP students, he combined them with a general overview of political and international trade aspects as well as technical and even medical implications, such as the difficulties encountered in the use of pneumatic technology in tunneling and underwater foundations. This was the case in the talk he gave on the drilling of the Mont Cenis tunnel to the Association Polytechnique on January 17, 1864 (Perdonnet 1864, pp. 27–48).

On the other hand, his assertiveness could also manifest as aggressive, as in his long-winded introduction to the previous lecture he gave to the same audience on January 10, in which he discussed the new the Rhine Bridge at Kehl. The bridge used a novel building process, in which pneumatic technology played a key part in establishing the foundations. He knew the project well, as he had been one of the expert advisors to the engineers. He took half the time allotted to his exposé to accentuate his excellent connections by naming and lauding the previous speakers in the series, to welcome the varied categories of his listeners from workmen to EP students and academics, to discourse at length as always on the advantages of the spread of the railway, to postulate thereby a future betterment of tolerance and international relations, all while decrying the British focus on business and wealth that he abhorred[205] (while praising British industriousness), lambasting the historical Spanish intolerance of Jews like the Pereire brothers and Protestants like himself (while praising their rapid railway development), and chauvinistically celebrating a supposed French tolerance in comparison (Perdonnet 1864, pp. 6–14).

We do not know what matured him to the point that he learned how to deflect and mediate conflict situations, if not to mellow his character, but his trip with Dufrénoy and Élie de Beaumont in 1823 had definitely played a major role.

The trip to England had not been the only defining moment in his school years. The educational background that he had unwillingly experienced in Yverdon also influenced him profoundly, although the positive effect of these two early years with Pestalozzi only surfaced much later. Together, the two critical situations awakened his intellectual curiosity, self-worth, independence, and mentoring instincts. The years with Pestalozzi also influenced his egalitarian outlook and probably awakened his interest in others and in pedagogy.[206]

205. A reflection perhaps of the chauvinist disdain for 'a nation of shopkeepers'?

206. This included his later advocacy of women's education. Yverdon had added a girls' school in 1813.

Teaching through compassionate engagement in the development of the individual and mentoring as Perdonnet later displayed them were Pestalozzian traits that he had not demonstrated as a youngster, but that he had nevertheless succeeded in ingesting and applying to adult education. They were certainly uncommon within the caustic and aggressive intellectual stance often prevalent in French academic discourse, as we saw in Olivier's attitude toward those he called pure scientists.

After leaving the École des Mines, Perdonnet traveled to England again, this time accompanying a classmate Léon Coste[207] who was sent there by the *corps*. There the duo examined the refining of metals and visited the early railways. This expedition may also have been at the recommendation of Dufrénoy, who had just published the account of his and Élie de Beaumont's trip, *Voyage Métallurgique en Angleterre* as a treatise that year.

It was this second visit that awakened Perdonnet's passion for railways that would set the course of his career. While in England he met and befriended the twenty years older railway pioneer George Stephenson and his son Robert, as well as Nicolas Wood and Thomas Grainger (Coste and Perdonnet 1830.1, p. 5). His experience with the new mode of transportation obviously fascinated him so much that he completely reoriented his career plans. Upon returning to Paris, he began working on Flachat's Paris–Saint-Germain Railway in 1835 with Clapeyron and Lamé, who had both recently returned from Russia. He published an article with Coste on railways in the *Annales des Mines* in 1829 that was reissued as an offprint, *Mémoire sur les chemins à ornières*, the following year (Coste and Perdonnet 1830.1, p. 5), and they also published a second treatise on the industrial refining of ores that had been the reason Coste had been sent to England in the first place. Their *Mémoires métallurgiques sur le traitement des minérais de fer, d'étain et de plomb en Angleterre* (1830) was an expansion of the experience of his first trip with Dufrénoy and Élie de Beaumont whom they mentioned in the title. The fact that the second edition in 1837–39 even included the names of these men as authors is yet another indication of how influential the first expedition and these men had been on his development.

While Perdonnet was not the initiator of the passion for railways in France—a whole coterie of engineers had begun the development in the previous decade[208]—

---

207. Coste had been a regular élève ingénieur. He graduated the same year, with diploma.

208. For a concise overview of the early development of the railway in France, see Barbin 2021.

he soon made it his own. His enthusiasm for the new mode of transportation and his energy in pursuing that passion infected everyone around him. Among others, it influenced his father Vincent, who presided over the earliest committee for a railway in Switzerland in 1838 (Perdonnet *Traité élémentaire,* 1855, vol. 1, pp. 38–39). The driving force behind the proposed line from Morges to Yverdon[209] was William Fraisse (Fraisse 1838), another former classmate of Perdonnet and Coste at the École des Mines.

Perdonnet's strong but open mind, his varied educational background that included proof of his scientific qualifications, his international experience, and his industrial bias in railway engineering qualified him as a candidate for a position at the ECP. He was hired in 1831 to teach mining.

Teaching and publishing about railways

As a former student of the École des Mines it was natural for Perdonnet to be entrusted with a course in mining, but his focus had shifted, and mining and geology were no longer his primary interests. He immediately took the opportunity to insert ten lectures on railways into his semester. This was the first appearance of the railway in an academic course. From 1838 he gradually relegated mining more and more to the geological engineer Amédée Burat (grad. Mines 1832) and developed railway engineering into a separate course for the academic year 1837–38. This caused him to be regarded somewhat skeptically—not by his colleagues at the ECP, but by politicians and even some of the professional public at the time (see Figuier, *Merveilles de la Science*, vol. 1, p. 296). By that time, Perdonnet's was only the second academic course devoted exclusively to the topic.

Although he had introduced the subject in his mining classes in 1831, the first course devoted entirely to railways was Charles Joseph Minard's, which he published as *Leçons faites sur les chemins de fer à l'École des ponts et chaussées en 1833–1834.*[210] Minard's main source of information had been George Stephenson's close collaborator Nicolas Wood, whom he cites

---

209. Political troubles prevented construction until 1852–55, when it became part of the second Swiss railway.

210. Minard had apparently already recognized the importance that iron would gain in construction in 1806 (obit by Victorin Chevallier in *Annales des Ponts et Chaussées* 1871, 2e. sem, p. 4).

throughout his text and whose 1825 first edition of *A Practical Treatise on Rail-roads and Interior Communication* he followed closely. The course dealt primarily with track construction and therefore did not include much material on the mechanical aspects of the topic. Except for a brief treatment of inclined planes, he introduced locomotives only from the eighth chapter (of nine), but without discussing their construction.[211] The field was developing rapidly, and Perdonnet was able to teach a fully developed course four years later when, in contrast to Minard, he concentrated on the mechanical engineering aspects of railways, and of course on the use of iron.

Like Colladon, and as a Pestalozzi disciple, Perdonnet insisted on the visual and haptic aspects of learning and took his students on excursions so that they could see and touch what he lectured on in class. His dedication—and there are many accounts that testify to the enthusiasm he injected into his espousal of the railway and his teaching—soon vindicated his insistence that railways were to become the premier means of transportation for the future.

Except for the very brief earliest version in 1831–32,[212] Perdonnet's course seems not to have survived in transcript form. However, two plate volumes testify to what he taught: an *Album des chemins de fer* with 60 plates issued by his *répétiteur* Germain Cornet in 1849 (later extended to 78 plates by 1853), and the *Nouvel Album des Chemins de Fer* with 53 plates published by his *préparateur* (demonstrator) A. Jacquin a decade later, in 1864. His teaching and professional activity did give rise to several multi-volume treatises that must also have served as texts for the students. From the introduction to the second volume of published plates in 1864 it is clear that the student notebooks were full of highly detailed drawings:

> The railway course at the École centrale consisted of only ten lectures in 1831, during which I drew all the figures on the blackboard myself.... Later I recognized that the students used up most of the lecture time in drawing and paid less attention to the oral explanations.... It is undeniable that the

---

His pioneering work was mentioned by Perdonnet throughout the first chapters of the first volume of his *Traité élémentaire* (1855), and in *Notions générales* (1859), bibliography, p. 377.

211. Wood's treatise would be expanded in 1831/32 and again in 1838. No subsequent update of Minard's course seems to have existed. The second edition and the Italian translation, both of 1840, only reproduce the original 1833–34 text.

212. The only transcript of this course appears to be in the national archives.

> students profited most when the professor drew the diagrams during the lectures, but as the program grew, the number of lectures stayed constant, and this method became totally impossible to maintain. (Jacquin 1864, Perdonnet's introduction pp. 1–2)[213]

This, and the fact that the ECP's *conseil des études* limited the number of lectures on railways, paring them down to 32 at mid-century, is why Perdonnet's assistants Cornet and Jacquin,[214] who had probably been the ones to draw all the diagrams on the blackboard (Perdonnet in Jacquin, p. 1), published the drawings in 1849/51/53 and 1864. Perdonnet, who stressed drawing as *la langue de l'ingénieur* (Perdonnet 1867, pp. 31–33), regretted the fact that this compromise was necessary and replaced the drawings that the students had formerly copied from the blackboard with special exercises:

> When we abandoned the sketches made by the students in their notebooks, I substituted measured diagrams of the most important machines and apparatus that they had to draft on graph paper with extensive notes to supplement the captions, as well as more or less detailed projects, and I increased the number of tests to check their progress. (Jacquin 1864, *Introduction*, p. 2)[215]

The total of 131 plates in the two volumes (1853 3rd edition and 1864) deal almost exclusively with iron. They demonstrate that Perdonnet concerned himself with iron construction at a machine scale, an intermediate scale represented by railway

---

213. Le cours des Chemins de fer à l'École centrale n'était en 1831 que de dix leçons, pendant lesquelles nous faisions nous-même au tableau toutes les figures.... Plus tard nous ne tardâmes pas à reconnaître que les élèves, passant une grande partie du temps de la leçon à dessiner, portaient moins d'attention à l'enseignement oral.... Il est incontestable que, les figures étant faites pendant le cours par le professeur, la majorité des élèves en tirent un meilleur parti; mais si, le programme s'accroissant, le nombre des leçons reste invariable, ce mode de procéder devient tout à fait impossible.

214. Cornet is not recorded and probably did not enjoy a particularly distinguished career. He was Perdonnet's *répétiteur* for the third year for an entire decade from 1845 to 1855 (Comberousse 1879.1, p.A11), then worked for the Chemin de fer de l'Est, and died before 1889 (*École centrale 1889*). Jacquin is probably André Auguste Jacquin (1822–95). On p. iii, vol. 2 of the second edition of *Traité élémentaire des chemins de fer* in 1860, Perdonnet refers to Jacquin as his secretary and he thanks him warmly again in the foreword to the third edition of 1865 (vol. 1, p. vii).

215. Nous suppléons du reste à la suppression des figures faites sur le cahier par les élèves en exigeant d'eux des dessins cotés faites sur le papier quadrillé, des machines ou appareils les plus importants, des notes étendues qui complètent les légendes, ainsi que les projets plus ou moins détaillés, et en multipliant les interrogations qui constatent les progrès.

accessory structures, and to a lesser extent at a building scale. Although the plates that present buildings do not include details in contrast to those that deal with railway cars, locomotives, and machinery, they do show the structural systems. The fact that Perdonnet did include all three scale levels in his lectures, however, confirms that—together with his colleagues in machine construction—he played a comprehensive role in the presentation and discussion of the new material during the transition from the experimental to the industrial phase of iron construction. The crossover that his lectures generated between the fully developed industrial use of iron in railway and locomotive construction and the material's experimental use in building structure, enabled the students to see its potential for bridges and buildings.

The drawings that Cornet published in 1849 and in the expanded editions in 1851 and 1853 represent station plans, track, rolling stock, and related iron machinery. An exception is one masonry viaduct (pl. 6 in all versions) that however serves to show embankments and cut-and-fill. There are medium-sized turntables in iron (1853 pls. 27–30), fifteen plans of railway stations (1853 pls. 33–39, incl. 36 bis and 39 bis and ter), and three cross-sections that show roof structures (1853 pls. 40–42), whereby only two of these (pls. 40 and 41) are in iron. Both designs show the truss type Polonceau had developed, although the 1849 hall of the Embarcadère de Strasbourg that still stands as the Gare de l'Est (pl. 40) was engineered by the ENPC engineer Pierre Cabanel de Sermet. The descriptions suggest that these roofs must have been discussed in detail during class, and the dimensions of the iron members are clearly marked on the plates and discussed in the descriptions:

> 40. Framework of the great hall covering the tracks of the Strasbourg Line's passenger terminal in Paris. It consists of rafters [i.e., frames] made of rolled iron sections that form triangulated voussoirs; these main ribs are spaced 4 meters apart and are interconnected by means of round bars for the posts and compression members. The roof form is a flat arch.
>
> 41. Roof truss of the right-bank Saint-Germain to Versailles Railway Line. The rafters [frames] and purlins are made of I-sections interconnected by iron lintels [fishplates] bolted together out of several members. The compression members are in cast iron as are the columns and abutments. The posts and tension members are round bars. Sheet metal stiffened by angle-iron was the appropriate material used to cover the 10-meter trusses spaced at 8 meters over

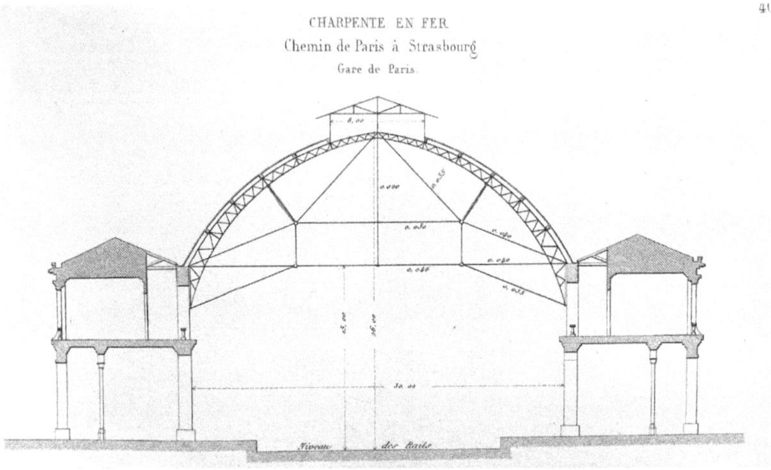

Fig. 23. Cross-section of the hall of the Gare de l'Est, then known as the *Embarcadère de Strasbourg*. The iron truss is derived from Polonceau's design, and it was built in 1849 by Pierre-Sophie Cabanel de Sermet. Perdonnet published and described the image in great detail, which demonstrates his concern with building-sized structures in iron. (Source: Cornet, *Album des chemins de fer*, 3rd edition 1853, pl. 40)

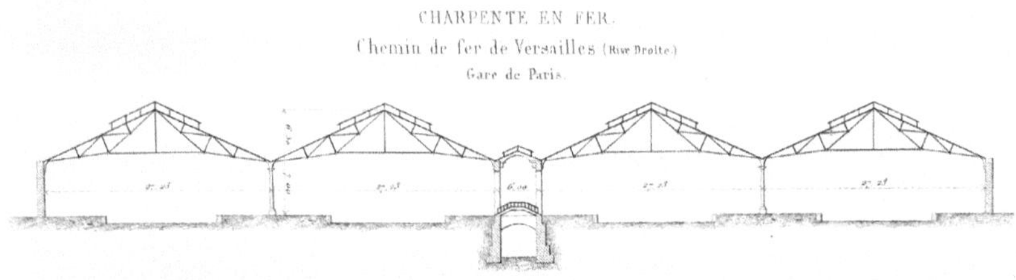

Fig. 24. Cross section showing Polonceau trusses that spanned the tracks in the Paris station of the right-bank railway from Paris to Versailles, the first construction phase (1842–46) of the third Gare Saint-Lazare (1842–53). Polonceau had been Perdonnet's student and became his collaborator. (Source: Cornet, *Album des chemins de fer*, 3rd edition 1853, pl. 41)

> that part of the station covering the Western Line. (Cornet 1853, p. 10)[216]

---

216. 40. Charpente de la grande halle recouvrant les voies du chemin de fer de Strasbourg, gare des voyageurs (Paris). Elle se compose d'arbalétriers en fer méplat, affectant la forme de voussoirs décomposés en triangles; ces grandes nervures espacées de 4m,00 sont consolidées au moyen de tiges en fer rond, formant poinçons et faux-entraits. La forme du comble est une ogive surbaissée.
41. Comble en fer du chemin de Saint-Germain et Versailles, rive droite. Les arbalétriers et les pannes sont en fer double T, consolidés aux joints par des plates-bandes en fer boulonnées entre elles; les contre-fiches sont en fonte ainsi que les colonnes et entablements, les poinçons et tirants sont en fer

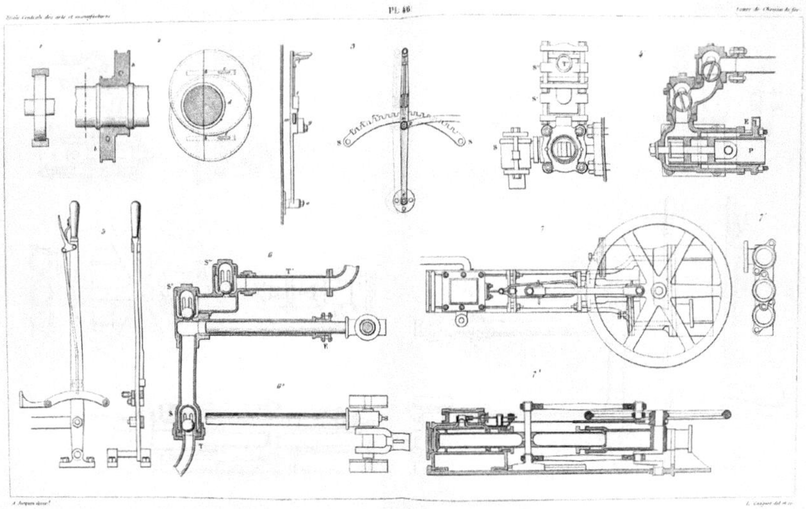

Fig. 25. Details of parts for locomotives, railway cars and machinery in iron. They demonstrate Perdonnet's concern with the small-scale use of the material. (Source: Jacquin, *Nouvel album des chemins de fer*, 1864, pl. 46)

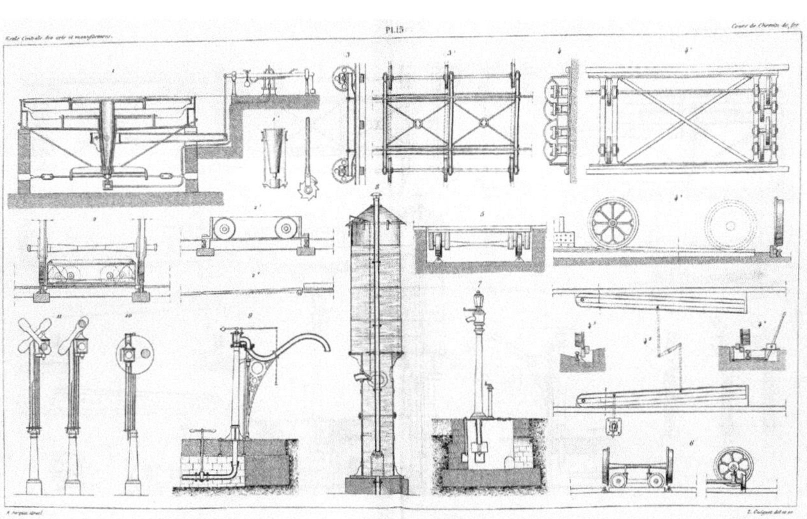

Fig. 26. Several plates show the structure of medium-scale *accessoires de la voie* (auxiliary structures) in iron. (Source: Jacquin, *Nouvel album des chemins de fer*, 1864, pl. 15)

rond. On a fait une judicieuse application de la tôle consolidée par des fers de cornières aux fermes qui recouvrent la partie de la gare affectée à la ligne de l'Ouest, et dont la portée est de 10 mètres sur 8 mètres d'écartement.

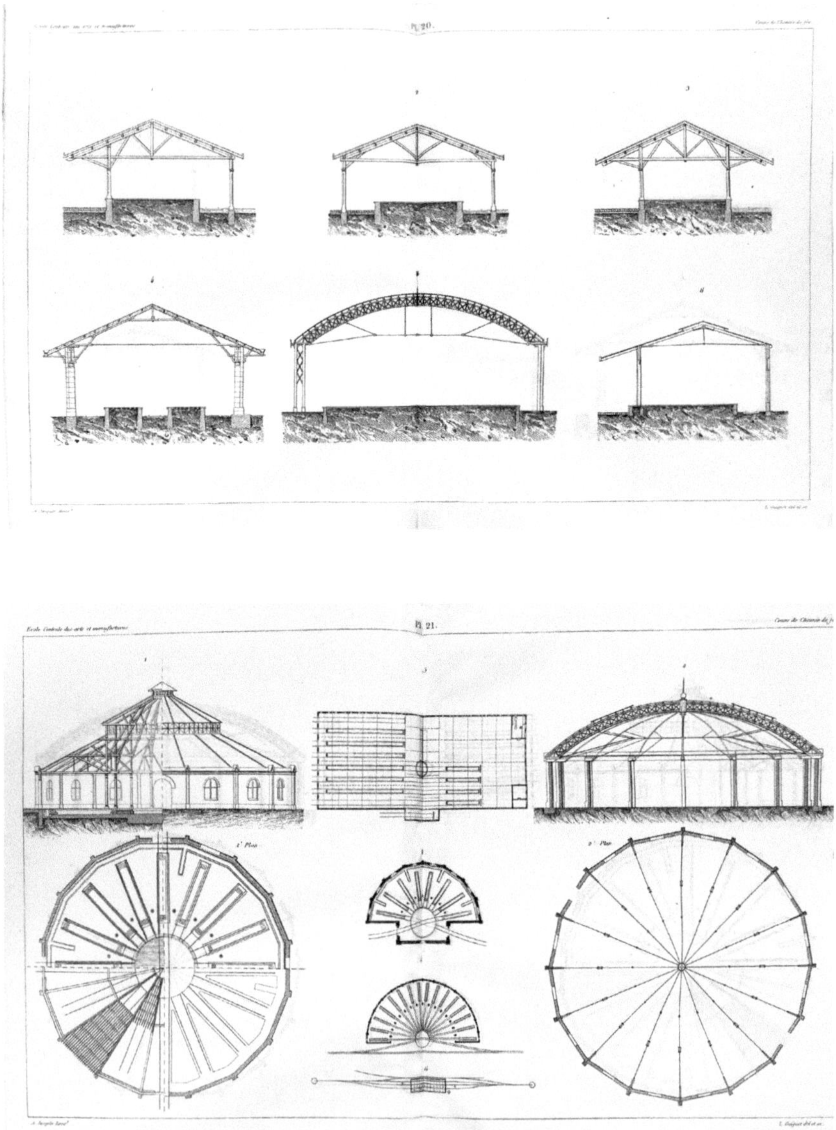

Fig. 27. The iron station roofs that Perdonnet published in his later album were far less detailed than in his first. This indicated that industry had adopted the material and taken over the design and manufacture of large-scale iron structures between 1850 and 1860. Even the few special structures that he did show after 1860 no longer provided measurements or explained details. (Source: Jacquin, *Nouvel album des chemins de fer*, 1864, pls. 20 and 21)

Most of the drawings in Jacquin's later volume of 1864 were also devoted to locomotive and track construction, but several dealt with iron bridges, water towers, workshops, locomotive sheds, and stations as well. These inclusions indicate that the lectures continued to treat iron construction seamlessly from a focus on smaller-scale machinery (for example pls. 40–48), to medium-scale iron *accessoires* that were to be erected for servicing the railway, and to building structures.

In contrast to Cornet's 1853 volume, however, Jacquin does not describe the structures, nor do his images show dimensions.[217] This less detailed treatment of iron roofs in the *Nouvel album* suggests that details were beginning to be relegated to the industrial workplace as iron construction had begun to mutate from a tentative and experimental phase into its industrialized phase in the decade between 1853 and 1864 (see the quotation from Jolly in the chapter 'Teaching about iron: from experiment to industry' further on). By the 1860s, the visual correlation from the small machine and medium-scale to the large structures would enable the students to make the intellectual leap from the scale of machinery to the larger scale of building structures—especially when they were supplemented by the courses on machine construction and the requirement to document buildings in the architecture and drawing classes. Together with the mechanical courses and Bélanger's organizational *vulgarisation* of theory, the railway course formed an interdisciplinary link that would later enable the ECP's graduates to transform what they had learned and manually practiced in drawings at a small scale to the large scale of building structures and bridges.

Over the years, as Perdonnet's railway course evolved, his lectures were supplemented by his copiously illustrated treatises. The *Portefeuille de l'ingénieur des chemins de fer* in six volumes and one paper (1843–55, 3 vols., atlas, and three addenda, co-authored with Camille Polonceau) had 143 plates, including the second addendum with eighteen additional plates. These contained descriptions and images of many major railway station sheds in iron: Birmingham, Bristol, Leeds, London, Manchester, Orléans, Paris, Saint-Germain, Versailles, and Vienna.[218] It was clear that Perdonnet had studied the topic in detail. This major

217. Cornet's earlier 1849 edition also did not have descriptions.

218. Pages 97–180 of the volume *Légendes explicatives des planches* of this five-volume work, document many stations, both large and small on all the important lines of the period and in detail as well as one bridge and one crane.

work was followed in 1857–59 by *Nouveau portefeuille de l'ingénieur des chemins de fer* with 168 plates, and the two-volume *Traité élémentaire des chemins de fer* (1855–56) with 520 figures and fifteen plates in which he credits his relative[219] and former student Gustave Bridel (grad. 1847) with substantial contributions.[220]

The field developed so quickly that Perdonnet greatly expanded this last treatise in 1858–60, and by 1865 the final edition had doubled to four volumes with a total of four portraits, six maps, five folding tables, and fifteen plates (the last included in the total of 1061 figures). In fact, he devoted a great deal of energy to this popular account of the development up to 1865, collecting (with Jacquin's help) all the information he could amass from his many contacts, from manuscripts, and from published accounts. He credited his sources fully (3rd ed. 1865 vol. 1, p. v), and he included all ongoing developments and experiments in Europe, the Americas, Asia, Africa, and Australia. The fourth volume contains addenda to the material in the other three volumes[221] that he continued to add at the very last moment before publication. He ambitiously planned yearly supplementary volumes (vol. 1, p. vii) that never appeared, however, as he died a year and a half later, in September 1867.

Perdonnet's extraordinary effort in the *vulgarisation* of his field in the service of public education[222] paralleled what in those years France's premier *vulgarisateur* Louis Figuier was undertaking in a more comprehensive manner

---

219. Perdonnet's mother was born Jeanne Françoise Georgette Bridel (1769–1802). Her precise connection to Gustave Bridel is not documented, but she may have been his aunt or another relative of the unidentified Dutoit-Bridel who had supported Perdonnet's admission to Pestalozzi's school in Yverdon.

220. Like Polonceau, Bridel was Perdonnet's protégé, and he was so self-effacing that he declined to be named co-author. This modesty that the school characterized in the evaluation of his diploma as eminently reliable—that is, regular in his habits, orderly, intelligent, and diligent (Type de la régularité, de l'ordre, de l'intelligence et du travail, Bridel 2001, p. 55), led to his name being forgotten as the author of his many pioneering hydraulic installations, railway works, and tunnels, and particularly his critical role in excavating the Gotthard Tunnel (1872–82) through the Alps.

221. For example, an account with two figures of Stansilaw Kerbedz's new bridge over the Vistula in Warsaw that was then under construction (1865, vol. 4, pp. 385–389) as an addendum to the brief paragraph (pp. 503–504) in his long chapter on iron bridges in vol. 1 (pp. 489–524).

222. See Sandras 2020 for Perdonnet's complex involvement with the Bibliothèque des amis de l'instruction and the Association Polytechnique and his collaboration with Évariste Thévenin on the popular *Almanach général des chemins de fer* that was published full of anecdotes for the least intellectually demanding segment of the public.

for the entire spectrum of science and technology. Perdonnet made an outstanding effort to enhance public awareness of railway construction and management, and his treatise remains the best and most comprehensive early history of the railway. It has the distinct advantage that it was compiled and commentated by a pioneering participant who knew many of his sources personally. As Figuier admiringly wrote in his evaluation of the third edition:

> We have impatiently awaited the third edition of this key work.... No one is better placed to obtain precise and authentic information than the learned director of the École centrale des arts et manufactures. His rapport with railway engineers, some of whom had been his students, allows him to draw [his information] directly from the source. (Figuier, *L'année scientifique et industrielle*, vol. 10 1866, p. 525)[223]

However, this treatise and the short general *Notions générales* that he wrote for the less educated public in 1859 did not present much in the way of iron construction at a building scale. Both works discussed small-scale carriage, locomotive and machine construction, and their detailing. The first volume of the *Traité élémentaire* in 1855 also treated mid-scale accessories at great length (pp. 426–477), and large-scale iron bridge construction to some extent (pp. 352–372), but it only presented a few bridges in detail.[224] The second volume that he published a year later contains multiple plans of station buildings but only two roof structures, one in Cincinnati[225] and the Baltimore and Ohio Railroad's first terminus in Philadelphia (pp. 98–99).[226] In the second edition of that volume in 1860 he added a sketchy cross section of the Great Northern Railway's new terminus, King's

223. La troisième édition de cet ouvrage capital était attendue avec impatience.... Personne n'était d'ailleurs mieux placé que le savant directeur de l'école centrale des arts et manufactures pour se procurer sur toutes les questions des renseignements exacts et authentiques. Ses rapports avec les ingénieurs des chemins de fer, dont une partie ont été ses élèves, lui permettaient de puiser toujours à la source.

224. This section was, however, greatly expanded with more detailed drawings, including the new technique of pneumatic foundations in the final 1865 edition, vol. 1, pp. 524–548, and vol. 4, pp. 217–229.

225. This terminus was located either on the Little Miami (1837–48), on the Cincinnati, Hamilton, and Dayton (1851), or the recently begun Ohio and Mississippi Railroad in 1854.

226. The location of this station is unknown. It could have been a predecessor of the Philadelphia, Wilmington, and Baltimore Railway's terminus at Broad Street.

Cross station built by Lewis Cubitt 1851–52 (p. 40) and one very detailed drawing of a roundhouse on the Chemin de Fer du Nord, a *remise polygonale des Ardennes* (p. 48).[227] And in the third edition in 1865 he added only three more: a cross section of the roof of London's 1860 Victoria station by Robert Jacomb-Hood (vol. 2, p. 253), a truss of the Gateshead Maintenance Depot's two 1852 locomotive fueling sheds of the Northeastern Railway (p. 295), and a simple 1849 canopy over the platforms at Épernay on the Paris-Strasbourg Line (p. 363). Clearly, he had decided by then to relegate structural detailing in iron at a building scale to specialists and to the industry. This bears out the observation on the differences between Cornet's and Jacquin's albums.

The profound influence that Perdonnet and his many publications exerted internationally has never been highlighted. August Crelle, the eminent Prussian mathematician who designed the first Prussian railway, the Berlin-Potsdam Line in 1837, wrote his theoretical treatise on the use of pneumatic energy for railways in 1846 in French. As he wrote:

> The author of this memorandum, having written it in a language that is not his own, hopes that one will take this into consideration as to the style and the difficulties that he had in composing it, and in this regard, he solicits the reader's indulgence. (Crelle 1846, footnote, p. 1)[228]

Crelle did not mention the reason for going to this trouble, but it was self-evident to his readers. He wrote it in French because Perdonnet and his colleagues, the preeminent theoreticians of railway development who would read it, were all French. This is also the reason that the cover states that the treatise that was published in Berlin was also available from the publisher and bookseller Bachelier's shop in Paris.

The influence that Perdonnet's strong personality, his enthusiasm for the railway as a harbinger of modern civilization, and his pedagogy exerted on those of his students who would later become structural engineers and innovators in iron construction, have all been underestimated. He communicated both his excitement for the field that he was co-inventing as he taught, and his true respect for the self-made engineer, especially for George Stephenson, whom he

227. This was probably on the Hautemont-Mons Railway that opened in 1858.

228. L'auteur de ce mémoire, l'ayant écrit dans une langue qui n'est point la seinne, a besoin d'espérer qu'on tiendra compte, pour le style, des difficultés qui lui ont présentées cette rédaction, et, à ce titre, il sollicite l'indulgence des lecteurs.

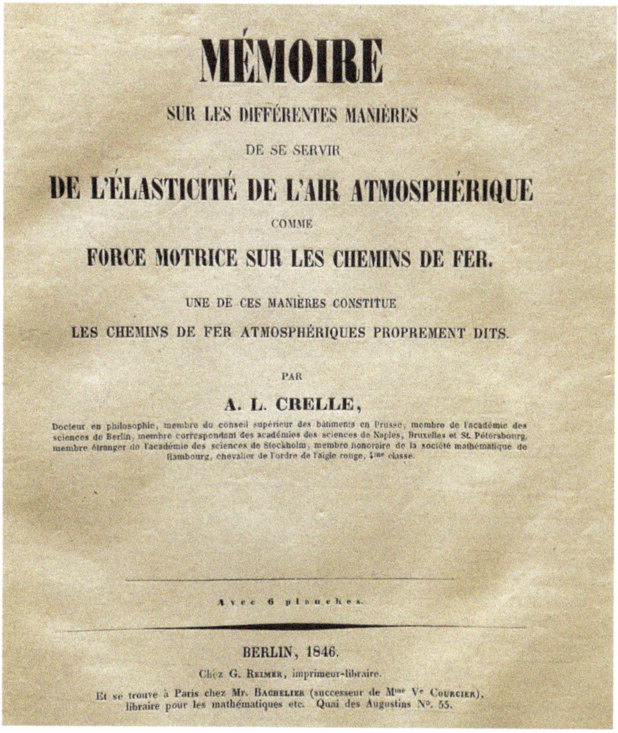
MÉMOIRE
SUR LES DIFFÉRENTES MANIÈRES
DE SE SERVIR
DE L'ÉLASTICITÉ DE L'AIR ATMOSPHÉRIQUE
COMME
FORCE MOTRICE SUR LES CHEMINS DE FER.
UNE DE CES MANIÈRES CONSTITUE
LES CHEMINS DE FER ATMOSPHÉRIQUES PROPREMENT DITS.
PAR
A. L. CRELLE,
Docteur en philosophie, membre du conseil supérieur des bâtiments en Prusse, membre de l'académie des sciences de Berlin, membre correspondant des académies des sciences de Naples, Bruxelles et St. Pétersbourg, membre étranger de l'académie des sciences de Stockholm, membre honoraire de la société mathématique de Hambourg, chevalier de l'ordre de l'aigle rouge, 4me classe.
Avec 6 planches.
BERLIN, 1846.
Chez G. Reimer, imprimeur-libraire.
Et se trouve à Paris chez Mr. Bachelier (successeur de Mme Ve Courcier), libraire pour les mathématiques etc. Quai des Augustins No. 55.

Fig. 28. Crelle's treatise on compressed-air driven locomotives was published in Berlin in French in 1846 and sold by Bachelier in Paris because the experts in the field were all French. (Source: Peters Collection)

repeatedly lauds in his many writings right down to his very last publication in 1867. This admiration testifies to his down-to-earth approach to technology and his non-elitist attitude toward technical education.

The Protestant and Swiss connection

As I suggested in Perronet's case, the influence of Protestant or Swiss-educated men, chiefly but not only from Calvinist Geneva,[229] pervaded the development of

229. As already mentioned, Protestant and Freemason groups of industrialists in Alsace were also involved in the support of public education, and as Dupin admiringly noted (*Voyages,* vol. 3, 1824, tome II, pp. 176–178), the pioneering textile manufacturing center Protestant Glasgow was particularly active in the area of free public education for everyone. Ce résultat est admirable, et nous explique en grande partie le progrès des arts industriels, au sein de Glasgow. (The result is admirable,

popular public education as well as engineering education in France. The Swiss connection spans from Rousseau, via Pestalozzi, Perronet, Perdonnet, Dumas, and Colladon, as well as Perdonnet's former colleague at the École des Mines, William Fraisse,[230] and the Protestant connection includes Vicat and Seyrig among other prominent engineers. Dumas, who had been sent to Geneva at the age of sixteen to study science, was the only one among all these who was not of Protestant extraction. Most had close ties with both Saint-Simonianism and the French form of non-religious Freemasonry.[231]

Although Saint-Simonianism under Barthélemy Prosper Enfantin, who bought *Le Globe* in 1830, had grown increasingly religious and fundamentalist in nature and was quickly suppressed in 1832 as an anti-national sect, Henri de Saint-Simon's original ideas continued to influence industrial and technical development in France. His doctrine was popular among French railway entrepreneurs, financiers, and engineers. Virtually all of them leaned more toward liberal ideals and did not follow Enfantin.

It was not only the engineers, scientists, and industrialists, but also Protestant politicians who fostered education. The interim Minister of Agriculture Adrien de Gasparin, who was a close friend and political collaborator of Guizot, had supported Lavallée when Vatimesnil had left the government in 1829. Through his intervention in the Chambre des Députés on March 17, 1830, Guizot had provoked King Charles X to dissolve the legislature, which precipitated the three-day 'July Revolution', the so-called *Trois Glorieuses*. This of course greatly enhanced his political influence. Lavallée had been acquainted with him from his time as investor in *Le Globe* in the mid-1820s, and he, too, supported Lavallée—first in his role as Minister of the Interior, and then as Minister of Education from 1832. Although Grattan-Guinness credits the brief one year that Vatimesnil served as Minister of Education (1828–29) and his contact with Joseph Fourier with the idea for universal public education (Grattan-Guinness

---

and it explains to a large extent Glasgow's progress in the industrial arts.) He was equally impressed by education in Birmingham, which he touches on briefly (ibid., pp. 224–225) and on secondary education in Manchester (ibid., p.168).

230. Along with the Genevan engineer Guillaume-Henri Dufour, Fraisse was one of Napoleon III's teachers at the military academy in Thun, Switzerland.

231. The influence of Swiss Protestant thought in French engineering education mirrors the influence of Protestant thought in wooden bridge truss construction in eighteenth-century German-speaking Europe. See Peters 2015.

1990, vol. 3, p. 1113, and Colladon 1893, p. 188), there were many who lobbied for the idea at the time, particularly among those surrounding *Le Globe*. It was, however, Guizot, as King Louis-Philippe's education minister from 1832 to 1837, who was intimately involved with the actual drafting of the 1833 law on universal primary education. Perdonnet mentions Guizot's influence as having been responsible for the introduction of popular education in France, and he also credits his sponsorship of the fledgling Association Polytechnique for adult education. He considered Guizot responsible for that institution's survival and influence (Perdonnet 1867, p. 11). Guizot played a similarly influential role in the development of technical education in France to that of Playfair in Britain. He was certainly a friend of workman's education, as he had himself apprenticed as a cabinetmaker in Geneva.

Pedagogical interests

Perdonnet concerned himself with technical education rather than with primary schooling, but he was not its only advocate. Charles Dupin, who held the founding chair for *la mécanique appliquée aux arts* at the Conservatoire National des Arts et Métiers from 1819, may have been thinking of an expansion of this institution when he lobbied unsuccessfully for an École Polytechnique de l'Industrie Civile after the revolution of 1830. He may not have been aware that Lavallée had just organized a school for precisely that purpose (Pothier 1887, p. 69 and Weiss 1982, p. 19) or, like Sainte-Preuve, he may have preferred to see such a school founded as a public institution. There had also been sporadic courses for what were called the industrial classes in Metz and Mulhouse (Grattan-Guinness 1990, p. 1112). Like Perdonnet, Dupin had been involved in worker education along with his advocacy of education for other previously neglected fields.[232] In this he had been influenced by Monge's example, which had inspired a whole generation of his students to devote themselves to educating young professionals. Dupin had subtitled his book on geometry in 1825 'à l'usage des Artistes et des Ouvriers, des Sous-Chefs et des Chefs d'ateliers et de

232. Dupin had helped found the Académie Iolienne in Corfu in 1808 and was, among his other educational endeavors, a member of the Conseil de perfectionnement of the École Supérieure de Commerce that had been founded in 1820. See also Fourcy, 1828, (1987 ed.) Annexes par Jean Dhombres, p. 162.

manufactures' (for use by artists, workmen, foremen, and workshop managers).[233] While Monge and the other teachers at the EP had geared their courses to the education of future state employees, Prony had initiated the interest in expanding into the education of laborers. As a result, the following generation of teachers—among them Dupin, but also Bélanger, Coriolis, and Mary at the ECP, as well as their colleagues Péclet, Perdonnet, and Olivier—directed their courses toward practical application in industry (Pothier 1887, p. 219). As the title of Dupin's work indicates, the range of their clients spanned from the new industrialist class to simple workmen.

There were differences in intent, however. Dupin, a mathematician and naval officer, appears to have concerned himself with the education of workmen at a distance, while Olivier's background was technocratic in a Saint-Simonian sense. Perdonnet, who espoused a Protestant Swiss sense of work ethic and egalitarian democracy, was also more in tune with the non-elite segment of the population. He genuinely admired and fostered autodidact technologists who came from simple backgrounds. Physical work and an egalitarian attitude had been part of his education in Yverdon. He made a point of stressing this in what may have been his very last public talk to the convalescent workmen in the Asile Impérial de Vincennes,[234] where he described in 1867 how a simple workman had entered the ECP in 1835 and had become a successful industrialist:

> Around 1835 we admitted a cabinetmaker to the École centrale who was already 30 years old. Today he is the director of a large factory on the banks of the Rhine and has placed his two sons at the school. This year one of them has been ranked third and the other fifth in the transition between the second and the third year. (Perdonnet 1867, p. 21, footnote 1)[235]

Charles Pronnier (grad. 1851) had been another prominent example. He had been an ironworker before studying at the ECP (J.-F. Belhoste 2008, p. 66). The ability

233. Perdonnet admiringly mentions this concern of Dupin's in *De l'Utilité de l'Instruction*… 1867, p. 10.

234. The series *Conférences Populaires faites à l'Asile Impérial de Vincennes sous le Patronage de S. M. l'Impératrice* published talks from many fields that were conceived as contributions to popular education at the time. According to Sandras, Perdonnet was an active member of the committee that proposed the lectures.

235. Vers 1835, nous faisons entrer à l'école centrale un ébéniste âgé déjà de 30 ans. Il est aujourd'hui directeur d'une grande fabrique sur les bords du Rhin et a placé à l'école ses deux fils, passés cette année, l'un le troisième, et l'autre le cinquième de leur promotion, de seconde en troisième année.

of workmen to better their lot financially and socially was a constant theme with Perdonnet. In an earlier talk in 1866 at the same institution, he wrote about the steam locomotive, saying:

> The honor of its invention is split between a simple English miner, George Stephenson, and a French civil engineer, Marc Seguin. (Perdonnet 1866, p. 10)[236]

Stephenson, originally a simple mechanic in a mine and the upper-class industrialist Seguin were both Perdonnet's personal friends. Like his colleague Flachat, they were self-made engineers. In the introduction to his 1859 book *Notions générales sur les chemins de fer*, his popular—as opposed to professional—treatise, Perdonnet clearly expressed his concern for the education of workmen and the public generally. As he said in a lecture that he gave to workmen at the Association Polytechnique on January 10, 1864:

> ...it is not to scholars that I speak, but to those who do not know, to those men who have not had the fortune to enjoy a stellar education in their youth and who, as adults, are eager for enlightenment, eager for scientific knowledge, to these men who have my entire sympathy. (*Première conférence. Le pont du Rhin* in: Perdonnet 1864, p. 5)[237]

From the very personal tone of his two talks at the Asile Impérial, it emerges that Perdonnet had a forceful character. He was evidently an opinionated man who was extremely proud of his accomplishments and thought a great deal of his own capabilities. It is also clear from many anecdotes that he had a good sense of humor; he must have been an enthusiastic teacher and a stern but inspiring mentor. He claimed in the 1866 talk that in his very first lecture on railways at the ECP in 1831 he had categorically and aggressively stated that the invention he was propagating would be comparable in its impact to the revolution caused by the invention of printing—i.e., movable type (Perdonnet 1867, p. 16–17). This assertion shocked one of his auditors, and he knew that his was then very much a minority opinion. Both the eminent scientist François Arago (Perdonnet 1867, p. 16–19)

---

236. L'honneur de l'invention doit être partagé entre un simple ouvrier mineur d'Angleterre, Georges [*sic*] Stephenson, et un ingénieur civil français, Marc Seguin.

237. ...ce n'est pas pour les savants que je parle, mais pour ceux qui ne savent pas, pour ces hommes qui n'ont pas eu le bonheur de recevoir une brillante éducation dans leur enfance, et qui, devenus adultes, sont avides de lumières, avides de science, ces hommes qui ont toutes mes sympathies.

and even the Minister of Public Works, the historian Adolphe Thiers, were dead set against railways. Perdonnet reported that Thiers had said after returning from a visit to the newly opened Liverpool-Manchester Railway:

> Gentlemen, I have just returned from examining railways in England, and I think that I have understood them well, and I tell you that they are not what many people suppose them to be. As far as I am concerned, I would be entirely satisfied if France were to build 16 kilometers a year. (Perdonnet 1866, pp. 17–18)[238]

After his visit to England in 1827, Perdonnet's passion for the new means of transportation had become all consuming. He strove constantly to convince investors, the profession, the government, and later his students and the general public of its importance. Time and trade were on his side—and he succeeded in his endeavor, helped by the sheer force of his personality. As he wrote in 1859:

> The railway question has indeed become one that concerns governments in the highest degree. The public is as impassioned today by the railway question as it was by religious issues in the Middle Ages. (Perdonnet 1859, p. 2)[239]

The new mode of transportation found immediate favor with many of France's financiers like the Pereire brothers, James Mayer de Rothschild, or the anglophile Charles Laffitte. They all supported the development. By 1866, a generation later, and despite serious official and financial opposition from the canal lobby, there were 12,000 km of railways in operation (Perdonnet 1866, p. 24). This was certainly in large part due to Perdonnet's incessant lobbying. His imposing presence impressed his enthusiastic auditors as an anonymous report from his oration at the graduation ceremonies of his first school, the Collège Sainte-Barbe in 1861 testified (wonders-of-the-world.net - Eiffel Tower, accessed online 05.12.2018). He used his social position, his fame, and his teaching as a platform for promoting its spread.

---

238. Messieurs, je viens de voir en Angleterre les chemins de fer; je croix les avoir bien vus, et je vous déclare qu'ils ne sont pas ce que bien de gens supposent. Pour mon compte, je m'estimerais bien heureux si, en France, on en construisait 16 kilomètres par année. See also his *Traité élémentaire* first edition, vol. 1 1855, pp. 20–21 for a slightly different version of Thiers's statement.

239. La question des chemins de fer est effectivement devenue l'une de celles qui préoccupent les gouvernements au plus haut degré. Les populations se passionnent aujourd'hui pour la question des chemins de fer comme elles se passionnaient au moyen âge pour la question religieuse.

The role played by the imperial couple

The railway was one branch of Perdonnet's concern, and egalitarian technological education the other. His 1831 comparison between the spread and influence of printing and the railway was typical of his mindset. The conflation was not random: it was personal and evidently tied to his close relationship with the unschooled railway pioneer George Stephenson. He repeatedly referred to his admiration for the autodidact Stephenson in many of his writings. They had met on his second visit to England with Léon Coste in 1827, and that visit had at the same time reoriented his professional focus from geology to railway development. Perdonnet found support for his pedagogical interest at the highest level of society, in the French imperial couple whose influential views on worker education ran parallel to his.

The role that the French Emperor Napoleon III and Empress Eugénie de Montijo played in the development of popular education in France was an important one. Especially the empress was an active proponent of public education and women's rights.[240] The Asile Impérial de Vincennes, at which Perdonnet's later talks were presented and that had such an influence on workmen's education had been founded as a political, and more precisely, a public-relations gesture, which is why Empress Eugénie had it extensively photographed by the pioneering photographer Charles Nègre. It was one of two model convalescent hospitals that she had apparently initiated and that Napoleon III established by decree in 1855. Its purpose was to rehabilitate workmen who had been injured or had fallen ill during the major rebuilding of Paris under George-Eugène Haussmann from 1853. This rebuilding was part of a major socio-economic push to make Paris a modern center of empire along the model of London, where the future emperor had lived in exile for some years. The empress participated in his passion, and in competition with the seven years older Queen Victoria (soon to be Empress of India herself) whose Crystal Palace had caused an international sensation, she also pushed her husband to build the Halles de Paris as an iron structure and followed its development actively.

240. She unsuccessfully lobbied the Ministry of National Education to grant the first baccalaureate to a woman (the first was later Julie-Victoire Daubié in 1861) and the Académie Française to elect George Sand as its first female member. She also presented the first Légion d'honneur awarded to a woman, the painter Rosa Bonheur in 1865.

Fig. 29. *Carte de visite* of the emperor Napoleon III and the empress Eugénie de Montijo in 1865. Photograph by André Adolphe-Eugène Disdéri, the inventor of the type. (Source: Wikimedia Commons)

Paris counted just over 1½ million inhabitants before the surrounding municipalities were incorporated in 1860,[241] and around 400,000 additional workmen were imported to work on Haussmann's projects. The influx added well over 25% to the existing population, which provoked serious organizational problems.

241. En 1856 l'enceinte des fortifications renfermait 1,525,942 habitants (Du Camp 1875, vol. 1, p. 12 footnote). In 1856 the fortifications encompassed 1,525,942 inhabitants.

In the course of this development, the empress—who was intensely concerned with bettering the social position of women—also involved herself as patron in the second institution, the Asile Impérial de Vésinet, which was reorganized in 1858 to become a companion convalescent home for female workers. Empress Eugénie was a strong, independent woman who served as her sometimes-vacillating husband's conservative political advisor and as regent several times during his reign. Her forward-looking interest in emancipation extended to the education of workmen, likely through a feudal sense of responsibility. She also founded and personally financed the series of three free lectures a week by eminent men from all fields to the workmen in Vincennes—but not to the women in Vésinet, although that institution did boast a substantial library.

Among the many lecture offerings at Vincennes were two lectures by Perdonnet, *Les chemins de fer* on railway construction in 1866, and *De l'utilité de l'instruction pour le peuple* on universal education in 1867. Empress Eugénie's cousin, Ferdinand de Lesseps,[242] also gave a lecture there in 1868 on *Le percement de l'isthme de Suez.* The empress later represented France by opening the Suez Canal in 1869, an endeavor that had been spearheaded by de Lesseps.

The personal interest that the emperor and empress took in the free public education of workmen and the political risk that they accepted in doing so sometimes surpassed that of their subjects. In 1844, during a period of his political imprisonment, the future emperor had written a popular treatise on the empowerment of the worker, *Extinction du pauperisme.* Like the two unnamed *cadets* in Perdonnet's EP studio in 1821, as a youth the emperor had been involved in *carbonariste* activities in Rome in 1830, but in his case with the *carbonari,* not with anti-Bourbon motives like the French, but for the political reasons of Italian unification, and against the Austrian influence in northern Italy and the secular power of the pope.

Despite their own religious and political conservatism, the imperial couple's interest in social welfare and popular education was genuine. It may have been influenced by the future emperor's early schooling in Constance and Protestant Augsburg and his brief military training in Thun, Switzerland (1830) under the Genevan engineer Guillaume-Henri Dufour. The later Swiss commander-in-

242. Ferdinand de Lesseps's mother, Catherine de Grevigné y Gallegos, was a great-aunt of Empress Eugénie through the Grevigné and Kirkpatrick lines. Eugénie de Montijo was thus Ferdinand de Lesseps's first cousin once removed.

chief and co-founder of the Red Cross became his mentor and friend. Dufour was well connected in military, engineering, and scientific circles in France,[243] and the future emperor stayed in very close contact with him through 1836 and remained on friendly terms with him thereafter. Perdonnet, whose views on popular education even more clearly stemmed from a Swiss sense of democratic egalitarianism, documented the imperial couple's advanced social views when he wrote about the course offerings of the Association Polytechnique:

> Political economy is a completely new science.... We hesitated long to teach it, fearing that it only offered an opportunity to tackle political questions that prudence as well as the rules prohibited us from dealing with. It was only last year when the emperor and empress, as well as the head of the Department of the Seine encouraged us to do so that we included it in our curriculum. (Perdonnet 1867, p. 42)[244]

Both the empress and Perdonnet, however different their motivations might have been, supported free and compulsory education for the masses, including education for women and girls (Perdonnet 1867, pp. 46–47).[245] This makes one speculate that a possible personal relationship may have existed between the imperial couple and the innovative pedagogue. Perdonnet was after all a member of the empress's committee to choose lecturers for her *conférences populaires* (Sandras 2020). He may have been closer to the empress than to the emperor because Per-

243. Through his education Dufour (grad. EP 1809/Metz 1810) was at home in French engineering and scientific circles. He was well connected in part through his fellow Genevan Colladon, but also through his own experience. He had, for instance, served at Corfu (1810–14) during the same time as Charles Dupin (1808–11). Despite the fact that Dufour was army and Dupin navy, they interacted at the Académie Iolienne as fellow polytechniciens interested in teaching mathematics, and Dupin later referred to Dufour as "Mon ami le colonel Dufour" (Dupin, *Voyages dans la Grande-Bretagne,* vol. 3, 1824, tome I *Voies publics,* p. 287, footnote).

244. L'économie politique est une science toute nouvelle.... Longtemps nous avons hésité à l'enseigner, craignant qu'elle n'offrît l'occasion d'aborder des questions de politique que la prudence aussi bien que les règlements nous interdisent de traiter. L'année dernière seulement, encouragés par l'Empereur, par l'Impératrice et par M. le Préfet de la Seine, nous l'avons comprise dans les programmes de notre enseignement.

245. Sandras 2020 suggests that Perdonnet's support for women's education possibly originated in his wife's working-class background, although Perdonnet as a child of his time insisted that the male of a family take the lead in intellectual matters, at least outwardly. There had been a girls' school at Yverdon when Perdonnet was there.

donnet's egalitarian opinions were well known, and Napoleon III was suspicious of Freemasonry with its egalitarian doctrine. Although it is not known whether Perdonnet was in fact a Freemason, many of his colleagues were, and he had indeed once been accused of *carbonariste* sympathy if not membership.[246]

Whether or not the emperor harbored reservations about his motivations, Perdonnet's conviction of a democratic rather than a patronizingly feudal approach to general education had influenced the organization of engineering education a decade before the imperial couple began their efforts in the same direction. Together with Bélanger and Mary, Perdonnet was one of the most influential thinkers in the development of French and international engineering education at the time, and politically and socially he was the best connected among them.

His egalitarian convictions also manifested in his involvement with the Association Polytechnique pour le développement de l'instruction populaire that he had helped found in 1830 and of which he was president at the time of his death in 1867. The Association was a workman's free education society, similar to but more organized than the ad hoc lecturing and writing for the working classes that involved John Ruskin, William Fairbairn, and Robert Willis at that time in England. Perdonnet gave regular lectures there, and not only on railways. In the *Entretiens populaires* for 1861, for instance, he published a lengthy article: *Des Grandes Inventions au point de vue des services qu'elles ont rendus ou qu'elles peuvent rendre,* (On great inventions from the standpoint of their current or possible future public service, Perdonnet 1862), in which he described the development and construction of military ordnance (both guns and cannon). This is further proof of his concern with iron as an industrial material.

Perdonnet's non-elitist attitude worked both ways: the academic had much to offer the workman, but a worker could provide the academic with valuable experience and techniques too. His and Colladon's conviction that artisans could even impact innovation in engineering matched the sentiment that Rankine would later express in 1856 (p. 8). In Perdonnet's case, the desire for the intellectual enfranchisement of the working class was both his personal attitude given his marriage with a working-class woman, and a Swiss contribution to the development

246. *Carbonarisme* and Freemasonry are ideologically related. Napoleon III had forced the installation of Maréchal Bernard Pierre Magnan as Grand Master of the Grand Orient de France by decree in 1862 to supervise and suppress any hint of opposition to the regime. See Denslow 1957, vol. III, K-P 'Bernard Pierre Magnan', accessed online 10.12.2018. Grand Masters were normally elected, not appointed.

of general education in France. It certainly did not originate in the basic tenets of the French Revolution. There were many opponents to compulsory education at the time in France, both among conservative Catholic groups and curiously also among those who supported the Revolution. The latter group argued that it violated the Frenchman's freedom of choice to be educated or not. Perdonnet discussed the issue with Johann Konrad Kern, the Swiss ambassador to France. Kern, a founder of the Eidgenössische Polytechnikum (the present ETH Zurich) in 1854–55 and the first president of its board of directors from 1857–60, assured him that

> In our country we definitely do not consider mandatory primary education to be a violation [of our rights], but rather a guarantee of freedom. (Perdonnet 1867, p. 49)[247]

Perdonnet shared his compatriot's view that compulsory and free education established the basis of an informed democratic system.

In the early years, his publications had originally been geared exclusively to professionals, but with time he gradually became increasingly concerned with informing and educating the broader public. In 1859 he wrote:

> I published a treatise in collaboration with Camille Polonceau with the title *Portefeuille de l'Ingénieur des chemins de fer*, and then, on my own another called *Traité élémentaire des chemins de fer*.
>
> The first of these works was purely technical, comprehensive, and costly, and it addressed specialists exclusively.
>
> The second was affordable to a larger public and was written from both a technical and an economic standpoint. But there, too, the technical part still occupied such an important place that reading it could sometimes present difficulties for those totally unfamiliar with engineering.
>
> So in this new work today, I have tried to make myself intelligible to everybody. (Perdonnet 1859, pp. v-vi)[248]

---

247. On ne regarde nullement chez nous l'instruction primaire obligatoire comme une violation, mais plutôt comme une garantie de la liberté.

248. Nous avons publié, en collaboration avec M. Camille Polonceau, un ouvrage intitulé le Portefeuille de l'Ingénieur des chemins de fer, et, seul, un autre ouvrage sous le nom de Traité élémentaire des chemins de fer.

Le premier de ces ouvrages, purement technique, est d'une grande étendue, d'un prix élevé, et s'adresse exclusivement aux hommes spéciaux.

Coriolis, Péclet, and especially Colladon's contributions in the initial period

While Bélanger concentrated on the relationship between analytical theory and construction, and Perdonnet on the pedagogical dissemination of technology and railway construction, others in the faculty were consolidating the curriculum by establishing cross-field relationships in their teaching. Virtually all of what they contributed involved the introduction of iron and understanding its industrial potential.

The three courses—*mécanique*, *théorie des machines*, and *construction des machines* and their various sub-sections like *turbines* or *théorie physique des machines à vapeur*—were central to this. Content and titles changed several times between 1830 and 1865 as mechanization grew in complexity and in importance for the industrial world (Comberousse 1879.1, pp. 60–61). Machine technology was a rapidly growing field, especially in the earlier decades. The hectic development had made Bélanger's decision to split machine technology in 1838 an urgent issue. The theoretical part that Bélanger took over became mechanics, as we saw, and the practical part was reformulated as construction. Each became a field with its own methods and defined boundaries. The distinction between the two was new and simple: mechanics analyzed structural behavior and construction generated objects. This newfound clarity made both mechanics and construction useful to the ECP graduates who would later develop iron structures.

Benoit had originally suggested a unified course comprising both *mécanique* and *construction des machines* in 1829, but his idea was quickly abandoned. Even in the early period, Benoit's co-founders had viewed his proposal negatively. They considered it too technological, meaning that it was organized as an encyclopedia by categories of practical details, and that it was not scientific enough because it included little or no theory. In other words, it lacked the didactical aspect of *vulgarisation*. A lot of material had been missing in Benoit's proposal, and it simultaneously duplicated material that was more appropriately relegated to other courses (Comberousse 1879.1, pp. 53–54, and Weiss 1982 pp. 98–99). Perhaps this led to Benoit resigning when he did, and at any rate,

---

Le second, en même temps technique et économique, est à la portée d'un public beaucoup plus nombreux. La partie technique y occupe néanmoins une place si importante, que la lecture peut en devenir parfois difficile pour les hommes entièrement étrangers à l'art de l'ingénieur....
Nous avons cherché, dans le nouvel ouvrage que nous livrons aujourd'hui à la publicité, à nous rendre intelligible pour tout le monde.

Fig. 30. The mathematician Gaspard-Gustave de Coriolis taught the mechanics course at the École Centrale 1830–31. He was founder Eugène Péclet's brother-in-law and closely associated with the school in the early years. His chief contribution to the pedagogy was that he recommended his assistant Jean-Daniel Colladon as his successor, and later his former classmate Louis-Charles Mary to teach architecture and construction. (Source: engraving by Zépherin Belliard, Wikimedia Commons)

he never actually prepared his course (Colladon 1893, p. 177). When he left, the material was split into machine theory, machine construction, and theoretical mechanics. The two theoretical parts that Bélanger would later organize were taught for the first year unsuccessfully by the mathematician Narcisse Didiez, Dupin's assistant at the Conservatoire and the designated co-teacher. He was succeeded by Péclet's brother-in-law Coriolis, who taught them as a single course for another year. Coriolis also conceived the subject in purely mathematical terms as he had recently formulated it in his treatise *Calcul de l'Effet des Machines*.

Coriolis had been Poncelet, Bélanger, and Mary's classmate at the EP and for a year at the ENPC. He was a theoretician who would write a famous mathematical treatise in 1835 on projecting the trajectories of billiard balls. Perhaps the inspiration for visualizing this concept had been his teacher Monge's tracts on a

Fig. 31. The experimental/theoretical physicist Jean Daniel Colladon as a young man. He developed the haptic approach to engineering education at the École Centrale in the early years of the program. (Source: © Bibliothèque de Genève, inventory number 009, public domain)

card trick or the movement of knights in chess? All three studies were certainly serious papers, but at the same time the choice of vehicle also implies a sense of the practical world and a certain sense of humor and playfulness in their authors. Coriolis was known as a lively lecturer, although some criticized his approach as too abstract. However, it fulfilled what the founders intended in rendering the field of engineering more exact—that is, in their terms more scientific.

Coriolis was not a *vulgarisateur*, but when he left the school at the end of the second term in 1830, he recommended that Colladon, who was both a theoretician and an inventive constructor, should replace him. Not only that, he also later recommended Mary to teach architecture in 1833, thereby contributing substantially, although indirectly, to the success of the program.

Jean-Daniel Colladon was an excellent choice. He had been intimately involved in planning the original didactical concept and he was fully in tune with the school's goals. He had taken part in the earliest meetings of the founders in

1828 (Colladon 1893, pp. 176–177), and had served as the go-between with the Minister of Education Vatimesnil. He supported the school in other ways too. It was his initiative, as well as the financial help of Pierre Laurans, (Gourisse n.d., p. 54) that saved the ECP during the cholera epidemic of 1832. At the time, Lavallée, who was himself ill and had just lost an infant daughter to the disease (Dibos, personal communication), almost gave up and proposed liquidating the school (Pothier 1887, pp. 85–87). Colladon was very much a part of the original program and he quickly showed himself to be an enthusiastic teacher. Although he only taught for a very few years, his hands-on approach to theory exerted a far-reaching influence on the general development of the curriculum. Colladon was what we would today describe as a combination of an experimental and a theoretical physicist. He matched in practicality what Bélanger would soon develop in analysis.

In the school's first year, Colladon had already helped the overworked Péclet by taking over the section on the transmission of energy through steam as adjunct teacher (*professeur adjoint*) for physics and mechanics (Comberousse 1879.1, p. 50). As part of this teaching assignment, he also developed a section on steam engines (Colladon 1893, p. 179).

The chair professor Péclet was a rotund, amiable, modest, and spontaneously enthusiastic man (Faure in Comberousse 1879.1, p. A77). Having lost his only child at an early age, he took a paternal interest in his students (Dumas in Comberousse 1879.1, p. A71) and always inquired at the end of his lectures whether they needed more explanation. He also habitually deprecated his own research contributions (Pothier 1887, pp. 54–55). He was easy to get along with, and the students both loved and respected him (Comberousse 1879.1, p. 51). His interest in *vulgarisation* stemmed, according to Dumas, from his years of teaching at the Collège royale de Marseille,[249] in a city with what he called a pragmatic industrial flavor (Dumas in Comberousse, 1879.1, p. A72). From 1820 on he had also held popular public lectures on applied physics and chemistry at the school. At the ECP, Péclet saw it as his task to translate theoretical physics into what he named *physique industrielle.* This was a major task, a conceptual shift from pedagogical to didactical *vulgarisation,* similar in effort to what would become Bélanger's more narrowly focused transformation of mechanics. Péclet was obliged to delegate part of the subject matter to his assistant Colladon while he himself retained the sections on

249. Despite its name, the school was a lycée, not a collège.

Fig. 32. The founder Eugène Péclet was a physicist who was particularly interested in making physics comprehensible to the public. (Source: Guillet, *Cent ans de la vie de l'École centrale*, 1929, pl. 4, opposite p. 13)

the industrial and technical applications of heat and light. These were courses that were of particular value to the students (Comberousse 1879.1, p. 52).

Like Perdonnet's emerging course on railways, Colladon's course on steam and prime movers was unprecedented as an academic subject. No one had ever attempted anything like it before.[250] The only comparable precedent had been Nicolas Clément's industrial chemistry courses at the Conservatoire from 1819 that had treated steam as energy. And like Perdonnet's, Colladon's course also demanded flexible on-the-job learning from the teacher (Pothier 1887, p. 75). Teaching also proved to be quite a challenge in another respect, as Colladon was only twenty-seven when the school opened in 1829. He was therefore younger than several of his students (Colladon 1893, p. 194).[251] Perdonnet by comparison was a full three years older when he began teaching in 1831.

250. The second course on steam engines appears to have been introduced in 1838 at Durham (Chevallier and Johnston 1838, p. 5).

251. In 1831, forty-eight students were over 21 and a few of them over 30 (Lavallée/Dumas, 1831, p. 11 footnote).

At Lavallée's urging and on Coriolis's recommendation, Colladon inherited the mechanics course when Coriolis left the following year. As his interests lay equally in experimental and in theoretical physics, he agreed to take it on and incorporate his steam-engine segment with the proviso that he would receive funds to transport whole engines to the school. This would allow him to teach the theory of mechanics in a more practically oriented fashion than Coriolis had done.

An anecdote in Colladon's obituary written by his friend Louis Figuier in 1893 shows how unusual and effective his dramatic teaching style was:

> The early graduates of the founding years of the great École centrale still remember how Daniel Colladon taught physics and mechanics: he used to perform a dramatic demonstration in front of them each year. He loaded a rifle with a bullet and holding his hand tightly over the mouth of the barrel, he fired the gun. The air compressed inside the barrel that was firmly blocked by his hand, and this prevented the bullet from leaving the weapon. He then displayed his undamaged hand to the applause of the students. (Figuier, *La Science Illustrée,* 1893, 2nd sem., pp. 150–51)[252]

According to Colladon (1893, p. 180), there were only about 100 steam-driven machines in all of Paris at the time, and he was familiar with virtually all the builders.[253] This allowed him to bring full-sized machinery and parts into the school that the students could familiarize themselves with, take apart, and examine. Colladon knew that to understand something is not necessarily to *know* it. Haptic experience helps one to know. The *feel* of a material, its weight, relative size, surface quality, heat conductivity, and so forth all contribute to technical knowledge.

---

252. Les anciens élèves de l'École centrale, où Daniel Colladon professait la mécanique et la physique, au premier temps de la fondation de cette grande école, se rappellent encore que le professeur faisait devant eux, chaque année, une expérience assez émouvante. Il chargeait à balle un fusil et, tenant la main fortement appliquée contre le bout du canon, il faisait partir le coup. La compression de l'air dans le haut du canon, exactement fermé par l'apposition de la main, empêchait le projectile de sortir de l'arme, et l'opérateur montrait sa main intacte, aux applaudissements des élèves. See also Colladon's technical and less dramatic account in *Mémoires et compte rendu,* 1882.2 pp. 191–192.

253. In comparison, the textile manufacturing center Glasgow with just over a fifth of the population of Paris in 1821 (147,000 to 786,000) had around eighty such machines installed in the early 1820s (Dupin, *Voyages,* vol. 3, 1824, tome II, pp. 173 / 181 footnote).

This experience gave the students practical construction experience and it allowed them to analyze what they discovered and propose design improvements. They apparently even built some machines themselves (Comberousse 1879.1, p. 48). Colladon was of course aware that France lagged behind Britain in factory mechanization. He therefore also took his students into the mechanics' workshops and let them work there (Colladon 1893, pp. 198–199). Whether he had been inspired in this by Leblanc is not clear, but he went further by actually involving his students in mechanical work. This appears to be the first time that physical work of this nature had entered any academic curriculum, engineering or otherwise—at least in France. It was also the first time that theory was taught and *vulgarisé* through objects at any professional school.[254] This can be seen against the contemporary background of full-scale public experiments with technology (Chatzis and Fontanon 2011), but it was unique in an educational environment. Colladon told the manufacturers that getting to know the sons of industrialists at the school was to their advantage, while explaining to Lavallée that contact with the burgeoning manufacturing scene in Paris was also to the school's advantage. This argument convinced Lavallée, who pledged additional funds from his own pocket (or perhaps from his father-in-law's) should the amount that the founders' council provide prove insufficient (Colladon 1893, pp. 200–201).

Supplementing the external experience, the school also set up its own workshops where the students could observe model building, glass blowing (Pothier 1887, p. 42), and the fabrication of the machine parts that they had designed in class. Later they would test and build these parts into the machines that they had analyzed (Comberousse 1879.1, p. 48). Colladon taught only briefly, but his conflation of theory and practical work became the *leitmotif* of the school and the chief pedagogical vehicle of *vulgarisation*.

The initial prospectus that Lavallée put out in 1829 stated that it had been Benoit's proposed course that had suggested the revolutionary idea of actual student participation:

---

254. There is at least one analogous contemporary instance, but not in a professional school. The 'Moravian star', a three-dimensional illustration of an exercise in Monge's descriptive geometry, was developed as a haptic exercise by a teacher in the Moravian (Herrenhuter) Unitäts-Knabenanstalt, a boy's school in Niesky, Saxony, sometime between 1821 and 1830. In its original form it was geometrically an 'augmented rhombicuboctahedron'. Developed as a pedagogical tool, the representation of the Star of Bethlehem soon evolved into an industrialized artifact. (Peters 2015, p. 120).

> The students have to assemble and disassemble a certain number of machines. They will run them and evaluate their usefulness. (Pothier 1887, p. 39)[255]

We cannot know if this had really been Benoit's suggestion, but it was Colladon who first implemented it as a down-to-earth, hands-on approach. This haptic element had previously been lacking in Coriolis's theoretical version of the course, entirely based as it was on mathematics.

Steam engines were important for the future field of construction. Their invention had prompted the industrial introduction of iron for machinery. Initially they were used for pumping out mines, and later for a broad range of other functions: as agricultural tractors, steamboats, locomotives, and all the early railway infrastructure. Bolts and rivets appeared for connecting boilerplates that could withstand high pressures. These connectors necessitated the development of lugs and flanges to connect structural members to hold machines up and for simple sheds to house them. Colladon's practically oriented course provided the first academic link between iron and construction. It presaged what Perdonnet would establish in his fully developed railway construction course from 1837, and what entrepreneurs like Antoine Castor would put into practice around mid-century with their steam-driven machinery that opened new possibilities in construction, like bucket-chain excavators or pneumatic caissons for underwater foundations.[256]

Apart from the section on steam engines, Colladon also incorporated the section *résistance des matériaux*, which Péclet had not had time to treat in any detail, into a new course, *analyse géométrique et mécanique générale*. It was important to him that he illustrate every theoretical principle visually and haptically with practical examples of machines and structures. He detailed how he did this at

---

255. Les élèves devront monter et démonter un certain nombre de machines; ils les feront fonctionner et évalueront leur effet utile.

256. The steam-driven bucket-chain dredge was invented in Britain, first implemented in a primitive form by John Rennie in 1804 and 1808 (Weale, *Quarterly Papers,* vol. 1, 1844, part 1, 'On the dredging machine', p. 5). Betancourt had designed a dredge around then and used it in 1810 in Kronstadt (but it was patented after 1812 by another), and these first examples were gradually improved and used in many situations until their perfection around 1844 by several firms (Weale, *Quarterly Papers* vol. 1, 1844, part 2, 'On the utility and construction of the dredging machine'). The dredge was simultaneously perfected by Castor in France in 1844 based on an early model he had used in the Seine in 1841 (Castor 1869, p. 10), but it would be only patented by Alphonse Couvreux in 1859 (F pat. 40260) and as an excavator in 1860 (F pat. 44893).

great length in his autobiography, describing how he had the students draw the steam engines and the parts they had disassembled at full-scale (Colladon 1893, pp. 201–206). The hands-on approach to teaching was quickly picked up by his colleagues and it continued as one of the basic themes of the school's pedagogy for decades after he left. This form of experiential learning contributed a whole other level of knowledge to the students' education.

When Colladon left the ECP in 1833 to take up a professorship in mechanics at the Académie de Genève (today's University of Geneva), he requested to be allowed to continue teaching the steam-engine segment as a separate module at the end of each academic year. He continued to do this through 1835 (Pothier 1887, pp. 32/119), and at the same time he also took his yearly visit to Paris as a convenient opportunity to use his students as free labor for his steamboat experiments on the Seine.[257] In this way, and with Lavallée's blessing, he anchored his practical and experiential approach to knowledge even more firmly in the mindset of the school.

On the other hand, when he left, Colladon's friend and successor Joseph Liouville, who had served as his *répétiteur* since 1831, openly admitted that he was not at all interested in practice (Pothier 1887, p. 120). He reverted to pure mathematical theory in the main course, *analyse géométrique et mécanique générale*, for five years with no success at all.[258] Liouville then moved on to a professorship at the EP in 1838, which suited his abilities and the abstract nature of his interests better, and that was when Bélanger began to enter the picture.

The interaction between Bélanger, Walter, Faure, and Ferry

The attempt to balance theory and practice characterized the development of the school during the formative years. Colladon had already proved that it was possible to integrate them elegantly. The school hired the practitioner Walter to teach one part of what would later develop into mechanical engineering as *théorie des machines* (1830–38). However, Walter's largely practical bias was not considered a satisfactory solution.

---

257. This is reminiscent of Wilhelm von Humboldt's proposal to involve university students in their professors' research as a pedagogical technique.

258. It appears that Liouville relied on the 7th edition of Monge's *Traité élémentaire de statique, à l'usage des écoles de la marine* (1835), edited by Hachette, in his course. A copy of that edition exists bound together with the prospectus of the school for the years 1834 and 1835.

The *conseil des études* that had succeeded the founders' council as the decision-making body in 1832 was not convinced by Walter's teaching. Bélanger resolved to rectify the situation by taking over *théorie des machines* from Walter and reconfiguring it. At the same time, he took over Liouville's course and developed it in three stages with changing titles. For the first phase (1838–41) he kept Colladon's original title, *analyse géométrique et mécanique générale*. He then split off the geometrical analysis in 1842, which the *conseil* delegated to Émile Martelet (a specialist in descriptive geometry and eventually Olivier's successor). Then Bélanger concentrated on *mécanique générale* for five years (1841–46), finally reuniting the sections in a completed transformation as *analyse et mécanique générale*. Martelet taught this too from 1847 to 1853 whereafter Hippolyte Sonnet replaced him. After 1846, Bélanger was free to concentrate his efforts entirely on *théorie des machines* for the remainder of his career at the ECP.

When Walter relinquished his version of *théorie des machines* to Bélanger in 1838, he took on the practical side in *construction et établissement des machines* from Ferry, who had been teaching it from the beginning. Walter retained Faure as his assistant, while Ferry kept his second course, *métallurgie du fer*, that he had been co-teaching with Walter, also assisted by Faure. This was the roundabout way that the separation between theory and construction finally found its mature expression.

The various constellations of the field of machine technology formed the core of courses (in addition to railway construction) that dealt exclusively with the new material, iron. They were in constant flux and expanded rapidly. All three practice-oriented teachers, Walter, Ferry, and Faure were responsible for researching and developing their course content. Bélanger coordinated their efforts and provided stabilizing leadership, corresponding to the way Péclet had previously guided Colladon's development.

Walter, the initial holder of the chair in machine technology, was a graduate of the École Spéciale Militaire de Saint-Cyr. However, instead of following the usual military career path, he became director of the Fonderie et Ateliers de Construction in Vienne (Isère) and built several foundries. Based on his practical experience in iron metallurgy he joined the faculty of the ECP in 1830, where he proved to be popular with the students and successful as a teacher. Between 1835 and 1838 he published his treatise *Métallurgie pratique du fer: Description méthodique des procédés de fabrication de la fonte et du fer* that was based on his practical experience and his course. The more scientific-minded lobby in the *conseil des études* criticized that, as a former military engineer, he overstressed the practical side of metallurgy to the detriment of its theoretical basis. This is quite possibly the reason that he gave the metallurgy course over to Ferry in 1838. According to Pothier (1887, pp. 213–214), neither Walter nor Ferry were especially versed in theory. However, as a graduate of the École des Mines (grad. c.1808), Ferry had as thorough a theoretical background as Perdonnet. Both Walter and Ferry had the ECP-trained Faure as their assistant who, by virtue of his education (he had studied under Colladon and graduated in 1834), was well able to combine the two aspects. The discussion went back and forth between those who supported innovation—in this case, science—and those who championed practice.

It was significant and highly unusual that Walter's transcript credited his *répétiteur* Auguste Faure as editor even though Faure then held no tenured position at the school. In fact, given his ability both in practice and theory, Faure's contribution as assistant was possibly even more influential than the titular professor's. He also took over Walter's *construction et établissement des machines* and a subset called *turbines* in the years 1847–51 for the ailing Walter, who died shortly after the summer term ended in the last days of July 1851. This is documented by two student notebooks: *1ère Année d'Études. Cours de Construct[io]n de machines. Prof. M. Faure* and *2e Année d'Études. Cours de Turbines. Professeur M. Faure.*[259]

In contrast to the military engineer Walter, Faure was a product of the ECP's graduating class of 1834. He had been associated with the school as instructor from 1838 on, officially as *répétiteur* and then *suppléant* (substitute) for *construction et établissement des machines* (for Walter) and for *métallurgie du fer* (for Ferry) from 1847 to 1850.[260] However, he was not promoted as Walter's successor because he was still too young and had no major publications or outside

259. The first is by the student Ignacio Mendiola, and the second by Antonio Montero. A series of eleven notebooks by these students—ten by Antonio Montero from Montevideo, Uruguay, who did not graduate, and one by Ignacio Mendiola from Puerto Principe (now Camagüey), Cuba (grad. 1852)—are in the Peters Collection.

260. In the brochure published by the school in 1847, (Lavallée/Dumas 1847), he is listed on p. 2, for 1847 as a *répétiteur* for *construction des machines et métallurgie du fer*, and on p. 23 as substitute teacher for *construction des machines: première partie du cours professée per M. Faure; suppléant M. Walter de Saint-Ange.*

professional experience. Ferry was senior and had qualified himself by previously having co-taught *construction et établissement des machines* with Walter from 1830 to 1837. Besides, by then he was already professor for *métallurgie du fer*[261] and *technologie mécanique* and an active and successful practicing engineer (Lavallée/Dumas 1847, p. 1; Pothier 1887, p. 61). Polonceau (grad. 1836) came from practice to take over Walter's course for the year 1851–52, followed by Charles Callon (grad. 1833) who had already co-published a book on steam-powered ships in 1846 with Ferdinand Mathias (grad. 1838). Callon taught machine construction to the combined second and third-year students alternating each semester with Bélanger's machine theory (Bélanger 1864, p. xxiv). This alternating system with the classes of combined second- and third-year students persisted until 1865, when Perdonnet altered the program and Bélanger left. From then on, Callon only taught the third-year students until his death in 1877.

The younger Faure had studied under both Walter and Ferry, and in recent years he had worked more closely with Walter than Ferry had. He was better versed in theory and especially in *vulgarisation* than either of them, so he was retained to teach the *turbines* segment (1847–52) as Ferry's *répétiteur* and backup in theory (Pothier 1887, p. 209) as we also see from Antonio Montero's student notebook. Although he had been bypassed in the transition, Faure was so successful and indispensable that he was appointed professor for a subset titled *transformation de mouvement* in 1851.[262] After the school had been transferred to government jurisdiction in 1858, Faure was asked to teach another new course, *application de la résistance des matériaux*. Faure then taught both courses until his retirement in 1862. During that period, he was also professionally active as director of the Mines de Pontgibaud (Puy-de-Dôme) (Guillet 1929, p. 340) and, curiously, also as an expert on small arms. He helped found the Société des ingénieurs civils in 1848. It was originally an exclusively ECP-alumni foundation, founded by several élèves-fondateurs[263] (albeit with the autodidact and 'crypto-ECP founder' Eugène Flachat as its first president), and held its inaugural meeting at the school

261. This was a new course that had been created for Ferry in 1831 to support railway construction that Perdonnet had begun to teach, and it included canals and bridges as well (Lavallée/Dumas, 1831 p. 2).
262. Pothier 1887, p. 214, not mentioned in Guillet 1929. Thales records him as instructor for this course from 1845.
263. The founders of the Société were Charles Callon (grad. 1833), Camille Laurens (grad. 1833), Léonce Thomas (grad. 1833), Michel Alcan (grad. 1834), Auguste Faure (grad. 1834), and William-Charles Priestley (grad. 1836). See Jacomy 1984 for the history of the founding of the Société.

(Pothier 1887, pp. 203–205). Eleven years later in 1859, Faure was elected the society's president, following Callon in that position.

The notebooks that Faure's and Ferry's students were obliged to keep provide insight into how they taught the machine courses.[264] Faure's version of the *Cours de construction des machines* followed five goals:

1. Study the component parts of machines reduced to their human equivalent in order to understand their movement
2. Study the means for modifying and standardizing the movement of a given system
3. Apply the formulas derived from calculations to determine the forms and dimensions of the parts
4. Calculate how to design machines
5. Study the construction details of the parts and the machines. (Mendiola, 1850, p. 1)[265]

This list indicates the subtle ways in which the three teachers interacted and contributed to one another's courses. Since all three—Walter, Ferry, and Faure—taught several permutations of the mechanical and metallurgy courses in the early period, it is not possible to neatly identify their individual contributions. They surely influenced one another with their varying points of view, and they reacted to the courses their colleagues taught as well. The goals numbered 1, 2, and 5 are practical rather than theoretical and perhaps show Walter and Ferry's influence. The various forms of transforming force documented in the student Mendiola's notebook (rectilinear, both direct and indirect, rotary, alternating, etc.) are explained in 158 carefully detailed sketches that show the long-lasting impact Colladon exerted on the curriculum as implemented through Thumeloup's drawing course. They occasionally

264. Four of these are in the Peters Collection: Ignacio Mendiola, Cours de construction des machines 1st year 1849–50 (Faure), Antonio Montero, Cours de métallurgie (du fer) 1851–52 (Ferry), Jules Carimantrand, Cours de métallurgie 1856–57 (Ferry), and Antonio Montero, Cours de turbines, 1852 (Faure).

265. 1° Étude des organes des machines réduits à leur état humain[?] pour transmettre le mouvement 2° Étude des moyens unités [?] pour modifier et régulariser le mouvem.t imprimé à un système donné 3° Application des formules fournies par le calcul, pour déterminer les formes et les dimensions des organes 4° Calcul de l'établissement des machines 5° Étude de détails de construction des organes et des machines.

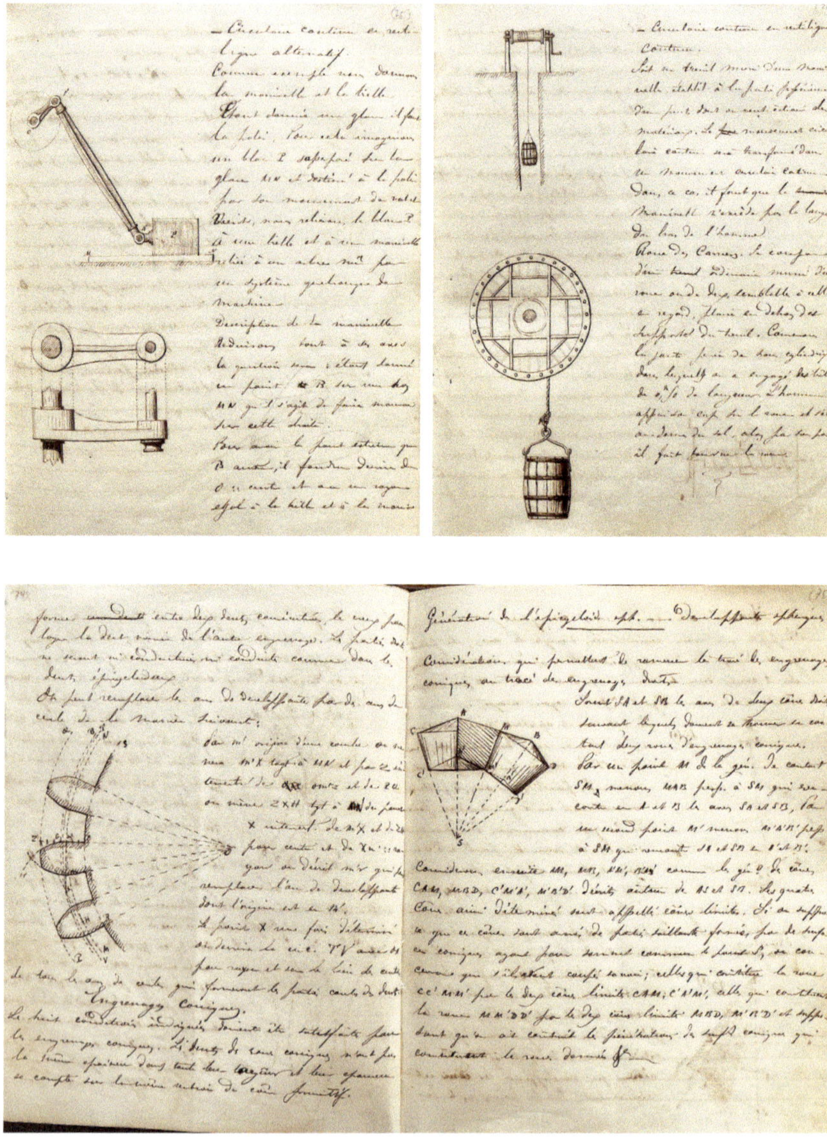

Fig. 33. Manuscript notes and drawings from four pages of the student Ignacio Mendiola's notebook from Auguste Faure's course *construction des machines* (1849–50), pp. 25, 21, 74–75. The sketches show the influence of the drawing teacher Thumeloup and the descriptive geometry teacher Olivier's courses. (Source: Peters Collection)

include dimensions and are often accompanied by geometrical calculations of gearing and diagrams that owe everything to Olivier's descriptive geometry.[266]

Voici comment on peut concevoir le mode d'ajustage et d'établissement de cette excentrique (this is how one can design the manner of adjusting and the construction of this eccentric gearing), Mendiola writes on page 31 of his notebook next to the detail of an eccentric rotary gear.

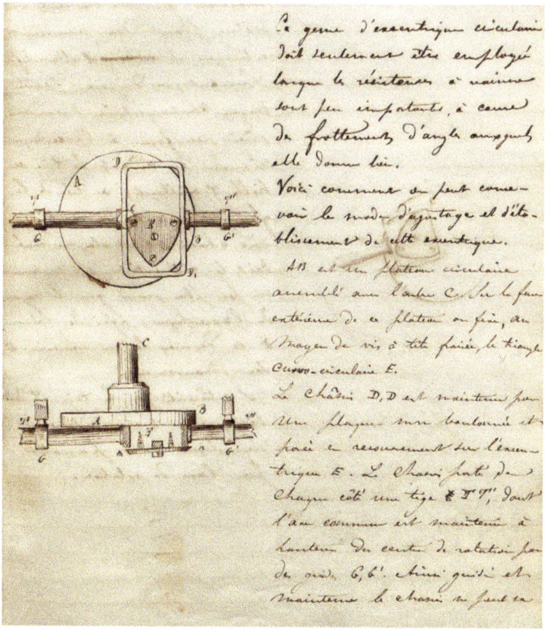

Fig. 34. Manuscript notes and drawings from page 31 of the student Ignacio Mendiola's notebook from Auguste Faure's course *construction des machines* (1849–50). The image and description reveal Colladon's legacy of detailed machine analysis. (Source: Peters Collection)

Teaching about iron: from experiment to industry

The students learned how to use iron in the courses devoted to railways, metallurgy and manufacture of iron, and machine construction. The difference between these ECP courses and analogous ones taught at the EP[267] or lectures

---

266. Olivier had begun publishing on the topic as early as 1827 and had published his comprehensive treatise in 1842 (Olivier 1842, foreword, p. iv).

267. Two courses taught by José Maria Lanz, and those by Betancourt and Hachette published in 1808, Dupin's course of 1826, in which he treated machine design as a geometrical problem, Jean

given at the Conservatoire National des Arts et Métiers is that they were not heavily theoretical, but were balanced between theory and practice, and that they crossed the boundaries between these fields.

Louis-Charles Mary, who was responsible for the architecture course—where one might have expected iron to figure prominently—did not even consider featuring iron in what he taught. We shall see in Part II that he could not entirely avoid it, but iron construction at the time was still principally machinery, and not buildings or even bridges. 'Real' buildings and bridges were built of stone. Iron was rails, locomotives, and carriages, and then, almost as an afterthought, sheds, station halls, and minor bridges. None of these counted as real architecture.

The structural use of iron was the exception rather than the rule in the first half of the nineteenth century. Architects ignored it, although the new material was widely used in foundries, exhibition spaces, and hothouses. Royal and aristocratic patrons commissioned cast- or wrought-iron bridges as scenic follies and gateways for their parks, especially in Russia, the Duchy of Anhalt, and Prussia (Kosykh 2021). Iron was an exotic material and used experimentally rather than as an accepted alternative to stone or wood in building construction. The one exception appeared to have been in Russia where the many onion domes of churches had depended on iron armatures as early as in the seventeenth century (Kosykh 2021), but these, like the few iron roof structures built for their fire-resistant properties both in Russia and later in France, were hidden. The manufacture of iron parts was invariably a cottage industry geared to the manual casting or forging of specially designed elements for individual buildings, not an organized industrial process of standardized prefabrication—except in the railway industry.

The number and type of components, connections, and details in these experimental structures was chaotic. Each builder developed a random panoply of members and connections for each unique situation. Everything was experimental and modelled initially on timber or small-scale houseware or locksmith detailing, and then later derived from boiler-making and ordnance. There was no recognizable geometrical or organizational coherence to iron construction. We tend to forget this because of the intense focus that historians have placed on the exceptional early structures in view of what was to come. It was only

---

Victor Poncelet's two courses (the first published in 1828 and the second written in 1829 but only published in 1839/41), and Giuseppe Antonio Borgnis's texts that were used at the EP.

by mid-century that the experimental character of iron construction gradually began to change, but another decade was to pass before iron established itself in mainstream construction and became a common industrialized structural material worthy of serious consideration by architects. César Jolly, head of the firm César Jolly et Joly fils, one of the largest and most experienced French contractors of the day,[268] published a relevant study *Études pratiques sur la construction des planchers et poutres en fer avec notice sur les colonnes en fer et en fonte* in 1863, in which he wrote:

> Over the past ten years the use of iron in the construction of floors, beams and roof trusses has steadily increased, largely due to the ease with which French foundries have been able to roll this metal into a series of special shapes that, separately or in combination, and by means of intelligent application, allow the construction of very large spans while guaranteeing both stability and cost economy.
>
> We judge that this result has not always been achieved, because aside from the multiplicity of better or poorer combinations, very different iron shapes have been used for constructions that are identical in type and dimension. According to us, these differences can be attributed to three causes:
>
> First, the absence of defined standards in structural deadload per square meter
>
> Second, the sometimes considerable variation in evaluating the maximum load that iron can safely support
>
> Third, the admittedly rare use of types with a poor distribution of material and weak resistance in relation to their weight that should be eliminated from the foundry catalogues. (Jolly 1863, introduction pp. 1–2)[269]

---

268. The firm Constructions Métalliques d'Argenteuil had been founded by the serrurier Pierre-François Joly in 1824. It later became César Jolly et Joly Fils, then Théophile Joly, César Jolly et Delafoy, then in 1893 E. Nouguier, Kessler et Cie., with Eiffel's former designer Émile Nouguier as partner.

269. Depuis dix ans l'emploi du fer a pris dans la construction des planchers, poutres et combles, une extension sans cesse croissante, due en grande partie à la facilité avec laquelle les forges françaises sont parvenues à laminer ce métal sous une série de formes particulières, qui, employées séparément ou combinées entre elles, permettent, par une application judicieuse, d'obtenir de très grandes portées en réunissant la garantie de la solidité et l'économie de la dépense.

Nous pensons que ce résultat n'a pas été toujours atteint, car, à part la multiplicité de combinaisons plus ou moins bonnes, des sections de fer bien différentes ont été appliquées à des constructions de nature et de dimensions identiques. Ces écarts doivent être, selon nous, attribués à trois causes:

1° A l'absence de bases bien définies sur les charges permanentes à appliquer par mètre superficiel de construction;

In this book, Jolly refers to the treatise by the ECP engineer Georges-Henry Love (grad. 1840) on the strength of wrought and cast iron, *Mémoire sur la résistance du fer et de la fonte* (1852). In fact, he states (p. 3) that it, as well as Love's second edition of 1859 (p. 12), provided the incentive for his book, and this demonstrates the growing interest of the pre-industrial contractors in the ECP graduates and their theoretical as well as practical work. The comparison I made earlier between Cornet's (1849–53) and Jacquin's (1864) plate volumes for Perdonnet's railway course underscores this transition from experiment to industry and the recognized need for standardization in the decade 1850–60.

Until the early 1860s, both theoreticians and industrial metallurgists had their work cut out for them to catch up with the advances that the pragmatic contractors had made in the use of the material. They had to sort through the multiple forms and details and begin to standardize forms and types. Only then could iron be economically employed in structural work. And as far as architecture was concerned, there were a few well-known but diffident attempts to feature iron, more decoratively than structurally. It was not until the historian James Fergusson issued the fourth volume of his *A History of Architecture in all Countries, from the Earliest Times to the Present Day* in 1862 titled *History of the Modern Styles of Architecture* with a brief but remarkably perceptive chapter on 'Ferro-Vitreous Art', and Eugène-Emmanuel Viollet-le-Duc published the second volume of his *Entretiens sur l'Architecture* in 1872 that architects could grudgingly envision accepting iron structure officially into the canon of architectural materials. The breakthrough came with Henri Labrouste's two libraries in Paris in 1850 and 1868, which publicized the shifting perspective on iron. The details of how that happened and the ECP engineer Baudet's involvement in the development will be explored in the chapter 'Implementation: the architect Labrouste adopts visible iron structure.'

Until the 1860s it was still evidently Walter's, Ferry's, Faure's, and then Callon's job to examine and define the metallurgical and design issues involving the new material and encourage the students to use it (see also J.-F. Belhoste 2004,

---

2° Aux variations parfois considérables dans l'évaluation du plus grand effort que le fer peut supporter avec sécurité;

3° A l'emploi, assez restreint cependant, d'échantillons qui devraient être rayés du catalogue des forges, par suite de la mauvaise répartition de la matière et de la faible résistance qu'ils donnent, comparée à leur poids.

p. 67). Perdonnet presented practical examples in subordinate structures that had serendipitously appeared on the fringes of railway construction. It was not Mary's task to deal with iron structure in architecture or structural engineering. After Mary retired in 1864, Antoine de Fréminville introduced the course *constructions navales*, in which he taught iron ship construction in nine of 25 lessons from 1865 to 1869. However, his course was terminated after five years, and it was only decades later in 1897 that Duplaix, who had begun his career with Schneider et Cie. in Le Creusot, developed a course on iron construction, *constructions métalliques*, that was finally devoted entirely to the field of structural work in steel. By that time, however, the pioneering period had passed, and structural systems in iron were well established in both construction and architectural design.

The importance of drawing

Iron was a new material in the early years, and the students primarily learned how to use it through visual instruction. Courses in freehand drawing and mechanical drafting schooled them in visual observation and graphic communication. The faculty formed drawing into a tool for the development of new knowledge. Olivier and Bélanger followed Monge and Hachette, who had naturally stressed the importance of engineering drawing, and the former Pestalozzi pupil Perdonnet acknowledged the importance of manual knowledge too. To them, drafting was much more the language of the engineer than were the abstract mathematics that the EP had come to emphasize.

Visual instruction through images is an alternative form of cerebral communication. It has the advantage that it is non-linear, and it provides multiple levels and scales of information from detail to overview simultaneously.[270] When information is communicated visually, the brain can make immediate connections viscerally, and it lets one combine information in new ways to create knowledge. These combinations can be linear connections through ratiocination, but images can also stimulate sudden associations that lead to insight—the so-called 'eureka effect'—and thereby to seeing as we defined it in chapter 1, 'The conceptual issues'. Visual instruction through images is therefore an *active* and often instantaneous form of learning that differs from the more *passive* linear reception to lectures and reading, in which cerebral action occurs through ratiocination after the information has been received, and connections between facts or associations then follow. Active and passive learning are complementary. By combining these learning processes through simultaneity and sequence, information transference is intensified, and it provokes the establishment of knowledge.

---

270. For a discussion of the communication of visual information, see: Edward Tufte, 1983. *The Visual Display of Quantitative Information.*

The teachers all knew that manual recording was the best way to propagate the novel thinking that they explained verbally and with diagrams at the blackboard. Perdonnet implied as much in the introduction to Jacquin's album (1864, pp. 1–2). When the students recorded the teachers' diagrams and drawings, they produced 'virtually tangible' objects where no physical objects were available or even existed. The *répétiteurs* regularly checked and certified the sketches that the students copied from the blackboard into their notebooks, and this strengthened the awareness of their value. Creating information by drawing supplemented both the lectures and Colladon's system of the haptic use of physical machinery in class.

The earlier engineering academies had also stressed drawing: the ENPC more as a presentation tool with artistic and architectural connotations, and the École du Génie de Mézières as a graphic tool related to descriptive geometry, cartography, stereotomy, and ballistics. With the foundation of the EP, drawing became a mathematical discipline that Hachette then expanded to include machine design.[271] Drawing was no longer merely presentation, but also a design tool. The ECP then expanded and refined this tool as a general haptic and visual educational method.

The ECP's drawing program was intensive and comprehensive. It went far beyond the scope of a mere service course and included sketching and finished line drawing. Drawing served as an analytical tool for examining existing industrial buildings or machinery, as a useful method for making personal notes, and as a productive tool for creating industrial layouts and design projects (Comberousse 1879.1, p. 48). It could also provide a means of communication in professional interaction. As Perdonnet later wrote:

> One may justifiably claim that drawing is the *language of the engineer*. In fact, it is the language that engineers use to communicate their ideas, their projects, and to disseminate them to the public. In certain cases, it may prove quite difficult for an engineer, the superintendent of a workshop, a foreman, to make himself understood by a worker who has no notion of drawing, and unfortunately there are still far too many to whom this remark applies.

271. See Belhoste, Picon, and Sakarovitch 1990 for a detailed analysis of eighteenth-century drawing programs. For Durand's rejection of the traditional form of drawing, see Guillerme, *Sur la poïétique architecturale* 1975, reprinted in Guillerme 2008, esp. pp. 171–174.

> A dozen years ago, teaching drawing was rather neglected: when the last London Exhibition highlighted our superiority in the manufacture of tasteful luxury goods, the English believed that this superiority was due to our greater skill in the art of drawing, and they have since then devoted substantial resources to establishing a large number of drawing schools in London. (Perdonnet 1867, pp. 31–33)[272]

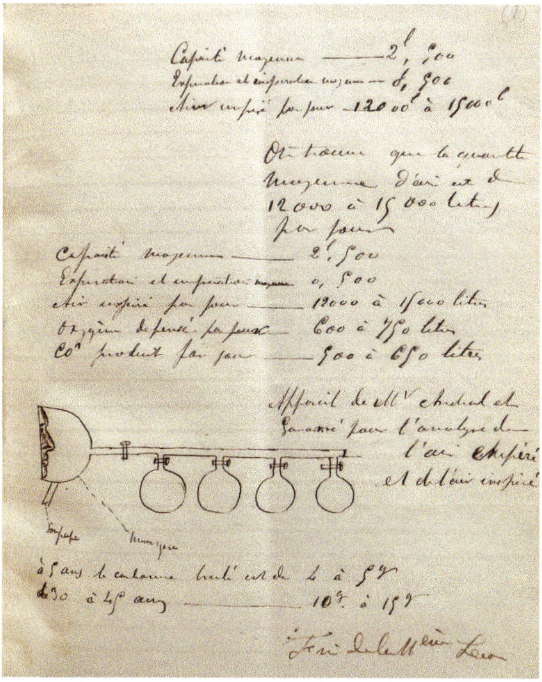

Fig. 35. A page taken from the student Antonio Montero's notebook from Louis Doyère's *Cours d'hygiène* (physiology) 1848–49, p. 91. It demonstrates how drawing permeated the entire program, even in the courses that were not primarily visual. (Source: Peters Collection)

---

272. On dit avec raison que le dessin est *la langue de l'ingénieur*. C'est en effet la langue dont se servent les ingénieurs pour se transmettre leurs idées, leurs projets, et pour les répandre dans le public. Bien difficile est dans certains cas, pour un ingénieur, un directeur d'atelier, un contre-maître, de se faire comprendre d'un ouvrier qui n'a aucune notion de dessin, et malheureusement il y en a encore beaucoup trop à qui cette remarque s'applique.

Il y a une dizaine d'années, l'enseignement du dessin était un peu négligé; la dernière exposition de Londres [1862] ayant mis en évidence la supériorité de notre fabrication des objets de luxe et de goût, les Anglais crurent s'apercevoir que cette supériorité tenait à notre habilité plus grande dans l'art du dessin, et ils ont, depuis lors, consacré des sommes considérables à créer à Londres un grand nombre d'écoles de dessin.

The students honed their drawing skills in courses ranging from architecture—where one would expect it—to *histoire naturelle industrielle* (Pothier 1887, pp. 40–41). They used these skills in Olivier's descriptive geometry and in Colladon's steam engine construction. From 1837 Perdonnet relied on drawing in his railway course too. No other engineering program was as intense in its treatment of the subject. Drawing permeated the entire school, from Louis Doyère's physiology (*hygiène industrielle*) to Auguste Cahours's basic chemistry, Achille Ferry's metallurgy, and Eugène Péligot's analytical (industrial) chemistry.

It was Olivier who had first directed the faculty's attention to the importance of drawing in a meeting of the *conseil d'études* in 1830 (Pothier 1887, p. 353). This was by no means a new interest on Olivier's part, as he had authored a memo on wash-colored drawing at the École de Metz a decade earlier, in 1818 (Péligot in Comberousse 1879.1, p. A60). The Bélanger-Olivier-Péclet committee then formulated what was to be taught in minute detail. We recall the stress that the three had laid on minutiae such as the choice of ink, line weight, and so forth. They stipulated that the drawing instructor was to school the students in sketching and drafting in an unconventional manner. The curriculum required him to treat the drawing course as visual analysis and communication rather than as art by concentrating on draftsmanship and detail. In a way it built on the graphical tradition of the *concours* projects of the ENPC (see Picon and Yvon 1989), but it went further, in a direction that the drawing professor César Leblanc had pioneered: as analysis rather than description. The goal was to achieve clarity in understanding problems as well as to communicate information. This was one of the key contributions to the civil engineering side of the curriculum and to the education of future iron constructors.

The program was consciously focused and well-coordinated, and it had a strong technological bias. It inculcated skills in design, observation, and analysis with a coherent complexity no longer practiced today. The means it used were buildings, details, and machines as models (Montero's 1851–52 notebook), not the classical statues and living human bodies that traditional art schools relied on. The instruction focused on scale and proportion—that is, on technical information—and not on modelling form by means of light and shadow, although that, too, was accommodated in the form of descriptive geometry for the purpose of additional volumetric and spatial clarification. Information was the goal, not art. Viewed on its own, the drawing class would have been what we would now term a service course. However, the *conseil d'études* placed drawing front and center in the curriculum,

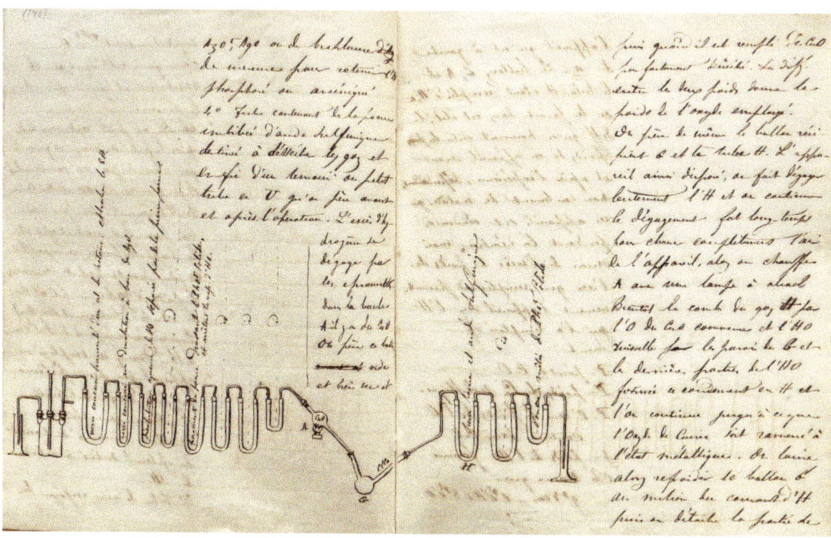

Fig. 36. Pages 49 and 140–141 from Antonio Montero's notebook from Auguste Cahours's *Cours de chimie* (1848–49). Drawings were demanded in all courses and graded by the drawing teacher Thumeloup. (Source: Peters Collection)

and even though there was no designated course as such, meaning that drawing was not accompanied by lectures, it took up a large portion of the first year and nearly as much in the second and third years (Lavallée/Dumas, 1831, p. 15). No engineering school since then has devoted such energy to visual literacy.

Fig. 37. Pages 44 and 80 from Antonio Montero's notebook from Eugène Péligot's *Cours de chimie analytique* (industrial chemistry, 1851–52). The students' notebooks are a valuable source of information on how information was communicated to the students and received by them. (Source: Peters Collection)

César Leblanc

The origin of this program owed much to the drawing professor César Leblanc. He was an engraver who had graduated from the first art class of the Conservatoire National des Arts et Métiers in 1806 and had then gone on to work as a map maker. Nine years later he was hired back by his alma mater as a draftsman on the basis of his expertise in engraving, and he gradually assumed the course in machine drawing as the institution evolved its technological focus in 1819 on the recommendation of the Minster of the Interior, Comte (later Duc) Élie Decazes. Dupin was the first professor for mechanics, with the chemist Nicolas Clément for industrial chemistry, the economist Jean-Baptiste Say for industrial economy, and the specialist in machine construction Gérard-Joseph Christian as director. Parallel to Christian's attempts to standardize machine parts (Guillerme 2008, p. 309), Leblanc devised visual methods for his novel course to explain machine function and construction, and is therefore considered one of the founders of the discipline of technical drafting. In 1830 he published *Choix de modèles appliqués à l'enseignement du dessin des machines*, in which he explained his didactical intentions:

> My aim is to enable the representation of machinery, to popularize knowledge of their function, so to speak, and to show how the various parts work together. [I want] to offer clear and direct methods to those who wish to devote themselves to this kind of study, not only how to draw existing machines correctly, but also how to conceive construction designs for all sorts of combinations or for machines that have only been proposed. (Leblanc 1830, unnumbered p. 1 of the introduction)[273]

It was because of this expertise that he was hired by Lavallée to teach drawing and develop the program, and as its subtitle states, his book was published as the text for his course.

Some accounts confuse César Leblanc with Adolphe Leblanc, who was also a mechanical draftsman and engraver and was employed as César Leblanc's assistant with the title of *maître du dessin des machines* for the initial year (1829–30). Both only taught for the first year (Pothier 1887, p. 63). César Leblanc's withdrawal from the program must have been sudden, which may be why his participation in the curriculum has been neglected. However, given what he wrote, it seems probable that he was either involved in designing the curriculum or that he at least influenced the course that the Bélanger-Olivier-Péclet committee put together.

In the foreword to his book, Leblanc explained his pedagogy and the reasoning behind it. His system was complex and fully developed by 1830, the fruit of a decade of teaching experience, so we may suspect that it predated the founders' curriculum. Had he therefore—like Gourlier and Colladon—also been part of the discussions that preceded the formulation of the school's goals? No source

---

273. Le but que je me propose est de faciliter la représentation des machines, de populariser, pour ainsi dire, l'intelligence de leurs fonctions et du jeu de leurs différentes parties, et d'offrir aux personnes qui se livrent à ce genre d'étude des moyens sûrs et prompts, non seulement de dessiner avec une entière correction, d'après des machines déjà exécutées, mais aussi de tracer à l'avance des épures pour la construction de toutes sortes de combinaisons ou d'appareils qui ne seraient encore qu'en projet.

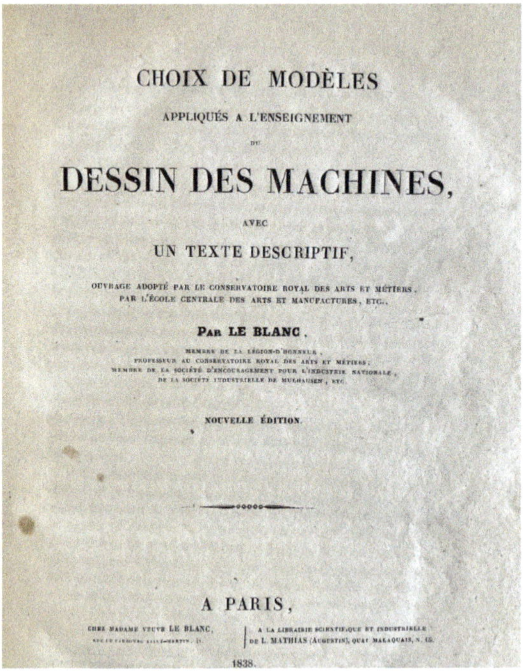

Fig. 38. Title page from a later edition of Leblanc's 1830 treatise *Choix de Modèles* stating in the subtitle that it was intended for use for his course at the ECP: *ouvrage adopté par le conservatoire royal des arts et métiers, par l'École centrale des arts et manufactures, etc.* Leblanc taught only for one year and then disappeared from the history of the school. His text was never used. (Source: Peters Collection)

mentions his participation. Or did his course at the Conservatoire that he had been teaching since 1819 serve as a model for the founding group? The architect Gourlier was also an engraver and may have been in contact with Leblanc or influenced by his work. Although we may suspect contacts and influences here, they cannot be substantiated. What Leblanc wrote as the school opened its doors for the first time was certainly in perfect accord with the ECP's didactical goals and its pedagogy. In fact, what he wrote is a complete and beautiful manifesto of the entire program. It contains the germ of what the graduates would accomplish in iron construction throughout the rest of the century. Leblanc's intentions meshed perfectly with Olivier's, Bélanger's, and Colladon's interests. Even before Colladon, he had taken his Conservatoire students to workshops and had them draw the machines there. He may even have influenced Olivier's subsequent research on gearing, as he stressed it in his introduction:

In order to achieve this goal, I first had to set [the students] a number of problems in geometry as preparatory exercises, the solutions to which are essential for this type of drafting. I followed these with examples of projections and interpenetrations of various solids, taking care to prune anything that did not immediately relate to machine drawing. I separated out the mechanical components, choosing the ones best suited to demonstrate their action by linking them with simple axial lines and their forms by many graphically coordinated details. This was the only way to represent whole machines, including all their parts. I spent time on how to draw gears, as these form an essential element of mechanics. I also sought to teach how to draw plans, elevations, and sections of a machine from measured sketches, and I devoted several plates to the installation of large machinery. Finally, after dealing with the principal examples of shadows and applying them to the most common parts of machines, I finished with several washes and colored examples of various installations.

All the objects I chose as examples or models stem from experience. The forms collected in this treatise were never arbitrarily chosen, and it is necessary to stress this so that the students are made aware of the truth of the fact that forms are always necessitated, in fact driven, either by their function as parts of a machine or by the material of which they are made. Thus, for example, the curvature of gear teeth is fixed and cannot be altered without increasing friction and therefore detracting from the mechanical efficiency and the lifespan of the machine. One should not think, however, that the forms dictated by theory and by experience are not tasteful and beautiful. It is, on the contrary, the guide that rejects all useless appendages. Indeed, when all the parts of an object, however simple it may be, are so arranged as to contribute to the overall goal and are mutually supportive, the result is an appropriateness and a harmony that pleases the eye far more than useless and pernicious decoration. (Leblanc 1830, unnumbered pp. 1–2 of the introduction)[274]

274. Pour atteindre à ce but, j'ai dû présenter d'abord, comme exercices préparatoires, un certain nombre de problèmes de géométrie dont la solution graphique est indispensable dans ce genre de dessin. J'ai fait suivre ces préliminaires par des exemples de projections et de pénétrations de divers solides, en ayant soin d'élaguer tout ce qui n'a point d'application immédiate au dessin des machines. Je me suis attaché à dessiner séparément des organes de machines, après les avoir choisis de manière à faire comprendre leurs mouvemens par le tracé de simples lignes d'axe, et leurs formes par de nombreux détails convenablement coordonnés entre eux. C'était le seul moyen de parvenir à représenter complètement des ensembles comprenant tous ces organes. Je me suis étendu sur le tracé des engrenages, parce qu'ils forment une partie essentielle de la mécanique. J'ai cherché aussi à faire connaître la méthode de lever le plan d'une machine, et d'en tracer les élévations et les coupes

Leblanc contradicted much of what had previously been accepted as the purpose of architectural and engineering representation,[275] establishing drawing on a pragmatic basis, using descriptive geometry as his *vulgarisation* tool to establish a straightforward communication of technical information. His methods were so evidently similar to the ECP's approach to drafting that the parallel calls for explanation, but there appears to be no material available.

Leblanc had developed his ideas together with Gérard-Joseph Christian at the Conservatoire from 1819 on. In the foreword to the first volume of his own treatise in 1822, Christian documented their collaboration:

> M. Leblanc, whose merits are well-known and who has drawn and engraved the plates for this work, is planning to publish the construction drawings of the most common machinery as they will gradually appear in this treatise.
>
> This will prove of real use to artisan-mechanics, and I do not doubt that this work, for which M. Leblanc has already collected a great deal of precious material, will gain the approbation of all who are knowledgeable in the industrial arts. (Christian 1822–25, vol.1, 1822, footnote p. iv)[276]

---

d'après des croquis cotés; j'ai également consacré quelques planches à faire voir les dispositions qui sont nécessaires dans la pose d'une grande machine. Enfin, après avoir exposé les principales études d'ombres en les appliquant aux objets qui se présentent le plus habituellement dans les machines, j'ai terminé par plusieurs lavés et coloriés représentant divers ensembles.

Tous les objets que j'ai choisis comme exemples ou modèles, sont sanctionnés par l'expérience. Les formes de pièces réunies dans cet ouvrage n'ont jamais été prises arbitrairement; il est à propos d'en faire la remarque, afin que les élèves s'accoutument à cette idée vraie que les formes sont toujours nécessitées, et pour ainsi dire commandées, soit par la fonction que telle ou telle pièce doit remplir dans une machine, soit par la matière qui doit entrer dans la construction de cette même pièce; c'est ainsi que la courbure des dents, par exemple, dans les roues d'engrenage, doit être rigoureusement déterminée, et qu'elle ne peut être altérée sans accroître les frottemens, sans nuire par conséquent à l'effet mécanique et à la durée de l'appareil. Qu'on ne croie pas pour cela que les formes exigées par la théorie et par l'expérience manquent de la véritable beauté que réclame le bon goût. C'est ce guide lui-même qui, en pareil cas, rejette tout accessoire inutile. En effet, quand toutes les parties d'un ensemble, quoique simples, sont disposées de manière à concourir au but général, et à se prêter un mutuel appui, il en résulte une convenance, un accord qui plaisent aux yeux beaucoup plus que ne sauraient le faire les ornemens inutiles et dès lors nuisibles.

275. See Guillerme, *La Puissance du factice, Du lavis aux images de synthèse*, 1985, reprinted in Guillerme 2008, pp. 221–229.

276. M. Leblanc dont le mérite est connu, et qui a dessiné et gravé les planches de cet ouvrage, se propose de publier successivement les épures de construction des machines les plus en usage, répandues dans ce traité.

Leblanc and Christian can thus perhaps both be considered the éminences grises of the ECP's drawing program.

However, after only a single year of teaching, Leblanc withdrew. It may be entirely coincidental, but Christian left the directorship of the Conservatoire within the same year, both of them apparently under somewhat obscure circumstances. The reason for Leblanc's dismissal or withdrawal is unknown, and no academic appointment was ever made at professorial level in drafting again. His contribution to the ECP's pedagogy was never recognized and is not recorded. Whereas Gourlier is mentioned, and he apparently continued to be connected to the school with occasional lectures, Leblanc simply vanished, and his contributions are completely ignored by Comberousse, Pothier, and Colladon, who didn't even mention Leblanc's name in his autobiography. However, Leblanc must have maintained close contact with several founders and teachers because he engraved the plates for the second edition of Péclet's *Traité élémentaire de physique* in 1830, the first volumes of Dumas's eight-volume *Traité de chimie appliquée aux arts* (1828–46), and Walter's *Métallurgie pratique du fer* (1835–38). He references Hachette's work in his book (1830, footnote p. 101), which may indicate a link to Olivier, who was Hachette's former student and friend, especially as Leblanc refers to the construction of gearing teeth: Olivier later published his treatise on that topic in 1842. Leblanc also thanks Walter for information on machinery (1830, footnote 1, p. 170). Did Leblanc leave in 1830 because the founders revised their concept and decided that drawing would be taught in the technical courses rather than as a separate subject? The brochure of 1831 seems to indicate so. The way that the passage is formulated suggests there was some sort of conflict that may have precipitated Leblanc's withdrawal:

> There is no special drawing course at the École centrale, and we judged one to be totally useless. The exercises in descriptive geometry, machine layouts, and the many drawings that the students have to do in the courses on mechanics, chemistry, physics, construction, etc. are sufficient to educate excellent draftsmen. (Lavallée/Dumas, 1831, p. 38)[277]

C'est un service réel qu'il rendra aux artistes-mécaniciens ; et je ne doute pas que ce travail, pour lequel M. Leblanc a réuni déjà beaucoup de matériaux précieux, n'obtienne les suffrages de tous les hommes éclairés dans les arts industriels.

277. Il n'y a pas à l'École centrale un cours spécial de dessin, et l'on conçoit qu'il y serait tout à fait inutile. Les épures de géométrie descriptive, les levers de machines et les nombreux dessins qu'on

The inconsistency among Leblanc's role in 1830, the subtitle of his book, and the apparent curricular about-face the following year demands explanation. I was unable to access any relevant archival material that might illuminate what happened. There are also many coincidences surrounding Leblanc's role and disappearance that raise questions: Leblanc's practice of taking his students to draw in machine shops before Colladon did, his interest in gearing teeth that was reflected in Olivier's *Cours de 1831–1832. Engrenages*, but that predated his 1842 research publication, the fact that the subtitle of the 1830 treatise stated that it was to be used by the ECP, and his introductory pedagogical explanations before the ECP's concept was fully formulated. There is also his connection to Christian at the Conservatoire and their more or less simultaneous withdrawal, as well as his work for Péclet, Dumas, and especially for Walter that continued after he left the school.

What happened to the drawing course in the five years after Leblanc left is unclear, but it seems that mechanical drawing was subsumed in this period into machine construction, as the 1831 brochure had stated. In yet another coincident event, it was not until 1836, following Leblanc's death at the end of 1835, that the architect Nicolas-Auguste Thumeloup was hired as the new coordinator of drawing at the school and charged with teaching a drawing course too. Evidently a designated course was by then no longer considered to be totally useless.

Nicolas-Auguste Thumeloup

Thumeloup was never named professor. His title was *chef de dessins*. While this indicates a downgrading of the position that the *conseil des études* accorded drawing in the hierarchical organization of the school, the faculty had evidently embraced Leblanc's ideas after all. Their basically identical goals were endorsed and detailed by the Bélanger-Péclet-Olivier committee and were energetically implemented in all the courses.

Thumeloup did not share the broad overview of his predecessor or his mechanical bias. He was a graduate of the École des Beaux-Arts[278] and had won the second grand prize of the Prix de Rome at the age of thirty in 1834 (*Journal des Savants* 1834,

fait exécuter aux élèves dans les cours de mécanique, de chimie, de physique, de construction, etc. suffisent pour former d'excellens dessinateurs.

278. He had studied under Auguste Guénepin, city architect of Saint-Denis, who had also previously been sous-inspecteur on the construction of the Montmartre slaughterhouse under Louis Bruyère.

p. 626). This, and Olivier's sponsorship (Pothier 1887, p. 353), were what had led Lavallée to hire him in 1836. Professionally, Thumeloup designed neo-Renaissance interiors of *palazzi* in northern Italy, and he taught traditional architectural history at the school as well. Where Leblanc had concentrated on machine construction, in which iron was naturally the preeminent material,[279] Thumeloup focused more on buildings, which gradually also began to include iron structures.

Many specialized teachers assisted Thumeloup in some form of drawing instruction. As far as it is possible to tell from their titles and their professional background, most of them appear to have been more associated with machine construction than with architecture. They would have provided a balance to Thumeloup, whose focus was apparently on formal issues. Over the years these supporting teachers had two titles: *chefs des travaux graphiques* or *dessinateurs des machines*, and their responsibilities were not clear. Curiously, the latter was often an architect.[280]

In accordance with the integrative concept of the curriculum, first Leblanc and then Thumeloup and his assistants were responsible for supervising the graphic components of all the classes, checking students' notebooks, and grading the results (Comberousse 1879.1, p. 56). Each drawing in each of the courses was evaluated on twenty points. This drilled the importance of drawing into the students

279. He had published several more treatises and collections of plates on the subject, including co-authoring the atlas to Walter de Saint-Ange's treatise.

280. Comberousse (pp. A9–24) gives the list of those associated with drafting and drawing. His list is, however, not complete, as Albert Fernique is noted elsewhere as being a chef des travaux graphiques in 1862 for the machine course in the first year (*École Centrale 1889*, pp. 183/184). Originally there had been the professeur de dessin César Leblanc (1829–30) whom Camberousse erroneously assigns to the years 1836–41 (p. A17); his replacement, chef de dessins Thumeloup (1re année, 1836–54); and as maître du dessin des machines for 1829–30 the draftsman and engraver Adolphe Leblanc, who was to engrave some of the plates of Charles Eck's second volume in 1841. Then, as chef des travaux graphiques: professeur-adjoint de géométrie descriptive Baillio-Lamotte (1re année, 1830–31), formerly of the École des Mines (1831 prospectus); Chavonhet (2e & 3e année, 1832–36), Charles-François Nouvian, who is described in the 1843–44 prospectus as *ancien dessinateur attaché à l'École de l'Artillerie et du Génie de Metz* and a printer (1re année, 1840–46); as dessinateur des machines Amable Tronquoy (2e & 3e année, 1846–67); Gastelier (1re année, architecture, 1846–69); the architect Jules-Fréderic Bouchet (1re année, 1854–60); the architect Jean Ferdinand Deconchy (1re année, architecture, 1860–71), seconded by Édouard Duchâtelet (1re année, adjunct architecture, 1855–71), and then fully responsible from 1871; Komarnicki, a graduate of 1863 (2e année, 1866–71); the architect René Demimuid (grad. 1858), later professor in Trélat's school and then at the ECP (2e année, 1869–71); both followed by the architect Jules Denfer (2e & 3e année, 1871), who succeeded Demimuid as professor.

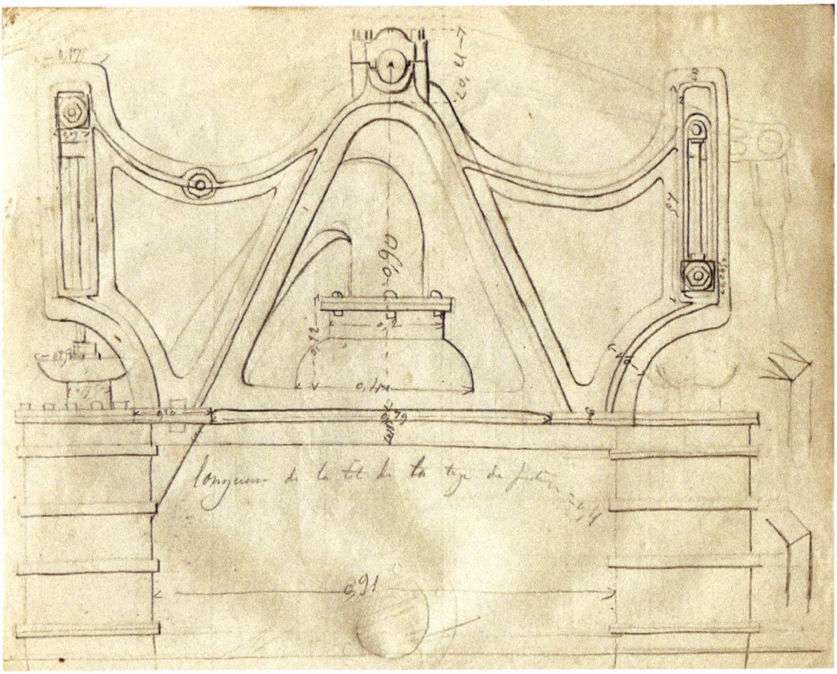

Fig. 39. Page 44 from Antonio Montero's notebook from Nicolas Auguste Thumeloup's course *croquis à la main* 1849–50. Although Montero was not a particularly good draftsman, as his other notebooks demonstrate, he learned to take great care in observing, registering, measuring, and detailing machinery. This fulfilled the pedagogical objective of visual learning and haptic skill needed in the profession. (Source: Peters Collection)

at every turn. The intensity of this engagement of a manual skill in the acquisition and communication of knowledge was one of the key contributions of the ECP's pedagogy to the formation of the students.

Comberousse was not at all good at drawing and never achieved even half the possible points in any of the drawings recorded in the ECP archives (Grelon 1994). Perhaps that is why he turned to teaching mathematics after a few years in railway construction under Flachat and then Émile Vuigner, rather than practicing engineering. Eiffel, on the other hand, who complained bitterly in letters to his mother that this requirement caused him no end of trouble (Weiss 1982, pp. 139–140), evidently became sufficiently visually literate and three-dimensionally (if perhaps not manually) proficient, which was the ultimate goal of drawing instruction at the ECP.

By 1839 the curriculum had matured. After ten years it was sufficiently tested and adjusted, and only a few new ideas were instituted, among them the adjustment of the balance between the basic courses, the advanced courses (1839–40), and the *travail des vacances* that required the students to complete an independent project during the summer break. This took the form of a complete project of a building or a machine by the first-year students and a notebook with sketches and comprehensive notes on a traineeship in a factory or workshop by the second year (Pothier 1887, p. 191). Paul Terrier's (grad. 1862)[281] *cahier de croquis* (sketchbook) of 1861–62 (albeit with later annotations) is his diary of the *travail des vacances* with its drawings and rough sketches on graph paper alongside copious notes, and it provides an example of what this looked like.

The content of what the students learned in their several drafting courses far exceeded what Monge had envisioned and what their own teachers had experienced in their studies. The students sketched and measured machinery and the masonry and iron structures rising around them in Paris, both during term time and in their required holiday projects. The Palais de l'Industrie for the second International Exhibition (1855) was a prominent example that Eiffel and Jenney's classes, for instance, could record and use in their own school projects.[282] Its iron structure had been designed, calculated, and erected by two graduates of the school, Alexis Barrault (grad. 1836) and Gustave Bridel (grad. 1847), working for the English contractor John Oliver York.[283] Many such structures found their way into the sketchbooks in Thumeloup's course and in the technical courses.

After Thumeloup died in 1854, the year after Olivier, and with Bélanger fully occupied with teaching mechanics, the drawing course seems to have declined in importance. The first-year course was taken over by Jules-Frédéric Bouchet, a student of the architect Charles Percier, who probably also inherited the organizing role. He was succeeded by Jean-Ferdinand Deconchy when he left to become the

---

281. Terrier later became the translator of Antonio Favaro's 1877 treatise *Lezioni de statica grafica* in 1879.

282. This structure stood until it was replaced in 1897 by the Grand Palais, erected by Moisant for the Exhibition of 1900.

283. York had apprenticed with an engineer and worked as manager—principal engineer—of the Horsley Ironworks (Aaron Manby's old firm) before moving to France in 1848 as manager of Brassey's ironworks in Evreux. He later built a railway in France and a gasworks in Spain before erecting the Palais de l'Industrie (1853–55), for which he was awarded the title of chevalier of the Légion d'honneur. He later built railways in Italy and England (obit. *Min. Proc. ICE* 1888). J.-F. Belhoste 2004, p. 71 mentions that part of the Palais de l'Industrie was apparently built by Ernest Goüin, which is possible, perhaps as subcontractor.

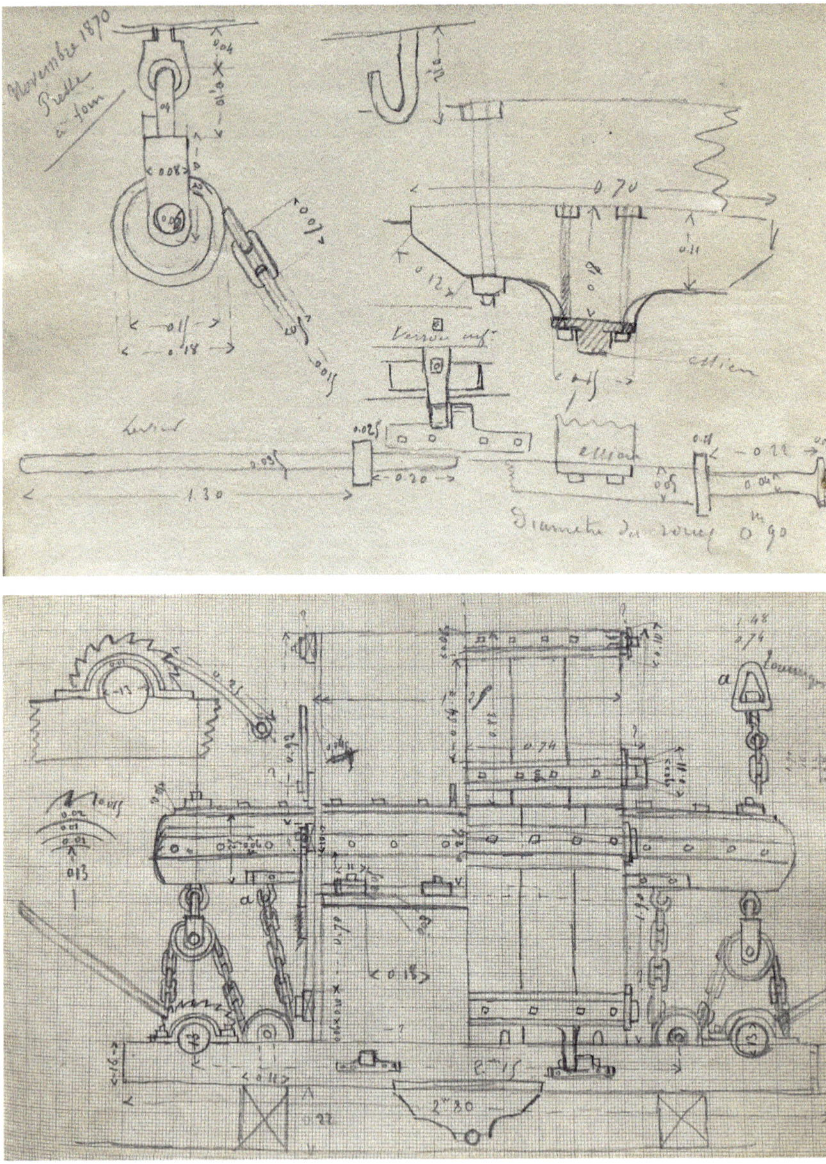

Fig. 40. The second-year *travail de vacances* demanded a detailed report on an industrial plant from each student during the summer recess. The student Paul Terrier's measured rough sketches and initial drafts of part of an installation on graph paper in his sketchbook *cahier de croquis* were probably done for Charles Callon's course, *construction et établissement des machines*, and accompanied by the drafting assistant (dessinateur des machines) Amable Tronquoy. They illustrate the process the students went through in fulfilling this obligation. (Source: Peters Collection)

Paris municipal architect. Both were assisted by the architect Édouard Duchâtelet, who subsequently took on the course himself in 1871. There appears to have been a hiatus in designating a person responsible for drawing in the two higher years during this time, but Deconchy or Duchâtelet may have fulfilled this role in the interim.

The synergy between Leblanc, Thumeloup, Olivier, Colladon, and Mary

The central position of drawing in the curriculum taught the students to think three-dimensionally in design and in the use of materials. It influenced their attitudes to construction and aesthetics. The intentional overlap of the two approaches—manual practice in sketching and descriptive geometry—gave them confidence to judge their work and to recognize its social value. Leblanc's original course (Lavallée/Dumas, 1829, p. 6) had been geometrically based while Thumeloup used traditional architectural sketching. Both schooled the students visually by training their powers to observe three-dimensional objects and translate them into two dimensions. Olivier's course in descriptive geometry recreated the method that Leblanc had used but in a more analytical form. The tandem approach was analogous to the way Renaissance artists had introduced perspective in drawing. It had never been combined pedagogically in construction. As Olivier wrote:

> Descriptive geometry must be considered as both an *art* and as a *science.* (Olivier 1843/1844, vol. 1, p. vi)[284]

The combination was a further manifestation of the didactical concept of *vulgarisation,* and it gave the ECP students confidence in their intuition and in calculation while it also schooled them in non-linear, multileveled visual information. In 1831, the brochure stated that no special drawing course was envisaged, but that drawing was to be taught in every course. This statement appears to conflict with both the 1831 curriculum (Lavallée/Dumas, 1831, p. 16), in which drawing took up a large part of the day in each year, and the *Cours de croquis à la main* later taught by Thumeloup. Perhaps the founders meant that there was no laborious copying of classical statues or drawing of live models as

284. La géométrie descriptive doit être considérée comme un *art* et comme une *science*.

in the École des Beaux-Arts, but that the subject matter in these classes was the technical, mechanical, and structural content of the other courses. This interpretation is borne out by surviving notebooks from the drawing course. However, that would have meant that Leblanc had left for another reason, and it does not explain why no drawing teacher was ever again appointed at full professorial level. The resolution of this conundrum must await the discovery of more documentation than is presently available.

Drawing had always been practiced by builders. Mary, for example, would use it extensively in his construction course. The new science of descriptive geometry had become central to engineering education ever since Monge had begun teaching the subject at the EP, and it had clearly influenced Leblanc's course. The combination of sketching taught by Thumeloup, geometrically based technical drafting, and descriptive geometry enabled graduates working in the field to confidently and unselfconsciously manipulate both structure and form. These skills can be observed in practitioners like Barrault with York in 1855 or Baudet with Labrouste from 1865 onward (see chapters 'Engineers and aesthetics: visual competence' and 'ECP graduates join the entrepreneurs de serrurerie').

The act of drawing—the coordination between eye and hand in the manual execution of images—is an important element both of observation and of intellectualization. It is a skill that is almost invariably neglected in engineering education today (Wetzg 2004) and increasingly in architecture too. Perdonnet, Bélanger, Colladon, and Mary were all fully convinced of its importance. As a companion to Bélanger's and Perdonnet's sentiment that drawing was 'the language of the engineer', Olivier stressed the visual aspect, stating *la géométrie descriptive est l'écriture de l'ingénieur* (descriptive geometry is the *handwriting* of the engineer) (Weiss, p. 127, Pothier, p. 54). Monge had also referred to descriptive geometry as a language even earlier (Picon 2000, p. 27; Monge 1799, p. 2), so they were all aware of the concept of drawing and drafting as a language with all the sensory implications of such a stance.

The insistence on sketching and drafting as visual means of understanding and developing thought, combined with the culture of practical workshop apprenticeships in industrial settings and with descriptive geometry as a haptic learning process, tied intellectual understanding (supported by Bélanger's linguistic precision as imparted in classroom lectures) to professional practice. The intermingling and overlapping of these three ways of learning and thinking—visual, haptic, and intellectual—was one of the strongest

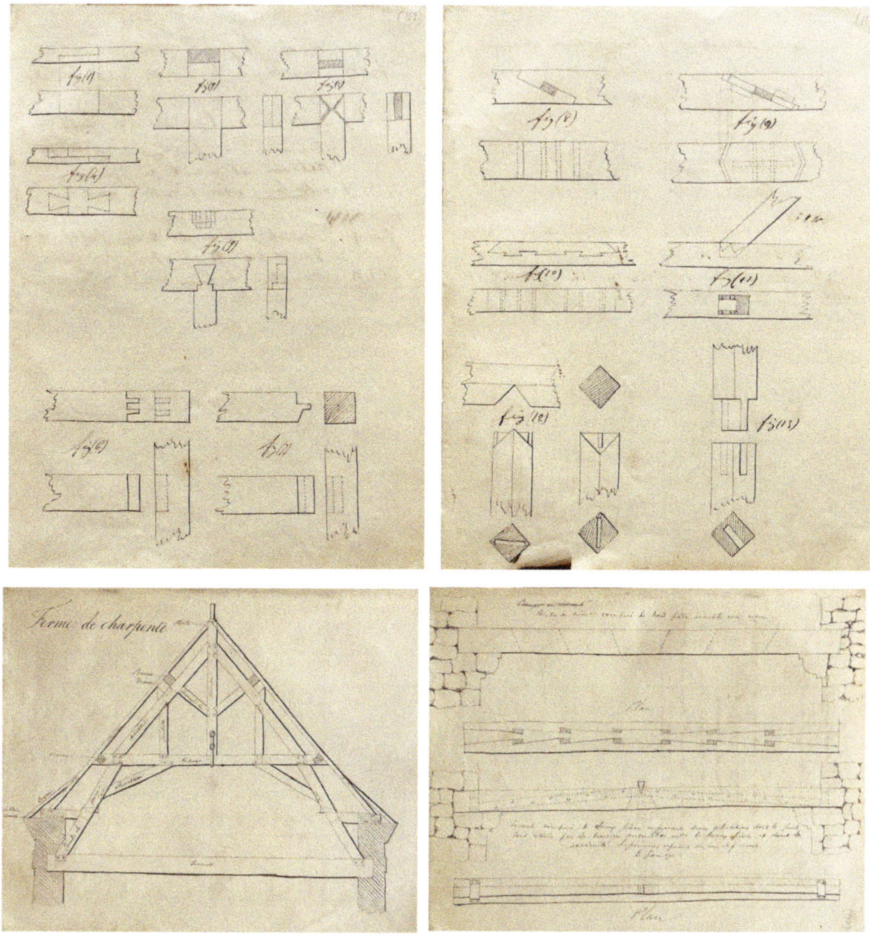

Fig. 41. A series of four drawings relating to wooden construction from pages 155, 157, 159, and 161 of Antonio Montero's exercise notebook in Olivier's *Cours de géométrie descriptive* 1848–49. These drawings that relate more to Louis Charles Mary's construction course demonstrate the typical overlap that occurred between the courses at the school. (Source: Peters Collection)

distinguishing elements of the ECP's pedagogy. The founders had intended this linkage from the very beginning:

> It is evident that no one today contests that oral lessons are insufficient to study science; they need to be accompanied by frequent exams, by numerous experiments or demonstrations, and by graphic exercises carried out under the

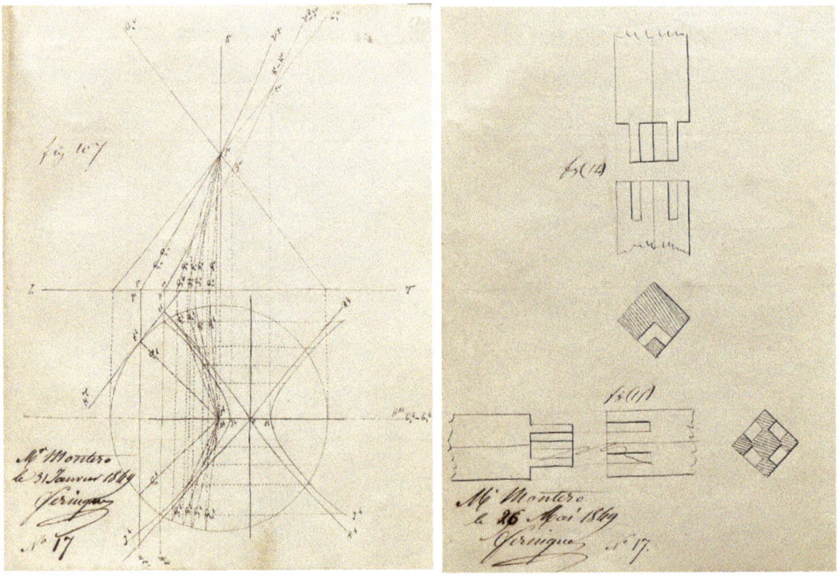

Fig. 42. Montero's exercise notebook from Olivier's course, page 81, dated January 31, 1849, and page 163, dated May 26, 1849. Both are validated by Olivier's assistant (*répétiteur*), the mathematician Clément Fernique. They prove conclusively that both the geometrical construction and the timber detailing stem from the same course. (Source: Peters Collection)

> professors' supervision. These several means of instruction must be taught in combination to form a complete teaching system. (Lavallée/Dumas, 1831 p. 6)[285]

Like all the others, Olivier fully supported these overlaps, and he even incorporated construction drawings from Mary's field in his course.[286]

Olivier's overlapping interest is proven by a closer look at two pages taken from Antonio Montero's notebook (1848–49). Page 81 shows a standard descriptive geometry exercise, and page 169 a carpentry detail from the above

---

285. Il est évident, et personne ne le conteste aujourd'hui, que dans l'étude des sciences appliquées, les leçons orales ne suffisent pas; elles doivent être accompagnées de fréquens examens, de nombreuses expériences ou manipulations, et de travaux graphiques exécutés par les élèves sous les yeux des professeurs. Ces divers moyens d'instruction doivent encore être combinés entre eux de manière à former un système complet d'enseignement.

286. Montero's notebook also contains measured as well as rough sketches of buildings, similar to those in his notebook from Thumeloup's class, and what seems to be an industrial layout, too, but these may have been incidental.

Fig. 43. An example of Albert Fernique's photographs of the construction of Eiffel's Statue of Liberty in Paris. It illustrates the engineer's interest in process more than in form. (Source: Fernique photographs 1881–83. Original in both Library of Congress (LoC) and BnF. Wikimedia Commons)

series. Both are validated and dated by Fernique, Olivier's *répétiteur*, who was the mathematician Clément Fernique.

Fernique's son Albert, who was seven years old when his father signed Montero's notebook, provides an example of how the stress on drawing and visual communication at the ECP could influence seeing and professional development.

Upon graduating as an engineer in 1862, Albert Fernique first became *chef des travaux graphiques (machines)* and *répétiteur* for the course *construction des machines* for the first year *(École Centrale 1889,* pp. 183/184). He then went on to become a well-known photographer. In 1873 he suggested 'magic lanterns' (an early form of image projector) as a way to use photographs in teaching (*Mémoires et compte rendu*, 1873, pp. 251–252). His best-known work is a series of photographs of Eiffel's construction of the Statue of Liberty. This series is an example of the ability to see and the visual acuity that the students gained from the ECP's border-crossing pedagogy. Although not unique as a documentation of a construction process,[287] the photographs demonstrate the ECP engineer's concern with the process, as opposed to the object-focused art of construction.

Olivier's influence

Olivier's course played a major role in the visual side of the curriculum. He had studied under Monge at the EP (1811–14, taking a year longer than usual to graduate for health reasons). He had then continued his graduate education (1815–18) at Metz, after which he immediately became an adjunct professor, staying until 1821. That was the year the Swedish king, Karl XIV Johan, Napoleon's former general Jean-Baptiste Bernadotte, called him to Stockholm to reorganize education at the Military Academy Karlberg and to tutor Crown Prince Oscar. Olivier returned to France five years later, where he made himself unpopular as a critic of the direction the EP had taken. He was ripe to participate in the foundation of the ECP.

Olivier was an ardent Bonapartist, a convinced Saint-Simonian, and an enthusiastic supporter of Monge's border-crossing approach to teaching. This led to Olivier's lifelong friendship with his teacher Hachette, who was keenly interested in the application of descriptive geometry to the arts and machine construction. We do not know whether Leblanc also influenced his interests. Olivier was a dynamic teacher who *élevait la géométrie descriptive à la hauteur d'une religion*, (elevated descriptive geometry to the level of a religion) (Comberousse 1879.1, p. 50). He often ended his lectures with an informal chat session, in which he dispensed advice to his students (Pothier 1887, p. 54).

---

287. The first such comprehensive series of photographs was a documentation of the demolition and re-building of the London Crystal Palace in 1852–54. There had been several more since then. (Peters 2018).

Fig. 44. Théodore Olivier, a convinced Bonapartist and Saint Simonian, was a firm supporter of visual learning through graphical means. His descriptive geometry course was one of the mainstays of the ECP program, and he researched ways of extending the technique into the industrial world. (Source: Guillet, *Cent ans de la vie de l'École centrale*, 1929, pl. 3, opposite p. 12)

Socially he appears to have been a charming man, although somewhat pompous (Pothier 1887, p. 54). Intellectually he was confrontational in his views, as we saw from the introduction to his treatise of 1851, and apparently quite rigid in pursuing his own interests (Dumas in Comberousse 1879.1, p. A67). He was known to have been vain, and he cultivated his remarkable resemblance to Napoleon down to the way he combed his hair (Comberousse 1879.1, p. 51). He may also not have discouraged the (false) suspicion that he was Napoleon's illegitimate son. Perdonnet was therefore not the only complex character at the school who tested Lavallée's patience and negotiating skills. However, the quirkiness of Olivier's character compromised neither his organizational competence and broad vision (Comberousse 1879.1, p. 73) nor his innovative teaching, and his students were devoted to him.

Although Pfammatter states that the didactical model of *science industrielle* superseded Monge's model of basing engineering on descriptive geometry, (1997, pp. 123–125), the field in fact adapted to the new model under Olivier's guidance by including practical applications like the graphical calculation of gearing or sun angles as part of the mathematics of descriptive geometry. Olivier expanded what Leblanc had intended in his drafting course and adapted the theory, supplementing what Bélanger was accomplishing on a non-visual level in mechanics. The relationship between oral and visual learning was important to Olivier, as was the relationship between sketches as a form of visual notation and finished drawings that translated them into solutions to problems. He explained this in the foreword to the second volume of his *Cours de géométrie descriptive* in 1844:

> In my view, there are two distinct parts to a descriptive geometry course: right from the very *outset*, that is in everything that concerns *the point, the line, and the plan*, there is both a *theoretical* and a *practical* aspect.
>
> The *theoretical* aspect only calls for *sketches*. These sketches help the reader or the listener to orient himself better spatially and then to follow the author or professor's point of view.
>
> The *graphical* aspect comprises a complete construction of the solution to the set problem—that is a complete *drawing*.
>
> A descriptive geometry course must therefore be divided into two different modes of instruction.
>
> The first is *oral*, and this is how the *theory* is explained in the lectures.
>
> The second mode of instruction is *manual* and consists of creating finished *drawings* in the studio.
>
> During the lecture the student should *sketch*.
>
> In the studio the student must make a finished *drawing*.
>
> I therefore consider instruction in descriptive geometry to be organized in the same way that chemistry is organized, and therefore, in order to be complete, to be taught as both *oral lectures* and *exercises*. (Olivier 1843/1844, vol. 2, pp. vi–vii).[288]

---

288. Pour moi, il y a deux choses distinctes dans un cours de géométrie descriptive : aussitôt que les *préliminaires*, c'est-à-dire tout ce qui est relatif *au point, à la droite et au plan*, se trouvent exposés, il y a la partie *théorique* et la partie *graphique*.

La partie *théorique* n'exige comme dessins que des croquis. Ces croquis servent à aider le lecteur ou l'auditeur à mieux voir dans l'espace et à se placer de suite au point de vue de l'auteur ou du professeur.

La partie *graphique* consiste en la construction complète de la solution du problème proposé, et ainsi en une épure complète.

Un cours de géométrie descriptive doit donc être divisé en deux enseignements différents entre eux.

Le premier enseignement est *oral*, et ainsi dans les leçons on explique la *théorie*.

Fig. 45. *L'entrée du cours*. Students in their decorated smocks entering the lecture hall or *amphithéâtre* carrying their notebooks in which they were also required to make sketches. (Source: Rousselet, *Nos grandes écoles* 1888, p. 289)

Fig. 46. The lectures were frequently accompanied by demonstrations that deepened the learning by visual means. Here the *cours de chimie*. This was not only true of the chemistry or physics courses but also of lectures like Olivier's for which he used demonstration models that he had invented. (Source: Rousselet, *Nos grandes écoles* 1888, p. 309)

To support this approach, Olivier also developed and built many transformable, three-dimensional demonstration models, originally for his public lectures at the Conservatoire from 1839 (Péligot in Comberousse 1879.1, p. A62),[289] and he brought some of them to class in the *amphithéâtre* at the ECP.

Olivier had spent several years in Stockholm. As a friend and disciple of Hachette, he had been particularly interested the use of geometry in machine construction and certainly knew Polhem's 'mechanical alphabet'. The celebrated eighty operable models of mechanical principles, as well as Polhem's automated cutting machine for cog wheels (the first automated machine tool) were, and still are a Swedish national treasure. At the time, these models were exhibited in the Royal Model Chamber that had belonged to the Academy of Agriculture since 1813, and they were actively used in teaching at the Mekaniska Skola. Olivier was possibly the only EP alumnus in Sweden. He was the reorganizer of teaching at the military academy and the royal tutor, and surely in touch with the Mekaniska Skola even if he might not have been actually involved in teaching there. The school was in the process of developing into the Royal Institute of Technology and took over the mechanical models in 1826.[290] It opened in 1827 after Olivier had left Stockholm, but had he been involved in the transition, and had the visually attractive Polhem models and their use in teaching inspired him to create his own manipulable models? Might Olivier's entire visual and haptic approach to teaching have been influenced by his knowledge of Polhem's late seventeenth-century example?

The idea of using transformable models in teaching was novel to engineers outside Sweden. Olivier certainly made quite a business of selling his models and is still well known as their inventor. The model-making firm of Pixii, Père et Fils

---

Le second enseignement est *manuel*, et ainsi, dans la salle de dessin, on fait exécuter les *épures*.

À la leçon, l'élève doit prendre des *croquis*.

Dans la salle de dessin, l'élève doit exécuter complètement une *épure*.

Je considère donc l'enseignement de la géométrie descriptive comme devant être établi ainsi qu'est établi celui de la chimie, et ainsi comme devant être composé, pour être complet, de *leçons orales* et de *manipulations*.

289. Olivier only began teaching at the Conservatoire National des Arts et Métiers in 1838, well after Leblanc's death.

290. The machine and other models remained at the academy and then eventually went to the Nordic Museum where they remain today, while the Institute of Technology's models are now in the National Museum of Science and Technology.

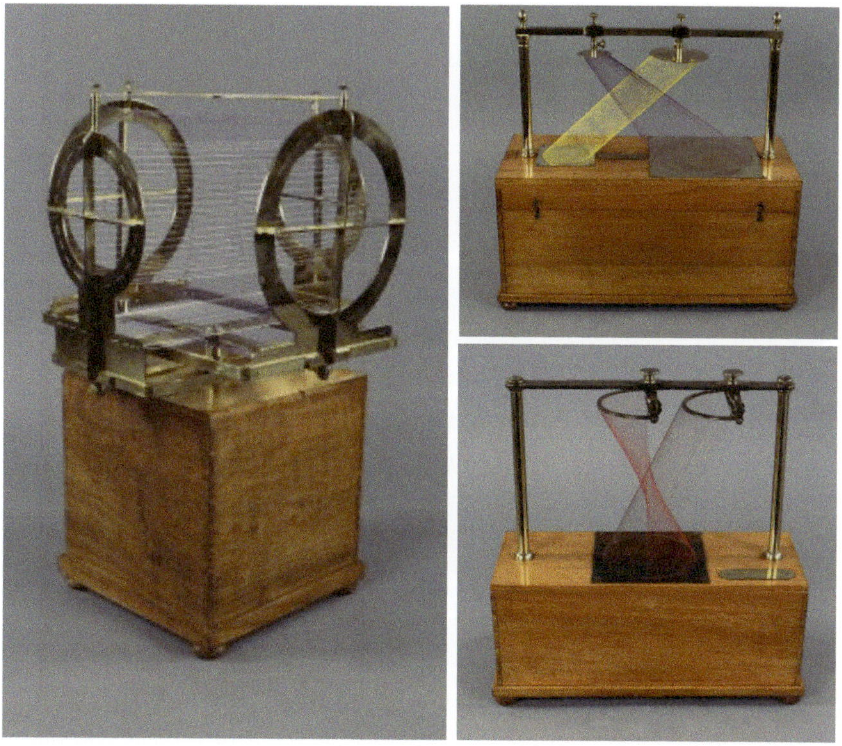

Fig. 47. Three of Olivier's original, three-dimensional models demonstrating the interpenetration of solid bodies from the collection of Union College. Olivier used these in his descriptive geometry classes to enhance the students' ability to visualize three-dimensionally. The models could be manipulated to change their form and thus their interpenetration. (Source: © Courtesy of the Union College Permanent Collection, Schenectady, NY)

manufactured them, and even after his death the firm's successor Fabre de Lagrange continued selling them at least until the 1870s.

About fifty of his models are still at the Conservatoire, and Olivier sold some to West Point and the Columbia School of Mines. Other collections exist at Cornell and in Lisbon, Porto, and Madrid.[291] William Mitchell Gillespie purchased forty-six originals from Olivier's widow, and after his death in 1868, Gillespie's widow sold them to Union College. Princeton had copies made, and Harvard as

291. Sakarovitch 1988, p. 329 footnote 116 / O'Connor and Robertson 2002 / Vierling-Claassen 2017. Curator Peggy Aldrich Kidwell lists more on the Smithsonian website.

well as the Science Museum in London and many others later purchased a large number from the manufacturer. This is yet another testimony to the influence that the ECP's pedagogy exerted on engineering education worldwide.

Like Polhem's models, Olivier's were three-dimensional visual clarifications. He could manipulate them by changing certain dimensions to display various interpenetrations. They differed from Colladon's machines that were both visual and tactile objects, ones that the students took apart, handled, drew, perhaps improved, and ultimately reassembled. Both types of objects served as additional teaching tools to supplement lectures and diagrams copied from the blackboard, and they expanded the students' intellectual understanding through the inclusion of different forms of sensory information.

Olivier used his models to help students visualize the problems he discussed, before giving them exercises in which these three-dimensional objects were projected and manipulated in two dimensions on paper. A powerful effect was the ability that the students gained, not only to visualize abstract geometrical problems, but to see them as physical forms in space, and the skill they gained to manipulate them as well.

Olivier integrated his passion into the pedagogical objectives of the school by demonstrating the practical applications of descriptive geometry to linear perspective, to the calculation and depiction of shadows thrown onto objects, to the art of stone-cutting (stereotomy), and to the design of complex wooden roofs. He also demonstrated descriptive geometry's ability to project information on the inclination of sunlight, to topography, and to the calculation and design of gearing teeth (Comberousse 1879.1, p. 52). Olivier's ideas on the practical uses of descriptive geometry had probably been incited by Hachette's passion and may have been influenced by Leblanc, too, but they were novel in their scope and in their relevance to engineering practice.

Despite his later disgust with those he termed *algébristes*, Olivier had not been fixated on graphic means of analysis in the earlier part of his career. In a rare private publication in 1836 that is generally not included in his publication list, he had studied the friction exerted by railway wheels on tightly curved track (Navier had studied the movement of wagons on curved tracks in the *Annales* the year before). Olivier's examination of the system proposed by B. Laignel was argued using analytical methods but with explanatory diagrams appended in the plates (Olivier 1836).

The four courses in railway construction, machine technology, descriptive geometry, and drawing converged pedagogically as they all used three-dimensional visual literacy, observation, and recording. This intentional overlap was another key component of the ECP program (related to the visual/haptic/intellectual overlap). Some have questioned whether descriptive geometry was particularly useful in machine construction (Picon 2000, p. 27), but as Olivier had clearly stated in 1844, it trained the students' three-dimensional visualization and contributed to their visual literacy. Its primary pedagogical value to the curriculum was the ability to visualize.

What the students learned from drawing with Thumeloup, from Colladon's approach in the intricacies of mechanical manipulation, and from Olivier's analytical descriptive geometry did not only have implications in Mary's construction and design. The ability to visualize spatially, to understand mechanical behavior, and to record machines and their parts by sketching them also served as a valuable technique in the ubiquitous nineteenth-century sport of industrial espionage. Visualization and recording proved to be very useful on study trips to England, where visitors to machine shops were strictly forbidden to take notes or sketch during their rapidly conducted tours. ECP graduates were trained to record what they had seen during the day from memory, by candlelight in their rooms in the evening. As Olivier reminded the students:

> Descriptive geometry is the handwriting of the engineer: he who can read spatially can visit a factory, a workshop, without taking notes, and then at home, draw up the tool and the machine that he has seen and understood. (Pothier 1887, p. 54)[292]

Perdonnet stressed the importance of this aspect of the students' well-trained visual and spatial skills in his obituary of Polonceau in 1859. Like the genuinely first-class teacher that he was, Perdonnet was proud that his students' abilities had surpassed his own:

> We visited factories in the way one is usually permitted to visit them—that is, by racing at breakneck speed through the workshops—but Polonceau saw all that was of interest with an incredibly rapid glance, and I was full of admiration to

292. La géométrie descriptive est l'écriture de l'ingénieur; celui qui sait lire dans l'espace peut visiter une usine, une manufacture sans prendre de notes, et ensuite dessiner chez lui l'outil et la machine qu'il a bien vus et compris.

> see him, upon exiting the workshop, draft a complete sketch of machines that I had barely glimpsed. (Perdonnet in Comberousse 1879.1, p. A120)[293]

The ECP continued to gain in influence as the program evolved, and the impact of visual thinking in engineering education soon began to impact structural analysis. Henri de Dion was one of the pioneers of visual structural analysis in the engineering construction course that he created for Émile Trélat's École Centrale d'Architecture in 1865.[294] From there it gradually spread to engineering schools, where it survived for almost a century through the medium of graphic statics.

Graphic statics

The impulse for graphic statics came in a roundabout way from French education. Aside from the early studies on the parallelogram of forces by Daniel Bernoulli, Isaac Newton, Pierre Varignon, and others in the late seventeenth and early eighteenth centuries, it was Monge's descriptive geometry, especially as developed by Hachette at the EP that was one of the initial impulses for developing a geometrical approach to structural calculations and design. EP graduates like Poncelet and others explored various possibilities. One of the most interesting, and possibly the earliest example of the *vulgarisation* of structural theory for innumerate technologists, was Prony's 1802 paper, *Instruction-pratique sur une méthode pour déterminer les dimensions des murs de revêtement, en se servant de la formule graphique* (Prony 1802.1). In this little-known twelve-page brochure he laid out a simple graphical way that allowed even mathematically untutored laborers to design retaining walls. He issued it in Germinal An X (March-April 1802) and expressly intended it for workmen. Prony thus stands at

---

293. Nous visitions les fabriques comme il est permis généralement de les visiter, c'est à dire en traversant les ateliers au pas de course; mais Polonceau saisissait tout ce qu'il y avait d'intéressant avec une rapidité de coup de l'œil incroyable, et j'étais émerveillé en le voyant, au sortir de l'atelier, faire un croquis complet de machines que j'avais à peine entrevues.

294. The school, now known as the École Spéciale d'Architecture, was founded in 1865 by ECP graduate Émile Trélat (grad. 1840) in reaction to the resistance to reform by some of the faculty at the École des Beaux-Arts. It stressed and developed the architectural aspect of the ECP's education, basing its pedagogy in part on the ECP, especially on Mary's design method, and it was supported by several ECP graduates and by prominent men in the building industries like Eugène Flachat, Ferdinand de Lesseps (associated in name only), and Eugène-Emmanuel Viollet-le-Duc.

the very beginning of both vulgarisation and graphical calculation, dimensioning, and design methods.

We can attribute Prony's concern to his education, which had been an unusual combination of practice and theory. On the one hand, he was Perronet's disciple and protégé and had worked as assistant site engineer on the Pont de Neuilly (1768–72) and the Pont de Sainte-Maxence (1771–86),[295] and on the other, he had attracted Monge's attention with a 1783 paper on thrust in arches that had been provoked by the settlement of the Pont de Neuilly after decentering.[296] As a result of his presentation of the paper at the Académie des Sciences, Monge had begun to tutor him privately (Dupin 1819, p. 14). Then, in the following decade, around 1791, he had also worked with Pierre-Simon Girard on geometrical problems (MacTutor, 'Riche de Prony').[297] He was thus uniquely positioned to initiate the crossover between theory and practice using geometry. In his 1802 paper, Prony referred to his treatise *Mécanique philosophique* that he had published two years before (1802.1, p. 3) and in which he had traced his thoughts to Euler, Lagrange, Laplace, Bossut, and Coulomb (1800 p. vii), and in 1802 he based his calculations on Coulomb's pioneering paper of 1773. He simplified them after discovering that the plane of maximum earth pressure bisects the angle between the back of a retaining wall and the natural slope of the soil (Gillmor 1971, pp. 106–107). This prompted him to design a simple diagram that allowed him to provide

> an easy method for designing retaining walls that can resist the lateral pressure exerted by the earth to those builders who have little or no knowledge of algebraic analysis (1802.1, p. 4).[298]

---

295. He would later gain even more experience as site engineer on the Pont de la Concorde.

296. This paper appears not to have been published by the Académie. It exists as MS no. 2215, *Sur l'application de la mécanique des voûtes au Pont de Neuilly*, at the library of the École des Ponts et Chaussées. Prony's first published mémoir seems to have been *Sur un nouvel instrument propre à mesurer les petits angles en général et au zénith* 1789, p. 19.

297. Coulomb, Prony, and Girard were among the very few practicing engineers at the time who had expertise in mathematical analysis. Prony's work with Girard was probably related to Girard's 1798 treatise that contains analytical diagrams (pls. 1–3). The study had begun in 1787 (Girard 1798, p. v) and Prony presented it together with Coulomb to the Institut national des Sciences et Arts on March 1, 1798 (11 Ventôse An VI) (Girard 1798, p. vii).

298. ...aux constructeurs peu ou nullement instruits de l'analyse algébrique, une méthode fort commode pour tracer les profils de murs de revêtement capables de résister à la poussée des terres qu'ils ont à soutenir.

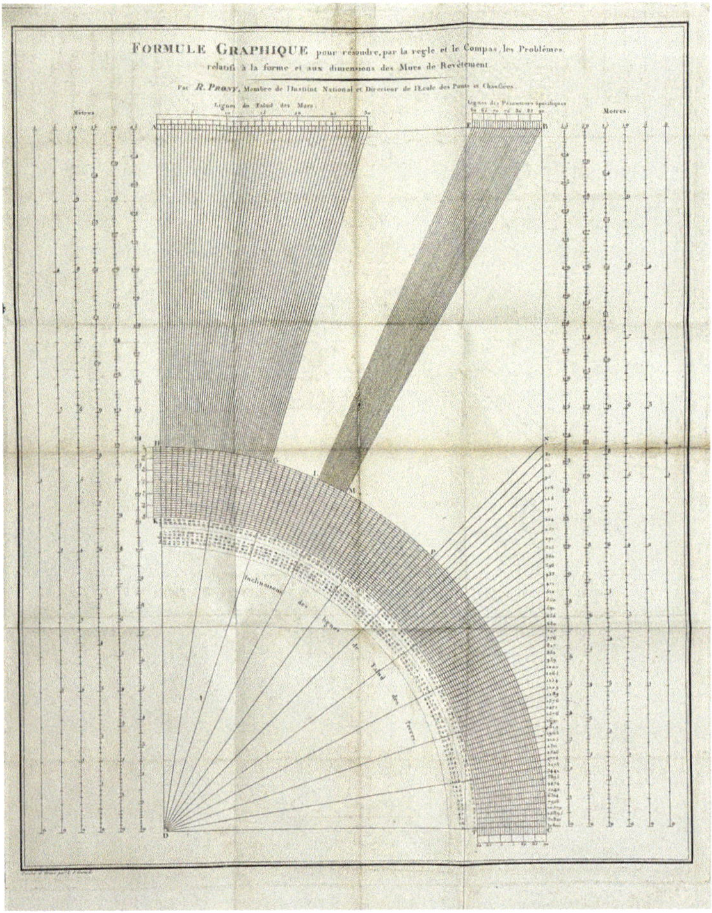

Fig. 48. Prony's large (59 x 47.5 cm) geometrical construction diagram for dimensioning retaining walls was intended for practitioners who were not numerate. The elegant diagram was to be mounted on a board and used as a template. Prony, *Instruction-pratique sur une méthode pour déterminer les dimensions des murs de revêtement, en se servant de la formule graphique,* 1802. (Source: Peters Collection)

Prony's geometrical construction owed everything to Monge's tutoring and way of thinking that had provoked the creation of descriptive geometry in the 1760s.

The paper documents the earliest known instance of an intentionally developed graphical solution to a problem in structural statics. It illuminates Prony's border-crossing concern with incorporating theory into practice, but it did not lead directly to the development of a new field of graphic statics. The reason lies in Prony's intention: his goal was to provide a scientifically reliable design method to innumerate builders, and not to simplify the process of calculation and dimensioning by graphical means for engineers. In the month following this brochure (Floréal = April-May), he published a more elaborate version that included algebraic analysis for his engineering students at the EP and ENPC (1802.2).[299] The sequence of the two versions tells us that he first sought to translate science to enable the laborer, and only then to present the analysis and explain the simplified graphical method to his students. The second publication goes into great analytical detail on all aspects of the problem, and in the midst of this analysis he lets his students know that

> Those who seek only practical construction can dispense with reading what follows and pass straightaway to article 34 and the explanation and use of a graphical method to solve, without calculation, the principal problems of forming and dimensioning retaining walls (Prony 1802.2 p. 26).[300]

Prony himself therefore considered analysis primary and the graphical method an addendum to how an engineer understands the problem. He supports this distinction in a footnote:

> However, this explanation [of the diagram] is only intended for laborers who have no theoretical knowledge of geometry or calculation: engineers will immediately recognize that these numbers represent the tabulated tangents of the angles marked on the arch TK (Prony 1802.2 p. 32).[301]

He supports this opinion again on page 34: 'voici comment on s'en servira pour suppléer aux formules analytiques dans la détermination des dimensions' (this is

---

299. *Recherches sur la poussée des terres, et sur la forme et les dimensions à donner aux murs de revêtement; Suivies d'une méthode pratique, à la portée des Ouvriers qui ont quelque habitude de se servir de la règle et du compas, pour résoudre très-facilement les principaux Problèmes relatifs à la forme et aux dimensions des Murs de revêtement... imprimé Pour l'usage de l'École polytechnique et de celle des Ponts et chaussées.* The sequence is proved by the date on the title page of the first paper and the contemporary sticker on Basel University's copy of the second treatise scanned by Google Books.

300. Ceux qui n'ont en vue que la pratique de la construction, pourront se dispenser de lire ce qui suit, et passer immédiatement à l'art. 34, où se trouvent l'exposition et l'usage d'une méthode graphique pour résoudre, sans calcul, les principaux problèmes relatifs à la forme et aux dimensions des murs de revêtement.

301. Cette évaluation n'est cependant que pour les ouvriers qui n'ont aucune connaissance théorique de géométrie et de calcul : les ingénieurs verront sur-le-champ que les nombres dont il s'agit sont les tangentes tabulaires des angles cotés sur l'arc TK.

how to use it to supplement the analytical formulas to determine the dimensions), and on p. 38 he comments how engineers will be delighted to see how laborers like fitters or carpenters will be enabled to use his diagram.

Just a few years later, Coriolis, Bélanger, and Mary would be among Prony's students. They would be the ones to integrate Prony's concern for *vulgarisation* into the ECP's innovative curriculum. Given Prony's preference, and following his own inclination, Bélanger chose the analytical approach to *vulgarisation* for his course and, unlike Olivier who based his intentions on Hachette and Monge, he regarded the graphical idea as an addendum, an incentive to 'see' structure through the medium of drawing.

It was only in the mid-nineteenth century that Carl Culmann at the Eidgenössische Polytechnikum began to coordinate all the geometrical ideas put forth by the French theoreticians in graphical form and develop them into a new field: graphic statics. His intention was not Prony's *vulgarisation* for laborers or a supplementary graphical version of theory for students, but harking back to Monge's original goal, Culmann primarily aimed at a graphical method that would *replace* algebraic analysis and simplify calculation and dimensioning.

Culmann had attended a *collège* in Alsatian Wissembourg from 1835 to 1836. He then prepared privately for the EP *concours* for a year under his uncle Friedrich Jakob Culmann, who taught at Metz. There he must have encountered Poncelet's thinking on projective geometry, although Poncelet himself had recently left, first for the Académie des Sciences and then the Sorbonne in Paris. Culmann fell ill and only began his formal engineering education a few years later at the Karlsruhe Polytechnikum, a school that still taught under the influence of Tulla's fascination with Monge's visual combination of engineering theory and practice (Maurer 1998, pp. 9/14–22). Culmann had thus encountered both Monge's and Poncelet's views on visual literacy and geometrical methods. This colored his understanding of the truss bridges he documented on a study trip to Britain and the United States (1849–50), and he began to explore graphical methods for calculating their member forces. Culmann may or may not have encountered Squire Whipple's treatise of 1847 or Herman Haupt's of 1851. Both were contemporary American books on truss bridges that contained primitive attempts at calculating bridge members and some simple graphical diagrams,[302]

302. Haupt had studied at West Point, but after Crozet had left, and Whipple at Union College before Gillespie initiated his engineering course.

but they cannot have given him much to work with. He was probably not aware of Prony's brochure or, if he was, he didn't consider it comprehensive, and it was Poncelet whom he credited with a first coherent graphical system. Poncelet had expanded Monge, Hachette, Girard, and Prony's work into a course that he taught military engineers in Metz, and although Culmann wrote about Poncelet that: 'das Zeichnen ist seine Sprache' (drawing is his language), he also stated:

> Poncelet's solutions were always merely translations of previously developed analytical expressions. (Culmann 1866, p. v)[303]

This, of course, was true of all the previous French developments except for Monge's original descriptive geometry, which had predated his algebraic treatment of the subject.[304] Later, as the founding chair of architecture and engineering at the Eidgenössiche Polytechnikum in Zurich, Culmann developed his ideas into a unified system of graphic statics in lectures from 1860 onward and formulated it in *Die graphische Statik* (1866). This treatise counts as the origin of the field, and in fact it really did fundamentally reorient the geometrical approach to structural understanding.

Influenced by Culmann's work, the geometrician Luigi Cremona in Bologna developed the field further. In *Le figure reciproche nella statica grafica* (1872) he traced graphical approaches to engineering analysis back to Fleeming Jenkin, his former classmate James Clerk Maxwell, an unknown draftsman named Taylor, and Rankine, neglecting the earlier French contributions. His second treatise, *Elementi di calcolo grafico: ad uso degli Istituti Tecnici* appeared two years later, in 1874. Cremona's work made it easy to vary forms and dimensions.

In France, de Dion had independently adopted a strikingly similar approach to Culmann's when he developed his visually based structures course for architects in Trélat's school around 1865. Because of his ECP education, especially Olivier's course and the general influence of the previous French development, de Dion was predisposed to use graphic means of calculation, and from the early 1870s he used what he had created for Trélat's students

303. Die Lösungen Poncelet's waren immer nur Uebersetzungen vorher entwickelter analytischer Ausdrücke.

304. The reason for this sequence had been the interdiction to publish the method that the Academy at Mézières had laid on Monge. He circumvented it by creating an algebraic version.

to develop an innovative approach to the calculation of curved structural members of variable dimensions and forms. It was based on deflection and replaced lengthy algebraic calculations. He presented his method to the Société des ingénieurs civils in 1875 (*Mémoires et compte rendu* 1875.1, pp. 523–526). Together with Seyrig he subsequently used his approach with success in the structural design of the Douro Bridge in 1876 and then in the calculation of the fixed frames for the 1878 Paris Exhibition for which he is best known. Unfortunately, de Dion died early, in 1878 at the very young age of forty-nine and before he considered his method ready for publication. If he had lived, his name might very well have ranked alongside Culmann and Cremona as one of the pioneers of graphic statics.[305]

Maurice Lévy had followed Culmann's lead when he published a compendium, *La Statique Graphique et ses Applications à des Constructions* in 1874 that incorporated Culmann's work and Cremona's first treatise, (erroneously dating it as 1870: p. ix), but evidently without knowledge of the method that de Dion was working on at the time. The bilingual Alsatian Lévy had apparently first envisaged simply translating Culmann's book into French, but he then opted for a comprehensive treatment of all aspects of the topic. He not only read German, but also English and Italian, and he quoted Culmann's, Maxwell's, Cremona's, and other authors' relevant texts in the original and provided the first historical overview of the development of the field in French (1874, pp. v-xi). His account included only a skimpy mention of prior French contributions (Lévy 1874, p. x).

Lévy introduced graphic statics in his course *mécanique générale* at the ECP from 1875 (Guillet 1929, p. 107), and published an expanded, four-volume second edition of his work (1886–88, accompanied by four plate volumes). This second edition spread the knowledge of graphic statics from the ECP to the French-speaking world,[306] just as Wilhelm Ritter's four-volume *Anwendungen der graphischen Statik nach Prof. C. Culmann bearbeitet* (1888–1906) spread it from the Eidgenössische Polytechnikum to all German-speaking engineers (Kurrer 2002, p. 478). Through Lévy and Ritter's amplification and *vulgarisation*, graphic statics

305. See Molinos and Seyrig in *Mémoires et compte rendu* 1879.1, pp. 427–433, 434, 444, 452.

306. Lévy was not alone in doing this. Terrier's classmate Paul Planat (grad. 1862), who had graduated well before Lévy began teaching, popularized graphical methods of calculation to architects in the advice section of his journal *La Construction Moderne* (personal communication from Santiago Huerta, 10.10.2020). Planat later served as director of the École Spéciale d'Architecture.

became the method of choice for the calculation and dimensioning of iron structure internationally, and it influenced design throughout the period of iron construction's most dynamic development in the last quarter of the nineteenth century.[307] It was only later that graphical methods of calculation and design gradually waned in importance as algebraic and transcendental methods of analysis advanced.

Engineers and aesthetics: visual competence

Apart from the role that the school's concentration on drawing played in the incorporation of visual thinking in observation, design, and analysis, the drawing course and the examples that the students recorded in sketches and drawings also helped establish their self-assurance in understanding iron structure, whether of machinery or of buildings. The competence it gave them formed their point of view that such structures were elegant in their own right, an opinion that Leblanc had already asserted in 1830 (unnumbered p. 2 of his introduction). This awareness, supported by the design method that Mary developed (see later under Part II), quickly spread into the French building world, and from there throughout Western culture. It would eventually influence the technologically based aesthetic propagated by the Modern Movement a century later.

Barrault proudly wrote in 1857 about his role as engineer to the contractor York, for which he completely redesigned the internal structure of the 1855 Palais de l'Industrie. He insisted on using visible iron structure rather than decorating it with historical forms or masking it with wood or plaster as the architect had envisaged:

> I adopted the opposite approach out of necessity and also out of principle: I did not want to try to reproduce a form of construction in metal that was conceived for stone, and also I did not want to use bad taste to hide the material that I used.
>
> In a word, I almost always showed the naked cast and wrought iron that I used following the laws imposed by science and the nature of materials, fully aware of the fact that I could not obtain the [visual] effect achieved through the use of ordinary construction [i.e., masonry], but considered that the simplicity of the forms and the grandeur of the interior would in themselves prove sufficient to

307. Several other treatises as well as a few translations in Austria, Germany, Italy, France, England, and the US contributed as well—among them Terrier's translation of Favaro—but these were the main sources through which graphic statics became known and used.

> endow it with a special character that conformed to its purpose, and consequently acceptable to the public and even to right-thinking artists. (Barrault and Bridel 1857, p. 2, cited in abbreviation by J.-F. Belhoste 2004, p. 74)[308]

The masonry exterior remained the architect Jean-Marie Victor Viel's, but the building's structure and the interior aesthetic were Barrault's. He had worked with Flachat, Petiet, and Jullien on several railroads, built multiple ironworks, and had many years of experience with iron production before he built the exhibition hall. Together with Flachat and Petiet he had published the *Traité de la fabrication de la fonte et du fer* (1842–46), and he was totally comfortable with the material in all its forms and conversant with all aspects of its production and use.

The separation between an internal visible structure in iron and an outer form in stone was new at the time. It mirrored what was then occurring in the design of major railway stations throughout Europe, where highly decorated masonry facades designed by academic architects fronted large, simple, and glazed iron sheds designed by engineers. Many of these engineers, on the continent, at least, were ECP graduates.

Barrault's rationale for the changes he made were all financial and deadline-driven, but his explanation for them was aesthetical. This was novel, and may have been the first time that an engineer so magisterially enunciated an aesthetical opinion, a judgment that had formerly been the exclusive province of artistically trained architects. No engineering graduate of any of the other schools would have dared to make such a forceful statement and with such aplomb. This self-confident aesthetical argument that Barrault made in explanation of his structure reappeared three decades later in Eiffel's well-known rebuttal to the protest of the artists against his tower.

Barrault's statements were strong, but in fact there were a few forms that were masked by decoration. Barrault used several types of cast-iron and wrought-iron

---

308. Par nécessité et au moins autant par principe, nous avons adopté un système opposé: nous n'avons pas voulu tenter de reproduire avec métal un mode de construction qui est fait pour la pierre et nous n'avons voulu non plus dissimuler par les artifices de mauvais goût la matière que nous avions à mettre en œuvre.

En un mot, nous avons presque partout montré à nu la fonte et le fer, employés suivant les lois imposées par la science et la nature des matériaux, sachant très-bien que nous n'arriverions pas aux effets obtenus dans les constructions ordinaires, mais en pensant que la simplicité des formes et la grandeur intérieure de l'édifice suffiraient pour lui donner un caractère particulier portant le cachet de sa destination, et par conséquent acceptable par le public et même par les artistes d'un sens droit.

trusses in the building. The visible cast-iron trusses had crossed diagonal bracing in the panels on the model of the 1851 Crystal Palace, but the built-up ones that were made of wrought iron were of two types: a plate girder and a type today known as a Vierendeel truss without diagonal panel stiffeners. The latter, the *entretoises B*, carried the roof above the upper story of the main space and were covered by cast-iron tracery. This may have been an aesthetic decision imposed by the architect rather than one made by the engineer. Barrault does not mention it in his treatise (Barrault and Bridel, 1857).

We may presume that Eugène Flachat, the railway builder with many railway sheds to his name and the first employer of many of the early ECP graduates, contributed to the self-confident stance on the aesthetics of structure. After Barrault died in 1865, Flachat wrote about his work:

> Using iron in construction permits no useless gesture: one must find any decorative aspect it requires in the lines of resistance, in the connections where these lines intersect. Nothing should appear contrived, and nevertheless each line reveals its purpose; structural logic must be attained without any element complicating the lines or distracting the eye.
>
> On viewing a structure, one instinctively looks for its significance, the purpose of all its elements. Anything that only serves as decoration is a distraction. Ornamentation, particularly if it constitutes an essential element of the structure, satisfies the eye and the mind.
>
> Iron is inappropriate for the small-scale work usual in masonry. It demands large bays and spans to be meaningful. The age of iron in building construction fulfills itself in a significant increase in size and height.
>
> Wherever the use of metal is limited to the scale of masonry work or timber framing in footprint or span it is overused or redundant. There is no reason for it.
>
> The interior of the Palais de l'Industrie responds to these demands with a grand harmony in its simple and generous lines, to which the eye responds with satisfaction. Ornamentation results from the natural forms of metal under the loadbearing conditions it is subjected to. The feeling of the scale of the forms is guided by a restrained taste and by the play of forces.
>
> (Flachat 1866, p. 537–538, the last sentence in part cit. in J.-F. Belhoste 2004, p. 74)[309]

---

309. L'emploi du fer dans la construction ne comporte rien d'inutile; il faut trouver le décor qu'exige l'art dans les lignes de résistance, dans les liens, dans les assemblages de rencontre de ces lignes.

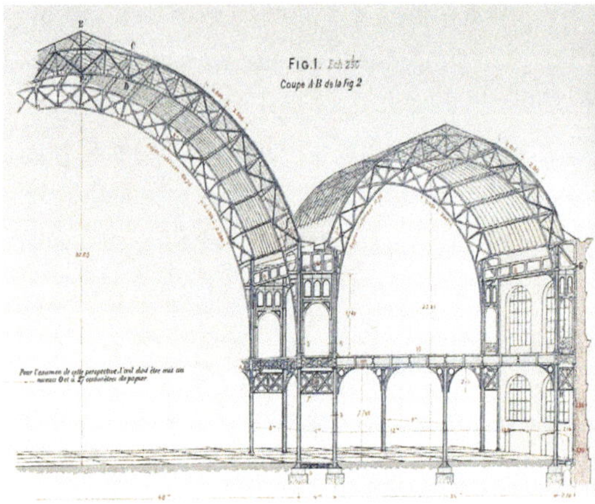

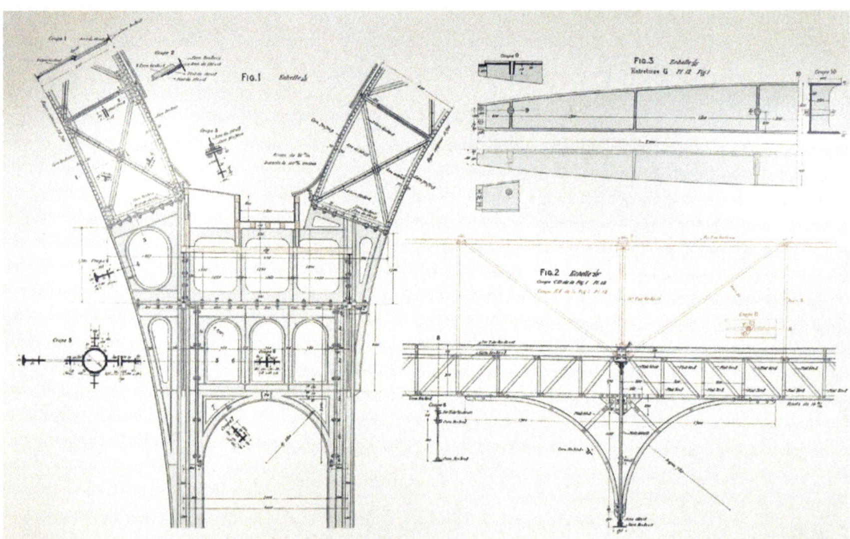

Fig. 49. The iron structure of the exhibition hall for the 1855 Paris Exhibition. While the architect dealt with the exterior, the façades, and the decoration, Alexis Barrault (grad. 1836) as designer and his assistant Gustave Bridel (grad. 1847) as site manager dealt with the visible internal structure for the entrepreneur. Barrault was particularly articulate about the aesthetic impact of his structure. No engineer had ever dared to voice such a clear aesthetic opinion before. (Source: Vierendeel: *La Construction Architecturale en fonte, fer et acier* 1897–1902, pls. 12 and 13) Photograph of the interior of the Palais des Machines in 1855. (Source: antique photograph, 1855)

Fig. 3. — Vue prise pendant la démolition du Palais de l'Industrie.

Fig. 50. Two views of the demolition of the building in 1897. (Source: photograph, 1897, Émile Gaillard [1877–1952], in: Picard, *Rapport général*, 1902, vol. 1, p. 324; and wood engraving by Louis Poyet, *La Nature*, 1897, p. 185).

This awareness, common only among ECP engineers, and certainly shared by Flachat himself, was forward-looking. Barrault's behavioral interpretation of structural form as the visualization of load paths contrasts with the purely geometrical considerations that the ENPC-educated Nordling advanced in his attempt at explaining the aesthetics of his trussed iron bridge piers in 1864 (Nordling 1864, pp. 12–16). This perspective would also be reflected in Jenney's attitude toward the engineering and architectural use of iron and the attitudes of the employees he mentored in Chicago. It explained Eiffel's self-confident pride in his tower, and it promoted a subtle shift in French architectural design as well, evident in the work of Henri Labrouste.

Il faut que rien ne sente l'effort et que, cependant chaque ligne révèle sa nécessité d'être; il faut aussi que l'homogénéité de la construction soit acquise sans qu'aucune pièce ne vienne compliquer les lignes et distraire le regard.

Instinctivement, chacun en regardant une construction, cherche à la fois la signification, la raison d'être de tous ses éléments. Ce qui ne sert qu'à l'ornement déplaît. Ce qui orne, surtout en constituant un élément essentiel de la construction, satisfait l'œil et l'intelligence.

Le fer ne s'arrange pas des petites dimensions que la pierre à créées. Il veut, pour revêtir des formes palpables nourries et significatives, de grandes baies, de grandes portées. L'âge du fer dans les édifices se révèlera par un accroissement considérable de leurs dimensions en surface et en hauteur.

Partout où le métal est, comme point d'appui et comme portée, restreint aux mesures que la pierre et le bois ont faites, il est prodigué ou il est grêle. Il n'y a pas de raison d'emploi.

L'intérieur du Palais de l'Industrie répond à ces exigences de l'art; il y règne une grande harmonie de lignes simples et largement entendues, où l'œil se repose avec satisfaction. L'ornementation y est le résultat des formes naturelles du métal dans les conditions de résistance où il est placé. Le sentiment de la mesure y est, pour les formes, guidé par un goût sévère et par l'entente des forces.

Implementation: the architect Labrouste adopts visible iron structure

Labrouste's transition from the primarily decorative use of iron in the twin vaults of the Bibliothèque Sainte-Geneviève's reading room (1843–50) to an innovative use of structure expressed as form in the stacks and the nine domes of the old Bibliothèque nationale de France's (BnF) study hall (1862–68) bears scrutiny. Architectural historians justly consider Labrouste a major figure in the development of modern architecture, and this judgment is historically tied to the design of the iron parts of these two buildings in Paris.[310] As Jolly wrote in 1863, the use of iron in construction developed from an experimental into an industrial material between 1850 and the early 1860s, and Labrouste's two library designs document this transition in exemplary fashion.

The Crystal Palace was designed in 1850, and the Bibliothèque Sainte-Geneviève was finished the same year. Both were innovative in different ways. The former marked the introduction of industrialized methods in iron fabrication and erection, and the latter marked the first quasi-structural visual use of iron in official architecture. Before that time iron structures had not been designed by academic architects and were therefore not considered to be works of architecture.

The Crystal Palace had been designed by the gardener Joseph Paxton and engineered by the railway engineer and contractor Charles Fox. The 18-meter rotunda inserted into the 1849 masonry London Coal Exchange was designed

---

310. This evaluation has since been revised and expanded, especially by Bergdoll in the chapter 'Historiography', in Bélier et al. 2012, pp. 25–41). However, here it is precisely this limited aspect that concerns us, not even the equally innovative technical aspects of gas and natural lighting, central heating, and ventilation.

Fig. 51. Interior of Henri Labrouste's reading room at the Bibliothèque Sainte-Geneviève (1850) showing the visible cast-iron arcs that are only part of the roof structure. (Source: Marie Lan Nguyen 05.04.2011 Wikimedia Commons - CC BY 2.0 FR) Interior of the study hall of the Bibliothèque nationale (1868) showing the structure of the roof in wrought iron. (Source: photo, Adelphilos, 2017, Wikipedia CC BY-SA 4.0)

Fig. 52. View into the wrought-iron stacks at the Bibliothèque nationale (BnF). (Source: wood engraving by Félix Poyet c. 1868)

and built by James Bunning, a Clerk of the City's Works.[311] The rotunda was a hybrid, and has been since considered architecture by historians, but it was really an object that had been inserted into a masonry building. It was decorated structure, but not in itself a building structure.

311. His title was subsequently upgraded to City Architect in 1847, but that did not reflect any formal training. Bunning had learned what he knew about building from his surveyor father and the surveyor and self-taught architect George Smith, and had worked as surveyor on Marc Brunel's Thames Tunnel around 1840. A large part of his professional career was devoted to infrastructure projects. See obituary in *Gentleman's Magazine* 1863, p. 799. Bunning later also designed a cantilevered widening of London Bridge in iron (p. 800).

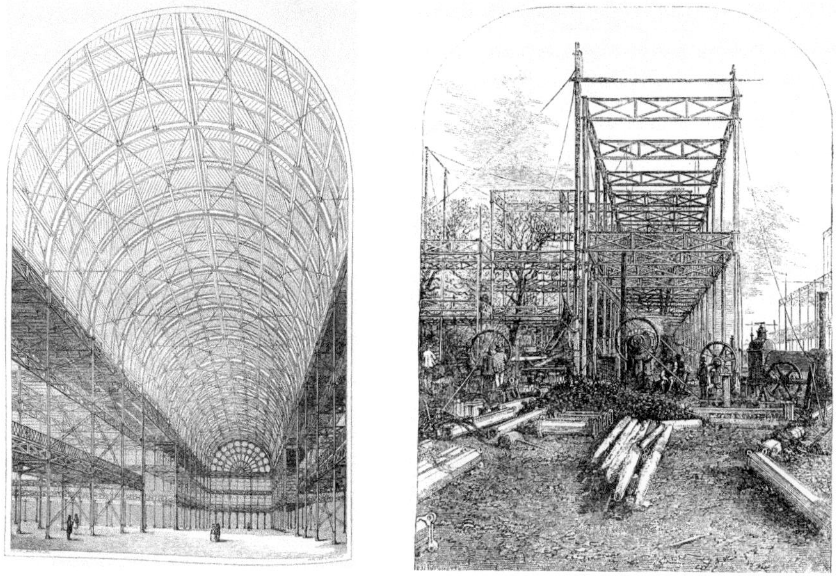

Fig. 53. Interior of the Crystal Palace for the London Exhibition of 1851. View in the transept looking towards the nave and the entrance showing the cast-iron and wood structure, and a view of the erection process. (Source: *Official Descriptive Catalogue 1851* v. 1, pp. 57/70)

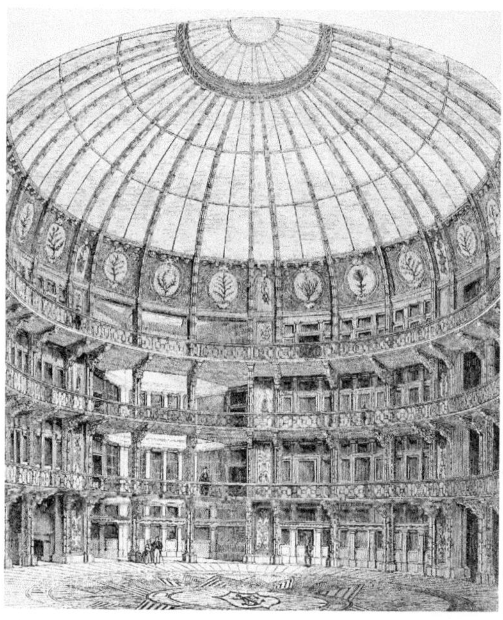

Fig. 54. Interior of the hall in Bunning's London Coal Exchange (1849). The building's structure was masonry, and this cast-iron construction was inserted into the building like an oversized piece of furniture. (Source: *The Builder,* vol. 7, 1849, p. 462, Sept. 29)

Fig. 55. A very early sketch by Henri Labrouste for the roof structure over the reading room at the Bibliothèque Sainte-Geneviève separated the vaults and the roof truss above. (Source: © Labrouste Archive, Bibliothèque Sainte-Geneviève, bsg MS 4273_0100)

The same can be said with reservations of Labrouste's first reading room, although its aesthetic influence was far greater. The significance of Labrouste's use of iron in this building lay in the fact that an academic architect had featured iron structure visibly in a prominent public building. The result looked structural, and it was self-supporting, but what the user saw from below was not really the roof structure, at least not in its entirety. What carried the roof was a complicated hidden truss-like configuration that was similar to those then common in French theaters. Labrouste's goal was clearly to present iron as an architectural material, but he had not yet achieved a congruence between structure and form.

Judging from the original sketch, the subsequent design drawings, and the earliest official published version of the cross section, Labrouste wanted to design a hybrid structure. But all versions are confusing. In what appears to be the latest drawing, the one in which the decoration of the arcs corresponds to the arcs as they were finally built, Labrouste fixed four of the verticals to what appear to be no more than connector straps that follow the curve of the two vaults. Two more posts rest directly on the apexes of the cast-iron arches. Point loads placed directly on brittle cast iron are not ideal.

Fig. 56. In Labrouste's many design versions, the cross section shows self-supporting cast-iron arcs that also helped carry the wrought-iron roof structure. Labrouste simplified the structure step by step, and the last two versions do look relatively simple, but the way the construction carries its loads remains unclear. The series shows how the architect was struggling to reconcile the vaulted form with a truss structure, and that he didn't quite know how to make it work. (Source: © Labrouste Archive, Bibliothèque Sainte-Geneviève, bsg MS 4273_0022r, bsg MS 4273_0023r, bsg MS 4273_0021r, bsg MS 4273_0024, bsg MS 4273_0020r)

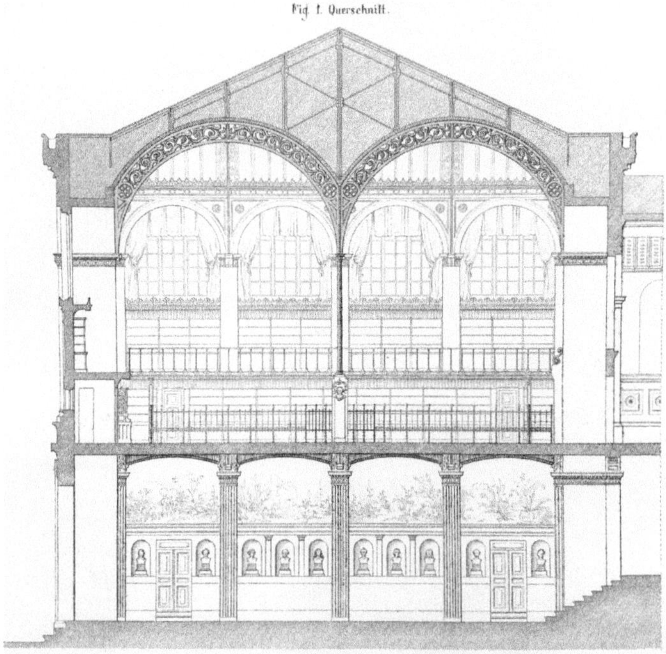

Fig. 57. The published as-built version of the roof structure over the reading room of the Bibliothèque Sainte-Geneviève is clearer than the many earlier designs, but the cross-section shows no connection details. This image was published in 1851 in the *Allgemeine Bauzeitung* Vienna, and is identical to that published by Gourlier in the official, government-issued publication of 1850. It is clearly stated to have been copied from an original drawing sent to the author of the article, so that the presumption is that Labrouste may have left the detailing to his contractor Roussel. (Source: *Allgemeine Bauzeitung*, vol. 16, 1851, pl. 386)

The as-built version of the design on the government-approved plans is slightly more logically organized. The influence of concentrated point-loads on the arches is mitigated, an improvement for which Labrouste's contractor Roussel may have been responsible. Here the cast-iron arches still appear to carry the load of six of the nine vertical posts, but via more substantial curved lower chords. In contrast to the first proposal, however, the final cross section fails to show any connection between the lower curved chord and the vertical posts on the image published by Gourlier, who had always shown himself to be interested in construction. The connections were also missing on the plan that illustrates one of the first articles to appear on the library, in the authoritative *Allgemeine Bauzeitung* in Vienna in

1851. Labrouste had sent it to Albrecht Rosengarten, who had worked for him in 1839, and who wrote:

> Thanks to the kind communication of Mr. Labrouste, I am able to provide several drawings to make my description clear. (*Allgemeine Bauzeitung* vol. 16, 1851, pp. 66 and pl. 386)[312]

We must presume that any information on connections was indeed missing on the originals. Why was this so? Had Labrouste simply delegated the detailing to his contractor?

Both the latest drawing and the as-built image show two more posts standing directly on the masonry, one on either side of the hall, and one more standing axially on the etiolated central column. The role of the cast-iron arches in supporting the roof construction is structurally ambiguous and unfavorable in both versions. In fact, the arches seem puzzlingly unnecessary. A straightforward truss supported on the lateral walls could have simply been built over suspended vaults as had traditionally been done in so many earlier theater roofs from the Baroque period in timber, right through to the use of iron in the early nineteenth century. It is abundantly clear that Labrouste had insisted on integrating construction with visible form but had not really been able to figure out how to make it work logically.

Robin Middleton called the ten-meter vaults that dominated the interior space 'suspended ceilings' (Middleton 1999, p. 34). They are actually self-supporting, hollow tile vaults (*Allgemeine Bauzeitung* 1851, p. 67), shells that carry no further load. The iron arches seem to be only secondarily responsible for their stability, if at all. Whatever the structural relationship between the arches and shells may be, this strange configuration and the unresolved relationship between the visible iron and the hidden structure above shows that academically trained architects were simply not yet equipped with enough of an understanding of structure, technology, and industry to integrate structure and form successfully—except for ECP border-crossers like Barrault who were just beginning to enter the profession.

No historian before Middleton had ever questioned Labrouste's technological competence in using iron in the Bibliothèque Sainte-Geneviève. Although

312. Durch die gütige Mittheilung des Herrn Labrouste, bin ich imstande, um meine Beschreibung anschaulich zu machen, einige Zeichnungen hier beifügen zu können.

Labrouste's interest in construction and materials was keen, it is doubtful that any non-mechanically schooled person would have possessed the necessary expertise to carry out innovative iron structures on their own, much less give them architectural expression. So how had Labrouste gained as much knowledge as he did, and on whom did he depend?

Even before undertaking his first library project, Labrouste was not entirely unfamiliar with iron structure. His construction teacher at the École des Beaux-Arts, Jean-Baptiste Rondelet—a mason and architect—had published an influential compendium on materials, construction, and related machinery, including the role of iron reinforcement in masonry and a few iron roof structures in his multi-volume *Traité théorique et pratique de l'art de bâtir* 1802–17. Presumably he discussed some of this material with his students in class.

After graduating in 1824, and armed with this knowledge, Labrouste spent five years as Grand Prix de Rome winner in Rome, in part together with the winners of 1823, Félix Duban; 1825, Louis Duc; and 1826, Léon Vaudoyer. This *bande de quatre* of Romantic architects concentrated on a precise and controversial documentation of the antiquities—in Labrouste's case, it was Paestum. After returning home he worked as site architect (*inspecteur*) for his friend and former classmate Duban on the Palais des Études of the École des Beaux-Arts (1832–38). This building had simple iron roof trusses and iron-reinforced flooring (Middleton 1999, p. 38; Eck 1836, vol. 1, p. 71 and pl. 24)—all still hidden, however. They were unremarked and did not enter the architectural discussion in any way.

In the interim, Charles Eck had published a well-received treatise that provided more information on the use of iron as structure than had been previously available. Charles Rohault de Fleury, a contemporary of Labrouste's who had studied at the EP and then gone on to become a Beaux-Arts architect, had built several hothouses in the Jardin des Plantes (1834–36), two of which still stand. These were the first iron and glass structures by an academic architect in France. Rohault and Labrouste had studied together under Louis-Hippolyte Lebas in the Vaudoyer-Lebas studio, and Rohault had also studied under his own father Hubert, commissioner of roads and honorary member of the *Conseil des bâtiments civils*, who had himself studied architecture at the EP under Durand. Rohault had left school a year before Labrouste, but while they were both there, Labrouste may well have discussed the use of iron with his more broadly educated engineer colleague. He was certainly later aware of Rohault's hothouses that had caused a sensation, and Eck had published them in 1841[313] (vol. 2, 1841, pp. 72–74 and pls. 49–50). Nevertheless, all the iron bridges and railway sheds that ECP engineers were erecting at the time (as well as the factories and exhibition buildings by others) were still largely disregarded by architects, despite the fact that Eck had begun to include a few of them in 1841. They were what we now call facilities, not works of architecture. What caused the change in how architects viewed these iron structures?

Eck's treatise

Eck was an inspector of public buildings in Paris and an architect and designer of building machinery. He played an important role in the development of construction awareness through his two-volume *Traité de Construction en Poteries et Fer* in 1836 and 1841. The treatise included multiple plates on the use of hollow tile in construction, in part combined with iron floor and roof structures. His stated intention was to supplement Rondelet's lack of detailed information in this area (Eck 1836, vol. 1, unnumbered p. 1 of the introduction). The first volume had appeared too early for him to have been aware of the new breed of industrial engineer that the ECP was beginning to graduate.

The two volumes are historically as important as Rondelet's better-known treatise, but with a different focus. They marked a shift in the understanding of construction and a documentation of the role of prefabricated materials and structures. Eck's initial goal was to present hollow-tile construction and to include iron as a means to this end as a correlate. However, he went further than that by documenting the last phase of pre-industrial iron construction, including roof structures—among them Duban's Palais des Études that Labrouste had worked on. His treatise—which is overall fairly reliable[314]—describes the first tentative steps of iron construction's transition to its modern industrialized incarnation.

We know nothing about Eck's education. He was an *élève de M. Lenormand* (Guyot de Fère 1832, p. 223). This may have been Louis François Lenormand, the architect Louis Lenormand's father, as the famous son was Eck's contemporary.

---

313. Eck's illustrations are based on the plates that Rohault had previously issued in 1837.

314. Some of the information published by Eck (the roof of the Théâtre du Palais-Royale) has recently come under critical scrutiny by Katrina Chalvatzi 2020.

In the first volume of his treatise, Eck repeatedly referred to himself as *architecte*, and in the second five years later he signed himself *architecte, ingénieur civil* (vol. 2, 1841, unnumbered p. 3). He cannot be traced among the graduates of any engineering school, but he did collaborate with Pierre-Dominique Bazaine on research projects during the last years of Bazaine's life.[315] Eck was therefore certainly knowledgeable about construction and materials.

He lauded the use of clay products in construction because, as he wrote, the material is universally available and its products can be manufactured on site (vol. 1, 1836, p. 1). In this he remained, like Rondelet, conceptually within the tradition of pre-industrial on-site *construction* rather than industrialized site *assembly* of prefabricated parts. In a slightly modified form, this principle also held true for the iron structures that he documented. All of them were constructed of small series of uniquely designed, wrought-iron components and forged to order by hand or by water-powered trip hammer in small, local *ferronneries* (ironworks). These components were of course prefabricated and only assembled on site. Their production was serialized but not industrialized in our sense because they were manually produced one by one for unique structures. They were never as precise as later industrially manufactured units could be, and they frequently required adjustment on site. Components like these and their resultant systems were a form of cottage industry rather than an industrial production as we now understand it.[316]

Fire resistance had originally motivated Eck to include iron construction (vol. 1, 1836, pp. 34–38), as it later also inspired William Fairbairn to publish his *Application of Cast and Wrought Iron to Building Purposes* (1854). However, Eck discussed not only incombustibility, but also the structural behavior of the iron and he featured hollow-tile floors that Amédée Gréban was then experimenting with at the army's *manutention des vivres de la guerre* (commissariat, 1835–36).[317] He also documented composite iron and tile *poitrails* (lintel beams) and *murs de*

---

315. vol. 2, 1841 *Mémoire sur la construction de nouveaux planchers*, unnumbered foreword and p. 7 footnote 1; and *Expériences faites par M. Ch. Eck, sur la résistance que présente le bois de sapin*, pp. 29–32.

316. Compare Charles Babbage's distinction between making and manufacturing (1833, pp. 119–122).

317. Eck, who was the government's inspector for the 12th arrondissement of Paris, was not responsible for that part of the city in which Gréban's building was being erected. The rue de la Manutention where it stood lies between the present Avenues de New-York and Président-Wilson, a location that was then in the 1st arrondissement (of twelve, now the sixteenth of twenty).

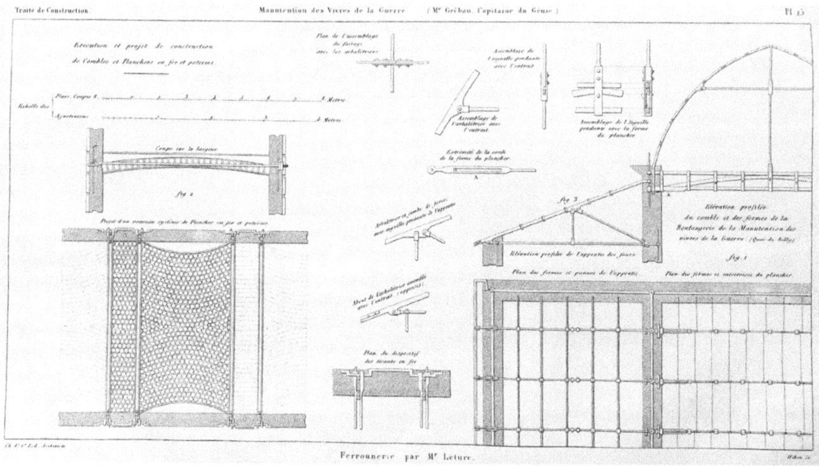

Fig. 58. An early proto-composite structure that Eck reported was a hollow-tile, plaster, and iron-rod floor, designed and built by Amédée Gréban as an experimental incombustible floor structure for the army's commissariat (*manutention des vivres de la guerre*, 1835–36). Although it was not a composite construction as we understand it today, Eck observed that the materials worked together structurally—possibly one of the first indications of the concept of composite behavior. (Source: Eck, *Traité de Construction en Poteries et Fer*, vol. 1, 1836, pl. 15)

refend (load bearing cross-walls). Where Rondelet had famously described the use of iron as *reinforcement* in masonry, Eck discussed the *composite behavior* of the two materials together and he warmly recommended the idea. This makes him a precursor to the conceptualization of composite structures or materials like reinforced concrete. He ended the main part of the first volume by discussing iron roof structures for palace buildings and theaters, with brief addenda on hothouses, suspension bridges, and iron fire curtains for theaters.

The inventiveness of the many solutions that Eck documented was inspiringly fascinating, but it stood in contrast to the industrialized standardization that would soon enable the success of structural iron. His treatise presents a broad panorama of innovative solutions, chiefly in wrought iron, but they were all missing the critical connection to the iron industry. This connection—which already functioned well in railway construction without ever being especially highlighted—was what the graduates of the ECP would soon apply to bridge and building construction, thanks to Perdonnet.

Professionals who had studied Eck's treatise were up to date in the use of iron in construction, especially when the second volume appeared in 1841. This volume expanded his purview beyond the examples by architects and contractors, and documented some of the facilities that French engineers were beginning to build at the time: market halls, railways, stations, storage facilities, factories, canal locks, caissons, and bridges, as well as roofs in Russia.[318] In some cases these began to show the tentative influence of industrially produced solutions that included a few minor rolled sections. They were not yet standardized, but they revealed a nascent awareness of the future potential of the iron industry in construction.

Eck's treatise was more general and useful to practitioners in this regard than William Fairbairn's slightly later *On the Application of Cast and Wrought Iron to Building Purposes* of 1854. Its purpose was also different, as Fairbairn's book concentrated on demonstrating the incombustibility of buildings in iron, focusing more on material behavior and detailing. However, Eck presented pre-industrial structures made of manually produced, forged iron, whereas Fairbairn—who published during the transition to industrial construction—included a section on industrially produced rolled iron (1854 Part II, pp. 73–113).

The 'entrepreneurs de serrurerie' and the mediating role of the Conseil des bâtiments civils

Labrouste was vitally interested in materials and construction—in fact, it was among his primary architectural foci, but he lacked the advantage of an additional EP education. Although he was not trained as an engineer, he was certainly familiar with his former classmate Rohault's structures and surely with Eck's treatise, too, especially since it discussed Duban's École des Beaux-Arts building, with which he had been involved. However, he still needed the help of a contractor who was accomplished in the experimental use of iron. *Serruriers* or ironmasters existed in Paris and formed an interconnected community.

Like the masons, who constituted a proto-democratic professional hierarchy

---

318. This information was probably communicated to him by or through Bazaine (Eck 1841, vol. 2, *Mémoire sur la construction de nouveaux planchers*, p. 10), although he mentions a savant étranger (foreign scholar) who gave him the information (vol. 2, 1841, main text, p. 108). Eck's choice of Russian structures raises questions, as none of the pioneering examples by Betancourt are among them like the 1819 triple iron dome of St. Isaac's cathedral in St. Petersburg that was nearing completion in 1841. (The architect Auguste de Montferrand collaborated with Betancourt and Charles Baird on this project) He does, however, show Carlo Rossi and Matthew Clark's complex theater roof for the Alexandrinsky Theater in St. Petersburg of 1833.

with its own legal and social status, the ironmasters seem to have formed a similar body that occupied its own social space within French society. As Robert Carvais writes, albeit about masons in the late eighteenth century:

> …despite certain misleading appearances, the strength of the building world lay in its ability to remain united at all times.… This true unity enables those involved to innovate in a delicate area: the representation of the profession through democratic elections.… This also demonstrates an exemplary political maturity which enabled the building field to pass without too much turmoil from trade to profession, and from guild to the freedom of enterprise at the dawn of the nineteenth century (Carvais 2021, p. 282).

Labrouste was thus surrounded by a cohesive group of inventive professionals[319] who did have the necessary experience, and he drew on the expertise of two of the most innovative among them: the *entreprise* Roussel for the 1850 building, and the Leturc firm for that of 1868.

The *entrepreneurs de serrurerie* (ironmaster-contractors) François and Théodore Roussel,[320] who built the roof and the floor structures for Labrouste's first reading room, and the iron founder Christophe-François Calla, who cast the arches (Roussel archive), certainly possessed the knowledge needed for calculating, detailing, casting, and erecting iron. Calla was primarily a founder of decorative art objects and street furniture, but he had also built complex automated flour mills on Oliver Evans's model (Eck 1841, vol. 2, p. 76 footnote), and his father had previously produced the hand-cranked screw propeller for Robert Fulton's submarine *Nautilus* as well. The Roussel firm was even more experienced in structural work. It had developed a flooring system and several truss roofs (Eck 1836, vol. 1, p. 51), and the Roussel brothers' father Jean-François had previously built the dome over the Halle aux Blés in 1811 (Eck 1836, vol. 1, p. 71 and pl. 36), as well as several very early iron and hollow-tile floors 1804–20 for the Assemblée nationale.

The Roussel dynasty was among the most innovative and experienced of these

---

319. Among these I have identified and named the Roussel dynasty (Jean-François, François, Théodore, and Henry), the Joly family (Pierre-François and Théophile Joly, and César Jolly) and the Mignon dynasty (Sylvain, Anselme, Charles, and Jean-Baptiste), as well as Achille and Paul Leturc, François Rigolet, and Jacques Travers. Eck 1841, vol. 2, p. 32 also names Fauconnier, Chavier, Albouy, Didier, and Fleuret as important names.

320. Originally traced via Geneanet.

early *entrepreneurs de serrurerie.* One can follow the full development of iron construction from the early experiments to the fully industrialized form in this firm's development alone.

It all began with Jean-François Roussel (1772–1821), who came from a family of

Fig. 59. The ironmaster-contractor (*entrepreneur de serrurerie*) Jean-François Roussel (1772–1821) was the founder of one of France's most innovative building dynasties. From the outset with the construction of the iron dome over the courtyard of the Halle aux Blés (1809–11) in the first years of the nineteenth century, his firm helped set the stage for the development of iron construction. (Source: Roussel family archive)

Fig. 60. The cast iron dome over the Halle aux Blés (1809–11). The architect was François Bélanger, the engineer François Brunet, and the contractor Jean-François Roussel. (Source: lithograph by Nicolas-Marie Joseph Chapuy 1838, [Musée Carnavalet] Wikimedia Commons). The demolition of the outer ring of the Halle aux Blés in 1887 exposed the dome. (Source: anonymous photograph 1887 - Wikiwand)

locksmiths, some of whom may have helped him found his business at the end of the eighteenth century. He was so successful, especially with the dome of the Halle aux Blés (1809–11), that he was accorded the title of *serrurier du roi* (royal locksmith) in 1816, which evidently conferred special privileges within the contractors' association.

He was joined and then followed by his son François (1799–1854). As François devoted most of his time to his military career as *officier de la garde nationale*, president of the Syndicale des Entreprises de Serrurerie (Locksmiths' Association), and judge at the Commercial Court, he took his younger brother, the general contractor Théodore (1801–73), into the firm as partner. Théodore may therefore have been responsible for many of the firm's structures in this period.
Théodore stayed on with the firm after his older brother's death in 1854 for

Fig. 61. The brothers François (1799–1854) and Théodore (1801–73) Roussel developed the firm further and built many of the iconic early iron structures in France, like the Bibliothèque Sainte-Geneviève. (Source: Roussel family archive)

several years to guarantee a smooth transition when François's son Henry (1826–97) followed as principal. The firm's projects in the period we are examining are thus not attributable to a single person.

Henry expanded and industrialized the firm as an *entreprise de charpentes métalliques*, a name change that reflects the industrialization of iron construction after the 1850s, when manually produced wrought iron began to be replaced by industrially produced rolled iron sections. He built part of Victor Baltard's Halles de Paris (1857–74; Joly built another part) and Jules de Mérindol's great slaughterhouse, the Halle aux Bœufs (now Grande Halle) of the Marché de la Villette (1865–67).

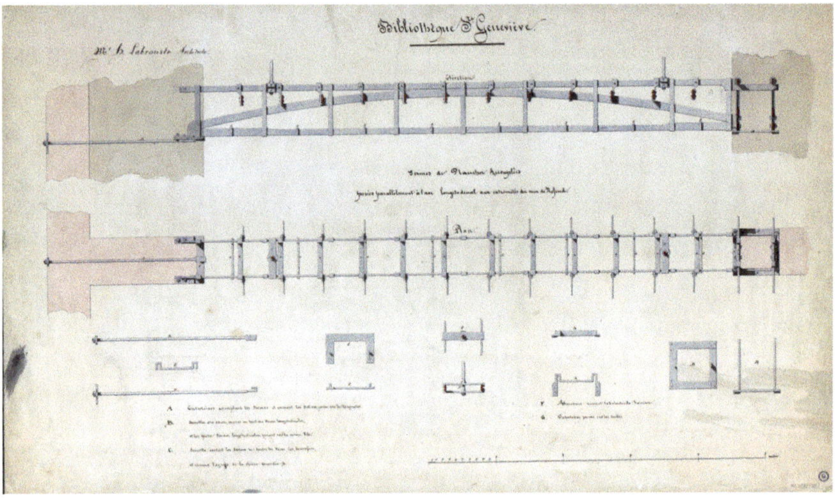

Fig. 62. An original drawing of the iron floor trusses of the Bibliothèque Sainte-Geneviève by Labrouste, probably from drawings by Roussel. (Source: © Labrouste Archive, Bibliothèque Sainte-Geneviève, bsg MS 4273_0019)

Henry headed the firm until 1880, when he sold it to the ECP graduate Jules Roussel (grad. 1869), who was not a relative (Roussel 2015 and private communication). Over the course of almost the entire nineteenth century, the four Roussels provided proposals and iron construction for many of France's most prominent structures.[321]

The consistent leadership from one generation to the next guaranteed stability in the development of the firm and a constant accretion of expertise. In the late 1820s or early 1830s, the firm built a double wrought-iron truss of 6.40 meters span on cast-iron columns into Théodore's house. This truss supported a 20-meter-high load

321. Aside from the iconic Halle aux Blés dome 1809–11 and the floor structures in the Assemblée nationale (1804–20), among the most important structures that the Roussel firm erected were Pierre-François Fontaine's planned modifications to the Palais des Tuileries (Eck 1836, vol. 1, pls. 9 & 10); the roof trusses for Jacques-Ignace Hittorff and Jean-François Lecointe's second Théâtre Favart (1823–25; Eck 1836, vol. 1, p. 66 and pls. 46 and 51) and for Lecointe's Théâtre de l'Ambigu-Comique (1826–28; vol. 1 pls. 23 and 28); the 17-meter dome of the chapel in Hippolyte Lebas's prison de la petite Roquette (1826–36; vol. 1, pp. 52–53, pl. 35); the roof proposal for Delespine's Marché des Blancs-Manteaux (Eck 1841, vol. 2, p. 41–42, pl. 32); Rohault de Fleury's long curved hothouse in the Jardin des Plantes 1834–36 (vol. 2, pp. 72–75 and pls. 49–50); two unsuccessful proposals for rebuilding the roof of Chartres Cathedral in the mid-1830s (vol. 2, pp. 32–35 and pls. 21–23); the vertical chains for Hittorff's Panorama des Champs-Élysées (1839, vol. 2, p. 62, pl. 47); the iconic Bibliothèque Sainte-Geneviève; parts of the structures for the 1878 Paris Exhibition; and the Halles de Paris and the Halle aux Bœufs.

Fig. 63. Henry Roussel (1826–97) represented the last generation of the innovative dynasty. He developed the firm into an industry, the Entreprise Roussel, and continued the importance of the firm, building part of les Halles de Paris (1857–74) and the Halle aux Bœufs (1865–67), now known as the Grande Halle de la Villette. He sold the firm in 1880, and it is now part of the currently largest French steel contractor Eiffage Métal. (Source: Roussel family archive)

Fig. 64. Les Halles de Paris (1857–74) was partly erected by the Entreprise Roussel (Source: Baltard and Callet, *Les Halles de Paris*, 1863 atlas, pl. 1). Halle aux Bœufs, now known as the 'Grande Halle' at La Villette. (1865–67; Source: Guillaume Vellot 2016, Wikipedia CC BY 2.0)

bearing cross-wall estimated to weigh 66.2 metric tons, and it testifies to the firm's structural daring and confidence (Eck 1836, vol. 1, p. 38 and pl. 20). The practical expertise of such contractors was tapped into and expanded with a knowledge of theory and material technology by the teachers at the ECP, and especially by the graduates. Their contact with empirical practitioners had originated with Colladon, and his successors continued the tradition.

The government's *Conseil des bâtiments civils* (Council for public structures) supported and participated in this transfer of expertise. Gourlier, who had been the first professor of architecture at the ECP and had co-published three volumes of public construction examples between 1825 and 1850, was an *inspecteur général* and secretary of the council, which had been responsible for reviewing and approving the plans for the Bibliothèque Sainte-Geneviève in 1843. The council existed until 1848, upon which inspection was transferred to other government departments. In the half-century of its existence, it united an influential group of professional builders, a sub-culture in the profession that was responsible for approving all non-military public buildings and their construction.

Originally, the members of the council were what in Britain are called quantity surveyors (project control managers or cost engineers in the United States), in charge of the *métré*. As it grew in complexity, the council went through ten iterations between 1795 and 1841 (Gourlier 1848/95, pp. 25–28), and its enhanced role included the general approval of public buildings and all aspects of their construction (Château 2008, pp. 111–112). Given the council's mission to assure the economic and proper use of public funds, its members were clearly biased toward construction, structural behavior, and technology. Eck, who apparently was not a member (at least not as recorded in the 1843 and 1846 volumes of the *Annales des Ponts et Chaussées, partie administrative*), cites examples of the innovative role played by the council in construction. He mentions the riveted joists designed by the engineer Louis Schwickardi (dit Aubert) that the council approved (Eck 1841, vol. 2, pp. 7–8/49–50/56), and the first known iron gussets that the architect Lucien-Tyrtée Van Cléemputte designed around 1830 in cast iron for connecting the members of the roof trusses over the archive depository of the Cour des Comptes (Court of Audit) (Eck 1836, vol. 1, p. 51 and pl. 32).[322]

322. These gussets were followed by complicated pseudo-gussets in which cast- and wrought-iron members were connected with a wrought-iron plate. They were designed by Alfred Neville for a small truss over the St. Denis Canal in 1845 (Mehrtens 1908, vol. 1, p. 534 / James 1986).

Gourlier was a prominent example of these technologically oriented professionals. Not only had he participated in developing the original concept of the ECP's curriculum, but he had also represented the council in the iron reconstruction of the (again hidden) nave roof at Chartres (1839–41). He was, however, not assigned to Labrouste's library.

As a typically bureaucratic organization, the council assigned a whole range of inspectors of various hierarchical designations to the library (*Annales des Ponts et Chaussées* 1846, vol. VI, 2e partie, *Partie administrative, Lois décrets … Personnel*, pp. 197–210). By the time Labrouste began building, the responsible *inspecteur général* and member of the council's governing body was Gourlier's colleague Auguste Caristie. Under him were an architect named Miller as *inspecteur*, a young ECP graduate named Laurent Courtépée (grad. 1845), Jacques Laudin and Dumas de Culture as *sous-inspecteurs des travaux*, and Gustave Dessalle and Monneau as *vérificateurs*. A Roussel functioned as *inspecteur pour entretien* of the site.[323] Caristie and Roussel had also been responsible for Duban's building at the École des Beaux-Arts on which Labrouste had worked earlier.

The detail plan for the iron truss construction of the floors, with the indication that the contractor was Roussel (whose details these must have been), were all individually annotated and certified by the recently graduated ECP chemical engineer Laurent Courtépée.

Labrouste was not the only builder interested in the new formal possibilities that iron presented. Unconventionally educated builders like the engineer-architect Léonce Reynaud were also intrigued by the new material. Reynaud had apprenticed as an architect and then, after having been expelled from the EP in 1822 along with Perdonnet, had studied under the pioneer of iron structure Louis Bruyère at the ENPC (see Part II). He was very much involved with the Romantic architects in their support for Duban's Palais des Études (Van Zanten 1978, pp. 169–173), and followed Durand as professor for architecture at the EP from 1837 and at the ENPC from 1841. By 1850 Reynaud had begun to recognize that the new industrialized form of iron implied new developments in construction as well, and he wrote in his *Traité d'Architecture*:

323. The name Roussel is not uncommon in France. This was neither a member of the metalworking family nor the later purchaser of the firm.

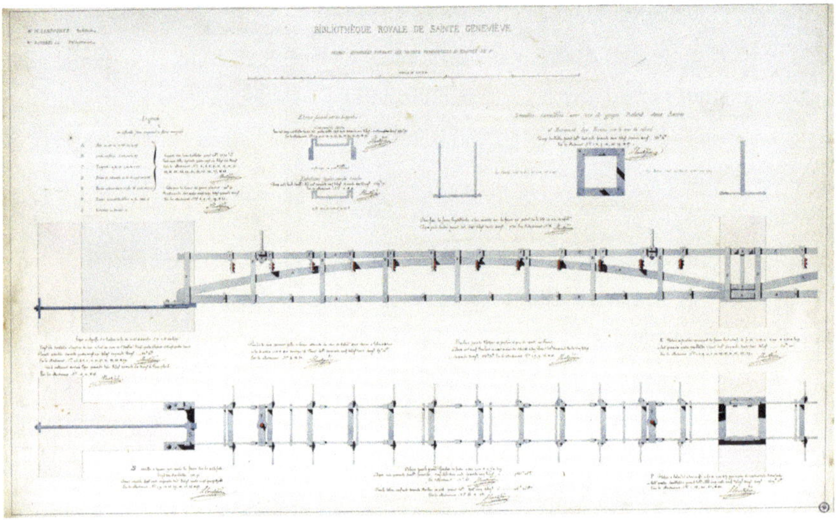

Fig. 65. Original plan of Roussel's wrought-iron floor trusses for the Bibliothèque Sainte-Geneviève, with fifteen comments and certifying signatures by L. Courtepée, *sous-inspecteur* of the *Conseil des bâtiments civils*. (Source: © Labrouste Archive, Bibliothèque Sainte-Geneviève, bsg MS 4273_0017)

> However, the new material that has now become available to us implies new forms and new proportions because it differs in essence from all those that have been used so far. What was suitable for stone is by no means suitable for iron. Although our industrial reality does not demand a total reorganization of the art [of building], it does present us with new elements and a new branch [of construction] that will doubtlessly develop strongly and to which we cannot assign any limit. (Reynaud 1850, vol. 1, p. 448, part quoted by Lemoine in Bélier et al. 2012, p. 188)[324]

Many builders were gradually feeling their way toward an awareness of the material's potential. Reynaud's approach was not yet fully industrial either. Werner Szambien thought that his use of iron was a modern modularization. He

324. Mais, à la nouvelle matière qui vient s'offrir à nous, il faudra de nouvelles formes et de nouvelles proportions, car elle diffère essentiellement de toutes celles qui, jusqu'à ce jour, ont été mis en œuvre. Ce qui convenait à la pierre ne saurait, sous aucun rapport, convenir au fer. Il y a donc, dans le fait industriel, le principe, non pas d'une rénovation complète de l'art, mais de nouveaux éléments, d'une nouvelle branche qui est, sans doute, appelée à prendre des développements considérables, et au progrès de laquelle il serait impossible d'assigner des limites.

attributed the repetitive structure of the iron-framed lighthouse at Walde near Calais 1857 to the influence of Durand's design method (1984, p. 327) that Reynaud had only briefly experienced. However, the lighthouse was in fact a prefabricated structure that he had imported in its entirety from England and that a British contractor had erected. It was modelled on Alexander Mitchell's screw-pile system as it had been used for the Wyre Light (1839–40) and Mapin Sands Lighthouse (1838–40), and these structures in turn were very similar to prefabricated iron bridge systems as they were then being developed by contractors in the United States. Although the lighthouses and bridges used more repetitive elements than Eck's examples did, they were conceptually also one-off, pre-industrial productions.

The non-academic architect Eugène-Emmanuel Viollet-le-Duc was formulating his innovative iron designs in lectures around mid-century, and would publish them in 1863. Viollet's astonishing designs—which some historians have credited to Labrouste's influence, or to his former employee Jean-Baptiste Lassus[325]—suggest that iron construction was attracting the attention of the more unconventional architects.[326]

Viollet's early designs in iron had simply (if dramatically) replaced stone compression members with cast iron, but his last one, and the only one he proposed in wrought iron, was a self-supporting frame for a house.

Although he only published the description and many details of his wrought-iron house in the chapter 'Dix-Huitième Entretien, sur l'architecture privée (suite)' of the second volume of the *Entretiens* in 1872 (vol. 2, pp. 303–351), his explanatory plate XXXVI of the building was published a decade earlier in 1863, and it presaged the system that Moisant and Saulnier were to develop in their Menier Mill (1869–72). Viollet wrote in 1872 that the Menier structure impressed him deeply (*Entretiens*

325. Lassus, who like Labrouste had studied under Lebas, and Félix Duban had participated in the restoration of Sainte Chapelle under Viollet in 1840 and in that of Notre-Dame de Paris (1844–64).

326. Corinne Bélier points out an essential difference between Labrouste's and Viollet's approaches to the use of iron in structure: Labrouste separated masonry and iron, while Viollet integrated them into a single system (Bélier et al. 2012, p. 228), coming closer to composite structure as Eck had described it in 1836. Viollet's concept resonated in Anatole de Baudot and Paul Cottancin's (ECP 1886) use of reinforced concrete in their church, Saint Jean de Montmartre (1894–1904). Baudot had studied with both Labrouste and Viollet, and had worked with Viollet on restorations and, like him was associated with Trélat's school, thus completing the circular connection to the ECP and its pedagogy.

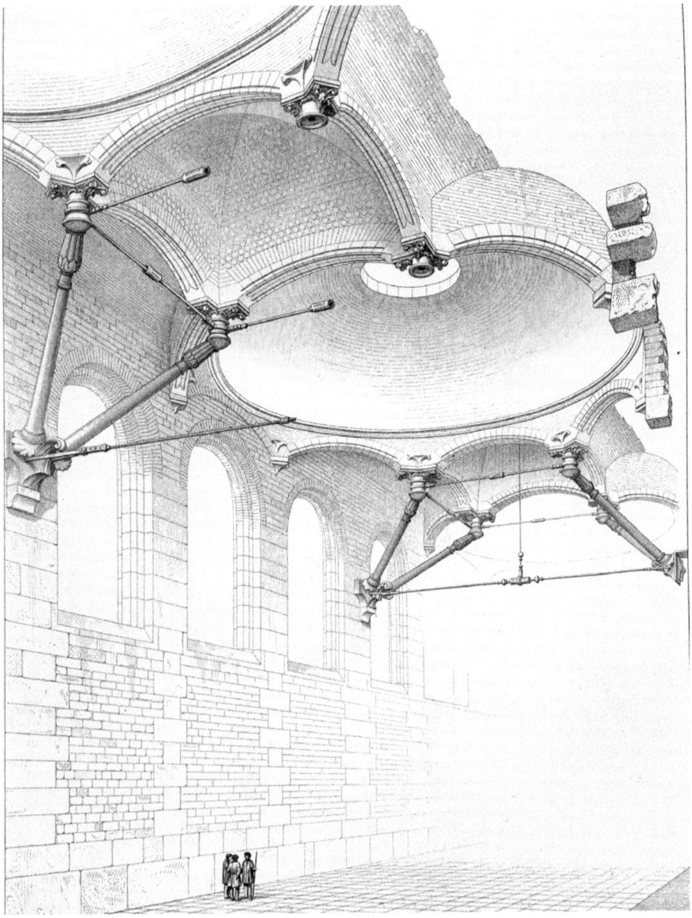

Fig. 66. In 1863 Viollet-le-Duc published his astonishing ideas for composite iron and masonry structures. He had derived them from his studies of Gothic structure, by replacing stone compression members with cast iron and adding some wrought-iron tension rods. (Source: Viollet-le-Duc, *Entretiens sur l'Architecture*, 1863 atlas pl. XXII)

vol. 2, p. 334 footnote),[327] and his note appears to be the first mention ever published of this pioneering structure. Viollet knew de Dion, and perhaps Saulnier and Moisant too. He interacted with de Dion, Émile Muller, and Trélat in the latter's school from 1865, and in the Société des ingénieurs civils, of which he became a member in

327. Viollet mentioned only Saulnier, but Moisant was the one who later developed and erected many pioneering iron structures. Saulnier, on the other hand, did not pursue iron construction further, barring the late and by then standard internal structure of the Menier family's Ferme du Buisson in 1880.

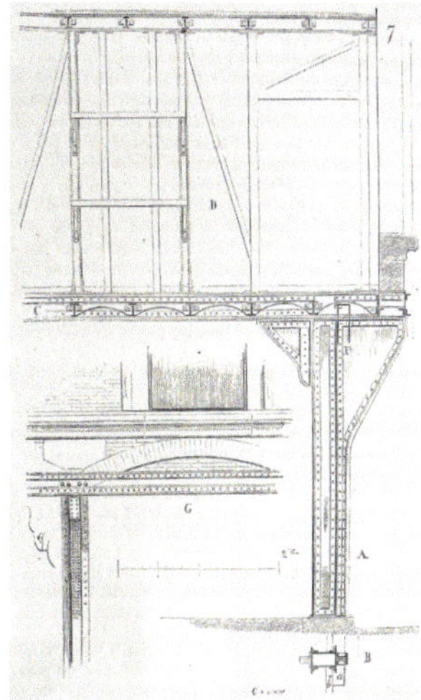

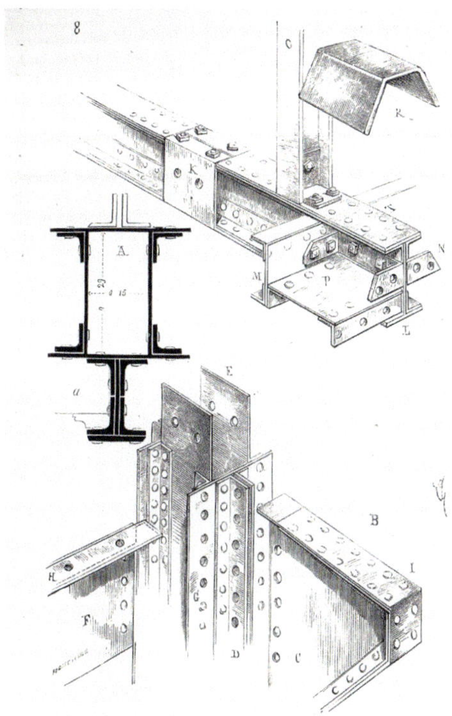

Fig. 67. Viollet also published a revolutionary proposal for a self-stabilizing wrought-iron framed house. (Source: Viollet-le-Duc, *Entretiens sur l'Architecture*, 1863 atlas pl. XXXVI) Viollet detailed the iron-framed house in the second volume of his work in 1872, the same year that Armand Moisant and Jules Saulnier finished a very similar structure for the Moulin Menier. (Source: Viollet-le-Duc, *Entretiens sur l'Architecture*, vol. 2, 1872, pp. 328 and 330).

1872 (*Mémoires et compte rendu*, 1872, p. 41), and even a member of its steering committee in 1873, when de Dion was vice president (*Mémoires et compte rendu*, 1873, p. 13). Given the 1863 date of Viollet's plate[328] and the fact that it so closely resembles the later Menier Mill, it is reasonable to presume that such ideas were discussed back and forth between interested builders at the time. It is, of course, not possible to determine who came up with the idea first and who influenced whom. These men did not work in a vacuum, and we may presume that they carried on some form of professional discourse. Viollet's ideas certainly paralleled those being explored by students and faculty at the ECP, and that is why he taught in Trélat's school (*Mémoires et compte rendu*, 1879.2 pp. 884–885). As he wrote regarding his wrought-iron house, also citing the industrial aspect as Reynaud had earlier:

> I by no means intend here to propose this fragment as a model for the rental dwelling of the future, as *the architecture of the future*, but as a precedent-free study of what modern industry provides us with right now regarding the building trades.... We must admit that here, at least, iron is not hidden, but presents itself openly. (Viollet-le-Duc, vol. 2, 1872, p. 336)[329]

Such was the lively experimental, proto-industrial professional world into which the enthusiastic ECP graduates entered at mid-century and upon which they were to make their mark in iron construction. It is also noteworthy that 27 years later Sylvain Périssé, an 1858 graduate of the ECP, would be the first to build a house on Viollet's model with visible iron structure clad in ceramic panels at 95 rue Montmartre between 1897 and 1899.

A paradigm shift around 1860

As Jolly had observed, iron manufacture and construction were changing around 1860. From 1855 to 1859 Labrouste had been in professional contact with de Dion in Eugène Flachat's office, possibly in the role of government inspector or architect[330] on the stabilization of the tower of Bayeux Cathedral (Saddy 1977,

---

328. The date of this plate in the atlas, nine years before the detailing of the project in volume 2, raises questions about the real date of this project. It is not possible to elucidate these questions here.
329. Je n'ai point cette prétention de donner ce fragment comme un type des constructions à locations futures, comme *l'architecture de l'avenir,* mais comme une étude sans réminiscences des moyens que nous fournissent les industries modernes touchant l'art de bâtir, des besoins du moment... Mais on voudra bien accorder qu'ici, du moins, l'emploi du fer n'est pas dissimulé, qu'il se montre franchement.
330. Labrouste was certainly in contact with his former student Lassus, who was then collaborating

p. 65). While he was designing the Bibliothèque nationale's study hall in 1858, he was already interacting with ECP graduates, and thus was familiar with their industrially focused approach to construction and their self-assurance in the use of iron and its formal expression. It is conceivable that Labrouste discussed its use and his ideas for the new national library with de Dion, with his assistant Louis Lasvignes (both grad. 1851), or perhaps even with Flachat himself, whose office was intimately involved in the changes (see chapter 'Flachat and the I-beam').

Whatever the impulse may have been, a lively development had taken place in Labrouste's mind in the years between designing the two reading rooms. Labrouste's understanding of structural iron had grown more sophisticated by 1858, when he began working on the new project. One cannot prove that his contact with Flachat's office and perhaps other ECP graduates had contributed to this development, but Labrouste's understanding of materials and the aesthetics of structure had matured along parallel lines with the thinking of the young graduates like Barrault or de Dion.

Eck's treatise, particularly the second volume of 1841, had also exposed architects, contractors, and his fellow government inspectors to an approach to building facilities that stressed materials and structure. Labrouste was evidently the first academic architect open enough to grasp the formal implications. One of the reasons for this openness was that he approached the act of designing like the ECP graduates did, as *organization* rather than *composition* (see Part II, the discussion of Mary's contribution). As Neil Levine perceptively wrote:

> Labrouste approached the library's design as a research problem rather than a compositional one. (Bélier et al. 2012, p. 172)

Labrouste chose contractors who were among the most experienced of the many who worked in Paris at the time. Not all the *entrepreneurs de serrurerie* were as

---

with Viollet-le-Duc. Viollet was involved, albeit probably peripherally, with the situation at Bayeux (Flachat's report in de Dion and Lasvignes 1861, p. 26). It is not quite clear that Labrouste was indeed in charge of this cathedral as inspector, as only the reports of two of the three inspecteurs généraux, Viollet-le-Duc and Léon Vaudoyer, exist in the national archives, and neither of these mentioned Bayeux. It may have been in the bailiwick of the third inspecteur général, Léonce Reynaud, whose report has apparently not survived. Centre Historique des Archives nationales, *Atlas des Inspecteurs-généraux des Édifices diocésains 1853* F/19/*/1822-F/19/*/1825, F/19/*/1940-F/19/*/1941.

experienced as the Roussels or Achille Leturc. Although the dome of the Halle aux Blés and the railway stations—with which ENPC engineers, or in the case of the latter, ECP engineers were involved—still stand as solidly as when they were built, unfortunate accidents did happen when scientifically based professional expertise was lacking.

Reminiscent of the fate of the Anthæum conservatory was the failure of the Marché Saint-Martin on December 10, 1879. However, the two cases are not

Fig. 68. Exterior view and interior of the Marché Saint-Martin on opening day in 1854. (Source: *L'Illustration, Journal Universel,* December 9, 1854, p. 396)

Fig. 69. Collapse of the Marché Saint-Martin roof under a load of snow on December 10, 1879. (Source: lithograph by Isidore Laurent Deroy)

truly comparable because the Anthæum collapse was due to a lack of construction experience, professional coordination, and structural knowledge, whereas that of the Marché Saint-Martin appears to have been simply weak structural detailing. The market hall had been built by Pierre Joly of the Constructions métalliques d'Argenteuil in 1854, according to his long experience with iron construction. Its span was 34.85 meters, which was standard for a Polonceau truss at the time. The truss may have been too slender for the span, and the purlins were more slender than usual and not sheathed in sheet iron (Mary 1862–63, pp. 226–227, pl. LIII). The result was elegant, but possibly weak. The collapse had several possible causes. According to a speculative note in the *Mémoires et compte rendu* (1880.1, p. 120), it may have been caused by contracting tension members due to extreme cold and the fact that the truss ends had been fixed in the masonry (albeit with only two bolts per side, see Fig. 98) rather than simply supported. The first half of December 1879 was unusually cold. Vierendeel (1897–1902, p. 498, cit. in Steiner 1984, p. 84) supported this hypothesis, despite his claim that the trusses were simply supported on one side, which may not have been the case. The roof had held for 25 years and collapsed under a load of snow. Joly hired his first ECP engineers a few years after the market hall had been completed and appears never to have lost another building.

ECP graduates join the 'entrepreneurs de serrurerie'

As manually forged and cast-iron parts were gradually giving way to industrially produced rolled sections and spans were increasing after mid-century, the changes introduced a level of insecurity into construction knowledge. Experience could no longer be reliably extrapolated, and the pre-industrial iron contractors and manufacturers began hiring ECP graduates. Barrault had been hired by York in 1855, and both Émile Baudet, who graduated in 1858, and Armand Moisant, who followed him a year later were hired by Joly, who was one of France's larger contractors at the time. The firm was headed by César Jolly (Joly's son-in-law) who had had some engineering training and would soon write on the transition between pre-industrial and industrial iron construction.[331] They entered the firm just a few

331. César Jolly had graduated from the École d'Arts et Métiers in Chalon-sur-Saône. The firm was among the most experienced iron contractors of the day and had built six Parisian markets, including the Marché Saint-Honoré, many bridges for the Chemin de fer de l'Ouest, the Gare de l'Est (1847–50), the second Gare Montparnasse (1848–52), the Polonceau girders of the great hall of the Gare Saint-Lazare (1853–54) and was then erecting part of Victor Baltard's Halles centrales (1857–74). Later

years after the Marché Saint-Martin had opened.

Having gained his initial practical experience with Joly, Baudet subsequently went on to a partnership with the *entrepreneur de serrurerie* Achille Mathieu Leturc[332] in 1865 (J.-F. Belhoste 2008, p. 71). According to Eck, Leturc was

> one of our most able ironmaster-contractors ... whose scholarly theory coupled with the consummate practice of his art, and also whose infinite imaginative resource has up to now contributed strongly to the advance of the iron industry. (Eck 1841, vol. 2, p. 8)[333]

Leturc (along with Schwickardi) had initiated the use of rolled iron for floors and trusses. He had apparently also built the first entire roof of out of iron in France,[334] for the slaughterhouse in Bourges (Eck 1841, vol. 2, pp. 56–57, pls. 44 and 45),[335] Gréban's experimental commissariat in 1836, and had, like Roussel, submitted several proposals for the reroofing of Chartres Cathedral in 1839 (Eck 1841, vol. 2, pp. 35–36 and pls. 24–27). The expertise of the two firms Leturc and Roussel was comparable.[336] The Conseil des bâtiments civils that had previously supported their experiments had ceased to exist well over a decade before the Bibliothèque nationale was under construction. No wonder, then, that Leturc would want to associate himself with the new breed of academic builders.

One of Leturc et Baudet's first projects—and the probable reason for their association—was the contract for the iron structure in Labrouste's new study hall, and perhaps also for the stacks. An earlier plan fragment of a simple iron floor structure, labeled *Leturc, Entrepreneur de serrurerie* can be seen on the BnF's

---

projects of the firm were the Paris Exhibition hall of 1867, the salle de fêtes of the Palais du Trocadéro in 1878, and the stations Saint-Lazare and Le Havre in 1888.

332. Information on Leturc from Wikipédie, 'François Charles Michel Leturcq' and Geneanet.

333. ...un de nos plus habiles entrepreneurs de serrurerie ... dont la savante théorie jointe à une pratique consommée dans son art, et dont, aussi, une infinité de ressources imaginatives ont, jusqu'à ce jour, puissamment contribué à l'extension des progrès de l'industrie ferronnière.

334. There were far earlier Russian examples, some of them dating back to the early eighteenth century (Kosykh 2021).

335. This structure was possibly built to the designs of an architect named Barthélémy Jullien in 1840.

336. Eck shows a double truss by Leturc of very similar design to the one for Théodore Roussel's house (vol. 2, 1836, p. 38 and pl. 20), and one more by Jacques Louis François Travers (p. 39 and pl. 21, identified on Geneanet) that indicates possible connections between the most active entrepreneurs de serrurerie. Daly discusses Leturc's innovative structural detailing in 1845/46 (*Revue de l'architecture générale,* vol. 6, columns 259–260).

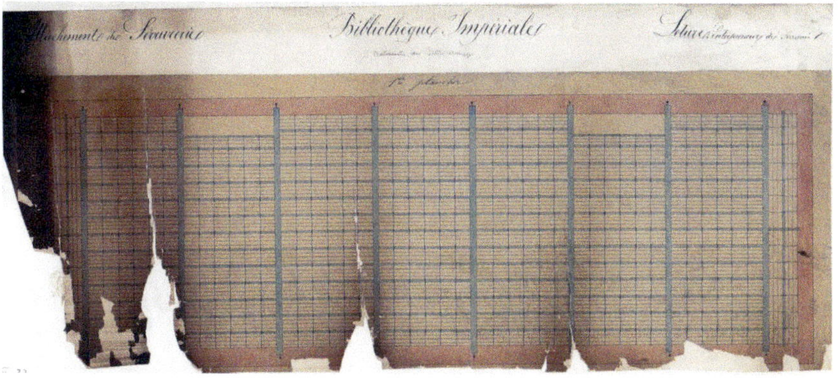

Fig. 70. An early and simple plan for the first-floor iron framing of the Bibliothèque nationale by the ironmaster-contractor Achille Leturc. (Source: © gallica.BnF.fr, HD-1019 (3)-FT6)

Gallica site.

The complexity of the final design makes it likely that Leturc approached Baudet early in the design process because of his expertise, and the young engineer must have contributed substantially to the structural design, detailing, and calculation of the domed structure. The knowledge base that Baudet's ECP education had given him in mechanics, iron construction, design, and descriptive geometry, as well as the ability that Olivier's models had instilled in him to visualize and manipulate three-dimensional structures had certainly prepared him well for the task.

Baudet soon took over the firm on his own and developed it further with the help of several *centraliens*.[337] When Baudet took his chief designer, Alfred Donon (grad. 1871), as partner, another ECP graduate, Jules Roussel (grad. 1869), replaced him, and in 1874 the firm introduced hydraulic riveting in France.[338] In 1878, Établissements Baudet, Donon et Cie., as the firm then called itself, was one of four contractors that erected 53 trusses each of the interior *galleries* of de

337. Aside from chief designer Alfred Donon and his replacement Jules Roussel, these were engineers named Kokocki (grad. 1864), Benoiste (grad. 1872), and Amblard (grad. 1881) (*École centrale 1889*, pp. 75/80/34). It is not known whether Leturc's son Paul André, who is also listed as an entrepreneur de serrurerie, remained part of the firm or not.

338. He apparently used Tweddell's portable hydraulic riveters for riveting warships in Toulon (Graces Guide to British Industrial History, 'Ralph Hart Tweddell', consulted online 06.02.2021).

Dion's Palais du Champ de Mars[339] at the Paris Exhibition of that year. Baudet assembled them on the ground and raised them by steam crane, which was a novelty at the time. Between 1880 and 1888, the firm built the complex iron structure of the Labrouste student Julien Guadet's Paris post office, now known as the Poste centrale du Louvre.[340]

Over the years, the links between the *entrepreneurs de serrurerie* and ECP graduates became ever stronger. Jules Roussel who had left the firm to take over the contractor Tribout in 1874 (see J.-F. Belhoste 2008, p. 71), bought out Henry Roussel in 1880. He amalgamated with Baudet et Donon as Baudet Donon Roussel in 1921, and they were then joined by Moisant Laurent Savey, and later Eiffel and others in 1960 to form CFEM (Compagnie Française d'Entreprises Métalliques) in 1966, today's Eiffage Métal.

The Bibliothèque nationale initiated Baudet's career. In contrast to the visually and structurally unclear iron structure of the Bibliothèque Sainte-Geneviève, the nine ten-meter ceramic domes of the new study hall with their visible armature of wrought iron are integrated elements of the structure. The rolled industrial sections and their iron connections demonstratively sport gilded rivet-heads. Form, structure, material technology, and the ornamentation of industrially produced structural parts melded at the Bibliothèque nationale to launch a new architectural awareness.

Over the quarter century between the initiation of the first library project in 1843 and the completion of the second in 1868—and possibly with input from the young Baudet—Labrouste had transformed his understanding of iron from apparent structure and *decoration* to actual structure and *ornament*.[341] Labrouste took a further step in the Bibliothèque nationale and boldly exposed the iron-framed stacks through a nine-meter high and prominently glazed archway axially opposite the entrance.[342] We do not know whether Labrouste knew of the three-tier cast- and wrought-iron archive shelving that Matthew Clark had manufactured and installed in Carlo Rossi's General Staff Building

339. These were manufactured by Fives Lille and Schneider-Creusot.

340. Guadet, a life-long adversary of Viollet's approach to history, became an influential Beaux-Arts teacher and theoretician.

341. For the transformation of detailing see Marc Le Cœur in Bélier et al. 2012, p. 193, and for the distinction between decoration and ornament see Part II, the discussion of Mary's contribution.

342. Neil Levine describes Labrouste's early and even more radical versions of this transparency (Bélier et al. 2012, pp. 176–178).

in St. Petersburg (1819–23; Fedorov 1992, pp. 70–72), but he apparently had seen the similar stacks in the circular reading room by Sidney Smirke (1854–57) in the British Museum on a visit to London (Middleton, p. 41 and Le Cœur, in Bélier et al. 2012, p. 136).

It was a daring architectural statement to visually link a culturally significant and academically iconic space to an engineering facility like Barrault's interior of the 1855 Exhibition building or Flachat's many railway station halls. Labrouste evidently did not suffer elitist qualms about learning from the design of facilities; a workshop he had seen and sketched in Saint Quentin in 1858 inspired the shed roofing for his stacks (Le Cœur, in Bélier et al. 2012, p. 163).

Like Eiffel's aesthetic statement with his tower twenty years later, Labrouste's architectural message was shockingly clear to the library's patrons. They saw storage stacks devoid of any decoration. It was pure structure, not an inserted and decorative piece of furniture like Bunning's rotunda and Labrouste's own Saint-Geneviève vaults had been. Barrault with York and Baudet with Leturc had facilitated this development.

Was the intentional public glimpse into the stacks—the iron 'guts' of a building that was officially classified as architecture—an official recognition of the aesthetic impact that ECP-educated builders were beginning to exert on their environment (and perhaps on Labrouste himself)? As a librarian said at the time, fully intending his remark to be pejorative, even the façade of the study hall facing the entrance court looked like a railway station (Le Cœur, in Bélier et al. 2012, p. 147). It did, and Labrouste's daring statement would influence many of the subsequent icons of early industrialized iron construction in and around Paris.[343]

Labrouste's gradual adoption of the technology and the aesthetics of iron as visible structure was a symptom of the cultural development of engineering. The ECP curriculum and border-crossing builders like Rohault de Fleury, Reynaud, or Barrault, and those who worked primarily in architecture like Trélat, Muller, Lavezzari, Demimuid, or Denfer disseminated a new concept of design and construction. The ECP's contribution was constitutive and based, as

343. Among the many were Moisant and Jules Saulnier's Moulin Menier (1869–72), Moisant and Boileau's department store Au Bon Marché (1872–76), and finally Eiffel's tower in 1889, and Moisant's structure of the Grand Palais des Champs Élysées in 1900.

Part II will explain, on Louis-Charles Mary's perception of design as a process and on Louis Bruyère's and Charles Eck's understanding of materials and structure. This was the future, and the future was iron. Academic architectural design at the Beaux-Arts, in contrast, remained based on Durand and Rondelet right up until the Second World War.[344]

344. The construction program, however, did develop a more modern format under Rondelet's successor Alphonse-François Marie Jaÿ from 1824 to 1863, and then especially under Édouard Arnaud (grad. ECP 1888) from 1920. See Brucculeri 2009.

France and Britain

Mutual influences in material studies

France and Britain were the two leading industrial nations in the first half of the nineteenth century, and they influenced one another. Thanks to the *entreprises de serrurerie* and the engineers graduating from the ECP, France was slowly drawing even with Britain in industrial technology toward the middle of the nineteenth century. Initially however, the flow of information had primarily been only one way, from Britain to France. The French government sent engineers like Dupin and Navier to England to report and spread the technical knowledge they had gained through their publications. When young French engineers like Goüin, Talabot, or Lavalley returned after working in England, they imported more information. Flachat, Perdonnet, and others spoke English. They had excellent contacts across the Channel and traveled to Britain too.

In the early part of the century, French engineers viewed England as the paragon of industrial and technological development, despite all the nationalistic reservations that bigoted men like Cachin expressed (see the chapter 'The first years'). In the 1830s, English had briefly even been a required course at the school.[345] The admiration of everything British went so far that early internal toilets, most probably a flushing variant of the classical 'commode', were known at the time as *anglaises* (see for instance Bruyère 1828, vol. 2, Recueil XI, pl. 1).

The French were certainly leading in mathematics-based engineering theory in the eighteenth and early nineteenth centuries, but what has been largely ignored is that British experimental research into material behavior had increasingly begun to inform the French theoretical approach. Engineering theoreticians

345. Due to curricular overload, however, the language requirement of 1832 became optional in 1834 and was discontinued entirely in 1836 (Guillet 1929, p. 136).

on the European continent had historically tended to use experimentation as a method to confirm their analytical derivations, rather than as a tool to create them.

The bias is of course not clear-cut, and there are overlapping forms of the two approaches. Pierre-Simon Girard's work *Traité analytique de la résistance des solides* that he published in 1798 is a prominent hybrid example. Nevertheless, we might characterize the two mindsets culturally as an analytical French 'material science' that works through theoretical derivation coupled with experimental confirmation, versus an experiential British 'material research' that works through a process of experimental empiricism—that is, trial-and-error leading to theoretical knowledge via deduction. The statement is admittedly oversimplified because experiments on timber and then on iron were of course carried out in both countries. Materials were chiefly studied in military environments, most often in naval shipyards, and experimentation always served practical considerations of shipbuilding, ordnance, construction technology, and performance. But French scientists, as opposed to their colleagues in Britain, tended to think of experiments as an adjunct to their primary goal: the establishment of theory. The bias in France was toward science, and in Britain toward application. The difference exemplifies one of the basic distinctions between scientists and engineers.

In 1817 Peter Barlow explained his logic for the distinction in the forward to his treatise on deflection in timber beams, *An Essay on the Strength and Stress of Timber, founded upon experiments performed at the Royal Military Academy*. Even though he was primarily a mathematician himself, he did not follow the French tradition, but derived the physics of the structural behavior of timber exclusively from experimentation:

> Being thus well aware of the defects of former theories, and of the danger of building a new one upon other physical assumptions, which, though more plausible, might be equally erroneous, I resolved to introduce no data but such as I obtained from actual experiment; and it is by this means that I have arrived at my general theorems. (Barlow 1817, p. x)

Barlow's statements were therefore not theoretically derived and then experimentally tested to confirm them like Girard's. Girard had developed his experiments on bending beams from 1793 based on what Lamblardie had planned to do in the Le Havre shipyard (Girard, p. v). He used his experiments to test the hypotheses that he had based on the theoretical work of Mariotte, Leibnitz, Jacob Bernoulli, Euler, and Lagrange (Girard, p. xlix). He also used Buffon's tests, as the tables at the end of his treatise show. Girard does, however, recognize that there is a value in using experimentation to modify theory—if not to create it:

> Only experiments can suggest modifications to these formulæ. (Girard, printed marginalia p. 6, cit. in Tardini 2014, p. 35 footnote 3)[346]

Other writers of that period, like Rondelet and Jean-Henri Hassenfratz also shared the view that experimentation should be used to confirm theory (Tardini 2014, pp. 38–43). Todhunter points out (vol. 1, 1886, pp. 103–105) that both Girard and Barlow had struggled with the identification of the position of the neutral axis under loading, which had led to problems in their theories. In their histories of the field, Timoshenko and Todhunter tend to downplay the importance of British experimental work in the strength of materials. Their reasoning was that the British researchers' proficiency in theory—an essential complement to experimentation—was modest compared to that of the academically educated French. Therefore, in contrast to his colleague Barlow half a century before, the mathematician Todhunter had a hard time accepting the influence that pragmatic experimentation exerted on the further development of theoretical work in France. Despite his own reservations, the theoretician Timoshenko admitted that

> A considerable amount of experimental work was completed [in England], and the results that were compiled found use not only in England but also in France. (Timoshenko 1953, p. 99)

He thereby implies that there was indeed value in Barlow's progression from experiment to theory. We note the same stress on experimentation in Eaton Hodgkinson, who was also researching the strength of beams in those years. In his paper 'On the transverse strain and strength of materials' in 1824, he writes that

> our knowledge of the action of the fibres or particles of bodies during their flexure, chiefly perhaps for the wont of sufficient experiments, seems still to be imperfect. (Hodgkinson introduction 1824 cit. in Todhunter 1886, vol. 1, p. 21)

Barlow was delighted to find that Dupin had confirmed his results on the bending of timber beams in *Expériences sur la flexibilité, la force et l'élasticité des bois,*

346. L'expérience seule peut indiquer les modifications que l'on doit faire à ces formules.

avec des applications aux constructions, en général, et spécialement à la construction des vaisseaux (1813/1815; Barlow 1817, footnote pp. vi–vii).[347] The theoretician and naval engineer Dupin had adopted an experimental, evidence-based approach to material behavior. In fact, Dupin, whom Guillerme theatrically characterizes as a 'midwife' of French industrial technology,[348] had been aware of research in Britain and met or corresponded with Barlow, although he had not known of Barlow's work before his own publication. He would travel to England repeatedly, also visiting the Royal Military Academy at Woolwich where Barlow taught.

Dupin acted pragmatically by examining the materials that he tested in great detail: how long they had been seasoned and used and where he had obtained them (Dupin 1813, pp. 8–9). These were two issues that would not normally have occurred to an abstract theoretician. Girard had also considered taking similar variables into account but had abandoned the idea because of its complexity when applied to theory (Girard 1798, p. liii). The contrast between the two reports documents how the perceived value of an experimental approach to the development of theory had begun to grow in the years between 1798 and 1813, an attitude that paralleled the French admiration for pragmatic British technology.

Dupin drew upon a long French tradition of naval experiments in wood in his research.[349] He had been inspired by the examination of the behavior of timber that the botanist Duhamel du Monceau had undertaken for the navy over half a century before (Dupin 1815, p. 195). In contrast to both Barlow and Duleau, he recognized the neutral axis that Duhamel had experimentally determined well before Coulomb's analysis in 1773 (Dupin 1813, pp. 12–13 / 1815, pp. 166–167). Following Duhamel, Buffon, and Girard, he looked at bending strength or *résistance virtuelle*, as he termed it, which he considered to be more useful in shipbuilding than an examination of timber's ultimate strength (Dupin 1813, p. 11). Only toward the end of the expanded version of his paper in 1815 did Dupin

---

347. Dupin reciprocated this admiration and commented on the connection between their work and results in 1820, *Voyages dans la Grande-Bretagne,* vol. 1, 1820, tome II Force militaire, études et travaux, p. 92 footnote, erroneously dating Barlow's publication to 1818, and he documented their contact vol. 1, tome I, p. xv, footnote.

348. '…l'un des accoucheurs de la société industrielle française', in *Variations sur les rêveries du Baron Dupin,* 1976, reprinted in Guillerme 2008, p. 37.

349. For a concise history of the topic of bending in wooden beams and French research in this area, see Ruggeri 2011 and Tardini 2014, pp. 33–60.

apply what he had learned to the issue of ultimate strength, relying primarily on the results of Buffon's eighteenth-century experiments (Dupin 1815, pp. 194–199), but not undertaking his own.

Like Barlow, Dupin derived the relationship between load and deflection experimentally and only then formulated statements that he called *théorèmes* from the results. He demonstrates his research process in the expanded form of his paper, which is the version that Barlow had read after he had published his own results (Barlow 1817, footnote pp. vi–vii). Timoshenko recognized Dupin's work as an empirically based seminal treatise and a precursor of Navier's theoretical work on the strength of materials (Timoshenko 1953, pp. 80–81). In his book, Barlow discussed his agreement with Dupin in detail (1817 pp. 120–131) and expressed the goal of the experiments he undertook:

> My object, therefore, has been not only to furnish new series of data, but also to make such experiments as might lead to an explanation of the causes of those apparent anomalies which were observed in the results of preceding writers [ie., the French scientists], and which rendered the whole of them doubtful and uncertain.
>
> In this respect I have succeeded beyond my first expectation. (Barlow 1817, pp. vii–viii)

Barlow and Dupin's experimental approach contrasts with Duleau's as expressed in his 1820 treatise on wrought iron, *Essai théorique et expérimental sur la résistance du fer forgé*, although Timoshenko (1953, pp. 81–82) writes of Dupin and Duleau's reports as similarly experimentally driven. In fact, however, Dupin tellingly presents all his experimental results first and only then formulates *théorèmes* from them, while Duleau, following tradition, begins with a chapter on the theoretical derivation of material behavior under loading and then describes his experiments, repeatedly stating that they confirm his previously hypothesized theory:

> The following experiments all agree in confirming the correctness postulated by theory as applied to elastic bodies. (Duleau 1820, p. 22)[350]

350. Les expériences suivantes s'accordent toutes pour confirmer cette propriété, annoncée par la théorie comme appartenant aux corps élastiques.

It was not until mid-century that French engineers, and typically ECP graduates like the railway engineer and future president of the Société des ingénieurs civils, Georges-Henry Love (grad. 1840, a classmate of Trélat's), began to accept the advantages of experimentation in the creation of theory. Love indicates this in the title of his 1852 treatise, *Mémoire sur la résistance du fer et de la fonte; et emploi de ces métaux dans les constructions, basé principalement sur les recherches expérimentales les plus récentes faites en Angleterre* (Memorandum on the resistance of cast and wrought iron and the use of these metals in construction, based primarily on the most recent English experimental research).[351]

The study of timber and iron was not the only area in which French material science profited from English experience. Hydraulic cement—in which the French excelled through the work of Louis-Joseph Vicat—was another material that benefited from empirical research. Although Vicat began his research based at first on publications by the chemist Balthazar-Georges Sage and the geologist Barthélemy Faujas de Saint-Fond, his engineering inspiration came from the studies of John Smeaton, whose assumptions he corrected through experimentation. This led him to create his commercially viable artificial hydraulic cement.

The difficult relationship between experiential and analytical approaches to knowledge surfaced for example in the disagreement between Navier's theoretical study on bending beams that he published in 1826 and expanded in 1833, and the results of Vicat's practical experiments in 1833 (Foce 2009, pp. 483–487). The conflict between the two cultures of material science and material research would be reconciled in the pedagogy of the ECP from 1829 under the influence of Colladon, who was both an experimental and a theoretical physicist, as well as through the *vulgarisation* of structural statics by Bélanger.

Mutual influences in construction

British engineers and entrepreneurs influenced the development in France not only through their research but in practical construction too. Robert Stephenson, Manby, Locke, Mackenzie, Brassey, and York consulted, built, and manufactured

351. A preference for practical experimentation continued to be a British trait. See for instance Guillerme; *La maquette d'étude. Un détour et une tentation de la philosophie naturelle*, 1987, reprinted in Guillerme 2008.

in France, communicating what they knew to French professionals. Less is known about what they took back with them to the other side of the Channel. British professionals encountered wrought-iron structures, metallurgical theory, mechanics, and increasingly also ECP graduates. What did they learn from the young industrial engineers with their multiple capabilities in theory, practical machine construction, and industrial know-how? British educators—men like Barlow, Gordon, and Rankine—spoke French, and many of them were interested in the science behind the production of iron that was being developed at the EP. But it was only after 1850 that British engineers began seriously to study theory, thanks to the efforts of men like Prince Consort Albert or Sir Lyon Playfair (see Smith 1992, Humber 1861, and Sheilds 1861).

Engineers and scientists from both sides profited from each other and from the International Exhibitions of 1851 and 1855 in London and Paris. Even more international exchange was carried on illicitly through industrial espionage, as Perdonnet had proudly reported. This only obfuscated the influences and the transfer of information.

French ironmaking had traditionally concentrated on the production of wrought iron, whereas British industry had initially focused on the simpler production of cast iron. Under the influence of railway construction, British industrialists quickly shifted to the production of wrought iron as well (Sunderland). While the French had produced their iron by forging, the British invented and built rolling mills. This initially made iron production less expensive in Britain than it was in France, but France soon caught up (Dupin, *Voyages* vol. 3, 1824, tome I, p. 245). France was by no means solely on the receiving end of improvements, and soon, in a reverse relationship, the innovations the British had generated were perfected in France and re-exported abroad. There were even parallel developments like the invention of the steam hammer in Manchester by James Nasmyth and in Le Creusot by François Bourdon, both in 1839. The first to be built was Bourdon's *marteau-pilon.* It began working at the Schneider plant in Le Creusot in 1840.

French builders concentrated on the incombustible aspect of iron construction. As Eck showed, they became expert at building iron roofs for theaters and other public spaces. A very early French example was Jacques-Germain Soufflot's roof over a staircase in the Louvre (1779–81; Steiner 1984, p. 23). Russian builders had built even earlier ones (Kosyhk 2021), but no information has yet appeared

Fig. 71. François Bourdon's *marteau pilon* installed at Le Creusot in 1840. (Source: *Le Monde Illustré*, 29.06.1889, p. 432 – downloaded from gallica.BnF.fr) James Nasmyth's virtually identical steam hammer, designed in 1839 and built in 1842. (Source: Nasmyth, *Autobiography*, 1883, opposite p. 270)

to show that the French even knew about the Russian examples.[352] Theaters were especially vulnerable to destruction by fire because of their combination of combustible stage sets, opulent costumes, and open-flame footlighting.[353] Victor Louis's hollow-tile and wrought-iron roofs for the Théâtre du Palais-Royale in 1784 (Eck 1836, vol. 1, p. 71 and pl. 31) and the 1789 Théâtre Français (Chalvatzi 2020) became widely known.[354] By the 1820s, fire codes mandated iron construction for theater roofs, and they proliferated.

English builders knew what was going on in France, and this may have indirectly influenced their interest in fireproof construction. The mill owner William Strutt had purportedly attempted to obtain plans of the Théâtre du Palais-Royale in 1791 after watching the five-year-old steam-powered Albion Mill burn in London. The evidence is only anecdotal, and at any rate he was unsuccessful. France was in Revolutionary chaos and stood on the brink of war. So Strutt resorted to his own ingenuity and used cast iron, brick, and plaster in his mill at Darby (1792) and in rebuilding another at Belper in 1803.[355] By mid-century, the situation had evolved.

The 1853 fourth edition of Tredgold's *Elementary Principles of Carpentry* indicates what the British manufacturers, engineers, and contractors were learning from the French by the early 1850s. Tredgold's treatise had been posthumously edited and expanded by Peter Barlow, who began to include iron structures after 1840. In this edition Barlow added several recently built roofs. Among them are French-influenced, Polonceau-type trusses built by Lewis Cubitt for the goods sheds of the Great Western Railway in the early 1850s (pl. 50), Richard Turner's shed at the Glasgow docks[356] (pl. 51), his Galway Terminus of the Midland Great Western Railway (1851; pl. 53), and those for the smithy and the engine shop at

352. Although many late eighteenth-century Russian structures seem to have been built using western European expertise, these even earlier roofs were unknown in France and Britain; see for example Kosykh 2021. Eck shows later Russian examples in 1841.

353. Iron had previously been used in conjunction with wooden structures in theaters, especially together with stage machinery. See Soufflot's acquaintance Dumont, *Parallèle de plans…*, especially the 1774 edition.

354. Ango's floor structure that used a web-shaped joist c. 1782 (*Procès-verbaux de l'Académie Royale d'Architecture,* vol. IX ,1780–93, p. 79) and several roof designs were published by Rondelet (see Steiner 1984, pp. 23–24).

355. Strutt's work influenced the structure of Charles Bage's Ditherton Flax Mill (1796–97), which is considered a precursor of the iron building frame.

356. Built either at the Lancefield Quay (1844), the Finnieston Quay (1848), or the Springfield Quay (1850).

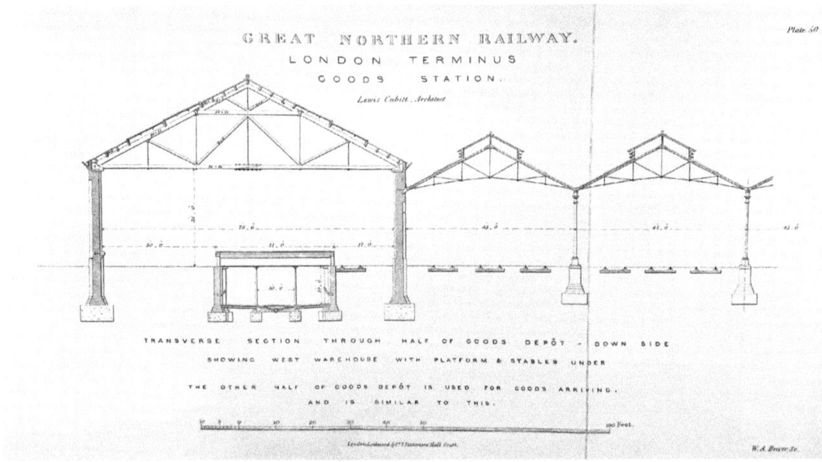

Fig. 72. Lewis Cubitt's Great Western Railway goods depot built in the early 1850s. (Source: Tredgold, *Elementary Principles of Carpentry*, 1853, pl. 50)

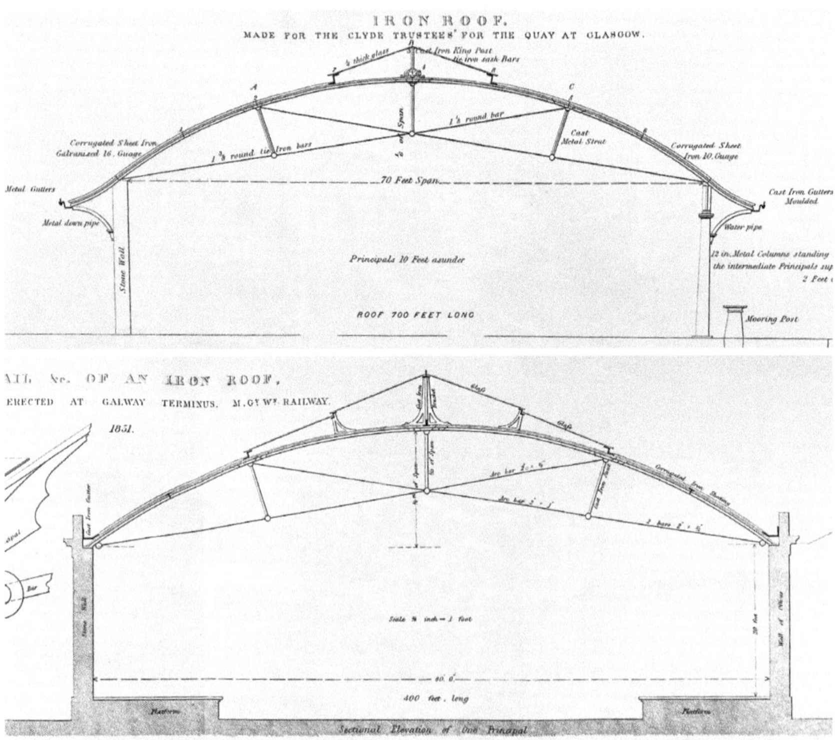

Fig. 73. Richard Turner's storage shed in Glasgow docks, probably on Springfield Quai 1850, and his station roof at the Galway Terminus of the Great Western Railway 1851. (Source: Tredgold, *Elementary Principles of Carpentry*, 1853, pls. 51 and 53)

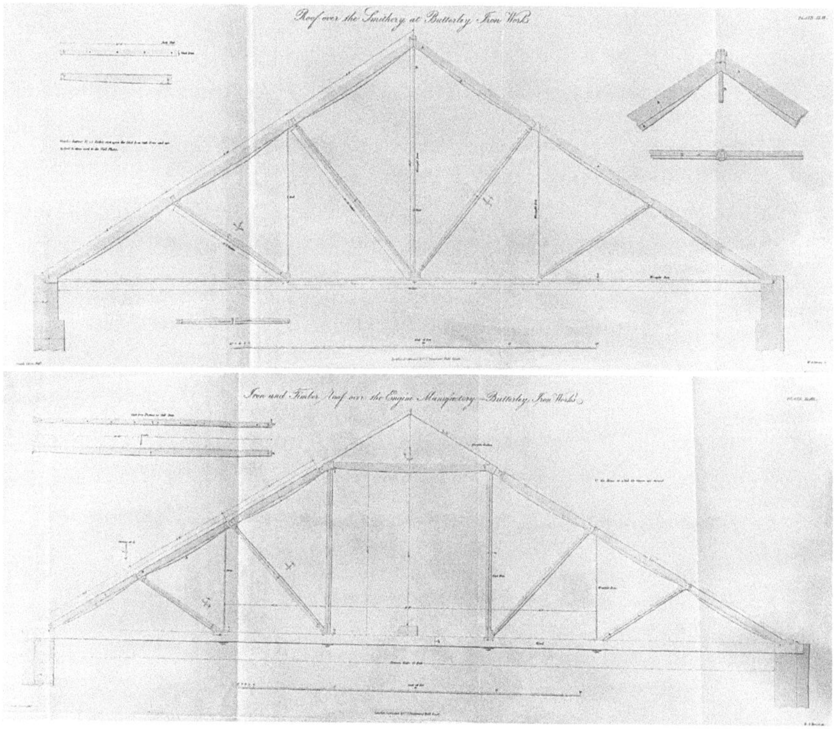

Fig. 74. Roofs of the smithy and engine assembly shed at the Butterley Ironworks, probably built around 1852 by Edward Reynolds. (Source: Tredgold, *Elementary Principles of Carpentry*, 1853, pls. 42 and 43)

the Butterley Iron Works—probably built to the designs of Edward Reynolds from 1852 (pls. 42–43).[357] Although Polonceau had almost certainly learned from the even earlier English trusses he had seen in Manchester (Fedorov, Heres, Lorenz 2022, vol. 1, pp. 274–278), these examples manifest a family relationship to Polonceau's famous design of 1837. Since they used some rolled members, they differed from the earlier French examples that had been entirely forged by hand.

Two English engineers originally enabled the industrialized rolling of wrought iron. Henry Cort patented grooved rollers in 1784 (UK pat. 1420),[358] and John Birkenshaw benefitted when he patented the first successful wrought-iron rail in

357. These roofs could also have been built under Joseph Glynn before 1850, although Glynn was chiefly known for his contributions to steam engine design, not roofs.

358. The first pieces to be industrially produced in large quantities were not rolled sections, but simple flat bars, the so-called 'strap rails' for railways in 1811 (Jewett 1967, p. 348).

1820 (UK pat. 4503).[359] The rail was roughly T-shaped, and a decisive innovation because it enabled the production of 15-foot (4.5 m) units. The previously used brittle cast-iron rails had only been useful in 3-foot (0.9 m) lengths. The new rail spurred railway development and, moreover, rolling them initiated a mechanized production process that led to a genuine industrialization of iron manufacture.

Aaron Manby immediately profited from the invention to roll deck beams in an approximate I-shape for building his first iron paddle steamer (1821–22) designed by him with his son Charles and Sir Charles Napier, who funded the project. He sailed it across the Channel and up the Seine to Paris. He had previously established himself in France in 1819 to avoid paying import duties on his steam engines, and he now moved there, establishing the technique of rolling iron sections in France.[360] Variant rail types and rolling mills followed quickly in England, France, Belgium, and the United States as the railways spread in the 1820s and 1830s. Rolled sections of many types soon appeared for other uses too. Boilermakers and shipwrights used them, and this gradually led to modern iron production.

Flachat and the I-beam

Following Manby's lead, builders in Britain used bulb beams, the so-called deck beams in their structures. They appeared for example in Richard Turner's sheds and hothouses, most famously in the Kew Palm House in 1848. But the modern I-beam, that iconic symbol of industrialized iron construction, evolved in France, where it achieved instant popularity as a floor joist.

Cast H- or I-forms had been known previously and were occasionally used in pre-industrial structures, but it was producing them in wrought iron that would bring about their industrialization. Flachat and the ECP graduates working in his office were involved in the development and introduction of the first rolled I-beams.

---

359. Mehrtens writes that an ironmaster named Nixon had previously rolled a rail in 1805, but it had not proven useful (1887, p. 32). However, no ironmaster of that name can be traced at that time.

360. Manby formed a firm in Charenton-le-Pont in 1819 to manufacture his patented oscillating steam engines, and in 1822 another, the Compagnie Anglaise, to manufacture lighting gas with his manager, the Irish chemist Daniel Wilson as partner. He purchased the Le Creusot ironworks in 1826, and it became the Société Anonyme des Mines, Forges et Fonderies du Creusot et Charenton. which apparently went bankrupt in 1833. The Le Creusot works were sold in 1836 to the Schneider brothers, and Manby returned to England around 1840 and sold the Horsley Iron Works in 1845. Wilson settled permanently in France and the Compagnie Anglaise continued to exist until 1847.

In 1841 Flachat developed a primitive type that he based on the railway rail. He used it in the station roofs of the Paris–Saint-Germain Railway (Halbou, pat. 8040 1849, pp. 8–9). Polonceau, who had probably worked on the designs for at least some of those stations that used the new beam in his truss, had taken on the directorship of the Strasbourg-Basel Railway in 1842 when construction on the largest project began, the Gare Saint-Lazare. Ernest Deligny (grad. 1842) replaced him in 1843 and became Flachat's collaborator for the railway line (Malo 1873, p. 19). In that capacity he was involved in the first (1842–48), and also at least in the initial design phase of the second stage (1851–53) of the station's construction (Vauthier 1898, p. 418).[361]

The I-beam might have become just another useful, specialized member in iron construction if its development had not been forced by a three-month-long Parisian carpenters' strike in 1845. Builders needed a replacement for their wooden floor joists, and iron manufacturers, who of course needed an appreciable volume of orders to retool and invest large sums in new plant, responded. The I-beam quickly evolved into its mature form over the course of less than a decade. By 1849 it was being rolled industrially (Jewett 1967, p. 362 / Mehrtens 1887, p. 41) and adopted by many contractors in their buildings (Steiner, p. 72). The development was fortuitously supported by Haussmann's rebuilding of Paris, which began in 1853. Boulevards were cut through the city fabric with hundreds of new multi-story buildings built to line them. The I-beam joist spread rapidly. Fairbairn reported on it to the British profession in 1857–58 (see later), and by 1859 it was being used as far afield as in the floors of the new Province of Canada's parliament building in Ottawa, then a settlement on the frontier of British colonial expansion. This may have been the first instance of the I-beam's use in a building in North America (Taite 2022, p. 26).

Flachat's initial I-beam preceded the strike. Because he needed a relatively small number, it must have been forged in a *ferronnerie* rather than rolled industrially. That is why the second volume of his magisterial treatise on iron manufacturing that describes rolling mills does not mention I-beams, but

361. Deligny, who had graduated as a metallurgical engineer, left for Spain in 1850 to design a railway, survey pyrite mines, and propose another railway line to service them. He returned to Paris in 1859.

only bulb beams, T-beams, and various rail forms. There is a *rail à double T* (Flachat et al. 1842–46, vol. 2, 1844, p. 657), which was the standard rail used in England and France and the ancestor of the modern railway rail, and this was certainly the origin of Flachat's 1841 I-beam design.

Flachat's experiment was a success. The new beam permitted straightforward connections like simple bolting or riveting to both the top and the bottom flanges, and moreover, the flanges provided convenient support for flooring systems too. Flachat's proto-I-beam obviously impressed Alphonse Halbou, director of the French branch of the Belgian firm the *Société Anonyme des Laminoirs, Forges, Fonderies et Usines de la Providence* in Hautemont (Département Nord). He recognized its potential, and its structural success inspired him to develop the form further. Halbou's experiments led to his patent of 1849 (F pat. 1BB 8040). In the description that formed part of his patent application, he reported what had motivated his research:

> Because of its special structural properties, the original rail that has been in use in railway construction for almost twenty years now provided the necessary data for initiating the experiments that we undertook. In fact, the problem was to determine the [most ideal] proportions so that one could modify it in order to make it as light as possible for the intended use while guaranteeing the maximum possible strength.
>
> The idea had occurred to the St. Germain Railway Company to use improved rails for the trusses of its station roofs.
>
> But like everything in an embryonic and experimental developmental phase, the improved rails invented by the St. Germain Railway Company are far from providing or being able to provide the results that fulfill the three conditions we mentioned above; [strength, lightness, and economy of construction], and the fact that they nevertheless did deliver the good results that we acknowledge was bought at enormous expense, due first of all to their great weight, their very high cost, and then by the structural effort required to supplement their lack of strength, a lack caused by their inadequate height and which had to be countered by intermediate supports. In other words, these iron elements can [only] be used in trusses where one has substantial financial resources and are entirely inappropriate for private construction, especially for building floors using iron members. The clearest proof of this is the fact that over the past eight years, ever since these iron members have been publicly accessible and known to architects and iron-contractors, they have never once been used again in floors, for which they

> are inappropriate due to their enormous weight and their enormous weight [*sic*] and their lack of strength. (Halbou 1849, pp. 8–9)[362]

We note that Halbou wrote that Flachat's beams had been used for eight years at the time of writing—that is, from 1841. His description is the only source to mention them and their use in the construction of the railway's stations, which obviously included the most important, the first construction phase of the Saint-Lazare station (1842–46). Although it does not state so specifically, the description implies that the I-beams were short and forged rather than rolled, Retooling an industrial plant merely to provide a new type of member for a single project would have been prohibitively expensive. Flachat's beams would probably have been manufactured in similar lengths to standard rails, which were in the range of 4.5 meters,[363] and two lengths spliced together (possible in wrought iron without a support from beneath) would have resulted in a span of 9 meters, but not more.

---

362. Le rail proprement dit, employé depuis près de vingt ans dans la construction des chemins de fer, est venu offrir, par sa construction particulière, les données nécessaires qui ont servi de point de départ aux essais que nous avons faits. Il s'agissait en effet, de donner à ce fer des proportions telles, que tout en lui assurant la plus grande résistance possible on pût, par diverses modifications, obtenir la légèreté indispensable à l'usage auquel on le destine.

Cette idée était déjà venue à la compagnie du chemin de fer de St. Germain, qui pour l'exécution de la toiture de ses gares, fit exécuter des rails perfectionnés qui furent employés dans la construction des combles.

Mais comme tout ce qui est à l'état d'enfance, & d'étude, les rails perfectionnés de l'invention de la compagnie du chemin de fer de St. Germain, sont bien loin d'avoir donné, & de pouvoir donner les résultats exigés par les trois conditions que nous avons annoncées plus haut; [solidité, légèreté, économie dans la construction] Et s'ils ont donné les bons résultats que l'on connait, cela n'a été qu'avec des dépenses énormes, nécessitées d'abord par leur poids considérable, leur prix de revient très élevé, puis par les travaux qu'il a fallu faire pour suppléer au défaut de force de ces fers, défaut provenant de leur peu de hauteur, & auquel on a du obvier en les faisant supporter par des points d'appuis; En un mot, ces fers qui peuvent être employés pour des combles, où l'on a la faculté de faire de grandes dépenses; sont tout à fait impropres aux constructions particulières et notamment à celle des planchers en fer; Et la preuve la plus évidente de ce fait est que depuis huit ans, que ces fers sont aussi connus que possible des architectes & serruriers, par l'endroit même où ils sont employés; On n'a pas essayé une seule fois de les appliquer à la construction des planchers, où ils sont tout à fait inadmissibles par suite de leur poids énorme & de leur poids énorme [*sic*] & de leur peu de résistance.

363. Rails were being rolled at lengths of 4.5–4.8 m (for the Paris-Rouen railway) in 1844 (Flachat et al. 1844, vol.2, p. 663). The lengths were limited by the needs of transportation to the construction site.

Halbou wrote (p. 12) that Flachat's beams were 120 mm high, but cumbersome because they weighed 20 kg/m, while his 120 mm beams weighed only 11 kg/m and carried more due to their better distribution of material. Moreover, he produced beams between 120 and 240 mm in height that could be rolled up to 14 meters in length.[364]

The year that Halbou registered his patent, several firms, including his own, began rolling them commercially (Steiner, p. 72). Unless I over-interpret the following statement, Halbou's comment on one of the two diagrams of his beams seems to suggest that he might have been open to licensing the roller forms that he had developed:

> The rollers needed for all the intermediate sizes, aside from those 120 and 160 mm high, are owned by the company, and their thickness can be varied between 6 and 20 mm according to the needs of industry. (Halbou 1849, p. 4)[365]

Whether such a license led to the beam's immediate proliferation (Jolly 1863, p. 13, credits the Forge de la Providence as the cause), or whether there was question of a patent infringement or parallel development has not been studied. The image that Halbou attached to his patent application in 1849 shows seven sizes ranging from 120 to 240 mm in height, with 45–68 mm flanges and a proposed variant with a stiffening flange in the middle of the web that he added to the original patent in 1850. Although he makes no mention of this, the range of profiles and profile thicknesses suggests that he had devised a system of exchanging or modifying rollers easily.

The predecessor of Halbou's employer, the Forges de la Providence, had been established by the iron manufacturer Ferdinand Puissant in 1832 in Charleroi, Belgium, under the direction of Thomas Bonehill who had arrived in Belgium several years before. A contemporary source (Valérius 1843, p. 61) reports that an 1835 Bonehill rolling mill (probably belonging to the Société le grand laminoir de la Providence that Bonehill had established with Puissant's widow that year)

---

364. Steiner cites variant dimensions and weights from the secondary literature (1984, p. 72 and p. 214 footnote 22). The logistics of transporting 14-meter lengths to a building site was another matter altogether.

365. Les cylindres nécessaires à l'exécution de toutes les dimensions intermédiaires outre les hauteurs de 120 à 160 mm sont en possession de la forge qui peut varier également leur épaisseur de 6 à 20 mm, suivant les besoins de l'industrie.

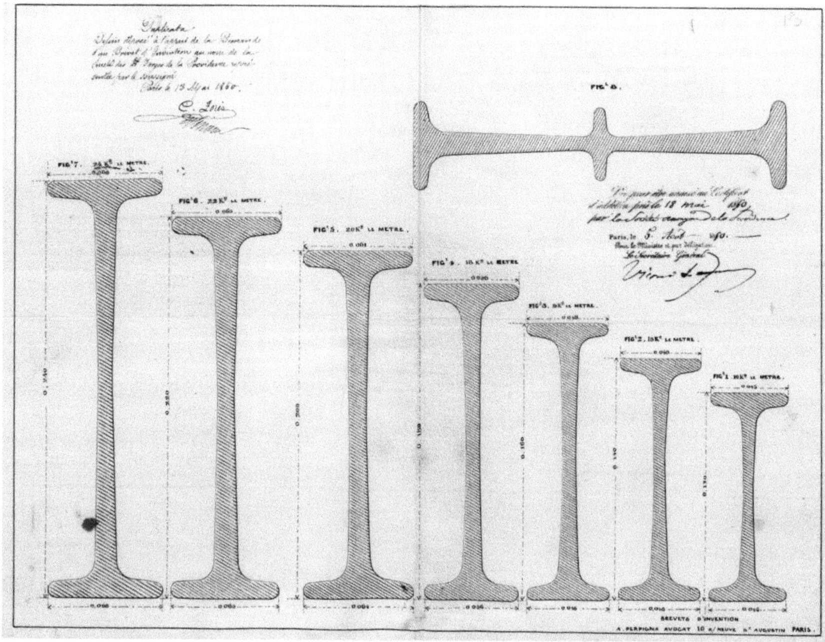

Fig. 75. The first series of industrially manufactured I-beams from the fifteen-year French patent application number 8040 dated March 1, 1849 by Philippe Alphonse Halbou for the manufacture of improved I-beams for the construction of floors, lofts, and trusses in iron. (Source: © INPI French Patent and Trademark Office - INPI Institut national de la propriété industrielle, pat. no. 8040 / 1BB8040)

was remarkable for the variety of sections it could produce, but no mention was made of adjustable mechanisms for rollers. Reiner Daelen, a German-Belgian[366] trained in Brussels in the early 1830s, installed what are recognized as the first adjustable cylinders, a so-called Universalwalzwerk in a rolling mill at Hörde near Dortmund in 1848 (Troitsch 2007). Whether there was any connection between Daelen and Bonehill or an influence of either of them on Halbou's 1849 development cannot presently be ascertained. The French branch of the 1838 Société Anonyme des Laminoirs, Forges, Fonderies et Usines de la Providence that had evolved from the Puissant-Bonehill collaboration was founded in 1843 to take advantage of the growing demand for rails for the rapidly expanding French railway network. However, Bonehill had left the firm in 1839 to become

366. To be precise, Eupen, where Daelen was born in 1813 was French between 1794 and 1815, when it became Prussian. It only became Belgian in 1920.

an independent consultant who built a total of over 40 ironworks in Belgium, France, and Germany over the course of his career. He may have influenced his successor Halbou's research although his patent application only mentions the Paris–Saint-Germain Railway as his inspiration.

In the course of this overheated industrial development, Belgian and French engineers and manufacturers often collaborated, influenced one another, or engaged in industrial espionage not only abroad but among themselves as well. It was therefore to be expected that several loosely interdependent developments would occur between 1841 and 1849, especially when the strike of 1845 opened up a new and potentially lucrative market. The year 1846 saw several variants.

According to Jolly (1863, p. 13), the first rolled (as opposed to forged) I-beams were manufactured at La Villette in 1846.[367] Deligny used them in the machine house of the Le Pecq–Saint-Germain atmospheric railway (1846–47; Vauthier 1898, p. 418 and Bordes 2010, p. 60), but we do not know whether Flachat was involved in their design. In the same year, however, Flachat ordered two more I-sections from a second forge located at Montataire (Oise), about 70 km north of Paris, and this time both Flachat and Deligny were involved in their design and manufacture (Malo 1873, pp. 19–20).[368] From Halbou's patent application and Léon Malo's account, we can surmise that both the original 1841 beam was used in the first construction phase, and some of the beams from 1846 were used for covering the Gare Saint-Lazare in the second phase (Malo 1873, p. 19–20).[369] These later trusses were all riveted rather than pin-jointed as the earlier ones had been (Steiner, p. 83). In 1853 Cornet described how the I-beams were used (see description to plate 41 shown on Fig. 24).

The construction of the station spanned a decade in two phases between 1842 and 1853, so it is not surprising that Flachat should have improved on his first heavy beams during that time. Although we do not know who had worked with Flachat on the 1841 beam design (Polonceau, Barrault, and Petiet were all in the office at the time), Deligny was definitely involved in developing at least one of

367. The responsible maître des forges was named Jules Lagoutte.

368. The Montataire beams were used in roofs of other buildings for the Le Pecq–Saint-Germain segment of the railway. Steiner names yet another mill at Fourchambault near Nevers (Département Nièvres) that rolled a third type in 1846 (p. 72), and Jolly in 1863 mentions I-beams rolled at Ancy-le-Franc (Département Yonne) and Ars-sur-Moselle (Département Moselle), however without dates (pp. 13–14). Flachat does not seem to have been involved in these.

369. Jolly's firm was responsible for the construction of these Polonceau trusses.

the 1846 rolled variants and in both design phases and one construction phase of the station.

All the early beams, both forged and rolled were, however, still manufactured in small production runs, and their price was consequently high (Steiner p. 72). That apparently changed with Charles Ferdinand Zorès's I-beam, because as Steiner writes (without explaining how), he was able to undercut the competition.[370] Flachat took advantage of this and used Zorès beams as well for platform canopies in the Saint-Lazare station, possibly also in 1846 (Mende 2006, p. 2159, citing J.-F. Belhoste 1996), although Steiner writes (p. 72) that the manufacturer only began to sell them directly to the public in 1848.

Flachat and his young ECP graduates were therefore intimately involved in the early development of the I-beam. They were in the vanguard of its design and application. In general, one might say that Flachat's Saint-Germain Railway served as a proving ground for a large part of the early organizational and technological development of the iron industry in France. As Malo wrote somewhat flamboyantly:

> This tiny Saint-Germain Line, that today is merely a speck lost in the folds of the Chemin de fer de l'Ouest, was, one can boldly assert, the veritable national school of railway construction. The seeds of almost all progress in this great industry sprouted there. It was a family affair: the hierarchy was benevolent. Everyone from the humblest employee to the most exalted, gave what he could in enthusiasm and ability in pursuit of the common goal. Science trumped bureaucracy. The goal was to create rather than to supervise, and here Flachat's inventive genius proved to be abundantly creative and active. (Malo 1873, p. 21)[371]

---

370. Steiner p. 72. Mehrtens suggested (1887, p. 32 repeated by Mende 2006, p. 2159 citing J.-F. Belhoste 1996) that it may have evolved with the help of a contractor named Chibon (Mende 2006) who had been experimenting with a composite, riveted I-section made of two rolled T-sections. This was probably Pierre Chibon, described as entrepreneur des couvertures d'édifices (roofing entrepreneur). Zorès is better known for his 1845 Zorès section, an inverted U-shape that he developed in multiple patents between 1852 and 1868. The Zorès section became popular in France and Germany and was used especially for decking in railway bridges.

371. Cette petite ligne de Saint-Germain, atome aujourd'hui perdu dans les replis du réseau de l'Ouest, fut, on peut l'affirmer hardiment, la véritable école nationale des chemins de fer. De ses bureaux et de ses ateliers sont sortis des germes de presque tous les progrès réalisés plus tard dans cette grande industrie. Tout s'y passait en famille ; la hiérarchie y était paternelle ; du plus humble employé

The time lag between Flachat's experimental innovation in 1841 that had inspired Halbou, the three or four early rolled forms in 1846–48, and the production of the I-beam on a larger scale as a standard rolled section from 1849, can be explained by the difficulty that manufacturers experienced in organizing the process, finding the capital to invest in new plant, and in developing the technology and training the necessary manpower to guarantee the precision of what their mills produced. Rolling was conceptually different from forging, which needed fewer workmen and comparatively simple machinery. Rolling demanded a large and complex machine park, a high level of specialization of the workforce, and an extreme division of labor including a full team of maintenance men to service and calibrate the machinery. Precision was an ongoing problem, and it was precision in manufacture that would enable industrialized construction. Barrault explained how critical calibration and maintenance were:

> When cylinders are poorly aligned, production becomes impossible. One can only produce bars of unequal thickness, cambered every which way, chewed at the edges, etc., etc., and commercially totally unusable. (Flachat et al. 1842–46, vol. 2, 1844, p. 656).[372]

Barrault, who first as Flachat's employee and then independently had spent years perfecting iron mills, was the lead author (Flachat 1866, p. 536) of the previously cited comprehensive treatise on every aspect of structural iron at the time. He published it with Flachat and Petiet in three volumes (1842–46), providing yet further proof of the fact that the railway industry spurred the development. The treatise predated the rolled I-beam, which is why it was not mentioned, but it provided a comprehensive analysis of the industry up to 1846, and it treated every aspect of the material and the organization of iron manufacture in great detail.[373] The section on rolling technology (Flachat et al. 1842–46,

---

au plus élevé, chacun apportait au but commun tout ce qu'il pouvait donner de zèle et de capacités ; la science y primait la bureaucratie ; il s'agissait moins d'administrer que de créer, et le génie inventif de M. Flachat s'y montrait merveilleux d'abondance et d'activité.

372. Lorsque les cylindres sont mal réglés, le travail devient impossible : on ne produit que des barres d'épaisseur inégale, cambrées dans tous les sens, mâchés sur leurs arêtes, etc, etc, et tout à fait inacceptables dans le commerce. The saws that cut the ends of rails, for instance had to be maintained after every 24 hours of use. (Flachat et al. 1844, vol. 2, p. 662)

373. The development of ironworking in France and the influence of political and economic factors and the availability and use of combustible materials is succinctly explained in vol. 1, 1842, pp. 23–28.

vol. 2, 1844, pp. 566–621) explains the complexities of the method.

The new beams were used chiefly in building structures, simply because of the limited size to which they could be rolled. They were used in small bridges, too, but the I-beams that were needed for large structures like the Langon Bridge (1854–55) or the Bordeaux Bridge (1857–60), both designed by Stanislas de La Roche-Tolay, were riveted together out of plates and angle irons. As late as 1870 Paul Régnauld (pl. 3) showed rolled I-beams that he described as *Fers spéciaux pour les ponts de 4,6 et 8.m 00 d'ouverture*. The largest he showed measured 300 mm in height and with flanges of 120 mm.[374] All I-beams larger than that were built-up members.[375]

While French builders immediately embraced the I-beam for their floors and roofs from the late 1840s, the contemporary examples of British roof structures that Barlow presented in his Tredgold edition of 1853 had rolled angle irons and bulb beams, but no true I-beam. Fairbairn appears to have been the first British professional to report on the new French advances in beam design and rolling technology. He visited the Paris Exhibition of 1855 and admired the size and the quality of the I-beams that French mills had been able to manufacture. In the second, much expanded edition of *The Application of Cast and Wrought Iron to Building Purposes* (1857–58, which he updated after he had visited the Exhibition, he pointed out that 'unless [in Britain] wrought-iron beams can at once be produced from the rolls similar to that represented on Fig. 37 [an I-beam], which is evidently the cheapest and best' (Fairbairn 1857–58, p. 88 and fig. on p. 86), it would be better to produce a composite one. He evidently found the French production of rolled sections to be far more advanced than what was then possible in Britain. Of their quality and use he wrote:

> ...we have yet to attain in this country a higher state of perfection in the manufacture, and a more extended application of wrought iron beams to the construction of dwelling houses. (Fairbairn 1857–58, p. 89, cit. from the 1870 edition by Steiner 1984, p. 73).

Therefore, by the mid 1850s, the problems of economy in tooling and precision—

---

374. According to Steiner, p. 73, Joly had used this size in a warehouse.

375. In 1863 (p. 23) Jolly mentions an unsuccessful attempt by the mill at Commentry (Département de l'Allier) to roll an I-beam of 500 mm in height. Even as late as 1880 in the United States the largest rolled I-beams appear not to have exceeded 15 inches (c. 380 mm) in height (Delaware Bridge Company's Illustrated Album, p. 10). See also Maw and Dredge 1872, who show very few examples of rolled I-beams in railway bridges.

at least to the tolerances required by structural work—were being competently tackled in France and the resultant products were appearing in French structures. It was precisely such industrially rolled sections that characterized the difference between the architectural expression of Labrouste's first library roof of the Bibliothèque Sainte-Geneviève and the second at the Bibliothèque nationale.

In many cases the French tended to concentrate on the export of expertise, both in the form of engineers and as knowledge, while the British principally focused on the export of goods. Of course, both did both (see Steiner p. 64 for the export of prefabricated iron buildings from France), but each culture had its own focus. The French exported individual structures while English manufacturers also actively developed standardized, prefabricated housing as a commodity for export, initially in wood as early as 1830, but then also in corrugated iron (Herbert, 1978). Joseph Paxton experimented with prefabrication and serial production for his 'Great Stove' hothouse at Chatsworth (1836–40). The British became experts in the design and construction of conservatories, increasingly in cast iron, and exported prefabricated iron structures of all kinds, as the example of Reynaud's 1857 Walde Lighthouse near Calais documented.

Fairbairn produced and sold not only iron ships and locomotives, but also what may have been the first totally prefabricated iron 'machine-building'. It took the form of a grist mill and was a hybrid structure that was both a building that housed machinery and a building-sized machine with space inside it for humans to service a manufacturing process. In a way, it was structurally analogous to an iron ship, and Fairbairn designed it for the Turkish Seraskier (defense minister and commander-in-chief) Halil Pasha in 1842 (Pole 1877, pp. 167–172 and Fairbairn 1864–65, vol. 2, 1865, pp. 119–121).[376] Innovative, iron-based exports increased rapidly, and by mid-century the Manchester industrialist Edward Bellhouse, whose father David Bellhouse Jr. had co-founded the Manchester Mechanics Association with Fairbairn in 1824, was exporting large numbers of

376. The europhile sultan Abdülmecid I, the first Turkish ruler to speak French, followed the international opening that his father Mahmud II had begun, who had contacted Fairbairn to consult on the production of ordnance and on industries for provisioning the army. Abdülmecid was later allied with France, Britain, and Sardinia against Russia in the Crimean War (1853–56). Damat Gürcü (the Georgian bridegroom) Halil Rifat Pasha, who had married into the imperial family, was then briefly commander-in-chief of the army, and on several occasions also of the navy. The issue at the time was the containment of Muhammad Ali, the de facto ruler of Egypt, in his aggression toward the Turkish empire in which the Russians, the French, and the British were all involved.

prefabricated, corrugated, and galvanized houses to California for the Gold Rush (*Allgemeine Bauzeitung*, vol. 15, 1850, pp. 184–185 and pl. 342).[377]

By mid-century the two countries had different but equivalent technology that they applied to industry in multiple ways. The structural iron industry and prefabrication technology were developing on both sides of the Channel, although Britain produced far more iron. By the time that Paxton and Charles Fox were designing the prefabrication and incremental component organization of the Crystal Palace in 1850, France was producing I-beams in bulk. How the reciprocal relationships developed and affected one another, and how influences flowed back and forth between these two diverse technological cultures and the other primary producers of iron in Austro-Hungary, Bavaria, Belgium, Prussia, Russia, and Sweden in the early to mid-nineteenth century is a topic for more detailed study.

---

377. Bellhouse also prefabricated and exported large-scale structures like the two-span lattice Puente Carlos Gomes near Rio de Janiero before 1860.

Perdonnet's administration

Lavallée retained the directorship for five years after turning the school over to the government in 1857–58. By 1862 he was 65 years old and tired, and the still dynamic, although only four years younger Perdonnet became the second director of the school, certainly on the recommendation of the emperor and probably by his direct appointment. Among his first moves was the installation of a *conseil supérieur de perfectionnement de l'enseignement secondaire spécial*. It replaced the original *conseil des études* and was chiefly made up of alumni, which had been a condition of Lavallée's gift of the school to the government.

Fig. 76. The aging Lavallée in the 1860's. He would retire as director of the École Centrale in 1862. (Source: *carte de visite* after 1860, reproduced in Guillet pl. 59) Perdonnet was only four years younger than Lavallée but still very active when he took over as director. (Source: Photograph by Nadar, BnF open source)

Fig. 77. Dumas photographed around 1850 after he had been appointed Minister of Agriculture for a year. (Source: Wikimedia Commons)

As director of the ECP, Perdonnet immediately put his stamp on the school, taking the opportunity to stress his educational philosophy—especially as regarded the education of workmen. He was not at all in agreement with Bélanger or Dumas in this regard (Weiss 1982, pp. 222–223), despite Dumas's background in Saint-Simonian thought, which supported egalitarian social views. Both Bélanger and Dumas thought that the school should primarily address its efforts to the progeny of the first generation of industrialists and only secondarily support those who could not afford the school's tuition.

However, Dumas had withdrawn from active participation in the school when the *conseil des études* was disbanded in 1862, and Bélanger soon retired in 1864. Perdonnet as director definitely called the shots. Not that he despised the paying student, far from it, but he considered the *boursier*, the student funded by grants, as equally worthy of support. It was clear that Perdonnet valued not only the potential in workmen's education but also their physical contributions to technical knowledge:

> Stone cutting and carpentry are not without their problems. There are craftsmen who, as 'professors of practice', instruct their co-workers in their solution. I do not disagree that they have an advantage over our 'real' professors, which is that they understand how to communicate well with workmen. But they have one fault, and that is routine that is so detrimental to progress. I wish that our masters of practice and our professors would collaborate to profit from one another! (Perdonnet 1867, p. 29)[378]

Perdonnet's statement underscores yet again his evaluation of the necessary relationship between theory and practice in civil engineering and construction technology and its border-crossing nature in education. It explains his support for the school's founding decision to introduce workshop apprenticeships for the students, and it documents his respect for the laborer's intelligence and contribution to the building process. This non-elitist attitude had always permeated the school, and it served the graduates well. Perdonnet ensured that it would continue.

We can see the effect of this attitude in the delicate and dangerous restoration of the crumbling tower of Bayeux cathedral (1855–59) when Henri de Dion explained the reasoning behind his careful choreography of the process and his appeal to the workmen's understanding and support as a matter of course:

> The complete success of this type of operation depends to a great extent on the precision and the intelligence with which the men who execute the process follow the instructions they are given. I have always been concerned to interest them in the success of their endeavors by explaining the goal we are trying to achieve. (de Dion and Lasvignes 1861 pp. 56–57)[379]

Such an appreciation of the laborers' collaborative contribution to the building process was certainly not usual in nineteenth-century building practice, and it still is uncommon today.

---

378. La coupe des pierres et le débit des charpentes ne sont pas sans difficultés. Il y a des ouvriers qui, sous le nom de professeur de trait, les enseignent à leurs camarades. Ils ont un avantage sur nos propres professeurs, je ne le conteste pas, c'est de savoir mieux se mettre à la portée de l'ouvrier, mais ils ont un défaut, c'est la routine, si nuisible au progrès. Que ces maîtres de trait et nos professeurs s'entendent pour se reformer mutuellement!

379. La réussite complète d'une opération de ce genre dépend beaucoup de l'exactitude et de l'intelligence avec laquelle les hommes qui l'exécutent suivent les instructions qui leur sont données, nous nous sommes toujours préoccupé de les intéresser au succès, en leur expliquant le but que nous nous proposions d'atteindre.

The end of the pioneering period

However, political forces were pulling the school in a different direction, and Perdonnet would make other decisions that were not so positive. The new *conseil* had urgent business. A report to the emperor submitted by the Minister of Agriculture, Commerce, and Public Works Eugène Rouher at the end of June[380] stated that France had not appeared to be as industrially competitive at the London Exhibition of 1862 as had been expected. His *Rapport à l'Empereur sur l'état actuel de l'enseignement professionnel et sur les moyens de le perfectionner* concluded that something needed to be done about the coordination of technical education at all levels, especially the education of workmen. Rouher proposed appointing Perdonnet as one of the committee members to examine the issue (Rouher 1862, p. 114).

The Rouher report led to several modifications in the curriculum that were made between then and 1865.[381] A few were positive, but most were not. One of the changes was standardized entrance exams. They replaced the earlier system of personalized interviews supplemented by essays that had looked for aptitude rather than previously acquired knowledge.[382] Others were the gradual curtailment of architectural drafting and an increase in machine drawing, as well as the doubling of some courses to avoid the huge classes in subjects that were taught in common to all second- and third-year students. The loss of architectural drafting—that is, the documentation of building-sized structures—would only be compensated by the introduction of Marcelin Duplaix's (grad. 1883) course *constructions métalliques* on steel structures in 1897—but that lay three decades in the future. Together with the earlier transfer of the ownership of the school to a bureaucratic government administration, Perdonnet's changes initiated the end of the dynamic phase of the ECP (see also Pfammatter 1997, p. 123).

---

380. Rouher was a former member of Guizot's cabinet and now head of the long-standing patron ministry of the school.

381. It also led to the official government recognition and endorsement of the Association Polytechnique's many free course offerings, albeit only in 1869, two years after Perdonnet's death.

382. As a result of this, Adolphe Lonchampt issued his *Recueil des Problèmes,* a thick volume of set problems in 1865. These were probably exams that the EP had already set over the previous years and that were to be expected for the ECP in the future as well.

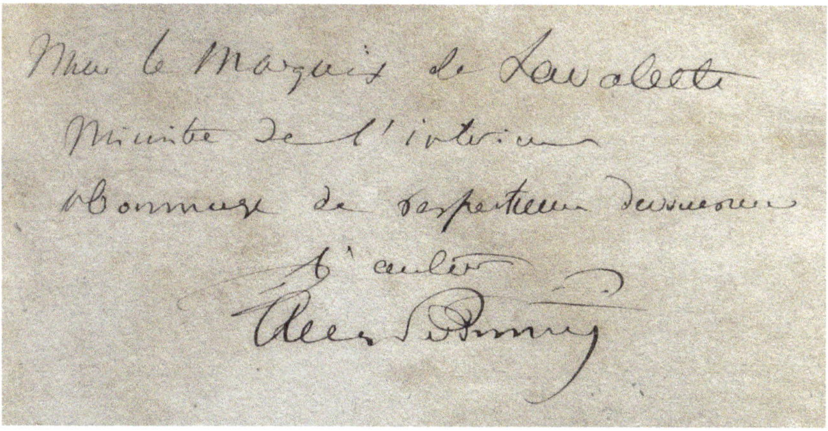

Fig. 78. Perdonnet's manuscript dedication to the incoming Minister of the Interior, Marquis Charles de La Valette, on the flyleaf of the first volume of the third edition of his *Traité élémentaire des chemins de fer*, 1865. (Source: Peters Collection)

Perdonnet tried to compensate politically for the loss of pedagogical vitality while at the same time actively continuing to push his interest in the development of railways. In the spring of 1865, he dedicated a copy of the new and final third edition of his *Traité élémentaire des chemins de fer* to the incoming Minister of the Interior, the Marquis Charles de La Valette, a supporter of Guizot and collaborator of Rouher who had apparently shown interest in railways at several instances over his career. The new minister was an anglophile and a Protestant. Perdonnet hoped for support.

The four volumes of La Valette's copy are in suspiciously fine condition. They display only very minor traces of shelf wear and none on the interior, and they even have a few uncut pages here and there. This suggests that the newly appointed minister, known as a *bon vivant* and social luminary (see 'La Valette' on the site www.senat.fr), may have been somewhat less interested in railway development than Perdonnet had hoped, or at least that he had other priorities to attend to at the time. Nevertheless, the dedication shows that Perdonnet had been an excellent choice for the directorship of the ECP. He was determined to leave no stone unturned and no possible contact unattended to.

The school had clearly grown too large to remain flexible, and was gradually evolving from an open and experimental institution into one with a more regimented format. The changes had been necessitated by the report, the new government ownership, and the finances of the school, and they sapped the élan of the early years. Playfair quoted Dumas himself, who criticized the growing intrusion of centralized government administration into education:

> If the causes of our marasmus appear complex and manifold, they are still reducible to one principle, administrative centralization, which applied to the University has enervated superior instruction (Playfair 1872, p. 9).

Although Germany also suffered under bureaucratic systems, the dispersion of schools and their regional rather than national organization prevented sclerotic ossification. The lively interplay of scientific and technological education continued to develop until the end of the century, while the ECP faculty and the graduates that the school produced gradually grew less adventurous in their professional experimentation. The boldly enterprising cycle was drawing to a close. There was no sudden change but rather a piecemeal narrowing of the educational purview, a misting over of pedagogical excitement, and it followed the professionalization of the fields of civil and industrial engineering. The fault was not—or at least not only—Perdonnet's. A sluggish and unimaginative government bureaucracy had replaced the dynamic and agile initiative and self-interest of the founders. Nevertheless, the change from a private to a public institution, with all the disadvantages that it entailed, allowed the school to survive and prosper.

In retrospect many of Perdonnet's changes were not supportive of the school's original concept. Some were forced by the sheer size of the student body in the 1860s or the growing diversification of engineering, while others were his personal decisions. He doubled the courses previously taught in common to the second and third years, relegating the introductory material to the first year and the more advanced subject matter to the appropriate upper-year sections. The increasing specialization of the engineering profession also necessitated the introduction of two new subjects. Faure's *résistance des matériaux* was the first that had already been introduced in 1858. The second, a specialized *constructions navales* was taught by Antoine de Fréminville, but only from 1865 to 1869. Especially the second ran contrary to Dumas's original concept of a generalist education. The proliferation of subjects paradoxically contributed to the waning of the school's leadership in engineering education and possibly alienated the few original faculty members and former élèves-fondateurs still left on the faculty.

Conflict

Several retirements at that time made the transition easier for Perdonnet. Mary left in the course of 1864, which allowed his course *constructions civiles et travaux publics* to be split between two chairs: Émile Muller (grad. 1844) took over *constructions civiles* from 1864, while the tunnel and railway engineer Adrien Ruelle became responsible for *travaux publics* the following year.

Mary's retirement from the ECP was probably neither voluntary nor amicable, especially since he had prepared and issued a new edition of his addendum to the construction course, *Détails pratiques sur la distribution des eaux*, headed 'Année 1865. École Impériale Centrale des Arts et Manufactures' in anticipation of his teaching the next year.[383] He was clearly not ready to retire, and he continued to teach *navigation intérieure* at the ENPC for another four years, through 1868 (Reynaud 1870). Mary and Perdonnet had most probably never been particularly close personally; they certainly did not see eye to eye on professional matters. Mary was as ardent a supporter of canal construction as Perdonnet was of railways (see introduction to Mary 1866, pp. 3–4). Lavallée had evidently managed to keep the rivalry subdued during his tenure as director, but when Perdonnet took over there was no one to arbitrate their different views.

After leaving the school, Mary updated his navigation course in 1866 and reissued the 1865 appendix *Détails pratiques sur la distribution des eaux* in 1868 for the ENPC. The reissue has reengraved plates and a rewritten text. The pagination is different from the 1865 edition, too, but except for the addition of a title page, the content is identical as far as my examination of both editions could tell. Mary must have copied it because he was unable to access the template he needed that must have remained the property of the ECP. This supports the assumption that Mary's separation from the ECP was not amicable. His transcript *Construction* was neither updated nor reissued by his former student Muller, while the other transcript, *Routes et ponts*, continued to be used and added to

383. The heading in my copy of the transcript is pasted on the top of the first page. This suggests that it may have been intended to serve both Mary's construction course at the ECP and his course *navigation intérieure* at the ENPC. Although Mary's name is not on the transcript, when discussing the water-supply of Besançon that he had designed, he writes on page 70: Appelé à Besançon en 1845, par l'administration municipale, pour examiner ces différentes propositions, j'ai été d'avis (When I was called to Besançon by the municipal administration in 1845 to examine these various proposals, I was of the opinion)

until 1873 under Mary's name by Ruelle, who was not an ECP graduate, and was therefore not associated with any in-house factions or political tensions.

There is no record to show that Mary's all-important design method (see the chapter 'Mary's method') survived his retirement. Muller did issue two excerpts[384] as well as two complete reprintings of Mary's 1865 *Détails pratiques*, but under his own name in 1871 and 1880, even retaining Mary's statement of being called to Besançon to design the water supply (1880, p. 68), but the transcript *Construction* vanished.[385] Coupled with the fact that the focus in visual instruction had shifted entirely to mechanical drafting, this signaled the end of Mary's influence at the ECP. Anyone who has ever been exposed to the vindictiveness of academic politics will recognize what all this signifies.

Bélanger, also a supporter of the canal system, left the school shortly after Mary. We do not know whether the circumstances of his retirement were as questionable as Mary's, but the two events do seem to have been coupled. We must remember that Mary and Bélanger had been classmates throughout their education, had worked together on the Canal de la Somme (1822–27) well before their involvement as members of the original ECP faculty. They had both been appointed in 1842 to teach at the ENPC by the then director and hydraulic engineer Jean Bernard Tarbé de Vauxclairs,[386] and they had collaborated on teaching water supply and drainage—Mary dealing with the construction side and Bélanger with the theory. There was clearly a close connection between the two men over and above their involvement at the ECP.

When Bélanger followed Mary in retirement, the pioneering drive that had particularly characterized his flexible mediation between theory and industry disappeared from the school too. He had taught *théorie des machines* simultaneously to the second- and third-year students, alternating with Callon's *construction et établissement des machines*. This tandem organization had perfectly matched the *vulgarisation* requirements of machine theory. As soon as Bélanger left, however, his *théorie des machines* was renamed *mécanique appliquée*.

384. One, with Jules Denfer, was on the use of iron in masonry and timber construction in 1872 and the other on pressure in gas conduits in 1873.

385. Muller did eventually issue a plate volume with 92 folding plates in 1882.

386. While Mary remained at the ENPC until 1869 (Gayant et al 1870, p. 5), Bélanger left in 1853, relinquishing his position formally to his assistant Charles Bresse in 1855. He left his other appointment at the EP in 1860.

The change in title indicates a shift in focus. It was a seemingly minute detail, but one that had serious implications. Bélanger had built his pedagogical approach to mechanics on a precise use of language and a careful definition of terms. His course was an integrated theoretical approach to physical objects, as the original title clearly stated, but the change mutated it into a dually focused course: an *application* of theory to practice, implying an asymmetry—that theory was of a higher order of value.[387] This contradicted Bélanger's careful balance between theory and practice, and it also ran contrary to Dumas and Lavallée's original *vulgarisation* concept that had stressed the unity of theory and practice. Had the government used the 1862 Rouher report to impose a pedagogical model that was similar to the one practiced at the EP and ENPC, and did Perdonnet really support what was happening? Dumas was certainly aware of a shift in educational policy when he had complained to Playfair of the nefarious influence of government administration in the university system.

Bélanger's newly debased course was split into two segments: Henri Tresca's second-year class and Édouard Phillips's in the third year. This further fragmented the material with two different foci: Tresca being more interested in material failure, especially plasticity, and Phillips in the dynamic effect of loading on springs within the elastic range.

Callon's course *construction et établissement des machines* underwent a similar fragmentation. Paul Lecœuvre (grad. 1838) split off *établissement* as a new *éléments des machines* in 1863 and then took over the second year for *construction des machines* in 1865, while Callon continued teaching the remaining portion to the third year until his death in 1877. Callon was a member of Perdonnet's new *conseil de perfectionnement,* so he presumably endorsed the change.

The reorganization of the curriculum and the doubling of instructors in Mary's, Bélanger's, and Callon's courses permitted a better educational sequence (Comberousse 1879.1, pp. 175–176), but at the same time the courses lost the experimental border-crossing excitement that the originators had brought to the material. They also lost the identification between the course material and the instructor and the coordinated overlap of interests between theory and practice that had been hallmarks of the success of the ECP's early

387. Addis described the history of this compartmentalization in the field. Bélanger also split his last three, post-retirement treatises (1864–66) into theory and application, but the division may have been one of convenience of presentation rather than of pedagogy.

Fig. 79. Perdonnet with his professors, photographed by Franck in front of the school's entrance early in 1864. Only six of the earliest faculty, Eugène Péligot (1), Perdonnet (3), Anselme Payen (4), Bélanger (6), Mary (7), and Léonce Thomas (11) are present. (Source: ECP archive © Archives nationales - 20170270/7)

pedagogy. Moreover, the change watered down the traditionally direct contact between teachers and students.

The difficulties these changes caused seem to be reflected in a photograph of the faculty taken in early 1864.[388] Photography was an expensive and complex process in the mid-nineteenth century. It required cumbersome apparatus, a complicated setup, and comparatively long exposure times of 20–30 seconds. Snapshots did not exist. That makes this photograph curious. The arrangement looks as though it were planned for an official group photograph, but the group is not set up symmetrically in front of the building entrance, the participants are not posed although they held their positions since only a few of them are slightly blurred. The men's clothes are somewhat in disarray, and many are caught in odd positions with candid expressions that they might not have wished recorded for posterity. The group certainly does not convey the impression that the men formed a team. It looks like an impromptu test, almost a mistake, and the unguarded facial expressions and body language seem to reveal an unrehearsed impression of the relationships between the men and the political conflict brewing in the period in which Perdonnet was instituting his changes in the curriculum.

The photographer François Gobinet de Villecholle, known as 'Franck', was the official photographer of all the *grandes écoles* at the time and an adjunct faculty member at both the ECP and the EP. He was known for artfully posed asymmetrical groupings of students that were often attempts at sophomoric humor. This image appears to be more randomly casual than tastefully arranged. It is certainly not humorous, and faculty were not normally portrayed casually. Since the original is mounted and marked with the photographer's name in the school's archive, Franck must have regarded it as a finished product. He knew the sitters and had photographed many of them for their official portraits. Is this image an intentional, perhaps tongue-in-cheek critique of the condition of the school and its faculty? Did Franck intend to portray a conflict situation? We cannot know, but as an old saying goes, 'They who understand will understand!'

388. The date of the photograph was determined from Bélanger (6), Mary (7), François Mantion (12), and Samson Jordan's (13) presence. As Émile Muller, who would replace Mary later in 1864, was not in the photograph, the image must have been taken early in the year. The men, both present and missing, were identified through photographs and the series of caricatures ('Portraits de professeurs de l'École centrale') in the Collections numérisées de l'INHA consulted online.

Seated from left to right are: (*1*) Eugène Péligot, the analytical chemistry teacher whose tenure dated from 1835 and who would leave the following year (1865). Next to him is (*2*) the élève-fondateur Charles Callon (grad. 1833). Perdonnet (*3*) sits in the middle, followed by (*4*) Anselme Payen, professor for industrial chemistry since 1834, when he had co-taught it with Dumas, and then from 1843 on his own, and whose participation was about to be downgraded through the appointment of a co-teacher in 1865. Last is (*5*) Paul Lecœuvre (grad. 1838), who had already taken over a segment of Callon's machine construction course and would teach the second year of *construction des machines* from the following year as well.

Bélanger (*6*) who would soon retire at the end of 1864 stands right behind Payen, looking skeptically at the photographer, and judging from the known caricature by Demimuid, Mary (*7*) is the man with the watch chain and hand in his pocket turned slightly away from the camera. The man to the left of Mary is possibly the descriptive geometry teacher (*8*) Émile Martelet, and then farthest left is (*9*) Amédée Burat, who had co-taught mining with Perdonnet from 1838 and who would take over the course on his own in 1865. The man standing with his hand resting on Perdonnet's shoulder is (*10*) Alphonse Salvétat (grad. 1841), who had been responsible for technical chemistry since 1844. Between Salvétat and Bélanger stands the faculty's second élève-fondateur, Callon's former classmate (*11*) Léonce Thomas (grad. 1833) who appears to be glaring into the camera. He taught *physique industrielle*, the course that Péclet had developed, as well as the steam engine course that he had taken over from Colladon in 1835. He would lose his first course in 1865. To the right of the lanky Bélanger, peering over his shoulder is (*12*) François Mantion, who had just been hired to relieve Perdonnet of the railway course in 1864, and above Bélanger stands (*13*) Samson Jordan (grad. 1854). He was Ferry's replacement as professor for *métallurgie du fer* and had also just joined the faculty in 1864. The man on the extreme right of the group looking skeptically into the camera may be (*14*) P. O. Delacroix, the professor for *législation et économie industrielle.*

The two last men cannot be readily identified. The one to the right above Mantion could be (*15*) Daniel, who taught *physique générale*, and the other immediately above Salvétat is perhaps (*16*) Louis de Mastaing (grad. 1851). He had been Ferry's *répétiteur* and had replaced Faure for *résistance des matériaux*, newly titled *résistance appliquée* in 1863—another significant name change.

This would leave only three men missing from the faculty active in 1864: Auguste Cahours, who taught general chemistry, Hippolyte Sonnet, who taught *analyse et mécanique générale*, and the faculty's third élève-fondateur Michel Alcan (grad. 1834), who taught *industries textiles*. Alcan's absence could be explained by the fact that his course had just been downgraded to a post-graduate elective.

This photograph is thus a significant document recorded just before the faculty purge of 1864–65. Perdonnet sprawls in the center holding his top hat and gloves or a scarf. Péligot and the person tentatively identified as Delacroix seem to question the situation. Mary, Bélanger, and Thomas seem unhappy, while Payen's tense posture suggests unease. We can understand Mary, Bélanger, and Thomas's chagrin because they would have expected to be seated in the first row in an official photograph alongside Perdonnet, Péligot, and Payen as the school's six most senior faculty. Of Lavallée's original team from the early to mid-1830s, only Péligot appears to be more or less comfortable with the situation, while Callon and Salvétat are even more at ease with the director. The year after this photograph was taken, 20% of those present in the photograph—Mary, Bélanger, and Péligot—had left the school, and the contributions of three more—Thomas, Payen and the absent Alcan—had been downgraded. All six were faculty from the early years. Perdonnet, who would only live another three years himself had cleared the decks and banished Lavallée's and Dumas's influence from the school.

The contrast between the faculty's regimented and fragmented perception of learning from the mid-1860s and their collaborative, experimental, and border-crossing approach in the 1830s under Lavallée's able guidance, could not have been more evident when we recall the enthusiasm of the young faculty, the background of an almost whimsical approach to scientific inquiry that Monge had manifested with his card trick and chess moves and Coriolis with his billiard ball study, Olivier's flexible models, Colladon's full-scale steam engines, and the playful drama that he had excited with his demonstration.

Part II: Pedagogical Implementation

Where it all came together: Louis-Charles Mary's contribution

A chief source of the program's original enthusiasm had been Louis-Charles Mary. He was the faculty member who in the early years had been responsible for coordinating the groundbreaking curriculum. While naming him as the instructor responsible for the construction course, historians studying the careers of Eiffel and Jenney in particular have neglected his contribution as a major influence on the program and discounted his influence on the future development of iron construction by *centraliens* or by architects like Labrouste. Curiously, the transcript of his architecture and construction course does appear—even upon a detailed examination of its content—to be entirely traditional, relatively uninteresting (Steiner 1984, p. 203 endnote 45, citing Turak 1970), and derivative (Nègre 2011). This dissociative view of Mary and his role has persisted despite multiple declarations of the veneration that the school and its graduates felt for him and his importance to the program, and it entirely misunderstands what Mary really did. We need to analyze the details of his background and his contribution in order to clarify this contradiction and explain why Pothier, Comberousse, and others rightly considered the self-effacing Mary to be one of the mainstays of the ECP's pedagogy, as Pfammatter was the first to point out.

Mary was a modest and hard-working man (Pothier 1887, p. 117), so modest that not even a portrait of him remains to testify to his appearance—aside from a hastily sketched caricature[1] in a copybook by René Demimuid (grad. 1858) and the above hitherto overlooked 1864 group photograph. Mary's intellectual background differed from those of his colleagues. He had not profited from the intellectual complexity engendered by a multi-focused education as Perdonnet had done, nor was he an innovator in the sense that his former EP and ENPC

1. See the sketch in the ECP archive in the Archives nationales Nr. 20170270/7.

classmates Coriolis and Bélanger were.[2] He did not develop new approaches to mathematics and mechanics or fight his way from abstract mathematics to applied mechanics to create a new field with a novel approach. Indeed, as a dyed-in-the-wool practitioner, the conflict between *innovateurs* and *vulgarisateurs* seems not to have preoccupied him overly much.

His border-crossing was of another kind: he concerned himself with reconciling and harmonizing an early version of Durand's theoretical compositional method as he had experienced it at the EP with Louis Bruyère's strong technological bias that he had evidently identified with at the ENPC. He tempered both with his own experience as a practicing engineer. His goal was to apply the result of this harmonization to a modern engineering education. In this he was eminently successful to a degree that has never been acknowledged by historians, and that he may not even have anticipated himself.

Coriolis, Bélanger, and Mary had all studied under Monge, Hachette, and Durand at the EP (1808–10). They followed this with study under the mathematician Eisenmann, the architect-engineer Charles-François Mandar, and the practitioner Louis Bruyère at the ENPC from 1810 to 1815.[3] The structure of their courses had been set by the director Prony[4] who, on the basis of his experience with Perronet and Monge, had explored the *vulgarisation* of theory for engineers and workmen using graphical means (Prony 1802.1 and 2).

The ENPC courses were not organized on the *vulgarisation* model that the later ECP would introduce, but we may assume that Prony's research interest was reflected in at least some of the teaching, especially since Bruyère had himself had studied under Prony decades before. The ENPC program would then have been the first point of contact that the three would have had with the ideas that would later lead to the development of the ECP method.

Coriolis left the school after a year and did not complete the course. He had concentrated entirely on physics. Bélanger had evidently preferred the direction

---

2. Bélanger and Coriolis (who only attended the ENPC for one year) had taken first and second prize respectively in *mécanique appliquée* as students at the ENPC in a competition to develop new theoretical approaches to mechanics.

3. Guillet 1929, p. 124 is mistaken when he claims that Mary had studied under Coriolis and Bélanger: they were contemporaries.

4. See Prony's *Plan d'instruction des élèves de l'École nationale des ponts et chaussées.* 1799 (Ventôse An VII) Paris: Imprimerie de la République.

Prony represented in tying together mathematics and construction, which may explain his later interest in drawing, although he concentrated on language in his own course and did not use graphics.

Mary had followed Bruyère, taking the technological route into construction and materials. He had practiced engineering for two decades (Pothier 1887, p. 291) when Coriolis recommended him for the ECP position in 1833.[5] While Bélanger had opted for practice in hydraulic engineering and then applied theory to his teaching, Mary, who had also worked as a hydraulic engineer, went into teaching without developing a theoretical approach of his own.

Even before Mary joined the faculty to teach the combined architecture and civil engineering course, Pierre-Charles Gourlier taught a course titled *constructions civiles et industrielles*. Gourlier had studied architecture in a traditional manner at the Académie Royale des Beaux-Arts[6] but had made his career in practice at the *Direction des bâtiments civils* from its inception in 1811 and then followed its merger into the *Conseil des bâtiments civils* from 1819. Despite being a traditionally educated architect, Gourlier treated architecture more as a profession than as an art. His course even included a section on the practice of both architecture and civil engineering for the third-year students.

Like other members of the *Conseil des bâtiments civils*, among whom was Viollet-le-Duc from 1838 (Barker, p. 4), Gourlier was actively interested in construction and had developed several forms of brick for which he received prizes at the national exhibitions of 1829, 1834, and 1839 (Eck 1841, vol. 2, *Aperçu sur l'art d'ériger les tuyaux de cheminées en briques*, pp. 5–6). Not only had he already co-authored the first of three volumes of public buildings, but he was also later to serve as the government representative and advisor to the iron-founder Émile Martin and the contractor-ironmaster Mignon's[7] reconstruction of the nave roof of Chartres Cathedral in cast iron (1836–41). Gourlier would also

---

5. In addition to teaching, Mary was appointed Paris municipal engineer in 1835 and advanced to director of that department in 1839 (Gayant et al. p. 4).

6. The Académie was only renamed École des Beaux-Arts in 1816.

7. The contractor responsible for the project was either Charles or Jean-Baptiste Mignon, the ironmaster and serrurier du roi Sylvain Mignon's second or third son. The oldest, Anselme, had died in 1834. Both the younger Mignons married daughters of their brother-in-law, the attorney Didier-Nicolas Riant, who had originally married Sylvain's daughter and later married Anselme's widow after 1834 (information from Geneanet). The firm appears to have been headed by Riant, who then became a railway entrepreneur, associate of Charles Laffitte, and competitor of the Pereire brothers.

serve as a member of the French committee sent to the London Exhibition in 1851 and would report on decorative and construction materials in 1855.

While Gourlier was responsible for teaching architecture—that is, the design and construction of buildings—as *constructions civiles*, he left after a year (and after remodeling the hôtel de Juigné into the ECP's original building). Antoine Raucourt was brought in to replace him and to include the public-works side of the curriculum as a new *constructions et travaux publics*. Raucourt mostly dealt with roads and ports (Guillet 1929, p. 124), which was the area of expertise that he had developed while working in Russia, but nothing is known about how he taught the architecture segment.

Neither the architect Gourlier, whose contribution to the curriculum would merit further examination, nor later the engineer Mary treated the architectural profession visually and aesthetically through the study of historical styles and spatial forms. The traditionally visual approach through history that the ECP felt obliged to include in the program eventually became Thumeloup's job, which he did in a service course consisting of only six lessons but accompanied by a thick volume. Gourlier and later Mary were free to regard architecture conceptually, the latter entirely as a technical subject, an engineering language. Comberousse later stressed that the technical focus in Mary's double course had from the very beginning constituted an important difference between the subject as taught at the ECP and a traditional architectural education (Comberousse 1879.1, p. 62).

There were two essential aspects of technical education that were not covered by Bélanger, Perdonnet, and the other instructors involved in mechanics and machine construction. One was the historical dimension, and the other was design method.

History

As far as history is concerned, the founders evidently felt that some link to previous experience was desirable in order not to detach the impetuous development of nineteenth-century technology completely from the past. Despite François Arago's several essays from 1828 on the history of steam engines in the *Annuaires du Bureau des Longitudes*, Charles Partington's 1822 account, Robert Stuart's 1824 history that had been translated into French in 1827, or Galloway's treatise of 1827, machine construction was all too new a field to fulfill that role. However, building construction had always profited from previously accumulated knowledge, and this was the model that served the ECP's pedagogy. Thumeloup's traditional architectural history course provided a brief introduction to the visual and aesthetical side, and Mary's course provided a longer technological link to history and, at the same time, a strong link to contemporary construction as well.

Mary felt obliged to pay lip service to the traditional masonry elements of architectural history in his course, but he left the detailed discussion of examples to Thumeloup. His text quickly veers off into a lengthier and more detailed discussion of the criteria for designing modern facilities, from market halls and factories to hospitals and prisons. Most of the examples he discusses were current, but some were historical, and he did not distinguish between examples based on a timeline. Mary used them all as case studies in method, and he discussed them as directly applicable to modern construction. He integrated his technological interest in history into his consideration of building methods without defining it as a discrete segment of the course. Since Mary dealt almost exclusively with traditional materials, the historical and modern aspects of construction blended seamlessly in his transcript. Earlier writers had also occasionally referred to historical construction techniques while concentrating on formal issues in architectural history, but Mary's focus was strongly dedicated to method and process. His non-antiquarian approach prepared a technological background for his students. Together with the construction and manufacturing methods that Mary's colleagues examined in their courses, this enabled them to grasp both the formal and the structural potential of materials, in particular iron.

Mary's technological approach paradoxically also prepared a foundation for what became eclecticism in architectural design. It surfaced in the antiquarian movement with the thoroughly researched Gothic construction that Viollet-le-Duc presented in his ten-volume *Dictionnaire raisonné de l'architecture* (1858–68) and in his innovative reinterpretations using iron in *Entretiens sur l'Architecture* (1863–72) (see Figs. 66 and 67). Although he had never studied architecture or engineering, and may or may not have known Mary personally, Viollet was a relentless supporter of the technological and functional interpretation of architectural form (see especially *Entretiens* vol. 2, 1872). He unsuccessfully attempted to introduce reforms at the École des Beaux-Arts in 1863, quit in frustration after a few months, and associated himself along with Mary's former students Henri de Dion and Émile Muller with Émile Trélat's innovative architecture school in 1865.

Mary's technological bias in history may have influenced Auguste Choisy's writings on Egyptian, Greek, Roman, and Byzantine construction from 1873 to 1904. Although the engineer Choisy only took *Navigation intérieure* with Mary at the ENPC from 1863,[8] he combined construction and history in his treatises and in his own courses at the ENPC from 1876 on. He had begun developing this theme when he returned from Rome in 1866, supported by his acquaintance with Viollet-le-Duc, who wrote:

> A young French engineer, M. Choisy, is about to publish a very complete treatise on the structure of Roman vaults based on [his examination of] the monuments. This compilation that I have before me, recounts in detail the various processes that these illustrious builders used. (Viollet-le-Duc 1858–68, vol. 9, 1868, footnote, p. 477)[9]

Neither Choisy nor Mary endorsed architectural historicism as Viollet did. Choisy explored historical determinism in architecture, while Mary only mined history for case studies, but it seems that Mary was wary of a possible danger of misinterpretation of his intentions when he wrote in the introductory chapter to his architecture course:

> ...and, as regards the case of France, don't we see that after Roman construction came Gothic construction, and later, after the Gothic was abandoned, the Roman returned mixed with Arab elements and one called it Renaissance style, and finally the Roman and Greek prevailed, and now, at this moment we try all of them. (Mary 1862–63, p. 2)[10]

---

8. Like Bélanger, Mary held dual appointments at the ECP and the ENPC.

9. Un jeune ingénieur français, M. Choisy, va publier prochainement un travail très-complet sur la structure des voûtes romaines, d'après les monuments. Ce recueil, que nous avons entre les mains, donne en détail les divers procédés employés par ces grands constructeurs... Choisy eventually expanded his interest from Roman to Byzantine and then to Egyptian construction, possibly reflecting on Durand's taxonomy.

10. ...et pour ne parler que de la France, ne voyons-nous pas qu'après les constructions romaines sont venues les constructions gothiques; que le gothique a été plus tard abandonné pour revenir au style romain mêlé d'ornements arabes que l'on a appelé le style de la Renaissance, que le Romain et le Grec ont fini par prévaloir; et qu'enfin, en ce moment, on s'essaye dans tous les genres.

Method

Design method was the second and, as far as iron construction is concerned, the more influential aspect of architecture that Mary explored, and here, too, he focused on process. He developed his method in a radically different way from composing forms as his teacher Durand had defined it by concentrating on the process rather than the object (Pfammatter 1997, p. 125). He stressed the organizational aspects of the results, not their formal expression, and developed what we now call design—that is, problem-solving rather than *composition.*[11]

Before beginning at the ECP, Mary published several reports on construction questions. During his years as a teacher, he authorized or issued multiple versions of his course transcripts, but he recorded nothing at all on what motivated him theoretically as he developed his method. No written trace of his pedagogical thinking appears to have survived. We can only surmise what motivated him from the changes that he made to Durand's method and how he drew on the technical material that Bruyère had published.

Mary was very much down-to-earth and not at all interested in abstraction, which explains the fundamental distinction between his and Durand's approach. He had been a practicing engineer for almost a quarter of a century when he joined the faculty. He collaborated with Bélanger on the enlargement of the Canal de la Somme (1822–27; *Mémoires et compte rendu* 1877.1, p. 67, obit Lepeudry), was a close friend of the bridge builder and concrete pioneer Vicat,[12] and had authored two papers on concrete construction in 1831 and 1832 and one on bridge construction in 1833.[13] His original work was well documented, and it was evident that construction and materials concerned him. He had been and continued to be a practitioner. His 1845–47 project for the water supply and drainage of Besançon (Mary 1850) that he built in 1850–54 appeared as a transcript that he used in class (Mary 1854).[14] He was appointed co-editor of the *Annales des Ponts et Chaussées* in 1842, became a member (along with Olivier[15])

11. Turak misses the distinction between Durand's and Mary's methods.

12. He wrote Vicat's obituary in the *Annales des Ponts et Chaussées* in 1862.

13. Several more reports are catalogued in the Archives nationales.

14. The Besançon project appeared even before that in Montero's notebook in 1851 before it was even finished (p. 45). It documents Mary's and the ECP's pedagogical commitment to tie every possible theoretical explanation directly to a practical, current example.

15. De Dion in *Mémoires et compte rendu* 1872, p. 123.

of the committee charged with reorganizing the EP in 1850, and advanced to the rank of Inspecteur général de première classe in the Ponts et chaussées by 1855. Mary acknowledged the shortcomings of abstract theory from the standpoint of the engineer, and was well aware of its limited applicability to practice. This intellectual position may have been why Coriolis, the theoretician of the physics of billiard balls, had recommended him to Lavallée in 1833.

Mary's devotion to method proved far more relevant to the school's pedagogy than his approach to history. He developed a way to organize function, structure, and materials through design that had broad implications for construction and industrial manufacture. This was his seminal contribution to engineering education. The material he chose to achieve this was derivative, but his contribution was novel. In order to resolve the apparent contradiction in this statement, I must explain his educational antecedents, and especially the influence that Durand and Bruyère, the two pioneers of system thinking in building, exerted on him. What were the goals that these two teachers had aimed at in their teaching and what did their intellectual environment look like?

The brevity of Mary's text on design method in the final version of his transcript (1862–63), a mere six pages of theory (pp. 26–31) followed by an explanation of its application (pp. 32–44), is misleading. It is completely disproportionate to its influence on the school's pedagogy and to its importance for the future development of iron construction. Small wonder that it has been—with the notable exception of Pfammatter—largely ignored in previous accounts. It was this brief methodological segment, supplemented by the bulk of Mary's course that he devoted to examples (pp. 43–103), that innovative iron engineers would need in their practice. The results of his teaching method—as minor a matter as it may appear to be from his course transcript—impacted every aspect of construction in which the new material, iron, appeared. His concept of design went far beyond architecture, and ranged from the design and construction of machinery, infrastructure, and buildings to the design and industrial production of components and their manufacturing processes. It formed the basis for Pothier's and Comberousse's judgment (Pothier, pp. 117–118; Comberousse 1879.1, pp. 75/267) that Mary's role had been central to the school.

Mary systematized the compositional process as it had been conceived by Durand in the previous generation and remolded it to suit the ECP's needs. He did this by combining it with the functional and material-based approach to architecture that Bruyère had taught and the attention to detail that Mandar had cultivated at the ENPC. Mary transformed what had been an ideology into a practical design process. He did not hesitate to appropriate Durand's method or even the plates that his teachers Durand and Bruyère had published, and this was what has earned him the epithet of 'derivative'; but what has not been sufficiently stressed is that he subtly altered what he borrowed and used it to fulfill another purpose. Valerie Nègre quite correctly pointed out that Mary did indeed take much of his theoretical material from Rondelet, Durand, and Bruyère (Nègre 2011, pp. 3–4),[16] but he reinterpreted it conceptually.

16. Mary 'lifted' six plates directly from Durand as well as paraphrasing his method, nine plates from Bruyère, and some of Rondelet's material in the chapter 'Principes fondamentaux de l'architecture.'

L.-C. Mary's teachers

Sganzin's construction course

Construction as it had been taught at the EP did not impact Mary's thinking. Joseph-Mathieu Sganzin, who had been drafted to teach in 1797 by his close friend, the director Lamblardie, taught construction as *application de la géométrie descriptive.* In 1807 he expanded it into what was intended as the practical side of the field in a *cours de construction* that he published in an abridged summary, *Programmes ou résumés des leçons d'un Cours de construction.* Mary and his classmate Bélanger were upperclassmen at the time and may not even have been fully aware of this change.

Sganzin had extensive expertise in construction. As an Inspecteur général des ponts et chaussées from 1803, he was responsible for many of the empire's harbor works. He called the new course an overview and based it on his own experience as an engineer, but he followed the traditional geometrically based approach. The original title indicated how Sganzin saw his educational brief. He focused on the graphical projection of three-dimensional design and added some material. The course was an incomplete introduction to the planning of *ouvrages d'art* and did not do justice to Sganzin's experience. It was certainly not a fully developed course on construction technology or the design of modern établissements (facilities), nor was it intended to be. He wrote:

> This course will henceforth be named 'construction course', and it will present the students with most comprehensive instruction possible on the subject by adding what was missing in the original course. (Sganzin 1809, p. 7)[17]

---

17. Ce Cours portera désormais le nom de Cours de construction, et aura pour objet l'instruction la plus complète qu'il sera possible de donner aux élèves sur cette partie, en ajoutant à l'ancien cours ce qui lui manquoit pour atteindre ce but.

However, he had qualified this ambitious goal in the preceding *Avertissement* by writing that it was simply a reminder of his oral explanations, not a complete exposé:

> This does not at all claim to be anywhere near a complete treatise on the field of construction, nor on that of the Ponts et Chaussées engineer; I only wanted it to remind the École Polytechnique's students of the most essential points that were covered in a lecture course. (Sganzin 1809, p. v)[18]

What Sganzin taught while Mary was at school was therefore just a service course, and the school dropped it in 1812. We only get a vague feel for the technological context Sganzin must have touched upon in this course and why he was so revered by his students from the posthumous fourth edition of the lectures (1839–40). Félix Reibell, who entered the EP the year Sganzin left and later studied under Bruyère at the ENPC, completely reworked and updated the text. He almost doubled it from 260 to 400 pages with three appendices instead of a single appendix, and he added a fourth appendix of 56 pages and plates by the theoretician Léon Lalanne.[19] He accompanied this text with an atlas of 180 plates instead of the original ten. This of course made it much more his own text than Sganzin's, but it nevertheless provides an indication of the material that the experienced Sganzin skimmed and that he might have taught if he had been given the opportunity.

Comparing what Sganzin appears to have done in expanding his geometrically based *application de la géométrie descriptive* with Olivier's later specialized application of descriptive geometry to the immediate needs of practitioners, especially with regard to the configuration of gearing teeth in 1842, it becomes evident that Sganzin fell short in this regard. It was only after another former Bruyère student, Léonce Reynaud, took over Durand's originally theoretical course in 1837 that architectural education at the EP, possibly under the influence of the ECP's growing success, began to move toward a more construction-based approach.

What is clear from the abbreviated published text of Sganzin's lectures is that he organized his course based on what was expected of a state engineer. The issue was the oversight of public roads and harbors, and certainly not industrial

18. Ce n'est point, à beaucoup près, un ouvrage complet sur l'art de bâtir, ni sur celui de l'Ingénieur des Ponts et Chaussées, qu'on a eu la prétention de faire; on a voulu seulement rappeler aux Élèves de l'École Polytechnique, les choses les plus essentielles d'un Cours oral.

19. Lalanne was later briefly director of the ateliers nationaux in 1848 and then the ENPC in 1876.

construction.

Sganzin divided his course into 30 lessons, ten that dealt with materials, of which only one very briefly touched on iron. The few pages devoted to the new material, pp. 90–96 in the 1809 second edition and pp. 75–80 in the 1821 third edition, merely discuss its physical properties, not how to use it. The second set of ten lectures dealt with roads and bridges, and the final ten with his own field of expertise, canals and ports,[20] with an addendum on graphical means of evaluating vault thrusts and earthworks. The course appears to have named many examples, but not to have included details of built structures. It was clearly an overview of public infrastructure and not an organized, practical approach to construction. His text continued to be used long after his retirement, both at the EP and by teachers abroad like Claudius Crozet[21] in the United States, but as far as any possible influence on Mary's method is concerned, Sganzin's course was insignificant.

Durand's background in aesthetics and geometry

Durand's influence on Mary, on the other hand, was a major one, despite the fact that he also taught a mere service course when Mary attended the school. That would change in 1816, when Durand was appointed full professor. The earlier students Mary and Bélanger cannot therefore have learned much other than what was present in his lectures and his text, the 1802–05 first edition of the *Précis des leçons*. Pfammatter (1997, p. 55 quoting Szambien 1984, p. 156) states that the original version of the course in 1799–1800 (An VIII) included 87 hours of drafting. This number likely reflected the total hours of the course: 30–35 hours of lectures, 13 hours of exercises that were probably spent by the students copying plans, plus the time devoted to site visits and exams. In any case it was drafting, not design, because until the reorganization of the EP's curriculum in 1816–17, when the students' actual design work was done outside class time,[22] Durand himself stated that the design part of the architecture course had consisted

20. He had been appointed directeur des travaux maritimes in 1795 (Base Léonor doc. 7/9) and subsequently inspecteur général.

21. Crozet had graduated from the EP in 1807, a year before Mary entered. Although he had not experienced Sganzin's expanded *Cours de construction* either, he used the summary as his text from 1816–23 at West Point and from 1839 at the Virginia Military Academy.

22. Compare the diagram of the school's curriculum, albeit that of 1818 when the content had been increased (Szambien 1984, p. 258).

in only four standardized drawings:

> Graphic design, as architecture students used to be taught it at the École Polytechnique, was limited to four drawings with the following content: first, the [classical] orders; second, doors and cross-ribbed vaulting; third, a room; and finally fourth, a staircase. This type of exercise left them with little scope for their ideas, and not very attractive ideas at that. Moreover, it took up so little of their preparation time for the *concours* over the course of the year that it was far more suited to turn them away from the field of architecture altogether by the ennui and boredom it generated than to enable them to develop their talents in this art. (Durand 1821, p. 1)[23]

The required drawings were therefore far too few and the subject matter far too primitive for Mary to have been able to gain any idea of Durand's compositional method. Given the content that Durand described, the drawings were boiler plate exercises and nothing more. The portion of the course that could have made it relevant to practice only came later, in 1816,[24] and it was only well after that reorganization of the program that the clearly frustrated Durand titled his introduction to the *Partie graphique* in 1821 *De la manière d'acquérir en peu de temps de vrais talents en architecture* (How to acquire real competence in architecture quickly). He had finally been permitted to replace the four drawings with an *assez grand nombre d'ésquisses relatives au mécanisme de la composition* (a fairly large number of sketches related to the technique of composition, Durand 1821, pp. 1–2). The EP graduates of the early years must have felt the lack, but in developing his own course, Mary nevertheless recognized the advantage of adopting the idea of Durand's theoretical system as a starting point and modifying it to serve the ECP's goals.

---

23. Anciennement à l'École polytechnique, le travail graphique relatif à l'architecture se réduisait, pour les élèves, à quatre dessins, qui avaient pour objets: 1° les ordres; 2° les portes et croisées; 3° une salle; 4° enfin un escalier. Ce genre de travail, qui ne présentait qu'un si petit nombre d'idées à leur esprit, et d'idées peu attachantes; qui en outre absorbait entièrement le peu de temps qu'on leur accorde pendant toute l'année pour se préparer aux concours, était bien plus propre, par la fatigue et l'ennui dont il les accablait, à les dégoûter de l'architecture, qu'à leur faire acquérir des talents dans cet art.
24. According to Dhombres in Fourcy, 1987, p. 162, the change occurred with the reorganization of the EP in 1816, although Durand had already taught there for twenty years, since 1796. For the sudden outburst of student projects resulting from the change, see the 30 projects documented in his *Choix des projets d'édifices publics et particuliers,* 1816.

Durand had come from a completely different intellectual background at a different time. The pedagogy that he developed and that Mary, Bélanger, and Olivier had experienced, had aimed at other goals than those that would interest the ECP's founders. Durand was teaching his final semesters in 1833, the year that Mary joined the new school's faculty.

Durand's education had been typical of his time. After working for a year in the office of a minor architect in 1775 (Szambien 1984, p. 15), the fifteen-year-old moved to Étienne-Louis Boullée's office and studied in the following years under several well-known tutors at the Académie d'Architecture.[25] Actual design instruction occurred in the private practice of the tutors and not at the Académie itself (Armstrong 2017, p. 5), where the students only attended lectures and discussions by the professors and other members. The classicist Ange-Jacques Gabriel was the director,[26] and the three professors were Julien-David Le Roy as the chair of architecture, Antoine-René Mauduit for mathematics, and Charles Bossut, Monge's mentor, for hydrodynamics. Jacques-Germain Soufflot appears to have taught construction and materials. He was not a professor, but like Boullée and other prominent architects and engineers, he functioned as a tutor. The academy members were subdivided into two degrees or classes, both of which could tutor students. The director and the professor of architecture were first-class members, as were prominent men like Soufflot and Perronet. Boullée was a second-class member until he was promoted to replace Soufflot after the latter's death in 1780 (*Procès-verbaux de l'Académie Royale d'Architecture,* vol. IX, 1780–93, pp. 30–31). Durand is nowhere mentioned as having been Boullée's official *élève* (pupil),[27] but he may have been by virtue of his employment in

25. The Académie d'Architecture had a school attached to it. It was the world's first officially organized architecture school and had been founded in 1692, disbanded in 1767 and reopened in 1771, suppressed in 1793, and later revived as part of the École des Beaux-Arts in 1816.

26. Above the director was the directeur général des bâtiments du roi, Flahaut de La Billarderie, Comte d'Angiviller. He was the link to the ultimate arbiter of taste, the king Louis XVI who, however, had other worries than infrastructure.

27. The only actual list of students and their tutors is dated January 1774, before Durand joined the Académie (*Procès-verbaux de l'Académie Royale d'Architecture,* vol. VIII, 1868–79, pp. 188–189). It names those students in the *élèves de l'Académie* category who were eligible to participate in the grand prix concours. Boullée is only mentioned as the tutor of Jacques-Pierre Gisors. Durand later belonged to this select group as he won a second prize in the concours in 1780.

his office. In 1779 Durand is recorded as having received a *prix d'émulation* as a pupil of the director Le Roy, who was responsible for devising 'visionary' projects for the student competitions, and in 1780 he won a *deuxième prix* in the student competition as Perronet's pupil.[28] Whether he remained based in Boullée's office while studying with these men is probable but not recorded.

The Académie, its members, and its schooling were bureaucratically organized and categorized. The three chairs were architecture (drafting and design), mathematics (geometry), and hydrodynamics, whereas materials and construction (timber and masonry), was relegated to tutors. This arrangement demonstrated where the focus of traditional architectural education lay: architecture dealt with *distribution* (site, form, and function), precedent (history), and *goût* (taste). Taste conferred the ability to judge quality. A cultivated taste was conditioned by classical proportions that were regulated by geometry,[29] and geometrical rules of proportion and their correct application were among the academicians' main topics of discussion. Hydrodynamics was the only technology that merited a chair because it was important for the government, which placed emphasis on the construction of canals for trade and harbors for defense.[30]

Despite the lack of emphasis on construction, the academicians did deal with innovations when they arose, as when Ango submitted his pioneering iron floor construction for evaluation in 1782 (*Procès-verbaux de l'Académie Royale d'Architecture,* vol. IX, 1780–93, p. 79),[31] but the Académie functioned chiefly as an advisory body of reviewers and did not treat innovations as novel approaches to construction with aesthetic implications for *goût*. Students were judged largely on the organization of their projects and the quality of renderings. They were trained more as draftsmen than as what we now understand as architects or engineers. Construction knowledge would come with practice, and except for large-scale

28. *Procès-verbaux de l'Académie Royale d'Architecture,* vol. VIII, 1868–79, p. 367, and vol. IX, 1780–93, p. 26.

29. The culture and preoccupation with defining 'taste' scientifically in France culminated in another field, in Brillat-Savarin's celebrated *Physiologie du goût, ou méditations de gastronomie transcendante,* 1825.

30. Bossut had been appointed as professor directly by the king in 1777, forced on the Académie by decree rather than chosen by the academicians (*Procès-verbaux de l'Académie Royale d'Architecture,* vol VIII 1768–79, p. 292). He was not ranked in the classification of first- and second-class members.

31. Eck (vol. 1, pp. 3–4) reproduced a 1785 report that the Académie authored on a series of hollow-tile flat arches, some of which were reinforced with iron bars. Boullée was one of the reporters.

projects like Perronet's bridges, for instance, engineers and architects relied largely on traditional craft knowledge.

After finishing his studies in 1782, Durand was therefore what we would consider a draftsman, and he stayed with Boullée, whose interests were two-pronged: architectural practice and hypothetical unbuildable projects similar to those that the architecture chair Le Roy developed for the student competitions.[32] This was the environment in which Durand had been educated. It was a very different world from that in which the founders of the ECP and Mary found themselves in the following generation.

Durand met his future architectural partner Jean-Thomas Thibault in Boullée's office, and when Thibault, who was known as a superb draftsman, returned from a stay in Rome (1786–90),[33] he and Durand joined forces. In 1792 they developed fifteen abstract projects that they submitted to the great national competition held in 1793–94 (An II). The architectural segment of this multi-faceted contest was intended as a replacement for the student competition normally organized by the Académie, which had just been dissolved. The intention was to identify a new and appropriate architecture for post-royal France. A total of 209 projects were submitted to the monumental sixteen-part competition, with one section devoted to architecture (Cirio 1992, pp. 80–81). Twelve of the duo's fifteen proposals were awarded prizes and one of them even took first prize (Cirio 1992, pp. 80–81, Picon 2000, p. 6). Durand's and Thibault's entries in the competition were based on Boullée's interest in idealized formalism. The success of their results was quite possibly due to this bias, as Boullée was an influential jury member.[34] The avowed objective had been to build the winning proposals, but this never happened, due at least in part to the chaotic political situation at the time. Robespierre's arrest and execution took place at the end of July 1794.

Durand's further professional development, despite several buildings of modest quality, sporadic collaboration with Thibault, and the consulting that Szambien documents, led him to theory and methodology.

---

32. Picon has suggested that Boulée's fantasies were motivated by his awareness of the conflict between architecture as art and the newly dominant scientific method (Picon 2000, p. 3), although Armstrong (2017, p. 21) points out that the 'visionary' projects had been introduced by Le Roy who had succeeded Jacques-François Blondel as professor in 1774.

33. Szambien (1984, p. 17 note 10) states that Thibault worked for Boullée from 1778 to 1788, other sources mention 1780–86 and 1790–99, and his stay in Rome as 1786–90.

34. The most prominent other members of the jury were Léon Dufourny, Claude-Nicolas Ledoux, Jean-Augustin Renard, and Antoine-Chrysostome Quatremère de Quincy.

Fig. 80 - Jean-Louis Nicolas Durand, a theoretician of architecture and the creator of the first design method at the École Polytechnique. He taught several of the École Centrale's teachers and influenced the school's design method. (Source: contemporary engraving.)

Durand's contribution: system and method

Like Monge himself a generation earlier in Mézières, Durand was hired as a draftsman at Monge's new EP in 1794. After a year he began teaching a course there in 1796–97 that attempted to elaborate an ideological, abstract 'science' of architecture. It was one of the seven weekly courses taught at the EP, but it was not technologically or practically oriented, and like Sganzin's construction course, it was not considered among the most important. Although the separation between architecture and structural engineering was fuzzy, Durand's course was directed primarily to the needs of civil engineers. He knew that they lacked a new form of education to fulfil their role in building public infrastructure, especially as far as designing buildings was concerned. In his introduction to volume two of the first edition of *Précis des leçons* in 1805 he wrote:

> Because engineers are asked more and more frequently to erect prominent structures, one highly recommends the study of architecture to the École

> Polytechnique's students, and at the same time one must give them the tools to do it well. (Durand 1805, vol. 2, p. 1)[35]

Picon pinpoints the dilemma that Durand faced. Like Boullée, he was on the one hand concerned with 'maintaining architecture as an autonomous discipline on the threshold of a world increasingly dominated by scientific and technological rationality' (Picon 2000, p. 3), while on the other attempting to devise a concept in reaction to his own background that would allow what had been the subjective art of architecture (despite all the Académie's attempts to define it objectively through mathematics) to fit into the field of the new 'objective' scientific value system. His solution, surely influenced by the methodology of Monge's descriptive geometry, was to teach architecture neither as an art nor as a profession, but by developing the first clearly enunciated compositional method supported by a taxonomy of plan types and forms.[36]

This was a very limited approach that could not satisfy building practitioners, nor was it intended to do so. The EP was an undergraduate institution, and those graduates who would go on to become civil engineers would then continue their studies at the ENPC. So, apart from two plates (pls. 3 and 4) in the first volume of the *Précis* that dealt superficially with materials and construction, Durand concentrated entirely on methodology. He stated very clearly in the chapter *Suite de l'introduction* in the reworked 1813 edition of the first volume (pp. 26–28), that what he described as the traditional professional preoccupation with the triad of decoration, distribution, construction was useless, and that method was the only possible way to learn architecture.

The aesthetic approach to architecture that Durand had experienced in his own education at the Académie was driven by rules of proportion. Lines on plans denoted proportions that had been derived from Greek sculpture and architecture, but the conundrum that the academicians were unable to resolve was that the strict rules were supposed to regulate the nebulous concept of 'good taste'. In the first volume of the *Précis* 1802 (pp. 10–11), Durand writes that he agreed with his teacher Le Roy that the Greeks themselves had not followed precise rules of

35. Les Ingénieurs étant chargés plus fréquemment que jamais d'élever des édifices importans, on ne saurait trop recommander aux élèves de l'École polytechnique l'étude d'Architecture, et en même temps leur faciliter les moyens de s'y livrer avec succès.

36. See Madrazo 1994, pp. 12–24 for a detailed analysis of Durand's course and its relevance.

proportion, which were in fact a classicist construct. This must have disturbed him because he replaced the traditional geometrical *rules* in his method with geometrical *logic*. He retained the highly structured character of the academic system and the proportional rules in detailing that was based on geometry[37] and contributed an innovation: a hierarchy of lines that were axes, along which forms were linearly strung, organized by an incremental process of composition.[38] Like the geometrical rules it replaced, Durand's method was formulaic. It began with the smallest unit, the (formal) component, and proceeded in clearly defined increments of scale to ever more complex compositions—that is, configurations of these components, and iterations of compositions. However, 'taste' never entered the argument, and instead his evaluation of what was 'correct' was dominated by an underlying sense of system. Durand's program was successful insofar as both his method and the encyclopedic taxonomy that accompanied it were accepted as a valid, supposedly scientific form of thinking.

After half a decade of development Durand published a first version of his method between 1802–05 in his two-volume *Précis des leçons d'architecture*, a book that became the required text for the course (Pfammatter 1997, p. 55). He republished it in 1809 and then expanded the first volume in 1813 as the '*Nouveau précis*' that became the expression of his fully developed system. A supplement followed in 1816 (co-authored with François-Tranquille Gauché) titled *Choix des projets d'édifices publics et particuliers*, and after two further republishings in 1817 and 1819, it was followed in 1821 with a new third volume, the *Partie graphique des Cours d'architecture depuis sa réorganisation.*

Durand's system featured a geometrical symmetry of repetitive, spatial and element types that apparently resembled the diagrams that Monge and Hachette taught in descriptive geometry (Picon 2000, p. 40). They were combined incrementally along hierarchical axes to generate ever more complex plans that are graphically attractive but ultimately non-functional. In the introduction to the first volume of the *Précis* (1802, p. vi) Durand cautioned against confusing drafting with architecture, but he failed to follow his own warning. As a result, his system was limited and selective: he rejected the culturally loaded 'signs and symbols' (Picon 2000, p. 31) that his forebears had relied on, but at the same time

37. *Précis* 1813 pt. 1 pls. 1–5/10, pt.2. pl. 4. This is the reason that Garric 2011 traces the origins of Durand's compositional method to the Académie.
38. *Précis* 1813 pt. 2, pls. 19/21.

he followed the academy's pedagogical approach by omitting concrete issues like site-specific criteria, materiality, or construction. Physical structure received even less attention in his system than it had previously at the Académie, and function did not interest him either. As Bernard Huet wrote, Durand's goal was rational, not functional architecture:

> ...the structures Durand proposes are never functionalist. They are rational. (Huet in Szambien 1984, p. 10)[39]

Durand's system was radically utopian in concept. Both Huet, who lauds Durand's limitation, and Picon who simply records it, point out its inherent pitfalls:

> The *Précis* is the first architecture treatise that takes architecture itself as its subject matter, to make architecture without referring to the art of construction. The only thing that counts for Durand is the founding act of architecture represented by the project. (Huet in Szambien 1984, p. 10, cited by Picon 2000, p. 55, footnote 3)[40]

Although he discussed traditional materials in his lectures[41] and showed and discussed rudimentary construction, particularly of roofs, in a single plate,[42] the bulk of the text is devoted to type and form, and most of the plates are geometrical and oversimplified. At the very end of the *Précis*, he did illustrate several real, non-symmetrical building plans that he had designed with an indication of materiality in wall thicknesses. He included them as examples of how architects should try to resolve the dilemma of creating symmetrical plans on irregular urban sites.[43]

Durand's method, and especially the importance of the 1813 plate titled *Marche à suivre dans la composition d'un projet quelconque* (Path to follow in the composition of any type of project) have been much stressed by architectural theoreticians and historians. Plate 21 of part 2 in the 1813 edition represents this stage in the development of this compositional method.

---

39. ...les édifices proposés par Durand ne sont jamais fonctionnalistes. Ils sont rationnels.

40. Le Précis est le premier traité d'architecture qui prenne pour objet l'Architecture même, de faire l'Architecture sans référence à l'Art de bâtir. Seul compte pour Durand l'acte instaurateur d'Architecture que représente le projet.

41. *Précis* 1802, pt. 1 pp. 25–64, and 1813, pt.1, pp. 37–51/101–119.

42. *Précis* 1802, pt. 1, pl. 3, and 1813, pt. 1, pl. 4.

43. *Précis* 1805, pp. 82–83 and pls. 25–26.

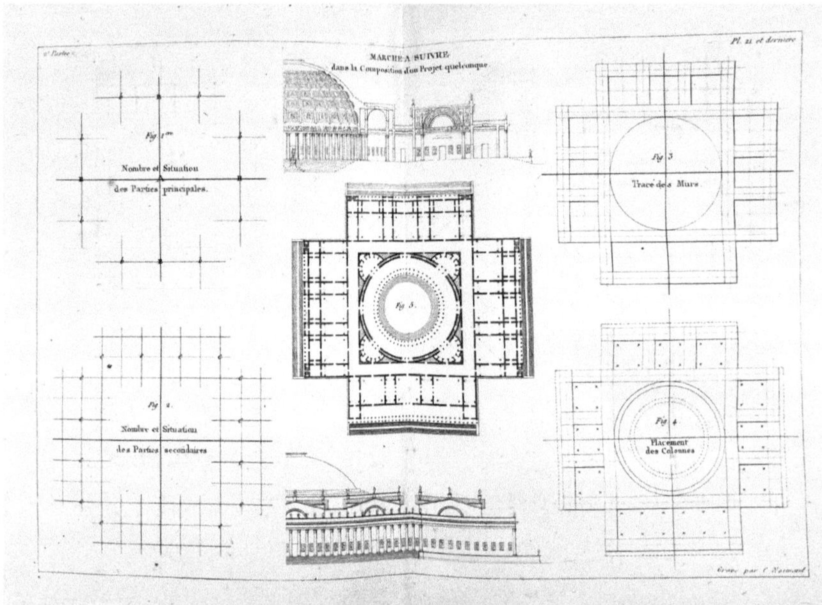

Fig. 81 - *Marche à suivre*, plate 21 in part 2 of Durand's treatise that first appeared in the third edition of volume 1, the *Nouveau Précis* of 1813. This plate encapsulated Durand's abstract design method and influenced architectural theoreticians throughout the Western world. (Source: Durand, *Précis des leçons*, 1817, part 2, pl. 21, 4th ed).

The *Marche à suivre* does not appear at all in the earliest edition, the one that Mary or the other ECP teachers who had graduated from the EP would have used while at school. Durand had previously included a sub-chapter with that title, but without the diagram.[44] The 1813 diagram made his 'path' clear: it illustrated his rigid formula for arranging spaces along a hierarchy of symmetrical axes. The 'path' had everything to do with symmetrical plan geometry and nothing at all with functional project design, let alone with the messy business of actual construction that must contend with material characteristics, site conditions, orientation, legal aspects, detailing, dimensions, costs, and the building process itself. The diagram is geometrically reminiscent of Andrea Palladio's plan for his late sixteenth-century villa 'La Rotonda' and is probably based on it, but Durand was not interested in any real building. He was only interested in the logic of the plan form and certainly not at all in the men, mud, and machines that would

44. *Précis* 1805, vol. 2, pp. 96–98.

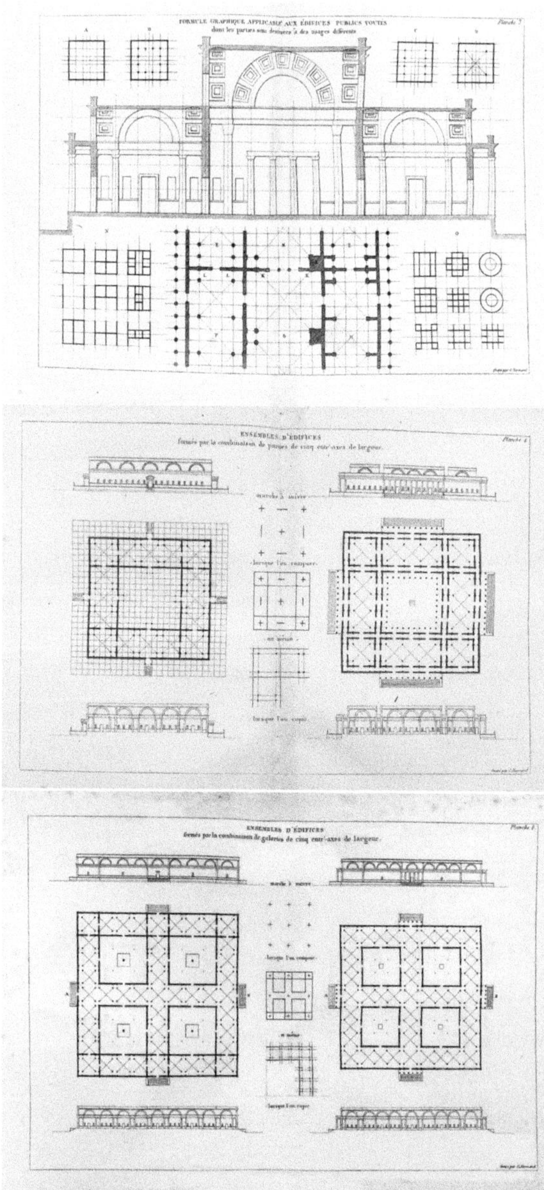

Fig. 82 - By 1821 Durand had expanded his system beyond the *Marche à suivre* with several further plates. Plate III demonstrated how the method was to be applied to the design of vaulted public buildings, plates IV, V, and the following plates presented examples of increasing complexity. All these were purely graphical exercises with no relationship to the functional use or construction of practical buildings. (Source: Durand, *Partie graphique*, 1821, pls. III, IV, V.)

become the lifeblood of Lavallée's 'pistons'. Together with an expansion of his diagram that Durand then published in plates III, IV, and V in the *Partie graphique des Cours d'architecture* (1821), it appeared well after Mary and the others had graduated from the EP.

In the introduction to the 1817 and later editions of the *Précis*, Durand states quite clearly (Madrazo 1994, note 3, p. 22)—much clearer than in 1802 (Durand 1802, vol. 1, pp. 22–23)—that his interests lie in the abstract and not in the particular. Durand demonstrated his bias in the projects that he illustrated in the 1805 second volume and even more clearly later in 1816 in *Choix des projets d'édifices publics et particuliers.* The latter is a portfolio of student work, and it was the product of the new drafting component of his course that followed the reorganization of the EP curriculum in 1816. It contains projects that are complex exercises in symmetry and composition rather than actual organized design solutions to architectural and functional problems.

Despite Durand's claim to the contrary (*Précis* vol. 1, 1817 ed., p. 1, footnote, referring, as proof to the *Choix des projets* 1816), very few architectural practitioners of any true originality graduated from the EP, even after the reorganization of 1816. Nevertheless, many architects throughout Europe were influenced by the distinctive geometry of his plans, of whom Picon names Leo von Klenze in Bavaria as the most prominent (Picon 2000, p. 7).

Because of his abstract and therefore unprofessional approach to building, the method seduced theorists even more than it did practitioners. Durand had created a rational, compositional design system, and he supplemented it with a catalog of all possible building forms. This encyclopedia, *Recueil et parallèle des édifices* predated the publication of the *Précis* by several years. It appeared in a first edition in 1799–1801 (Ans VIII and IX), and was reissued several times, each time documenting more and more building types.[45] Since

45. It began with two volumes, the first with 52 double-column pages of text by Jacques-Guillaume Legrand in 1799–1800 (An VIII) and the second with 92 plates in 1800–1801 (An IX). Multiple reprints followed, as well as variants: an edition of Legrand's text appeared on its own (posthumously) in 1809, a Belgian edition in 1830 had 10, and another in 1839 or 1841 had 20 plates added to the original 92. Szambien catalogs the sources of the plans and variant editions (pp. 99–108). By 1833, the year before Durand's death, the Venetian edition of the *Recueil* had grown to 196 double-column text pages with parallel columns in Italian and French, and 266 plates. It reached its largest form posthumously in 1857 with an addendum of 71 plates by Francesco Zanotto, totaling 337 plates.

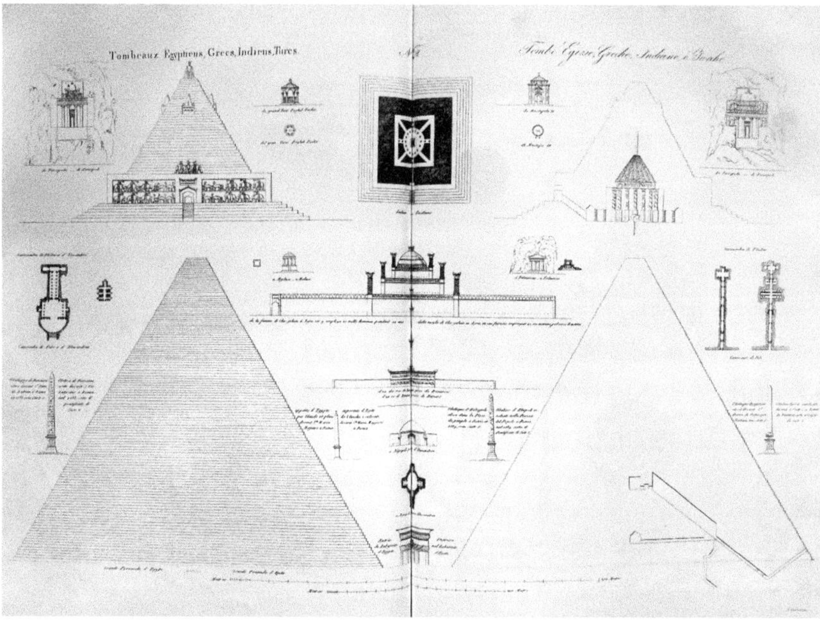

Fig. 83 - In his taxonomy *Recueil des édifices*, first published in An VIII and An IX (1799–1801), Durand published geometrically *corrected* plans and elevations to demonstrate geometrical design principles rather than a building's true form. Thus the Great Pyramid of Giza (Khufu's pyramid) is shown as an equilateral triangle, which it is not. (Source: Durand: *Recueil des édifices*, 1833, pl. 48.)

typology rather than actual buildings interested Durand, he 'corrected' plans or elevations to make them more geometrical than the actual structures that they represented. They were neither formally nor structurally correct. For example, plate 19 of the original plate volume (1800–01) and plate 48 of the 1833 edition show the elevation and section of the Great Pyramid of Giza as an equilateral triangle, which it is not.[46]

These changes were not errors. Durand purposely 'regularized' his plans to demonstrate type rather than to record actual buildings (Madrazo 1994, p. 13). Geometry and symmetry were the critera for his 'corrections'. Durand knew that the logic of his system did not require veracity, but example. Today, we might

46. The Great Pyramid of Giza measures 230.4 meters on each side at the base and the height was originally 146.5 meters, which makes the slope 186.37 meters in length, that is just over 19% shorter than the base and with an 81-meter lower apex than the image Durand published.

tend to consider this attitude a suspicious propagation of 'alternate facts' to support an untenable proposition. Durand also reduced the information contained in the plans he presented to their utmost abstract form:

> Function in Durand's case is a global concept that does not necessarily imply an unambiguous causal relationship between program and form. (Huet in Szambien 1984, p. 10)[47]

His neglect of actual forms, materiality, and function was, of course, pedagogically deficient. Neither architecture nor civil engineering can be reduced to simplified models. For Durand, everything was subservient to the abstraction that was his 'system'.

It was this lack of applicability that Navier as well as many other contemporaries deplored. The non-professional approach to architecture and construction as a hypothetical 'science' rather than as a professional activity underscores the problem that practicing engineers had overall with the EP's post-1804 and especially post-1816 pedagogical bias.[48]

Nevertheless, and despite his detachment from all professional reality in his teaching, Durand was still able to distinguish between structural form and non-structural decoration. He wrote about the problem in a sub-chapter on bridges, not discussing ornament, but strongly condemning decoration by comparing two examples.[49]

---

47. La fonction est chez Durand une catégorie globale qui ne nécessite pas l'établissement d'une relation univoque de causalité entre programme et forme.

48. Guillerme attempted to draw an intellectual parallel between Monge and Durand as regards the economy of their *formules graphiques*. (*Les limites de l'idéalisation* 1986, reprinted in Guillerme 2008, p. 66). His attempt is unconvincing since Durand's means were a formalist abstraction while Monge's mathematical method served the practical purpose of technical design, as Prony's, Leblanc's, and Olivier's use of it demonstrated. Guillerme also appears to implicitly recognize that Durand's method had limitations. See *Inventer l'éducation du génie inventif? Quelques étapes d'un parcours rétrospectif.* 1993, reprinted in Guillerme 2008, especially p.111.

49. The distinction between decoration (an overlay of cultural reading) and ornament (enhancement of structural reading) is my own. It differs a little from Robert Willis's, who differentiated between the two categories in another way: the first he called mechanical construction which was a straightforward stress of actual construction; and the second, decorative construction, which was a symbolical depiction of structure that might differ from the actual load-bearing reality (Willis 1835, pp. 15–17). See also Fergusson 1862, vol. 4, 'Ornamental engineering'.

> How should one decorate bridges?
>
> If the designer concentrates only on what is necessary for a composition, adding nothing useless, if everything necessary is designed most simply, in a word, if a bridge is designed according to convenience and the economic principles that should guide the composition of all types of structures, then bridges will be as beautiful as it is possible to make them. As a convincing example one can only compare the Neuilly Bridge with the previously discussed Louis XVI Bridge. This last was designed to be what one calls *decorative*, the other design was only concerned with construction. And as a result, the Neuilly Bridge is as satisfactory as the other is embarrassing and disagreeable. (Durand 1805, vol. 2, p. 26)[50]

This passage is significant, as it is one of the very few that mention contemporary structures in Durand's book, and the only one that discusses any of them.[51] The Pont de Neuilly, Perronet's mature masterpiece, was inaugurated in 1772. It was not decorated, while the Pont Louis XVI, soon renamed the Pont de la Concorde, was his last work in 1791. It has outsized, stubby columns surmounted by capitals fronting the piers. It also has massive blocks capping them, which for many years served as bases for statuary, and denticulated cornices under the parapets. In his own description of the bridge, probably originally written for the 1789 *Supplément* to his treatise, Perronet had stated the columns to be necessary structural masses. His description implies an aversion to decoration and insinuates his attempt to avoid it as much as possible (Perronet 1820, pp. 238–241).

---

50. Comment faudrait-il décorer les ponts?

Si dans leur composition l'on fait entrer tout ce qui est nécessaire, si l'on n'y introduit rien d'inutile, si tout ce qui est nécessaire est traité de la manière la plus simple; en un mot, si dans la composition des ponts on observe les principes de convenance et d'économie qui doivent guider dans la composition de tous les genres d'édifices, les ponts auront toute la beauté dont ils sont susceptibles. Pour s'en convaincre il ne faut que comparer le pont de Neuilli avec celui ci-devant de Louis XVI. Dans la composition de ce dernier, on a couru après ce qu'on appelle décoration; dans celle de l'autre on ne s'est occupé que de construction; cependant l'aspect du pont de Neuilli est aussi satisfaisant que celui du pont Louis XVI est pénible et désagréable.

51. The few others are Soufflot and Rondelet's Panthéon (1757–90; 1802 vol. 1, p. vi), and in 1805, vol. 2, in which he presents buildings, most of them are antique examples with the exception of a few markets (pp. 58–59), two abandoned hospital projects by Bernard Poyet in 1788 (p. 69), and 1799–1800 military barracks (p. 75).

Fig. 84 - Comparison between the details of the 1772 Pont de Neuilly (upper) and the 1791 Pont Louis XVI, now Pont de la Concorde (lower) (Source: anonymous photograph 1871 / from Structurae: postcard c. 1913 showing original condition before widening)

Durand was aware of Perronet's work and of his attitude to ornament and decoration—after all, he had been his student in 1779–80 and had designed a successful project under his tutelage. He had certainly also read Perronet's 1782 description of the Pont de Neuilly, in which he had explained his structural, functional, and material rationale for the forms of the arch, the novel *cornes de vache* (flaring from the crown to the springing), and the slenderness of the piers.[52] It was also common knowledge in professional circles that Perronet had only been able to finalize the design for the second bridge after many preliminary proposals and that he had been obliged to deal with the real-life issues of his client's *goût*. Durand must have known that the columns, capitals, and cornices were not gratuitous, but rather structure that had been shaped according to the representational demands of the responsible directeur général des bâtiments du roi.

The earlier Pont de Neuilly had been designed and built under Directeur général Marquis Poisson de Vandières, who had been Soufflot's mentor and a devoted supporter of classicism. He and perhaps the then king Louis XV had evidently preferred sparse detailing, whereas the 1791 bridge had been subject to the aesthetic preferences of his successor Flahault de La Billarderie, Comte d'Angiviller, the same directeur général (under Louis XVI) who had overseen the curriculum at Durand's school during his studies.[53] Szambien (1984, p. 22, note 4) questions whether an outrageous proposal by Boullée with ship prows instead of columns and an elaborate frieze might not have been a more appropriate target for Durand's criticism. But it seems that Durand was not taking issue with either designer under whom he had worked, both of whom had since died,[54] but was subtly criticizing the politically disgraced and exiled d'Angeviller's taste as it had defined royal construction and the Académie's aesthetic concerns at the time of his own education.[55] By discussing this example Durand appeared to be discrediting the issue of 'taste' generally and marking a clear distance to his own education. However, in contrast to his two mentors, and despite stressing the issues of function and economy in his critique, Durand's

52. Perronet 1782, vol. 1, pp. 1–2 / 1820 p. 3. Prony analysed the dimensions and forms in his 1783 paper.

53. In the first published proposal in 1783 (Peronnet vol. 2, pp. 25–27), Perronet had also written that Louis-Joseph de Bourbon, Prince de Condé, had been involved in at least part of the decision-making process as early as 1755. He was head of the royal household with the title of Grand Maître de France. He fled France in 1789 while the bridge was under construction.

54. Perronet in 1794 and Boullée in 1799.

55. D'Angeviller had been accused of squandering public funds and had fled France in 1791. He later died in exile in the then small Danish border town of Altona, later to become a suburb of German Hamburg.

interests really lay in geometry, symmetry, and the logic of composition, and not in structure or in any formal aspect.[56]

Durand's creation and elaboration of the first systematic compositional method was a significant contribution to architectural theory. His disinterest in 'taste' and the technology of construction or detailing, his overemphasis on the value of theory ('science') over practice ('empiricism') dovetailed with the mathematically driven focus that the EP pursued. This became especially apparent after the EP's curricular changes in 1816, changes that were irreconcilable with practice and the didactical concept on which Lavallée and his colleagues based their pedagogy.

The situation at the ECP in the 1830s was different. Mary knew that his own graduates would be building modest *facilities*, structures like factories, railway stations, and storage sheds, not representational military or civil *ouvrages d'art*.[57] Mary accepted Durand's dismissal of the issue of 'taste' that had conditioned the whole premise of both architecture and engineering up to that time, and at the same time he recognized the advantages and the shortcomings of Durand's method. The singular originality of his own contribution was that he modified the geometrical system to create a powerful tool for the design not only of buildings, but also industrial plant, machinery, and industrial production systems by understanding design, construction, and manufacture as processes. He recognized—theoretical baggage aside—that the strength of Durand's approach to design lay in its hierarchical logic and incremental organization. It consisted of well-defined steps that were easy to follow. His system approach to design was universally applicable. It was more than a simplistic understanding of design as art, as a mechanistic application of rules of proportion or as an inexplicable and indeed ineffable outpouring of genius. The generalization of design method was Mary's great contribution.

---

56. Madrazo 1994, p. 13 gives the example of plate 3 of the 1799–1801 *Recueil et Parallèle des Édifices* edition with its round temples as proof that Durand categorized by form and not by structure, but in fact it was geometry, not form, that concerned Durand: the two preceding plates also categorize Roman temples as rectangles.

57. Durand had not included factories or other manufacturing buildings in his classification. In the final 1833 edition of the *Recueil des Édifices,* the typology did, however, include a range of other facilities like baths, bazaars and markets, stock exchanges, barracks, theaters, cemeteries, schools, universities, customs houses, depots, various types of hospitals, hospices, and prisons. But most of them were historical examples that were only moderately applicable to modern needs, with the notable exception of a very few: several theaters, a hospital, a hospice, and one university.

The change that Mary made in Durand's method redefined the formal elements as structural and applied the hierarchical rules to function and construction. He reconfigured the method's focus to conform to the ECP's goals. It enabled the industrialization of material manufacture and led to the development of iron systems in construction.

Mandar's architecture course

The basis for the shift that Mary made from abstraction, geometry, and hierarchy to materiality and fabrication had been laid in the education that he had received at his graduate school, the ENPC. The contrast between how the 'building arts' were taught at the EP and the ENPC could not have been greater. After its post-revolutionary reorientation under Prony's leadership in 1798–99 (An VII), three courses were established at the ENPC (Picon 1988, p. 129), with the intention to supplement—but not duplicate—those taught at the undergraduate EP. *Mécanique appliquée*, which Prony taught together with Lagrange at the EP until 1815, was delegated to Joseph Eisenmann, and architectural design as construction was taught by Charles-François Mandar. Both Eisenmann and Mandar were appointees of Prony's predecessor Lamblardie and had been at the school since 1796. They were joined in 1799 by Louis Bruyère for stereotomy and carpentry. Prony, whose knowledge of practice had dated from his work as assistant on two of Perronet's earlier bridges and as site engineer on his final bridge, the Pont de la Concorde (1791), and whose interest in *vulgarisation* is demonstrated by several papers, developed the curricula for all three courses (Prony 1799). Eisenmann the mathematician concentrated on the analytical aspect that he then applied to concrete technological problems. As a pure mathematician he was unable to develop the subject as Bélanger would later do it. The two practitioners Mandar and Bruyère dealt directly with technology: Mandar with design and detailing in construction and Bruyère, who was originally hired to teach stereotomy (stone-cutting) and construction in masonry and timber, soon expanded his brief to include iron in his material technology as well as planning.

Sganzin's colleague and sometime collaborator Mandar taught from 1796 to 1820 (Picon 1995 and 2000, p. 29), and he had been involved, at least peripherally, in the pricing work of the *Conseil des bâtiments civils* (Château 2008, p. 106). His approach to the subject differed from Sganzin's geometrically focused course at the EP. He was a military engineer by training, and his first publication in 1801 showed where his design interests lay. His treatise dealt with the theoretical organization of fortresses, of which he was an acknowledged expert.[58] Picon reports that he gradually became convinced that this was a dead end (Picon 1995, p. 28), and his focus subsequently shifted in the direction of architectural design and construction. To fulfil the mandate that Prony had set for him, he developed and published *Détails de construction d'une maison* for his students in 1818. This was a full set of house plans that encompassed the entire scale range in design from overall configuration to construction details and materials, and it served as a model for his class. In 1826 he expanded this into *Études d'architecture civile*, which he published as an abridged transcript of his course, accompanied by an impressive 122 plates. This final publication was devoted to the formal composition and technical detailing of grand houses and their service buildings and grounds. His instruction was therefore narrower but more immediately applicable to practice than those of either Durand or Sganzin. While this was useful for a practitioner, it was not a conceptually designed course, nor was it a methodological treatment of the field of architectural design or construction. Nevertheless, Mandar's nit-picking attention to construction details and materials[59] did resurface in the three projects that his students were required to design over the course of their studies (Picon 1995, p. 33). His approach, as limited as it was, did therefore influence the students at the ENPC, and among them Mary.

Bruyère's system approach to construction and materials[60]

While Mandar taught architecture—or really design as construction—Louis Bruyère taught his course from another standpoint that included planning, and that we might oversimplify as materials and technology as construction. Bruyère was the least-known of the so-called Revolutionary Architects but the most active builder among them.[61] His influence on the development of iron construction was

---

58. Picon considers this work in its catalogizing of the development of fortification as analogous to Durand's *Recueil et parallèle* (Picon 1995, p. 28).

59. Steiner (1984, p. 22) points out that Mandar followed Rondelet's lead in inserting iron reinforcement throughout the structure.

60. A short overview of Buyère's innovations was published as a paper (Peters 2020).

61. Emil Kaufmann defined the term in 1952, based on a study in 1933 to describe the architecture of Claude-Nicolas Ledoux, Étienne-Louis Boullée, and Jean-Jacques Lequeu, essentially to distinguish the use of repetitive elements, antithesis of form or material, and multiple response (pattern) in

Fig. 85 - Louis Bruyère was both an architect and an engineer who concerned himself primarily with the functional and material approach to design. He taught construction at the École des Ponts et Chaussées and influenced the pedagogy of the École Centrale, especially through his student Mary. Bruyère also impacted professional design practice, and particularly the use of iron as an industrial material. Portrait by Elise Bruyère 1808–10. (Source: © ENPC Collection)

pervasive, but it has gone largely unacknowledged. As a young man he had begun his career in a traditional architectural apprenticeship in an office, not at the Académie. He then practiced as an architect until 1783, when he must have felt the lack of a structured approach to building, upon which he was admitted to the ENPC (Jullien 1832, p. 813). There he began his second round of education in an engineering academic context under Perronet and his protégé Prony, who had returned to the school that year as a teacher.[62] Since Bruyère was already a fully

the later eighteenth century in France. Later writers began to include Durand as theoretician and here I add Bruyère to the list. For a critique of Kaufmann's definition, see Garric 2011.

62. 1783 was an important year in engineering education: Monge became examinateur at the navy and Sage founded the École des Mines.

trained building professional, he was able to finish his studies in the minimum time of three years, and he graduated into the Corps des ponts et chaussées in 1786. He was then both a traditional architect/city planner who had learned about space, form, detail, and history in his traditional apprenticeship and also a modern engineer, schooled in the most up-to-date structural theory, material technology, and construction.

For the better part of a decade, Bruyère then worked as a minor state consulting engineer (*ingénieur ordinaire*), first in Tours and then in Le Mans from 1789 on canals, roads, and bridges, as well as designing a park in the center of town. He resigned that position in 1793 to return to private architectural practice. Bruyère was an experienced border-crosser between the mindset of the traditional and the new approaches to the building profession, which was unusual at the time. This characteristic obviously impressed his former teacher Prony, and in 1799 he began a very public career when the three-years older Prony, now director of the ENPC, hired him to teach *stéréotomie et l'art des constructions*. A promotion to ingénieur en chef and secrétaire du conseil général des ponts et chaussées followed in 1804, and in 1811, the former director-general of the ponts et chaussées, Jean-Pierre de Montalivet, friend of the emperor and now minister of the interior, recommended his appointment to the newly created position of directeur général des travaux publics, a combination of municipal engineer and architect in Paris. He filled this position, which brought him into contact with the *Conseil des bâtiments civils*[63] until 1820. In 1813 he apparently presented Montalivet with a *Collection des marchés de Paris avec projets*, an overall plan for the markets of Paris, with 127 drawings.[64] During his tenure he oversaw and influenced the design and construction of many public facilities: slaughterhouses, markets, a school, a stock exchange, and a wine depot, all of which he discussed and illustrated in his publication *Études relatives à l'art des constructions* (1823–28). This two-volume treatise exposes the full range of his own planning and construction experience.

---

63. Francearchives Portail national des archives. Iventaire-F13/10/A-F/13/139 Bâtiments civils, consulted online 17.12.2020. The position only became a ministry of public works as part of the interior ministry in 1830, and an independent ministry from 1840.

64. ENPC fol 486 (reference found in Guàrdia, Oyón and Garriga, 'Market and Market Halls', in Jon Stobard and Vicki Howard, *The Routledge Companion to the History of Retailing*, chap. 7, consulted online, 10.04.2020. It has not been possible to trace this document.

Although the designs were all by other architects, Bruyère contributed substantially to their concept and functional organization and possibly also to their construction. In a very real sense, he made them his own. As he wrote after he retired:

> I must confess that these types of structures were among those that interested me the most. (Bruyère 1828, vol. 2, *Recueil* XII, p. 3, footnote 1)[65]

A typical example of his personal concern with the function and organization of such facilities was the design of the prison bakery in Saint-Lazare Prison in 1819:

> A proposal for ovens to provide the prisoners with bread was once brought to my attention. They were to be installed at the Saint-Lazare prison in Paris and have a capacity of four thousand loaves of 24 ounces each. The design for these ovens and for the building to house them was by a very talented architect, but I wanted to reassure myself that the organization of the facility would satisfy all requirements, and I decided to consult a laborer in the warehouse who was kind enough to supply important and useful information that was then implemented in the realization of the project. One will find the plans and details of these ovens as built, with all the necessary explanations to understand them on plate 13. I limit myself here to say only that special conditions made it necessary to abandon the daily use of these ovens, but that their organization was nevertheless recognized to be the best possible one, thanks particularly to the advice and the information provided by the warehouse laborer we consulted. (Bruyère 1828, vol. 2, *Recueil* VII, p. 10)[66]

---

65. J'avoue que ces sortes de constructions étaient au nombre de celles qui excitaient le plus vivement mon intérêt.

66. On avait cru voire établir dans la prison de Saint-Lazare, à Paris, des fours destinés à cuire le pain des détenus en général, et dont la quantité devait s'élever à environ quatre mille rations du poids de 24 onces chacune. Le projet de ces fours et de l'édifice qui devait le contenir avait été fait par un architecte de beaucoup de talent; mais pour m'assurer que la disposition de ce projet satisferait à toutes les conditions, je crus devoir faire consulter un manutentionnaire qui eut la complaisance de fournir des renseignemens très-positifs et auxquels on se conforma dans l'exécution. On trouve, Planche 13, les plans et détails de ces fours, tels qu'ils existent, et les indications nécessaires pour en faire connaître toutes les parties. Je me bornerai donc à dire que les considérations particulières ayant fait suspendre l'usage journalier de ces fours, on a cependant reconnu qu'ils étaient disposés de la manière la plus favorable : ce que l'on doit particulièrement aux conseils et aux renseignemens donnés par le manutentionnaire consulté. The unnamed architect may have been Louis-Pierre Baltard (Victor Baltard's father) who was to publish *Architectonographie des prisons, ou parallèle des divers*

Indeed, the two-volume *Études*[67] that Bruyère published at the end of his career documents his concern with function as well as his erudition and inventiveness. His collection of studies demonstrates that his interests did not end with stereotomy—that is, the art of stone cutting—nor with the details of construction. Far from it. The two volumes encompassed all forms and scales of construction including city planning, dwellings, public facilities from warehouses to markets, farms, military hospitals, canals, river correction and pumping machines, the manufacture of tiles, centering for arches, leveling apparatus, bridges in stone and timber, and experimental designs in iron. The twelve chapters called simply *recueils* (collections) stress the process of construction and the manufacture and design characteristics of materials.

Both Bruyère and Mandar were clearly more grounded in actual building practice than their contemporary Durand at the EP, although Bruyère, at least, shared Durand's interest in systematization.[68] Mandar went into more detail in house construction than Bruyère, but Bruyère's concerns were those of a broadly experienced practitioner and far more generous in scope. He was intensely interested in the economy of materials and methods, and in the appropriateness of construction (Bruyère 1823, vol. 1, "Observations préliminaires", pp. 9–11). His contemporaries repeatedly stress that he was valued and even revered for his powers of observation and for his conscientious studies of function. As Bruyère himself had written:

> When, in the course of my job, I was required to concern myself with the public granaries then under construction in Paris, and with those proposed for Saint-Maur, my first move was to collect plans of the principal examples of the type, to study the various published reports on the question, and above all to consult those experts in charge of administrating the warehouses. I even

systèmes de distribution dont les prisons sont susceptibles in 1829, and who was to be commissioned to replace the old structure in 1824 and build the new prison hospital and chapel in 1834.

67. This work, in two volumes, is a collection of twelve groups of the more significant projects that preoccupied him, as well as some of his own, dating from around 1808 to 1828. It also includes a reprint of Agustín de Betancourt's great truss over the Moscow riding hall (1817) and a compilation, including the only published version of an otherwise influential report on model tests for his and Jean de Voglie's Fouchard Bridge in Saumur (1783), and another on the failure mechanism of masonry bridges by François-Michel Lecreulx; see Huerta 2015.

68. Mandar had also displayed a certain interest in systematization in his 1801 treatise on fortification, but this interest does not seem to have influenced his teaching.

> began a few experiments that circumstances and my infirmities however prevented me from completing. My studies remained unfinished, but others, more able than I, might continue them, and that is why I thought to present my few preliminary results in this study. (Bruyère, 1823, Recueil II, p. 3)[69]

Navier confirmed this quality in the speech he gave to the *Conseil municipal de la Ville de Paris* shortly after Bruyère's death:

> The structures built under M. Bruyère bear witness to his detailed study of the nature and purpose of each edifice. His markets, his slaughterhouses, his general wine depot form a collection of novel layouts that are perfectly adapted to the uses that these structures serve. Today these layouts are to be found replicated in all French cities, and they will serve as models for a long time to come. (Navier, 1832, cit. by Adolphe Jullien, 1832, p. 815)[70]

Bruyère compiled his *Études* from both the architectural and the engineering standpoints, and he translated[71] the concept of architectural design into the engineering world. He may have been the first designer who treated building technology as a core design discipline rather than as merely an adjunct to formal design. On the one hand, his designs were influenced formally by Durand's contemporary definition of types and his compositional methodology, and on the other they manifested his own distinct bias toward functional and construction issues.

Both men, possibly influenced by Prony's thoughts on systematization in *Mécanique philosophique* 1800, initiated the idea of repetitive systems in different

---

69. Lorsque j'ai été appelé, par ma position, à m'occuper des greniers publics qui se construisaient à Paris, et de ceux que l'on projetait d'établir à Saint-Maur, mon premier soin fut de recueillir les plans des principaux édifices de ce genre, d'étudier les différens écrits qui ont paru sur cette matière, et surtout de consulter les personnes instruites chargées de l'administration des réserves. J'avais même commencé quelques expériences que plusieurs circonstances et mes infirmités ne m'ont pas permis de compléter. Mon travail est resté bien imparfait ; mais d'autres plus habiles pourront le continuer, et c'est pourquoi j'ai cru devoir en présenter les faibles résultats dans ce Recueil.

70. Les construction[s] exécutées sous M. Bruyère prouvent une étude approfondie de la nature et de la destination de chaque édifice. Ses marchés, ses abattoirs, son entrepôt général des vins offrent un ensemble de dispositions nouvelles parfaitement adaptées aux usages auxquels ces monumens sont consacrés; on les imite aujourd'hui dans toutes les villes de France, et ils serviront long-tems de modèles.

71. Translation characterizes the ability to apply information from outside a field, while transformation denotes a modification or new application inside a given field (Peters 2000).

ways, Durand in geometry and form and Bruyère in construction. He incorporated taxonomy, form, structure, material production, and construction method in his studies, thereby communicating the interconnection of all these aspects. This border-crossing approach characterized Bruyère's teaching at the ENPC, and it influenced his students profoundly. About his teaching and the relationship to his students, one of them, the eminent railway engineer Adolphe Jullien wrote:[72]

> He was appointed to a professorship at the École des ponts et chaussées in 1799 where he established regular courses and defined precise rules to coordinate the haphazard collection of diverse observations and the many facts that had hitherto constituted almost the whole of engineering science. He gave the first lectures on construction to men who since have distinguished themselves by important works, and who all gratefully recall the benevolent and personal interest of their former teacher, and the zeal with which he explained all the details of his art, both in the courses he taught and in the ad hoc sessions he held with his students during which he had them draw up proposals for roads, bridges and canals. (Jullien 1832, p. 814)[73]

The key innovative projects that Bruyère published in his compendium and that influenced the future of iron construction were a series of wrought iron bridges. Most of them were never built, but one was. It was a true ancestor of modern iron construction, but because it was so small, it was never acknowledged by his contemporaries as the pioneering structure that it was.

The 'Pont sur le Crou' was a 12-meter, wrought-iron trussed arch. Bruyère erected it in 1808 and it stood for several years over the mouth of the tiny Croult

72. With the exception of this quotation and a few minor changes, Jullien's obituary closely follows a slightly shortened version of the text of Navier's speech of 1832. As Bruyère died on the very last day of 1831, the date of publication of Jullien's obituary in this volume of the *Revue encyclopédique, October* 1831, is misleading. It was evidently published early in 1832.

73. En 1799 il fut appelé à l'école des ponts et chaussées comme professeur; il chercha à y organiser des cours réguliers, et à soumettre à des lois précises les observations diverses et les faits nombreux dont l'ensemble constituait alors presque uniquement la science de l'ingénieur; il donna là les premières leçons de construction à des hommes qui, depuis se sont distingués par d'importans travaux, et qui, tous ont conservé avec reconnaissance le souvenir du bienveillant intérêt que leur portait leur ancien maître; et du zèle avec lequel il leur exposait tous les détails de son art, soit dans les cours qu'il professait, soit dans les entretiens particuliers qu'il engageait avec ses élèves, en leur faisant dresser sous ses yeux des projets de routes, de ponts et de canaux.

River in St. Denis.[74] Neither the bridge nor the river exists today. Bruyère found the theoretical basis for the form in wooden truss construction as exemplified in cross-braced centering designs for masonry arch bridges.[75] The novel aspect of this bridge was that Bruyère designed his truss in wrought iron with modularized, linear components and bolted connections. Although this was far from the first wrought iron bridge, it was most significant for the future development of iron construction.

Catherine the Great had commissioned several bridges in a combination of wrought and cast iron for the park of her palace in Tsarkoe Selo years before. Others had been built for Leopold III the Duke of Anhalt-Dessau in Prussia, and even one in the park of the Château de Raincy for Philippe Égalité, Duke of Orléans, in what is today the Département Seine-Saint-Denis. However, these had all been built as hand-crafted, decorative embellishments. They were park furniture or aristocrats' follies rather than working structures, while Bruyère's was conceived as a functioning road bridge. Vierendeel credits this bridge with the first use of triangulation in iron (1897–1902, p. 32), although the many earlier Russian examples and Victor Louis's roofs are also triangulated in a certain sense. The Croult bridge was built while Mary was Bruyère's student, and his teacher possibly discussed its concept in class or perhaps in the ad hoc seminars that Jullien mentioned.

---

74. There are two manuscripts on the bridge, one by Bruyère (ENPC ms. 2436) and one by Becquey de Beaupré (ENPC ms. 2134). The Croult Bridge was only published twice, once by François Thiollet in 1832 (Steiner 1984, pl. 9) and once by Cresy in 1847 (vol. 1, pp. 284–285). It was passingly mentioned in 1839 when Polonceau's father, Antoine-Rémi Polonceau, alluded to it in the preamble to his description of the Pont du Carrousel, calling it only *joli et bien combiné* (pretty and well put together). He was not at all a fan of wrought iron, criticizing it as *plus flexible, plus cher et plus oxydable que la fonte* (more flexible, more expensive and more prone to oxidize than cast iron) (A.-R. Polonceau 1839, p. 10). His son, Camille Polonceau, then may have referenced some of Bruyère's thoughts in his own composite truss design in 1837, or they may have been filtered through Mary's interpretation of them. Arthur Vierendeel and Frances Steiner have both mentioned the importance of the Croult Bridge (Vierendeel 1897–1902, pp. 31–32 and pl. 2; Steiner 1984, p. 32 and pl. 9). Picon and Yvon 1989, pp. 112–113 mention it. Mehrtens also mentioned it (1887, p. 36, and 1908, pp. 346/644, figs. 831–833 on p. 652) and recognized its importance, and he attributed the lack of continuity in wrought-iron truss design until the 1850s in France and almost until the 1870s in Britain to the difficulty of forging connectors at the ends of the members. This was only solved when riveted gussets became standard in the early 1860s (see Fig. 120 Plocq pl. I and Jolly, pl. VI as discussed in the chapter 'The result in iron…Eiffel').

75. Bruyère published designs and his theoretical thoughts on centering in 1828 (vol. 2, Recueil VII) along with Betancourt's 1819 report on the roof of the riding hall in Moscow. Although they were the same age, Betancourt had been a year his junior at the ENPC.

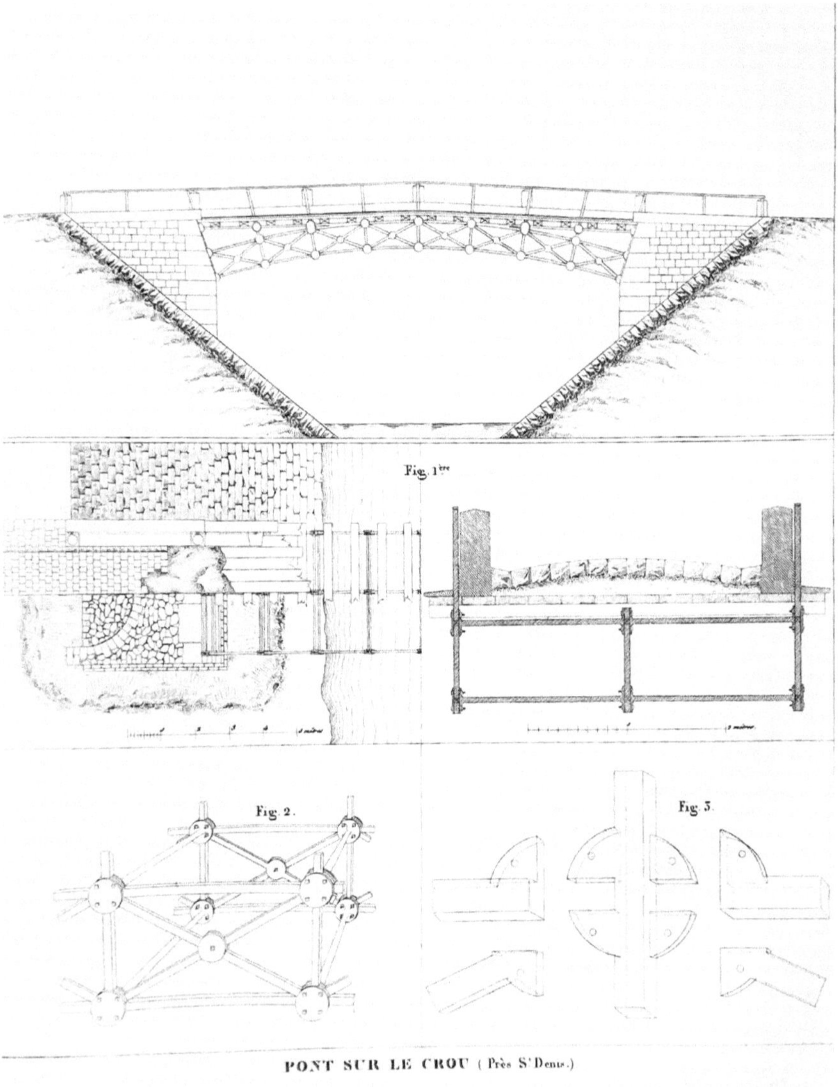

Fig. 86 - The 1808 Croult Bridge was a tiny, twelve-meter wrought iron arch that Bruyère built over a small creek in Saint-Denis, now part of Paris. Its importance lies in the fact that it was entirely made of wrought iron and conceived as an assembly of similar parts connected by means of a simple system. Since the material allowed all parts to resist any form of stress, it represented a conceptual advance over previous iron bridges in which those parts that were cast iron could only resist compressive forces. That, and the fact that the connections were standardized and easily dismantled and reassembled, make this modest structure the first modern iron system as well as the first true truss bridge. (Source: Bruyère, *Études relatives à l'art des constructions,* vol. 1, 1823, Recueil III, pl. 11)

The novelty of the structure was four-fold. First, all the parts of his truss were made of malleable wrought iron and none in brittle cast iron, which meant that the members could withstand all forms of stress equally well, not merely compression.[76] Second, Bruyère designed the bridge as a system—that is, as a systematic assembly of prefabricated and more-or-less standardized iron components. These two features alone made it the first true iron truss in the modern sense. In this Bruyère was earlier than Gérard-Joseph Christian, who apparently advocated standardized parts for machine construction in his courses at the Conservatoire (Guillerme 2008, p. 309).[77] The earlier Russian, German, and French decorative park bridges and the proto-trusses in churches, theaters, and other representative buildings—some of which became better known because Eck later published them—were not systematically or modularly conceived, although they were made of repetitive parts. Bruyère's design concept was possibly based on Durand's formal and spatial modularization and compositional method as well as on Thomas Paine's cast-iron voussoir system that he had presented to the Académie des Sciences in 1787. Although Bruyère had been sent by the *corps* to Tours, 240 km from Paris the year before, he may have been at Paine's presentation or at least heard about it at the time.[78]

The third novel feature was that Bruyère standardized not only the bars but also the connections between them. Although he did not yet design the connections as independent components or connectors as Van Cléemputte would do around 1830 (Eck 1836, vol. 1, p. 51 and pl. 32), his connection technology was standardized as simple bolting throughout, and he designed the connections in a

76. Bruyère was well aware of the dangers of fracture in cast iron. He had collected extensive material on the fractures that had occurred on the Pont d'Austerlitz, measures proposed by Girard to limit the cracking, and tests on the material (Bruyere, 1779–1810, notebook pp. 449–473).

77. I was unable to find an explicit reference to standardized parts for machine construction, but Christian expressed the principle in a general sense in *Vues sur le système général des opérations industrielles* (1819, pp. 70–75) and as an economy of means, referring to Smith's *On the Wealth of Nations* in the second part, *De l'économie manufacturière*, on pp. 99–117.

78. Thomas Wilson had realized such a system in his Sunderland Bridge over the River Wear in northeastern England (1796) based on a trial span Paine had built in Hyde Park, and Bruyère knew of this bridge (see Bruyere 1779–1810, pp. 371–379). However, Wilson's bridge and his smaller, short-lived Staines Bridge of 1803 (Bruyere 1779–1810, pp. 380–383) were in cast, not wrought iron, with all the disadvantages that this entailed.

form reminiscent of what would later become the bolted or riveted gusset plates of iron bridge and building framing.

Lastly, because of the way it was assembled, the system could be dismantled and reassembled. Shortly after it was completed, the Croult Bridge was in fact moved and rebuilt (Bruyère 1823 Recueil III, p. 5). As far as is currently known, this is the earliest recorded instance of planned disassembly in iron construction,[79] a characteristic that was to become one of the advantages and hallmarks of the new material. Whether or not Bruyère developed the concept of assembly and disassembly in iron as consciously as he did the idea of system, it was to become a feature of industrialized iron construction.

This was an advance over the individually designed and uncoordinated connection technology that was usual in cast- and wrought-iron construction at the time. Commonly, iron members had been connected with a medley of wedges, bolts, straps, interpenetrating members, and mortice-and-tenon joints, all derived from traditional carpentry, or in Russia, in part from locksmithing (Kosykh 2021). This, of course, inhibited disassembly. The Iron Bridge of 1779 exemplified this lack of system, or to a less chaotic extent, the Sunderland Bridge, erected based on Paine's ideas in 1796.

Bruyère was well aware of both of these bridges and of the other road bridges that had already been built in iron in England and in France. He published the most important of these in the first volume of the *Études* as a contrast to his own ideas (Bruyère 1823, vol. 1, Recueil III Ponts en Fer). All of the bridges were modularized in the sense that their parts were cast in series, but they were neither consciously conceived as demountable component systems nor were they in wrought iron. Their builders had been obliged to develop complicated and contorted ways of avoiding tensile, torsion, shear, and bending stresses to accommodate the inherent drawbacks of their material.

By conceiving his own designs solely in wrought iron, Bruyère circumvented the material limitations of cast iron and its related structural problems. This crucial simplification, together with his attempt at a systematic standardization of parts, provided the conceptual basis for the ECP's approach to industrialized structural design as Mary would later teach it. In the next generation this would rapidly lead to the modern development of iron and then eventually to steel framing.

---

79. Disassembly had been known in timber construction since the Middle Ages, for example house construction in Appenzell in Switzerland.

Fig. 87 - Before Bruyère's attempt to standardize iron connections, such structures had used a multitude of techniques and forms derived from wood construction. This chaotic condition is especially clear in the first successful iron bridge (1779) over the Severn in Coalbrookdale. (Source: View: Tony Hisgett, Wikimedia commons. Detail: 'Uncle Silver' partial, Wikimedia commons).

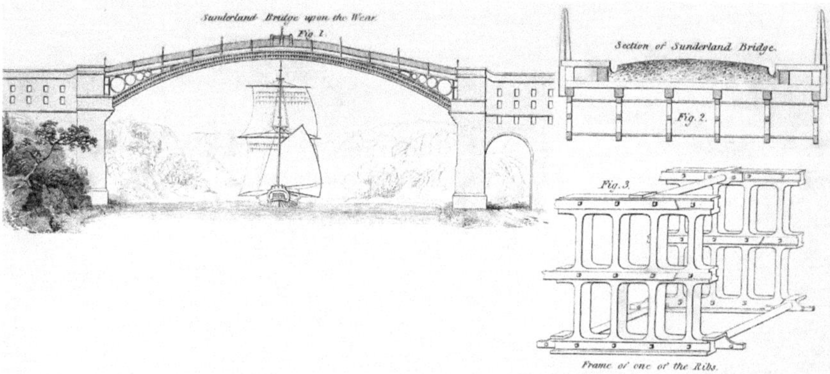

Fig. 88 - Gradually builders attempted to reduce the confusing number of special solutions. An example is Paine and Wilson's Sunderland Bridge over the Wear in 1796. (Source: Taylor, *Supplement to Nicholson's Operative Mechanic*, 1829, p. IV, partial).

As Jullien and Navier wrote, Bruyère had carefully researched the opinions of experts, and he was up to date on discussions involving the use of iron in structures and especially in bridges. One of the thirty-six volumes that comprise his collection of copies of reports, letters, and articles in the ENPC archive, labeled *mss.233 Tome 24BIS Ponts de fer*, contains memoirs and communications regarding the structural use of both cast and wrought iron in bridge building. They cover the entire previous French and British history of the topic from the late 1770s to the early 1800s (Bruyère 1779–1810). Aside from copies of several communications by Perronet and others, there is a series of long and detailed exchanges between the chemist Louis-Bernard Guyton de Morveau, one of the founders of the EP, an unidentified 'D. L. B.' (p. 121 of the notebook),[80] and the painter, inventor, and iron bridge pioneer Arnaud-Vincent de Montpetit, in which they discuss the relative advantages of wrought and cast iron extensively (Bruyere 1779–1810, especially pp. 101–116; see also Steiner 1984, p. 20).

While earlier builders—following the lead of boiler and steam engine mechanics—had sought to understand iron and its possibilities, and the bridge builders among them had fought with the constraints imposed by cast iron, Bruyère simply sidestepped the limitations.[81] He thus concentrated entirely on studying how to systematize, simplify, and serially prefabricate parts. Bruyère knew that wrought iron was still too expensive a material to use in large-scale structures, except perhaps in nationally prestigious projects. It had been used in the earliest French theater roofs at the turn of the century, but these were all modest-scale, one-off structures and they were hand made, not industrially produced. However, the use of the material was not the issue: his development of the concept of standardization and prefabrication was. The grand proposal he developed for a 114-meter, single-span, wrought-iron footbridge over the Seine from the Invalides to the opposite bank in 1810[82] was the direct result of his experimental Pont sur le Crou of 1808. He had designed it around the time that he was appointed city architect-engineer of Paris and the year that Mary graduated. The emperor Napoleon approved the project, but it was not built. Bruyère speculated that the

80. My search of the ENPC student listings between 1744 and 1930 posted online (Lacourt 2020) and of the online listings of the better-known graduates of the EP has not produced an engineer or scientist with those initials. My conclusion is to read D. L. B. as a cipher for 'M. le Directeur Louis Bruyère'.

81. Solving problems by avoiding them later became a standard technique of 'Yankee ingenuity' in United States iron construction and industrial design in the nineteenth century, the 'Rube Goldberg method' (Peters 1989).

82. See Bruyère 1823, vol. 1, pl. 12, Recueil III; Picard 1902, vol. 1, p. 406 and facing plate; also Picon and Yvon 1989, pp.112–113 (ENPC dessin D.III.C.11). No single span of that size was completed across the Seine in Paris until 45 years later, in 1856, when the wrought-iron Pont d'Arcole designed by Alphonse Oudry and built by Pierre-Félix Moreau opened.

Fig. 89 - In 1810 Bruyère proposed a 114-meter clear span as a wrought iron trussed arch over the Seine. It was based on his experimental span over the Croult two years before. It would have been exceedingly expensive but feasible as a prestige project of national importance, and Napoleon supported it. However, it was not built, and the first single-span arch over the Seine was only erected half a century later, in 1856. The reason for its abandonment is not clear. Artist's rendering of the arch. (Source: Picard, *Exposition universelle internationale de 1900*, 1902, vol. 1, pl. facing p. 406)

reason may have been because its curved deck would have blocked a direct view of the Invalides from across the river (Bruyère 1823, vol. 1, Recueil III, p. 6 and pl. 12), but a more likely explanation was that Napoleon needed all the iron that French industry could produce at the time for his war effort.

This was an exciting experimental period of iron construction, when the 39-meter cast-iron dome was built between 1809 and 1811 over Nicolas Le Camus de Mézières's 1767 Halle aux Blés to replace the wooden one that had burned. The dome had been designed by the architect-decorator François-Joseph Bélanger, the form calculated (not the structure, see Porrino 2201.2, p. 68 footnote 121) by the ENPC engineer François Brunet, who published a book on his work in 1809, and detailed, manufactured, and erected by Jean-François Roussel (Eck 1836, vol. 1, pl. 36 and p. 71). The structure was unprecedented. and we can detect Roussel's tentative attempts to adopt the development of bolts, lugs, rudimentary flanges, and the first gusset-like configurations like those that that steam engine manufacturers experimented with in the first years of the century, as well as those that Becquey de Beaupré and Lamandé had devised for their 1806 Pont d'Austerlitz.

The dome was the first large iron building structure in France and the largest in the world. Bruyère was not himself responsible for the design or construction. He inherited it at a late construction stage, when he took on the job of municipal

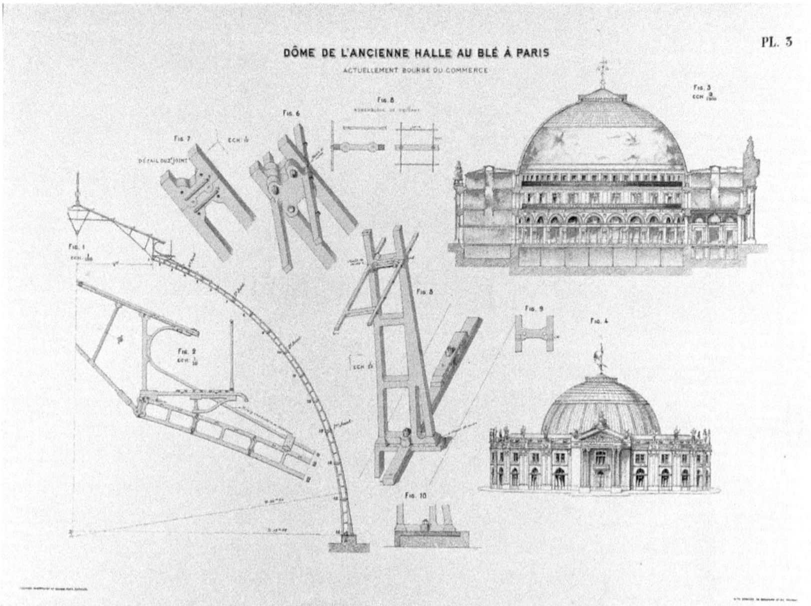

Fig. 90 - The dome over the Halle aux Blés (1809–11) was the largest cast-iron structure of its day, and it exerted a major influence on iron construction. It was erected by the ironmaster-contractor Jean-François Roussel. (Source: Vierendeel, *La Construction Architecturale en fonte, fer et acier*, 1897–1902, pl. 3)

architect-engineer in 1810. Still, it was the first major public project after his appointment, and he was therefore responsible for its success (Bruyère 1828, vol. 2, Recueil XII, p. 3, footnote 1).

The structure of this cast-iron building was backward-looking, if not yet entirely outdated. Bruyère's little bridge of 1808 had been conceptually much clearer than the Halle aux Blés project and had displayed an uncompromisingly systematic structural concept. The dome structure was typical of the early experimental period of iron construction, with detailing that still followed solutions that had been developed for timber structures. Each solution was designed to solve only one specific connection situation, albeit by then with a few innovative aspects that Roussel took from boiler construction. Viewed from the standpoint of Bruyère's 1808 Croult Bridge, it really was outdated in both its material and connection technology. Nevertheless, it dealt successfully with many structural, computational, manufacturing, connection, and geometrical issues. It was also the first structure for which the architect employed the services of an engineer as consultant and the two functions were not united in a single person, and it was a model for the future development of the building process in that regard as well.

The decision on the material and its advantages and disadvantages had been discussed back and forth at the beginning of the century, whereby the engineer, the Ponts et Chaussées as an institution, and Monge had all contributed in multiple reports to the discussion (Porrino 2021.1). By the time that Bruyère became involved, the decision to use cast iron had long been made for reasons of cost (Eiffel, Contamin, and Fouquet 1889, p. 5). Bruyère must have followed the manufacturing, detailing, and erection processes closely, and surely discussed the structural and construction issues that so differed from his own forward-looking views with the participants and likely also with his students. However, the difference between the design and building process of this structure and the ill-fated 1832–33 Anthæum twenty years later in Hove clearly illustrates the impact that science and scientists had in French iron construction, versus the purely empirical engineering approach of the early British builders. This was a cultural difference on which the founders of the ECP based their concept and on which Mary would build when modifying Durand's method.

Bruyère must not have liked the dome project because, although he promised (*Études* 1823, vol. 1, Recueil II, p. 13) to publish its details later in Recueil VII, he never did. This was in contrast to Eck who discussed it extensively in 1836 (Eck vol. 1, pp. 53–54 and pl. 36) and Mary who retained it right through the last edition of his transcript (Mary 1862–63, pls. IX and X). It is a pity that we cannot learn how Bruyère understood his role in the building process of this structure.

Curiously, very little of his novel thinking about iron appeared in the many buildings built during his tenure as municipal architect-engineer. The reason was almost certainly financial. Iron was still an expensive material and was needed at the time primarily for military purposes. After Napoleon's final defeat in 1815 and until Bruyère retired in 1820, France was battling to reestablish its industrial potential. The iron industry was in its infancy, and so the many public facilities that Bruyère documented were all designed as repetitive, modular, small-span masonry structures built with wooden roofs. The modular aspect may have been Bruyère's contribution to their design, but it was only in the decade following Bruyère's death in 1831 that the first iron *galeries* and market halls began to appear, iron roof structures were built again, and iron bridge building gradually gained momentum. The second volume of Eck's treatise (1841) documents the

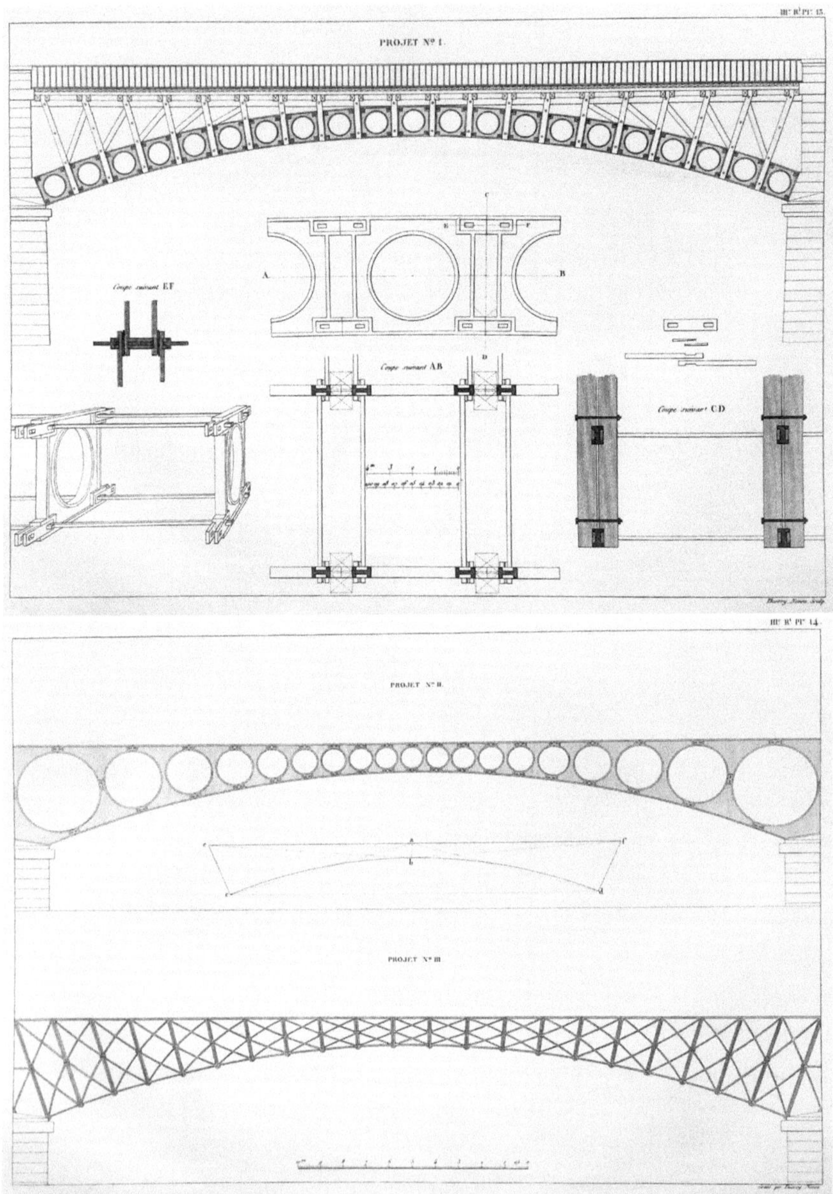

Fig. 91 - After building his first iron bridge in 1808, Bruyère continued to experiment with iron and composite modular systems. Although these examples were not built, his publication testifies to a preoccupation with prefabrication, modular construction, and the standardization of connection technology. He was ahead of his time, and in the next generation the École Centrale would complete his thrust with the critical connection to the iron industry. (Source: Bruyère, *Études relatives à l'art des constructions*, 1823, vol. 1, Recueil III, pls. 13 and 14).

proliferation that Bruyère did not live to see.

However, in the years that followed the Croult Bridge, Bruyère continued to research and teach his systematic approach to construction. He proposed three further long-span bridges that examined composite configurations in timber, cast and wrought iron, which would have been less expensive to build than purely wrought-iron structures (Bruyère 1823, vol. 1, Recueil III, pp. 6–7 and pls. 13–14). As Jullien wrote, Bruyère discussed his thoughts with his students, and these would carry Bruyère's innovative ideas from the classroom and into industry and building.

Together with his first successful little design, these later bridge proposals were revolutionary in the true sense of the word because they reframed how builders thought. They promoted rethinking construction as a configuration of modularized, repetitive parts with systematized repetitive detailing, and they conceptually presaged Perdonnet's work and his student Camille Polonceau's successful composite wood, cast-, and wrought-iron truss. Bruyère never saw Polonceau's truss, as he died on the very last day of 1831, but the systematic approach to construction and detailing that he had launched spread rapidly, and not only in France.

Polonceau's 1837 truss in France was not the only such structure (see also later in the chapter 'The result in iron: a brief characterization of six careers' on Polonceau). Squire Whipple in the United States patented a composite truss using only cast and wrought iron with repetitive, standardized detailing in 1841. Whipple had studied for a year (1829–30) at Union College in Schenectady, NY, a school that was aware of French developments and that would later establish the first American civil engineering program inspired by the ECP's pedagogy under Gillespie (Comberousse 1879.1, pp. 135–136). Although engineering at Union College only began as a degree-granting program in 1845, Whipple must have gained some expertise while there, or perhaps he contacted Gillespie after 1845. He was the first, according to Dorn (1970, p. 29, footnote 1), who actually analyzed a truss structurally in his treatise of 1847, albeit in a primitive fashion. He was obviously one of the rare builders who was enabled in both theory and practice at the time.

Whether or not Bruyère's approach to the modularization of structures had any direct impact on Carl Culmann's understanding of Whipple's and other trusses as iterations of panels is not known. Culmann first presented the idea in an article in the *Allgemeine Bauzeitung* (1851). It led to his graphic statics. Bruyère's work may have indirectly influenced his thinking.

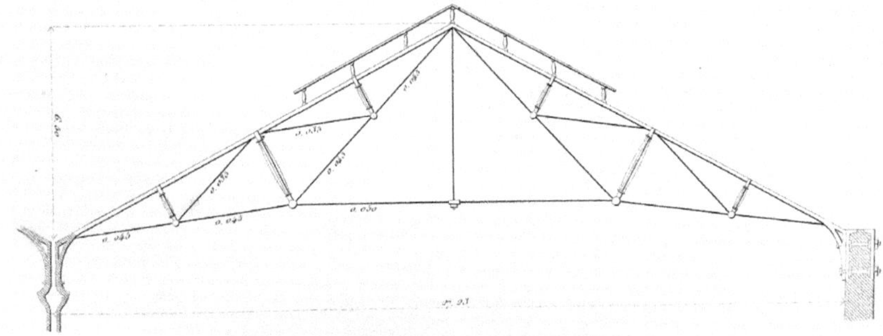

Fig. 92 - Bruyère's experimentation had a far-reaching effect. In France the École Centrale graduate Camille Polonceau created an economical composite truss made of prefabricated wooden, cast-, and wrought-iron members in 1837 that found widespread use. (Source: Cornet, *Album des chemins de fer*, 1853, 3rd edition, pl. 41 detail)

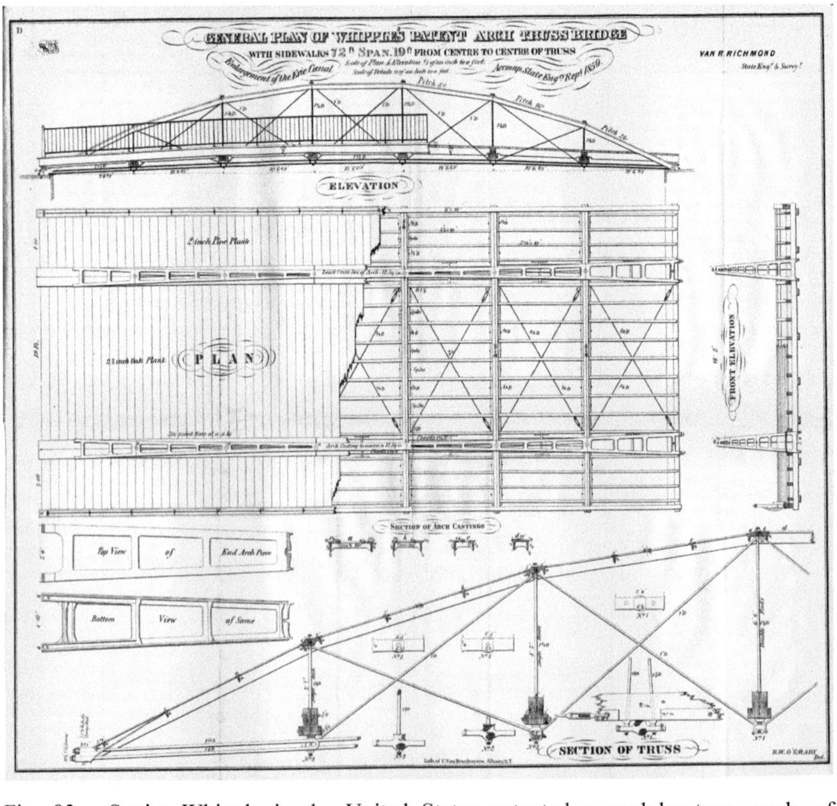

Fig. 93 - Squire Whipple in the United States patented a modular truss made of standardized iron components in 1841. (Source: © US patent no. 2064, US Patent Office)

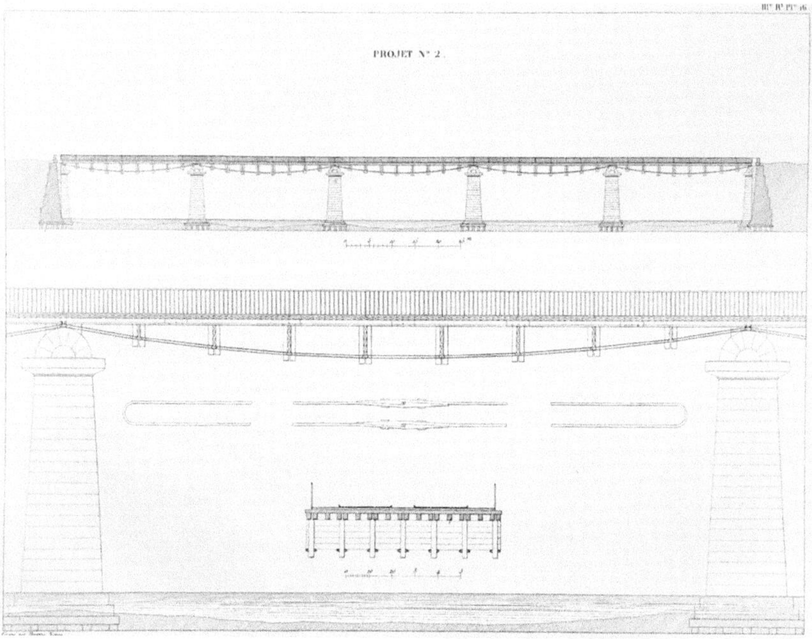

Fig. 94 - In 1823 Bruyère published the design for an underspanned chain bridge. The idea had originated with the engineer Jean-Félix Bélu in 1807 (Navier 1823, p. vi), and according to Bruyère's evaluation the design presented many problems and was far too weak. (Bruyere 1779–1810, pp. 513–524 of the notebook). Evidently the idea continued to intrigue him, and although he never built one himself, the image he published of its further development influenced other engineers. (Source: Bruyère, *Études relatives à l'art des constructions*, 1823, vol. 1, Recueil III, pl. 16)

Among other novel ideas, Bruyère had also studied the possibility of combining a composite construction in wood and iron with an underspanned chain (Bruyère vol. 1 1823, Recueil III, pp. 7–8 and pls. 15–16).

The idea, which Bruyère had adapted from a proposal by Jean-Félix Bélu in 1807, possibly inspired the lighthouse engineer Robert Stevenson's design for an underspanned cable bridge, the Crammond Bridge over the Almond in Scotland in 1821.[83] Guillaume-Henri Dufour adopted and credited Bruyère's proposal in his 1834 Pont des Bergues in Geneva.[84] Bruyère did not mention Stevenson's

83. Weale identifies it having stood on the Great North Road where it crosses the Almond between Edinburgh and Queensferry. (Weale 1843, vol. 2, p. clxxiv).

84. Peters 1987, pp. 197–198, quoting Dufour 1834, pp. 236–237, footnote. Dufour was an 1809 graduate of the EP and had studied for a year at Metz (1810) and evidently knew Bruyère's vol. 1 of 1823.

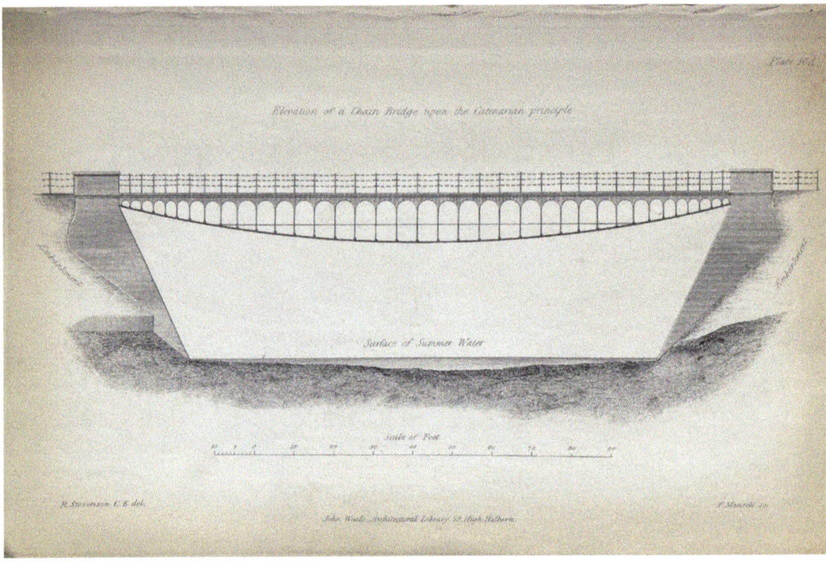

Fig. 95 - Robert Stevenson's underspanned wire cable bridge of 1821 in Scotland predated Bruyère's publication. (Source: John Weale, *The Theory, Practice and Architecture of Bridges*, 1843, pl. 104)

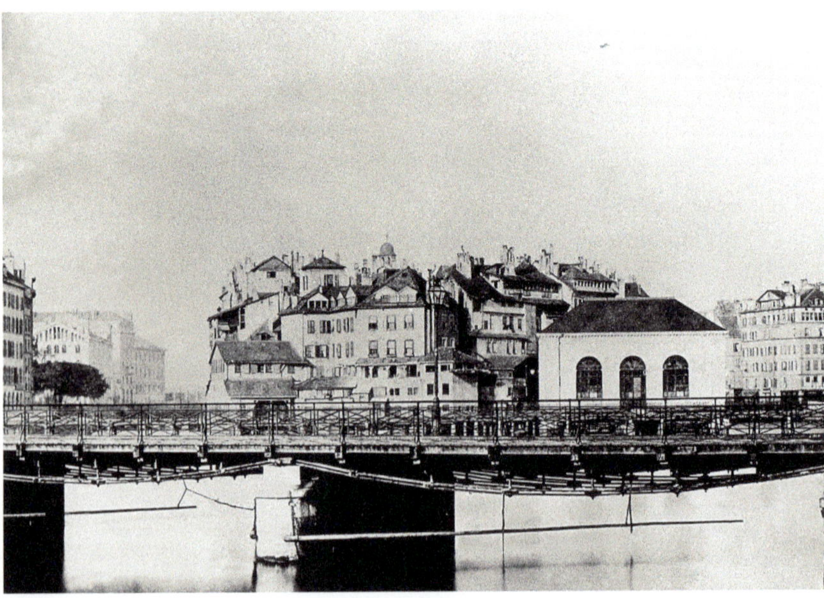

Fig. 96 - Guillaume-Henri Dufour credited Bruyère with the idea for his underspanned chain bridge of 1834 in Geneva. (Source: photograph before 1886 © Musée d'art et d'histoire, Service Vieux-Genève)

bridge in 1823, so we may suspect that his design (although not his publication) predated the Almond Bridge.

These are only a few examples, but they serve to indicate the extensive and yet hidden impact that Bruyère's experimental work had on the development of iron bridge building and facility design. Like Durand, Bruyère had been consciously moving in the direction of repetitive element configurations as system design. However, in contrast to Durand, Bruyère explored their technological and material implications in iron and in composite construction.

Bruyère's direct impact on Mary's methodology in architecture—and through him on the ECP's pedagogy—was profound. The ECP's faculty picked up where Bruyère had left off and completed the picture by inserting the last critical piece of the puzzle: the connection between iron construction and industrial production.[85] This is how Bruyère's exploration of iron as a material that could be used in an incremental, repetitive, and hierarchical configuration gradually permeated the construction industry and ultimately led to our modern understanding of construction as an industrialized system.

85. Although industrial production and organization was only introduced as a separate ECP course at the turn of the nineteenth to the twentieth century, the subject was inherent in railways, architecture, and metallurgy from the outset. Ferry was a recognized expert in industrial organization (Pothier 1887, p. 61).

How L.-C. Mary adapted Durand and incorporated Bruyère and Mandar

Viewed historically, Mary took the ultimate step to create a modern design method. It was focused, as far as his course was concerned, on building design. It was a long process that had begun in the late seventeenth century with the Académie's graphical understanding of building design and its evaluation through 'taste' and 'beauty' as codified in rules of proportion. It evolved via Durand's hierarchy, which he implemented and evaluated through geometrical logic, qualified by his and Bruyère's instinctive conception of 'system' in form, component manufacture, and construction, and it matured finally in the ECP's definition of design and building as a process.

Like Olivier in his approach to descriptive geometry, Mary resisted the EP's trend toward giving preference to abstraction and to scientific analysis over practice. He understood method as a sensible way to solve practical problems, and from the outset in 1833 his fellow members of the *conseil d'études*—the four founders with Bélanger, Perdonnet, Ferry, and Walter—supported his approach. They sensed the need to develop a practical design method, and they tied it to the conviction that drawing was the language of the engineer. Perdonnet later expressed this sense (Perdonnet 1867, pp. 31–33, previously quoted in the chapter "The importance of drawing"), and it was not entirely foreign to French academic engineers either.

The traditional *concours* at the ENPC also placed great importance on drafting in engineering, but the school originally taught it more as a method to communicate information than as a technique to create it. The ECP had stressed drawing as a learning and design tool from the beginning of the program in the early 1830s, when first Leblanc and then Olivier, Colladon, and Perdonnet included haptic and visual learning in their courses. The other teachers followed, and the student notebooks in all the courses are full of sketches. What the faculty recognized, possibly thanks to Leblanc's concept of drawing (based of course on Monge), was that the documentation aspect called for an accompanying methodical underpinning in creating new solutions to problems: drawing as a design strategy in planning, in site organization, in the building process, and in industrial manufacture. This realization underlay Mary's construction course, which he taught for forty-one years. In contrast to Sganzin's *Résumé* of his course at the EP with ten plates in its final version, six of which were devoted to geometrical calculations and the rest to simple infrastructure, the final form of Mary's transcript had fifty-eight plates of building plans, details, construction machinery, and manufacturing plant, which contributed a down-to-earth practical component to his teaching.

A simple adoption of Durand's method would not adequately have served his purpose. Mary had multiple sources at his disposal besides the earliest lecture-based version of Durand's compositional method:[86] he had absorbed the technical aspects of the building profession through Bruyère and the range of scales and the practical details of architectural design taught by Mandar. He was moreover familiar with contemporary French masonry technology represented by Rondelet who taught at the École des Beaux-Arts,[87] and with his colleague and friend Vicat's experimentation with artificial cement and underwater concrete construction. Mary was well prepared, and he expanded Durand's method with Bruyère's key understanding of technology as a full-fledged design issue. As a result, and in contrast to the corresponding programs at the EP and the ENPC, the ECP program for the *concours* for majors in construction was entirely grounded in technology. It encompassed the following elements:

> a complete heating system for a public building, an architectural design, a bridge or hydraulic engineering design, a railway project, a sketchbook, and a project in stereotomy. (Pothier 1887, p. 125)[88]

Together with the courses given by his colleagues, Mary's method supported an integrated understanding of the relationship between hierarchical planning,

86. Madrazo 1994, p. 22, footnote 2 citing Szambien 1984, p. 69. Bélanger was familiar with the same version, and in his brief stint at the EP in 1821, Perdonnet would have experienced at least part of the fully developed method.

87. Rondelet taught construction (stéréotomie) for eight years from the school's inception until 1824, nine years before Mary joined the ECP.

88. …un projet complet de chauffage pour un édifice public; d'architecture; de pont ou de navigation; de chemin de fer; le cahier des épures et coupes de pierres. Nègre, 2011, p. 2, footnote 3, erroneously citing Guillet 1929, p. 125, writes: …les élèves devaient aussi bien réaliser 'un projet complet de chauffage pour un édifice public' qu'un 'projet complet d'architecture'.

modular component manufacture, and realization through site assembly. The result was an early manifestation of an attitude that pointed the way to our modern concept of industrially based design.

Given our knowledge of the importance that the system-approach to manufacture and building later achieved, we can easily imagine that the direction Mary's teaching took followed a visionary objective. It would explain in large part the veneration with which Pothier and others regarded him. However, as Mary himself did not write on the topic, we will never know for certain whether his approach was clear-sightedly calculated to achieve the goal we recognize today, or whether it was, at least in part, an open-ended, fortuitous gamble. We also do not know what influence Mary's predecessor Gourlier may have had on his pedagogical approach. In this regard, Mary, his motivation, and his thinking remain something of an enigma.

When he was charged with developing the architecture course in 1833, Mary consulted the 1809 version of Durand's text (Nègre 2011, caption p. 3), which had appeared just after he had graduated from the EP, but not the later 1813 edition of the first volume with its *Marche à suivre* or the even later 1821 addendum with its more mature formulations of Durand's method.[89] The simpler version of Durand's theory, on the one hand, and Bruyère's practical modular structures, on the other, suggested an obvious connection, and they provided the pedagogical base to innovate.

Mary concentrated on industrial organization and construction (Pfammatter 1997, pp. 133–134). He preferred using contemporary *établissements* (facilities) or infrastructure, like his Besançon water supply, to Durand's grand historical and cross-cultural *ouvrages d'art* as examples. Among these examples were many of the Parisian facilities whose design and construction Bruyère had supervised: the Entrepôt des vins by Durand's collaborator François-Tranquille Gauché,[90] the Marché de Saint Germain, evidently erected between 1813 and 1817 by Louis-Adrien Lusson on Jean-Baptiste Blondel's plans, the Marché des Blancs-Manteaux

89. This is proved by the fact that Mary (plates V & VI) used elements from Durand's plates 20 & 9 of volume 1 part 2 that only appeared in the 1802 and 1809 editions, but not later (see Fig. 99 & Fig. 100).

90. Bruyère, *Études*, vol. 1, Recueil IV, p. 8 and Eck, vol. 1, 1836, pl. 4 & p. 69 name him correctly. Other sources spell the name as Gaucher.

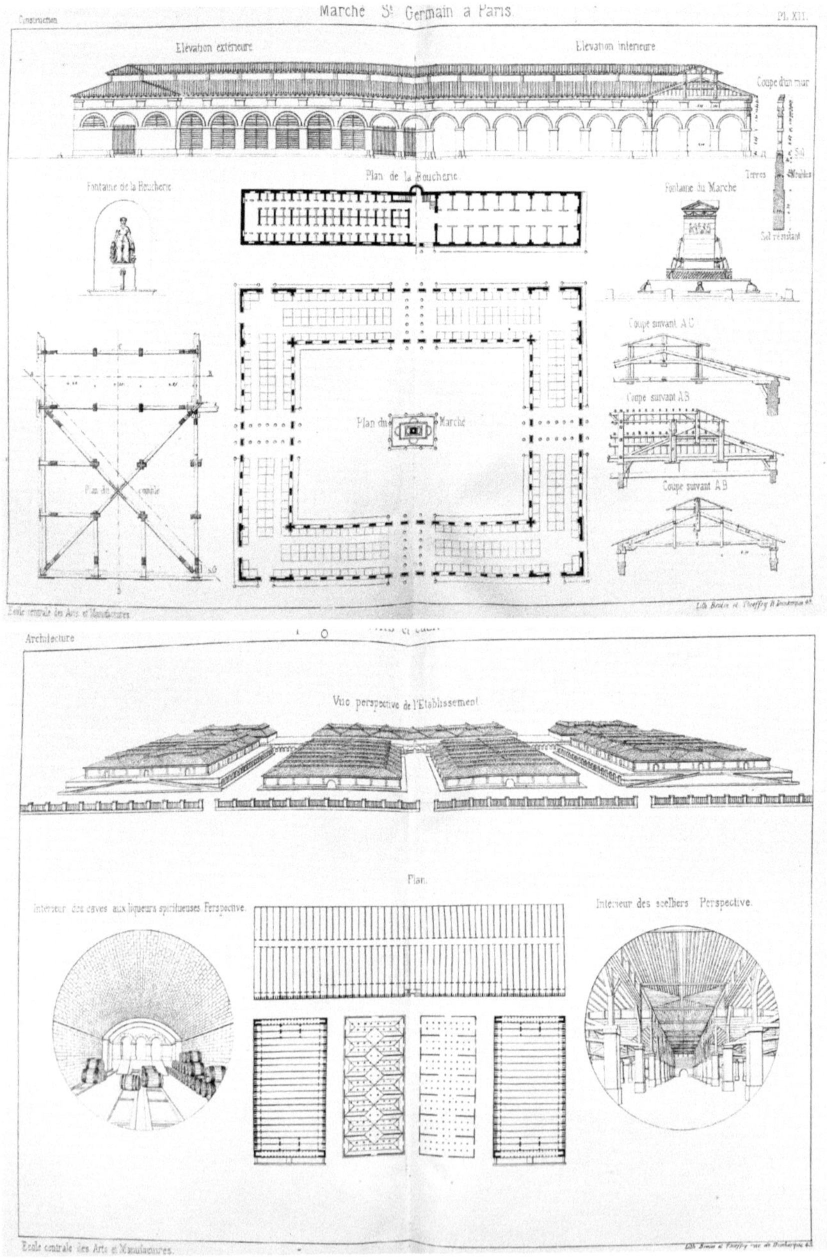

Fig. 97 - In his course transcript, Mary published many of the innovative facilities that his teacher Bruyère had overseen during his tenure as Paris municipal architect and documented in the *Études*. They showed that he preferred facilities (*établissements*) over representative buildings (*ouvrages d'art*) and how much he valued Bruyère's approach to

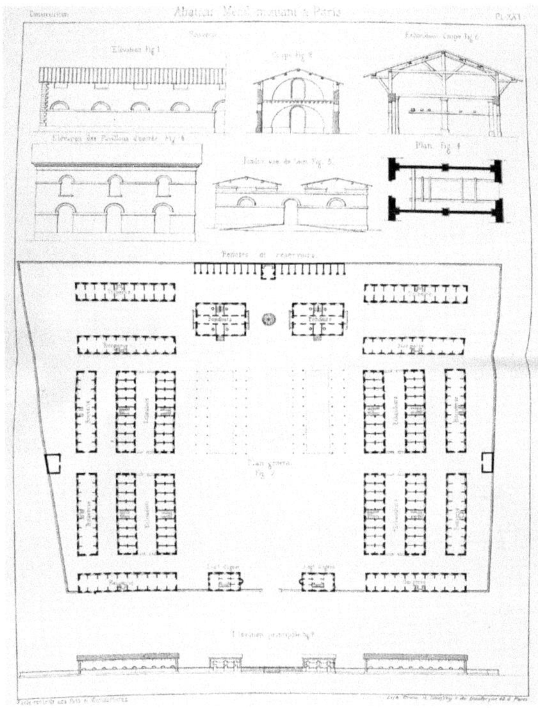

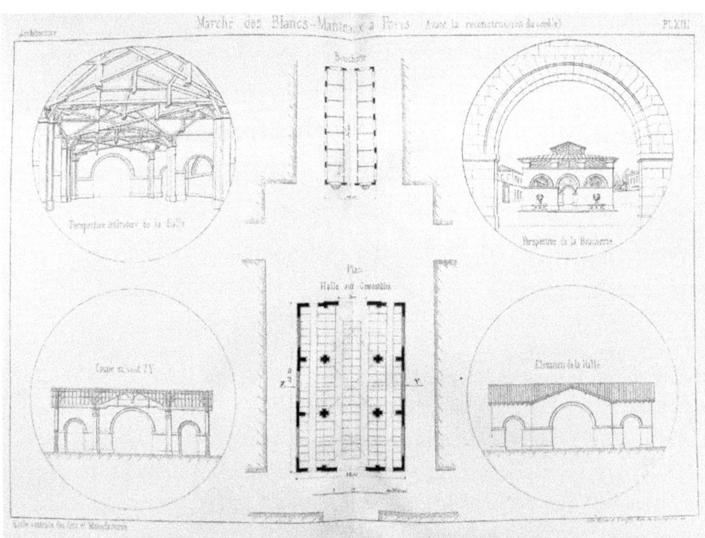

construction. Among these were Blondel's Saint-Germain market (1813–17), Gauché's wine depot (1811–13), the Ménilmontant slaughterhouse (Happe and Vauthier, 1810–18), and Delespine's Blancs-Manteaux market (1813–18). (Source: Mary, *Cours de construction*, 1862–63, pls. XII, XIV, XXI, XIII)

(1813–19) by Pierre-Jules Delespine,[91] and the Abattoir Ménilmontant 1810–18 by two architects named Happe and Vauthier, the latter of whom may have been the ENPC engineer Pierre Vauthier.[92]

Mary was following a well-established tradition in advocating for the primacy of construction and function over space and form. Picon explains the complex issue of functionality in late eighteenth-century architecture (Picon 2000, pp. 15–21). A highlight had been the original 1767 Halle aux Blés. This was an internationally influential facility that stressed functionality: even the circular form was due to its position at the confluence of six irregularly spaced access streets (Wiebenson 1973, pp. 262–263). Its architect, Nicolas Le Camus de Mézières, built nothing else of similar import.[93] He did, however, publish several treatises on wood and tile and a two-volume detailed guide for prospective clients in 1781 that concentrated on the organization and financing of building processes and on the costs of materials and construction.[94]

Mary mediated between the two poles of theory and practice using Bruyère's focus on materiality and construction to stress the Vitruvian precepts *firmitatis, utilitatis, et venustatis* (stability, utility, and beauty). He modified Durand's method from formal composition to an organizational process, retaining its hierarchical and incremental characteristics, which proceeded from the single *component* (for example the base or the capital) in stages of increasing complexity to the composite *element* (the entire column) to the organization of the *subset* (the portico) and finally to the building as a whole, the overall *system*. This last category is ambiguous: the term 'system' can apply in Durand's sense either to the whole object or to Mary's hierarchical process. As soon as Mary's graduates began to work in the field, this hierarchical approach would ensure the melding of industrial component production and structural design, leading ultimately to

---

91. Apparently with roof trusses by Antoine-Marie Peyre (Eck 1841, vol. 2, p. 41). Eck shows Roussel's unbuilt proposal for the roof in iron (1841, vol. 2, p. 41 & pls. 31/32). There is some uncertainty about Peyre's contribution as his name has not otherwise been associated with this structure.

92. Ménilmontant was one of five slaughterhouses all built between 1810 and 1818 under Bruyère's supervision, the others being Montmartre in 1810–18 by Poidevin and François-Joseph Bélanger; Roule by Louis-François Petit-Radel; Grenelle by Guy de Gisors and Alexandre Du Bois; and Villejuif, apparently designed by Nicolas Lenoir the year of his death.

93. He subsequently developed theories of the relationship between theater, music, and architecture.

94. Nègre (2009, p. 1092) enumerates one seventeenth- and several more eighteenth-century French books on construction.

prefabricated system design and, finally with Eiffel and his collaborators, to a radical 'kit of parts' concept in iron.

Mary's course matured around the time of the 1848 Revolution, when the title of his transcript (1852–53) changed from *Architecture* to *Construction*. The ECP's pedagogy was working as the founders had envisioned, and Mary had evidently grown more confident in his views. The effect that his students were beginning to exert on professional practice in general was unmistakable, as Labrouste's case would show. Mary began to stress construction even more exclusively than he had previously, and the effect that this had on the students who graduated after mid-century manifested as a concentration on the system aspects of design.

The organization and content of Mary's architecture course

Mary's double course was *constructions civiles et travaux publics*, and it used two transcripts: *Architecture*, which morphed around 1850 into *Construction*,[95] and *Routes et ponts*.[96] He taught a further course, *Navigation intérieure*, at the ENPC.[97] This was a division that corresponded to the tripartite organization that Sganzin had originally created at the EP.[98] The two-part course that he taught at

95. Nègre dates the earliest transcript to 1834–35 (Nègre 2011). WorldCat notes two versions: 1840–41 and 1844–45, and another is dated 1852–53. The edition used here is the final version of 1862–63, by then called *Construction*. The plates include a few of the newest examples of iron structures both in France and elsewhere. Some show details, including Polonceau's truss and a few structures like Victor Baltard's 1857–74 Halles de Paris. A comparison between an isolated plate volume labeled *Architecture* (1862–63) and a full copy of the course titled *Construction* for the same years show the plates to be identical. Thus, it is the same course.

96. In this course Mary also discusses iron bridges and goes into more detail in dealing with the new material, especially from the 1850s on. The section on iron bridges in the last, posthumous edition (Mary 1873) was especially comprehensive with 16 out of 52 plates (pls. XXXVIII–XLII, XLII bis, XLIII–LII; two are misnumbered) devoted to the topic (pls. XLVIII–LII are on suspension bridges). Virtually all these examples, however, date from the 1850s, with the addendum of 15 plates devoted to structures up to 1872 (pls. 1–11,15 on iron bridges). This update was posthumous: Mary had died in 1870. Mary's course went through a total of seven iterations according to WorldCat: 1837–38, 1841–42 (which included *navigation*), 1847 (for 1845–46), 1850–51, 1855–56, 1861–62, and 1873 (the text is also dated 1861–62, but with 15 additional plates).

97. He did teach it briefly at the ECP too, but apparently only in 1841–42 as part of *Travaux publics*.

98. Mary also published *Vitesses virtuelles* in 1841, *Opérations sur le terrain* in 1842 and *Détails pratiques sur la distribution des eaux* in 1854 and again in 1865 as addenda to *Construction*, and the last again in 1868 as an addendum to *Navigation intérieure* for the ENPC.

the ECP was relevant to the development of iron construction, not directly for its application to structures,[99] but to the organization of the construction industry. It especially influenced Baudet, Eiffel, and Moisant, who would later establish large-scale contracting firms.

Mary originally developed two of his transcripts for the ENPC: *Routes et ponts* in 1836 and *Navigation intérieure* in 1842.[100] As he only taught the first for a single year there, he reissued it with the date 1837–38 as a basis for the *Travaux publics* segment at the ECP. We might have expected *Routes et ponts* to deal heavily with iron bridges, but it did not contribute much relevant material until Ruelle's posthumously updated edition in 1873.[101]

The *Routes et ponts* transcript is curiously lacking in that regard. Mary did not mention Bruyère's proposals for wrought iron bridges, for instance. It appears from what he first wrote in the early 1850s and retained in all subsequent versions that he was uninterested in unbuilt projects (1873 reprint: 'Art. 2 Poutres en fer et en tôle, 1° Poutres droits', pp. 304ff). This could explain the lack. However, he even overlooked Bruyère's ground-breaking Croult Bridge, which he must have known about when he studied. He posited the origin of wrought-iron beam bridges in two 1840 experimental structures by Charles Antoine François Poirée (or perhaps one of his sons, Charles Antoine or Jules) and one by Vincent Maniel in 1849–50, and

---

99. Antonio Montero's 1851–52 student notebook that includes *Construction, Routes et ponts,* contains only three pages on iron detailing (pp. 14–16) and one general diagram of a suspension bridge (p. 228), but in the part of the later 1862–63 transcript that deals with taxonomy, several iron structures are included and shown in the plates.

100. Mary had been adjunct professor for *Routes et ponts* at the ENPC for a single year in 1836 and became professor for *Navigation intérieure* in 1842 (Reynaud 1870).

101. The lithographed transcript of 1873 postdates the course that Mary taught by almost a decade and his death by three years. The major part of the text, however, dates originally from 1861–62. It contains several words that are crossed out and some inserted corrections, especially of the plate and figure numbering. It also contains at least one error: a figure 27 mentioned on page 31 but missing from the corresponding plate VII. This suggests that the text, with the exception of the posthumous updates, had remained unchanged from a previous version. The first part, pp. 1–104, deals with roads, and the second, major part, pp. 105–397, with bridges, with added material right up to the construction of several bridges (Currières, Chenevières, and Champigny) dated to 1872. However, here, too, the major part was devoted to masonry bridges and only pp. 294–325 to arch and beam bridges in iron, pp. 327–388 to chain and wire cable suspension bridges, and pp. 374–88 to mobile bridges of all types. In this second part, Mary stressed the machinery used for manufacture and methods of erection. Machinery had been present but not a strong focus in his first course on architecture and construction, and evidently developed in the course of Mary's gradual shift in thinking around 1848.

he did credit Stephenson's work of 1848 as the real origin of wrought-iron bridge building (Mary 1873, p. 304), which in a certain sense it was. But he also missed the seminal contributions of William Fairbairn and Eaton Hodgkinson to Stephenson's work, as well as all the many early railway bridges by Flachat and others in France to which several of his former students had contributed.

Nevertheless, Mary's course texts make fascinating reading even today. In contrast to the typically dry transcripts or abridged summaries like Durand's *Précis* and Sganzin's *Programmes ou résumés*, Mandar's 1826 recipe book, Eck's two volumes of examples, or Bruyère's two volumes, Mary speaks directly to the reader. His texts are full of details and comments on his opinions, examples of projects, and work by his contemporaries that he finds interesting. He does not catalog or preach—he discusses. His curiosity and open character shine through his uncomplicated prose, and it is easy to see why his students appreciated him.

Of the three transcripts, it is Mary's architecture course that is of primary interest to us as the one that deals with method. Despite what one might think from its full title, *Architecture et construction des édifices*, it did not aim to educate architects. He expressed the idea of what he wanted to do in the introduction:

> When the founders of the École centrale included civil structures in the curriculum, they did so with no intention of educating architects, but they only wanted to give the students sufficient general ideas about architecture to provide them with guidelines for the study and organization of the industrial structures that they might be called upon to design.
>
> It is in this limited sense that the civil construction course was conceived, and one would be wrong to hope to find it a comprehensive exposé on architecture. (Mary 1862–63, p. 1)[102]

The second phase of this course, marked by the year in which the title morphed from *Architecture* to *Construction*, shows how Mary's thinking developed ever more strongly toward technology, and that increasingly involved iron. We can see

---

102. Lorsque les fondateurs de l'École Centrale ont compris les constructions civiles dans le cadre de son enseignement ils n'ont pas eu la prétention de former des architectes, mais seulement ils ont voulu donner aux élèves des notions générales d'architecture suffisantes pour les guider dans l'étude et l'établissement des constructions industrielles qu'ils pourraient avoir à élever.

C'est dans ces limites restreintes que le Cours de Constructions civiles a été conçu et on se tromperait si l'on espérait y trouver des notions complètes d'architecture.

the effect of the shift in interest in the difference between how the earlier graduates up to 1851 like Barrault, Polonceau, and de Dion worked—all of whom tended to approach design as the definition of and solution to individual problems —and how the later graduates like Jenney, Eiffel, Moisant, Baudet, and Contamin approached design more comprehensively as a system problem. Of course, part of the difference was due to the development of the building industries in that period, but these were in turn also influenced by ECP graduates as they joined the workforce.

Around the time of this shift, Mary issued an addendum to his course in 1854 as *Détails pratiques sur la distribution des eaux*. The topic of clean water supply and drainage was very much at the forefront of engineering interest, and had been ever since the cholera epidemic of 1832.[103] Mary concentrated on the planning of the system and the problems of construction. He described the history of water supply,[104] and devoted almost a third of the transcript (27 of 97 pages) to the network he had just completed in Besançon.[105]

In his introductory statement to the main transcript, which remains the most interesting for our study, Mary made it abundantly clear that form and construction are two sides of the same coin, but that construction was his focus:

> One can learn design by studying good models, but one can only become a builder through enlightened practice and by following building processes. Nevertheless, both sides of the field must influence one another in order not to fall into the trap of designing beautiful, unbuildable projects (Mary 1862–63, p. 2)[106]

103. In 1865 Mary updated the addendum. Whereas in the first version he had presented the practical issues of water supply and drainage in lengthy personal and practical explanations using Paris, but chiefly Besançon, as an example, the later edition is shorter and more compact. The development of practical water supplies had advanced, and the text presents a more balanced selection of examples from many sources, stressing the development in Paris, for which he was responsible, right up to the newest project of 1864 (p. 4), as well as citing Amiens, Besançon, Bordeaux, Dijon, Orléans, and a few British examples.

104. He cited Prony and Eytelwein (p.15), and Du Buat (p. 43).

105. However, he did leave current research and theory to Bélanger, who dealt with that in his *Notes sur le Cours d'hydraulique* that he had begun to publish for the ECP in 1850–51 and again in 1854–55. Bélanger had previously published these *Notes* for the ENPC c. 1842–50.

106. On peut apprendre à composer en étudiant les bons modèles, mais on ne devient constructeur que par une pratique éclairée et en suivant des constructions. Cependant ces deux branches de l'art ont besoin de s'éclairer réciproquement, afin que l'on ne soit pas exposé à faire de beaux projets inexécutables.

He explicitly stated that he would not address railway networks because that field had exploded into a separate subject that required its own specialized course (p. 1), and he blithely relegated the whole issue of iron as a structural material to his colleagues' courses as well. In the very first paragraph of the second part of the course, 'Construction des édifices', he explains that he will only address the traditional structural materials in buildings:

> The materials used in construction are wood and natural or artificial stone and metals. These last are described in the machine construction course because they are its particular purview. But wood and stone are of special interest to builders, and that is what we need to concern ourselves with primarily. (Mary 1862–63, p. 105)[107]

This made it clear that it was to be Perdonnet, Bélanger, Walter, Ferry, and Faure who were expected to provide the information on iron and how to use it for machinery and building structures. Nevertheless, Mary did discuss it, and often in detail. He was not averse to the material per se and was in fact knowledgeable about current experimentation in buildings. Iron of course played an important role in his water systems, with their pumps, conduits, and distributors too.

Pfammatter provides examples from the 1852–53 version of Mary's transcript (Pfammatter 1997, pp. 134–138), and Mary discussed iron sporadically throughout the text and included some of the most modern iron structures in his plates. From the earliest versions, the transcript had featured the dome of the Halle aux Blés in detail (p. 47, pls. IX and X). In 1862–63 he included sections on iron in floors: 'Emploi du fer dans la construction des planchers' (pp. 204–205, pl. LI). Among other interesting examples he cited Gréban's hollow-tile floor in the manutention des vivres de la guerre (commissariat) in 1835 (p. 207) that Eck had previously published (1836, vol. 1, pls. 15, 17, and 49), and he had previously illustrated it with a sketch on page 290 in the 1852–53 version (Pfammatter 1997, p. 135). He also discussed roofing in other metals like lead, copper, and zinc (pp. 248–250). He even had a brief chapter titled: 'Application du fer à la construction des combles' (The application of iron to the construction

107. Les matériaux employés dans la construction sont les bois et les pierres naturelles ou artificielles, les métaux. Ces derniers sont décrits dans le cours de construction des machines, parce qu'ils en sont les éléments; mais les bois et les pierres intéressent particulièrement les constructeurs, et nous devons nous en occuper spécialement.

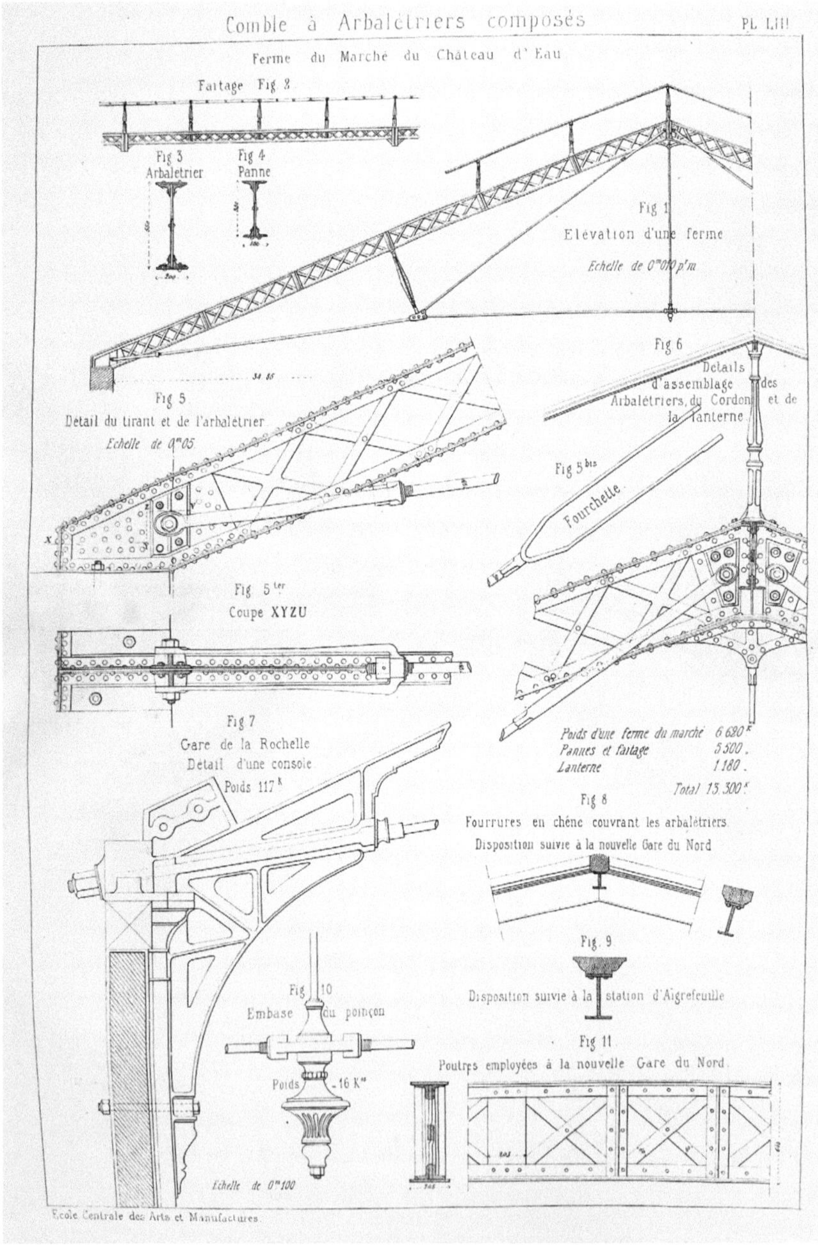

Fig. 98 - Mary's plate LIII showing the Marché Saint-Martin (a.k.a. Marché du Château d'Eau, 1854), the roofs of the La Rochelle station (1857), and details of Jacques Ignace Hittorff and Joseph Couche's new Gare du Nord (1861–65). (Source: Mary, *Cours de construction*, 1862–63, pl. LIII)

of roof trusses, pp. 224–234, pls. LI-LIII), in which he also incorporated some mathematical calculations—a rarity in his transcripts—citing Polonceau (pp. 224–226) and Paul-Joseph Ardant, military engineer and professor at Metz (p. 227).

It is true that Mary included few iron buildings in comparison to the number he presented in traditional materials, but those he did include always represented the most modern examples available. The newest that he included in 1862–63 were the station roofs in Bordeaux (1855), Niort (1856; pl. LI), Aigrefeuille (1857; pl. LII), La Rochelle (1857), the second Gare du Nord (1861–65), and the Marché Saint-Martin (1854), built by Joly, the one that later collapsed under a load of snow in 1879 (pl. LIII). These examples replaced other iron buildings that he had included in earlier versions of the transcript.

We can suspect the reason that Mary was reluctant to address iron on a par with wood and masonry in Jolly's explanation of the time lag between the experimental use of iron in building structure and its dominance in mainstream construction, which only really escalated after 1860. Construction as Mary wanted it to be understood was not experimentation: it was established practice. Mary's successors, who took over his courses in the mid-1860s, gradually shifted their focus from stone and wood to iron and concrete. These materials grew in importance as the engineering use of stone and wood declined. By the end of the century, the school shifted to courses devoted solely to the more modern materials in Marcelin Duplaix's *Constructions métalliques* in 1897, which became *Constructions métalliques et en béton armé* in 1924.

On the one hand, we must understand Mary's lack of focus on what was still seen as an experimental material in the context of his confidence that Perdonnet and the others would deal with it. On the other, his adaptation of the Durand method and his concentration on technology did support machine construction and the nascent industrialization of iron, the gradually emerging standardization of parts, and their organization into a system. The issue that concerned him was the industrial implications of modularized iron.

The question of plagiarism, a case for *vulgarisation*

Mary did not hesitate to cite others, but he made no mention of the fact that he developed his course concept based on Sganzin's organization, although that may have been a gradual evolution rather than a decision. He also freely appropriated visual material for his lectures from many texts, primarily from the three most influential treatises of the previous generation: Durand, Bruyère, and the prime classical text, Jean-Baptiste Rondelet's encyclopedic 1802–17 *Traité théorique et pratique de l'art de bâtir.*[108] We must remember that Sganzin was still alive, and the other three had only recently died when Mary took these liberties. The material was therefore contemporary, and he could expect it to be well known to the profession and the students without an explicit attribution—a lacuna that would be academically inadmissible today.

He borrowed structures as examples from Eck and Bruyère, and to a lesser extent he also 'lifted' material from other engineers who were still alive at the time: Mandar (pp. 97–98, pl. XXX),[109] Ardant (pp. 41/223/227, pl. XLVIII), Armand-Rose Emy (p. 220, pl. L), and several more. He names some of his sources, like Polonceau (p. 226) or his friend Vicat (pp. 136–146), but other less prominent ones he does not even bother to name. He also refers to his colleague Péligot's chemistry course (p. 141) and to obscure research by men like Jean Constant Petot (p. 147). The list of Mary's sources is long. It was not his intention to create new material, nor to steal and claim material for his own that he did not invent, but rather to organize the best modern knowledge that was readily available in a new way.

The reason for including a catalogue of Mary's use of foreign material here is to correct the implicitly dismissive attitude that judges Mary's teaching to be unoriginal and secondary to the pedagogy of the school. Some historians consider his adaptation of Durand, Rondelet, and Bruyère to be plagiarism, and have used it as proof that Mary was derivative. However, a closer examination shows that he used his predecessors' work as well as many examples by his contemporaries as raw material—in other words, he cannibalized it, and he purposefully reinterpreted the information that they provided in a practical, industrial context. By today's academic standards one might reproach Mary with neglecting to mention his sources, but his adaptive use of the material is a form of pedagogical *vulgarisation.*

One example serves to show how Mary's use of material by others aimed at a different goal: the Durand plates that he 'lifted' are straightforward catalogues of diagrams, but in Mary's versions they are hybrids—he reproduces some but not

---

108. Nègre refers in this regard to Françoise Hamon's unpublished PhD dissertation 'Construire pour l'industrie. L'enseignement de l'architecture industrielle à l'École centrale des Arts et Manufactures 1832–1914', 3 vols., Université de Paris-Sorbonne Paris IV, 1997.

109. The model house that Mandar had developed in his publication of 1818 serves as a simple example of a maison de campagne.

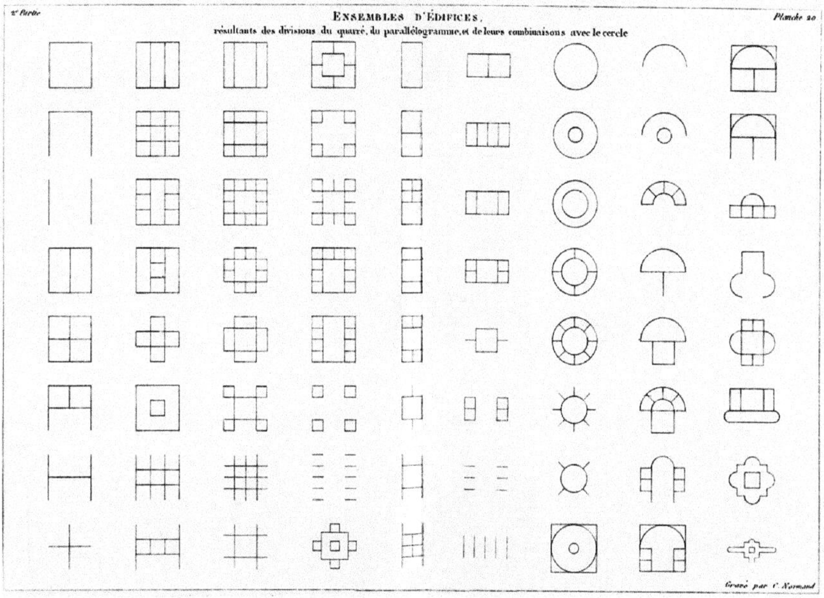

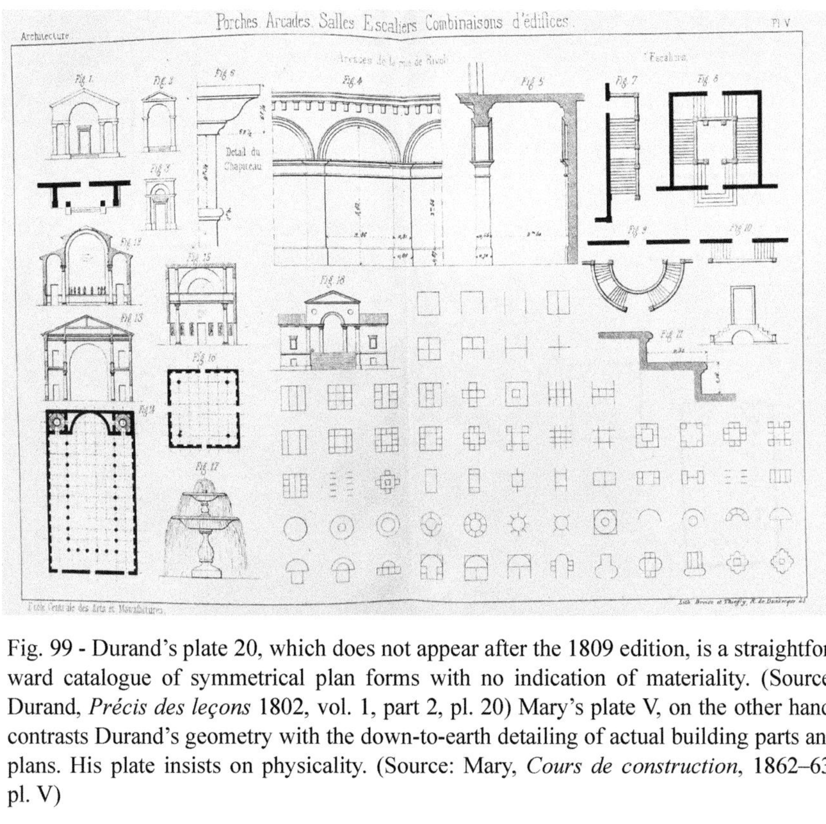

Fig. 99 - Durand's plate 20, which does not appear after the 1809 edition, is a straightforward catalogue of symmetrical plan forms with no indication of materiality. (Source: Durand, *Précis des leçons* 1802, vol. 1, part 2, pl. 20) Mary's plate V, on the other hand, contrasts Durand's geometry with the down-to-earth detailing of actual building parts and plans. His plate insists on physicality. (Source: Mary, *Cours de construction*, 1862–63, pl. V)

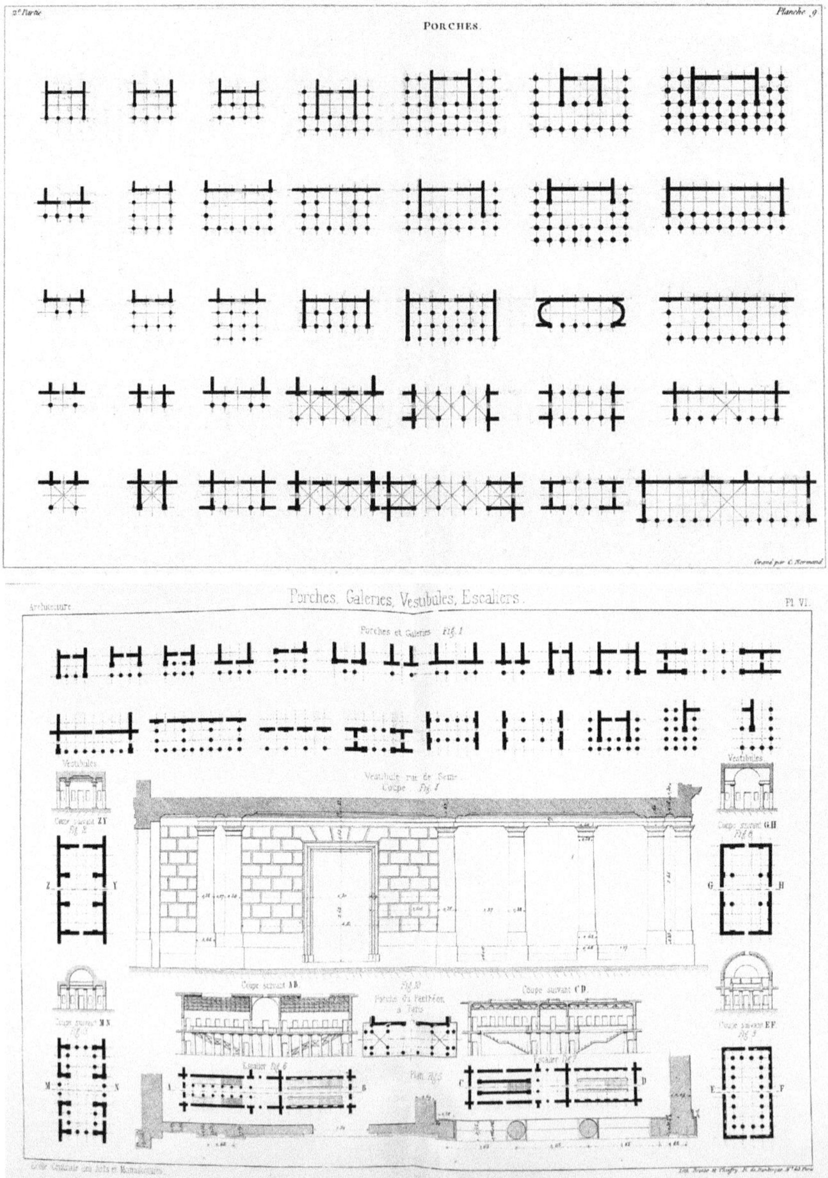

Fig. 100 - In plate 9, which also does not appear in this form after the 1809 edition, Durand does indicate wall thicknesses and the lines of vault ribs, but only in plan. They remain undetailed graphic indications. (Source: Durand, *Précis des leçons*, 1802, vol. 1, part 2, pl. 9). Mary uses a few of these diagrams in plate VI, and he illustrates them together with dimensioned wall details and with geometrical disturbances like staircases, pilasters, and embrasures, showing vertical sections with the structure and thicknesses of ceilings. The plate is dominated by a measured detail elevation. Again, physical reality intrudes on the abstraction of Durand's geometry. (Source: Mary, *Cours de construction*, 1862–63, pl. VI)

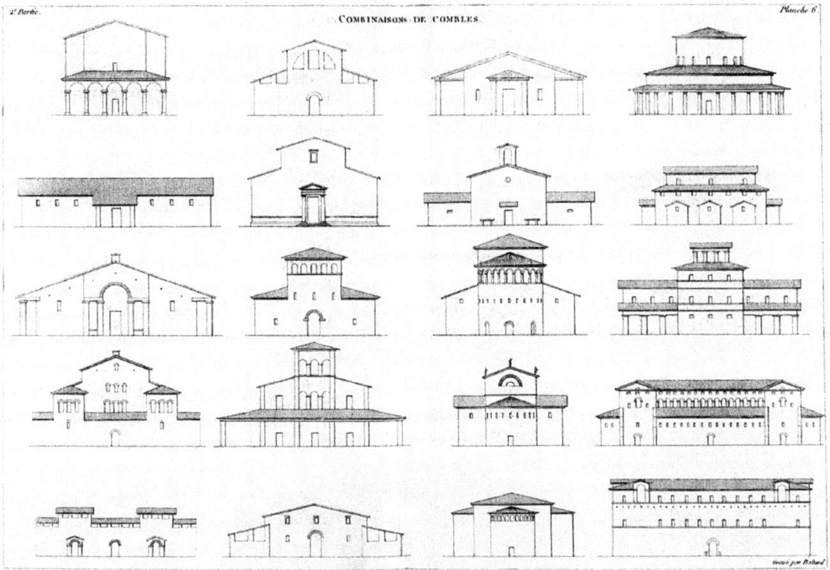

Fig. 101 - Durand uses the thin lines of geometrical abstraction in 1802 to show elevation types with no indication of materiality. (Source: Durand, *Précis des leçons*, 1802, vol. 1, pl. 6) Mary juxtaposes the elevations with reality by stressing the thicknesses and detailing that masonry demands with thick black lines in the lower left quarter of his plate. Mary's visual message is clear: geometry is all well and good, but construction demands more. (Source: Mary, *Cours de construction*, 1862–63, pl. VII)

all of each plate and always includes parts of plans and construction details from other sources at multiple scales. The resulting juxtaposition was not an editorial decision to save space: it was Mary's choice, and it underscores the focus that he refined step by step to achieve the ECP's practice-oriented goals. We might attribute the varying scales to Mandar's and perhaps also to Eck's influence, a perspective that was also mirrored in Perdonnet's treatment of small-, medium- and large-scale iron construction in the railway course.

'Principes généraux'

Mary divided the final version of his construction transcript into two roughly equal parts. He called the first 'Principes généraux', or general principles (pp. 1–105). It dealt with planning and design and included thirty-five plates that deal with architectural design—plans, facades, and details, as well as the few that included Durand's taxonomy and compositional theory. He followed this with a second part called 'Construction des edifices', or building construction (pp. 105–270), which treated construction technology and materials. It contained the remaining twenty-three plates with construction, detailing, site installations, building machinery, and materials.

'Principes généraux' begins with a brief traditional section on the formal elements of architectural history (pp. 3–26/31–38), for which Mary apologizes, repeating again that he will not educate architects, but engineers:

> Thus, we will skim the study of the [classical] orders of architecture, the design of columns, of capitals, and other ornamentation, which take up entire years of study for those who plan to become architects, because what our students need is to know how to organize and solidly construct a house, a shop, a studio, any kind of industrial facility. (Mary 1862–63, p. 3)[110]

The second chapter addresses 'Combinaisons des élémens des édifices' (Combinations of building elements), with a subsection '1° Règle à suivre dans

110. Ainsi nous passerons légèrement sur l'étude des ordres de l'Architecture, sur les dessins de colonnes, de chapiteaux, et d'autres ornements qui occupent pendant des années entières ceux qui se destinent à devenir architectes, car ce dont les Élèves ont besoin, c'est de savoir disposer convenablement et construire solidement une maison d'habitation, un magasin, un atelier, un établissement industriel quelconque.

la disposition des élémens entre eux' (The rule to follow in organizing relationships between elements). Instead of basing building design on formal components and elements like Durand, he bases it entirely on construction technology and materials.[111] To illustrate Mary's thinking, I quote examples that deal with columns from this section, but his text addresses other components in the same fashion:

> To design a structure, one needs to know its elements, functionally as well as formally. (Mary 1862–63, p. 26)[112]

Far from using this to initiate a traditional discussion of the classical column orders, Mary establishes design rules. He had already made it clear that

> columns cannot be reduced to dimensions that allow them to be crushed by the loads they are to support. (Mary 1862–63, p. 14)[113]

Where Durand had quoted Vitruvius and commented on the symbolic meaning of the various orders and their details without mentioning their loadbearing function (Durand *Précis* 1819 ed., vol. 1, pp. 11–14/42–43/5–76/79–80), Mary relates their aesthetic to their organization:

> Columns should be generally spaced equally apart, because one cannot deny that the principal cause of the pleasure one feels at the sight of a colonnade lies in its regularity, the proper relationship between the height and spacing of its columns, and the rightness of their proportions. (Mary 1862–63, pp. 26–27)[114]

---

111. In view of this distinction, Navier's 1813 comment that Durand concerned himself with problem-solving (Picon 2000, p. 29), does not apply in our sense, as Navier was apparently referring to Durand's logically traceable, supposedly scientific compositional method and not to the design of functioning buildings.

112. Pour composer un édifice, il faut en connaître les élémens, tant sous le rapport de leur destination que sous celui de leurs formes.

113. …les colonnes ne peuvent pas se réduire à une proportion telle qu'elles s'écrasent sous la charge qu'elles ont à supporter.

114. Les colonnes doivent en général être espacées également, car on ne peut se dissimuler que la principale cause du plaisir que l'on éprouve à la vue d'une colonnade ne soit la régularité de la disposition des colonnes, le rapport le plus convenable entre leur hauteur et leur écartement et la justesse de leurs proportions.

He then proceeds to explain the traditional spacing of the various classical orders in terms of their structural function and how their positioning relates to the other design components of a building, both physical and abstract: pilasters, outer walls, cross-walls, door openings, and lines of movement. The positioning is determined by function, not by geometry.

Mary's method

The structural and spatial explanation of component organization prepares the student for the innovative section of the text. Mary's method, his *règle*, defines the logic of how to set up relationships between components. He introduces it in six pages in the transcript (pp. 26–31), in which the paragraph 'Division en carrés par les axes' (Orthogonal and regular alignment by means of axes) is notable because it marks the first application of Durand's methodological ideas in a course outside the EP. This paragraph touches briefly on Durand's geometry of plan and elevation organization, the diagrams, and the use of *papier quadrillé*, (graph paper) to organize the design. Durand had only introduced graph paper himself in 1820 (Pfammatter 1997, p. 60).[115] Mary explains the logic of following the principle, first in plan, and then in elevation with the vertical organization of openings and the superposition of columns. Although Mary uses Durand's diagrams, his explanation for them is more complex than a simple coherence of form. He justifies them as an economical means to describe function and structure, and he has little use for Durand's rigid symmetry, because

> one must sometimes sacrifice regular, repetitive [or symmetrical] organization to satisfy functional demands. (Mary 1862–63, p. 29)[116]

Mary also treats the axes differently: where Durand used axes as a formal ordering principle and to regulate repetition, Mary uses them to define equal construction

---

115. However, the idea to use 'squared paper' as a guide to design was very much older in French architecture: see Peter Collins, 'The Origins of Graph Paper as an Influence on Architectural Design', *JSAH* 1962, pp. 159–162, in which he traces the origins back to Philibert Delorme and from there to the *Encyclopédistes* and their interest in the decimal system.

116. …on sacrifie quelquefois la régularité des dispositions pour mieux satisfaire aux exigences de la distribution.

modules and to determine structural spans.[117] In other words, he shifts his focus from the lines to the spaces between the lines: from the axes to the grid. He does not negate his own educational background when he refers to *goût* when writing about exceptions to the placement of vertical fenestration (p. 30), but he uses the logic of structure to argue his case, not 'beauty', and it is the principle of economy of means that resonates through the text as an underlying concept.

This brief foray into theory is easily missed, although it informs the methodological discussion throughout the entire transcript. It is the only expressly formulated remnant we possess of Mary's key conceptual contribution to the ECP curriculum and its subsequent proliferation worldwide. We cannot know for sure how much time Mary devoted to a detailed explanation of his *règle* in the lectures, but it was his most important contribution to engineering as well as to architectural education, and he explains its implications visually in plates V, VI, and VII. As simple and almost banal as this basic design principle may seem to us today, there was nothing self-evident about it at the time—no architectural or engineering theoretician appears to have expressed it in such straightforward terms before.

This organizational principle did not remain exclusive to Mary's construction course. Perdonnet and Charles Callon (who taught the course *Construction et établissement des machines*) also specified the use of graph paper for the drawings that the students made in their classes. Although this is a detail that is easily missed, the method, the graph paper, and the fact that Perdonnet and Callon used it, too, were small issues with major implications. The graph paper served two purposes: in Mary's case (and in Perdonnet's for station halls and railway sheds) it served primarily to control orthogonal grids of functional and structural organization, and in machine construction it functioned to control dimensioning and perhaps also to identify potential dynamic problems such as stress concentrations. The graph paper forged a visceral link between the three construction courses in the students' minds and therefore between the functional organization (that Mary called *distribution*), small-scale design, and iron in machinery and building structures. We find aspects of this technique demonstrated in Terrier's 1861–62 sketchbook (*cahier d'esquisse*) that was evidently intended as a preparation for the submission of his *travail des vacances*.

117. Without expressly saying so, when applied to iron construction, this would involve component sizing to permit off-site fabrication and easy transportation. My interpretation differs from Pfammatter's (1997, p. 134), who credits Durand with a structural interpretation of axes.

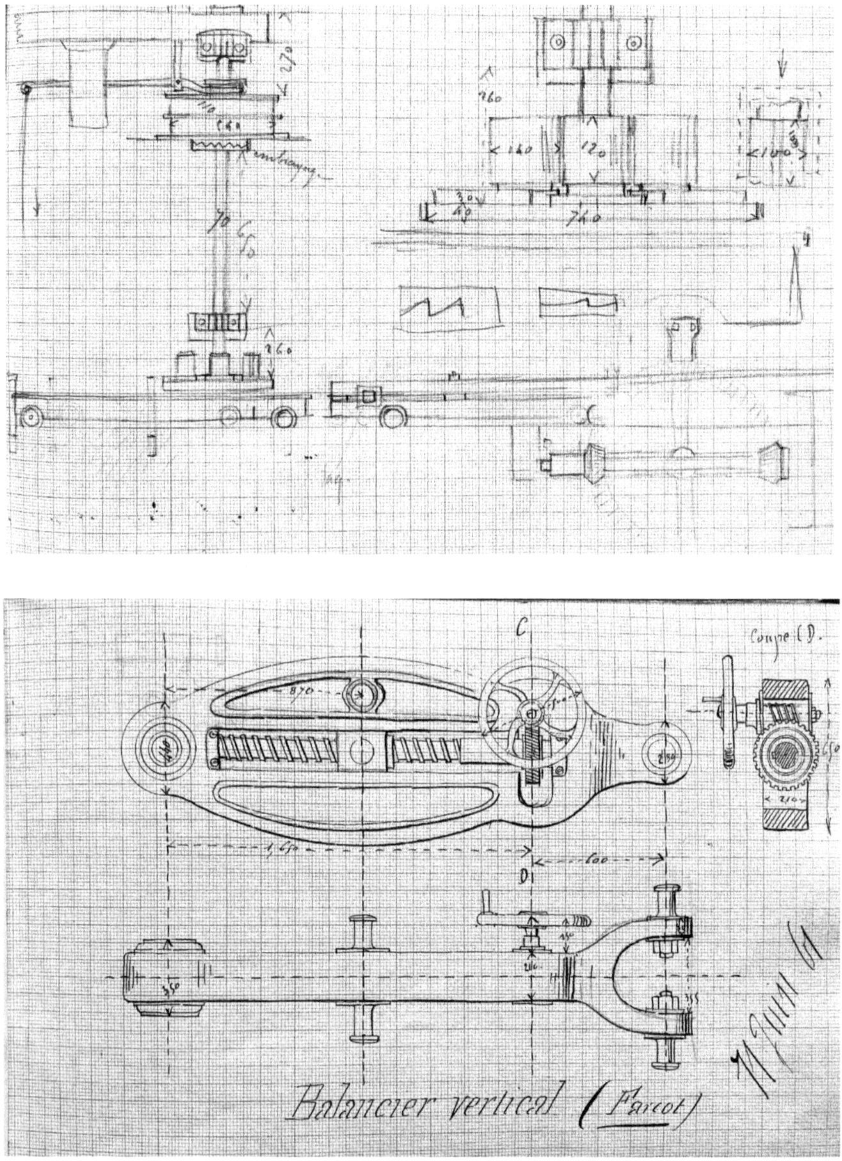

Fig. 102 - Graph paper was a useful tool for sketching and for finished drawings. Mary, Perdonnet, and Callon used it in their courses. It was a simple device that served to link the course content across disciplines in the minds of the students. This is illustrated by a rough measured sketch and a detailed machine part in Paul Terrier's sketchbook (*cahier de croquis*), probably for Charles Callon's course, for the second year's *travail de vacances* in 1861–62. (Source: Peters collection)

The next sub-heading, '2° Parties des édifices' (The elements that constitute buildings), is a systematic expansion of Mary's ordering principle as applied to a taxonomy of building parts. In this section he addresses combining components to reach the next higher order, the architectural element: 'Porches, vestibules, salles, escaliers, etc.' His elements also include functional details and services, like chimneys, lighting requirements, and water supply, always referring to examples that the students could see for themselves, and Mary explains their forms in terms of their functional parameters. Plates V, VI, and VII illustrate some of these issues with details that show thicknesses and dimensions, and the argument is always about function or construction. He uses his hybrid plates to contrast Durand's schematic diagrams with these pragmatic requirements to illustrate what he means, and his explanations situate the illustrations in a down-to-earth context that is far removed from abstract form.

Like Durand, Mary had no use at all for decoration, which he relegated to sculpture and painting. In his critique of decoration, Mary's intolerance exhibits a structural fervor that surpasses Durand's in the passage on Perronet's bridges. For example, after discussing structural elements he writes:

> This is the organization that one should follow as far as possible because it is how the most illustrious architects organized the buildings that they illustrate. Nevertheless, several architects of the past century lavished a multitude of attached columns on their buildings, columns pushed up against walls, jutting out or pushed back, or generally all sorts of very expensive devices for producing an unpleasing effect. If one has the money for decoration, it would be better to spend it on sculpture or painting which elevates the spirit and satisfies the eye far better than expensive, useless columns. (Mary 1862–63, pp. 28–29)[118]

In the third chapter, 'Composition des édifices, règles auxquelles elles doivent être assujetties' (The design of buildings, the rules that condition them), Mary provides

118. Telles sont les dispositions que l'on doit suivre autant qu'on le peut; parce que c'est ainsi que les Architectes les plus célèbres ont disposé les édifices qui les [*sic*] ont illustrés. Cependant plusieurs architectes du siècle dernier dans la seule vue de la décoration, ont prodigué dans leurs édifices les colonnes engagées, ou adossées à des murs, les avant et les arrière corps, et en général tous les petits moyens qui pour produire un effet peu satisfaisant entraînent à des dépenses considérables. On doit plutôt, si l'on a de l'argent à employer en décoration, l'affecter à des sculptures ou à des peintures qui élèvent l'esprit et satisfont les yeux beaucoup mieux que ne peuvent le faire des colonnes qui deviennent un ornement fort cher quand elles sont sans utilité.

more detail on his method at the even larger building and planning scale. The 'Moyen à employer pour faciliter le travail de la composition' (A method to facilitate design work), is a detailed qualification of the *règle* in terms of function and form:

> When designing a building for a given program, one must first check the area needed to satisfy the required functions. To do this one must know the space that is required for all the objects it is to contain: workshops, machines, workmen, inhabitants. When one knows the surface that the building or buildings will occupy on a given site, one needs to find the combination that will best satisfy the criteria. If there are to be several buildings, it would not be amiss to study the chief possible combinations of square, rectangular, or round structures. This would lead one to discover which are the best or those that could be appropriately modified. After making one's choice, one should test whether it is a good one, and for this, a first study on graph paper is appropriate, fixing the structural module that seems best to fit the programmatic conditions. Only after this quick and rough draft can one know whether the organization one has chosen is a good one. Then comes the detailed design.
>
> The principal combinations that one can make with rectangular or round buildings are shown in pl.V. (Mary 1862–63, pp. 43–44)[119]

Again, this is a very different interpretation than Durand had envisaged: Mary *combines* through functional organization, through *distribution* rather than by *composing* forms. In contrast to the philosophical approach of the EP course,

119. Lorsque l'on a à projeter un édifice quelconque, d'après un programme donné, le premier travail à faire consiste à se rendre compte de la superficie dont on a besoin pour satisfaire à ce programme. Pour cela, il faut connaître l'espace nécessaire pour loger tous les objets, ateliers, machines, ouvriers, habitans que l'édifice est destiné à contenir. Quand on connait ainsi la surface que le bâtiment ou les bâtiments doivent occuper dans le terrain dont on dispose, il faut chercher la combinaison qui satisfera le mieux aux conditions qui ont dû être imposées. S'il faut pour cela plusieurs corps de bâtiments, il ne sera pas inutile de connaître les principales combinaisons qu'il est possible de faire avec des corps de bâtiments carrés, rectangulaires ou circulaires. On verra quelle est celle de ces combinaisons qui conviendra le mieux, ou dont on devra se rapprocher. Quand on aura fait son choix, il faudra essayer si ce choix est bon, et pour cela, il faudra faire une première étude du projet sur papier quadrillé, en donnant à l'entre-axe la dimension que l'on aura jugée la plus propre à satisfaire aux conditions du programme. Ce ne sera qu'après avoir ainsi reconnu par un travail rapide et encore peu exact que l'on est arrivé à une combinaison satisfaisante, que l'on entreprendra l'étude détaillée.

Les principales combinaisons que l'on peut faire avec des bâtiments droits ou circulaires sont représentées Pl. V.

Mary's focus is physical and practical. Durand's goal in design was a formal composition—an object—but Mary's act of designing is a step-by-step process to fulfill a function (Pfammatter 1997, p. 139). This difference between *design* as an object and *designing* as a process summarizes the dissimilarity between the ECP's approach and those of the EP and the Beaux-Arts in a nutshell.

The logic of design method as a process appears so obvious to us today that we might easily mistake it for a cliché unless we contrast it with Durand's earlier static definition. It was novel, and Mary was revolutionary because he taught his students to think of design and construction in terms of step-by-step operations and systems. He was indifferent to formal and historicizing aspects, and thereby established an intellectual, cultural, and functional base for the modern design process that we today consider to be self-evident.

Mary's brief but critical exposé of method is followed by a longer section called 'Examen de divers édifices' (Analysis of various structures). This chapter describes the utility of the method by presenting a detailed discussion of the criteria for designing modern facilities, from market halls and many different types of factories to hospitals and prisons (pp. 44–105). Like Durand, Mary advocates type as an alternative to symbolic form, but his typology consists of actual structures, not of hypothetically corrected examples in the form of diagrams.

'Composition' redefined as 'organization'

So, what did the *règle* accomplish for the curriculum and how did it impact the concept on which the school was based? Instead of graphically—that is, visually *composing* plans and façades as all earlier practitioners had done, and to a large extent the ateliers at the École des Beaux-Arts[120] continued to do—Mary's students spent a great deal of effort on intellectually *organizing* functional programs and procedures.

I contrast *composition* as the formal manipulation of elements with the term *organization*, which defines their functional or structural relationship.[121] Mary

120. The École des Beaux-Arts did not exclusively concentrate on form as Durand had done. Material characteristics and construction were integrated into the program, for example by Rondelet, but the main goal of design was the resultant form, and solving functional or structural problems was a means to this end.

121. The distinction between composing and organizing is based on Karl Ludwig von Bertalanffy's systems theory as applied by Herbert Kramel to construction at the ETH Zurich around 1980.

himself related the concepts form, structure, and function in another way: he used the term *composition* to include both the formal aspects of decoration and the functional organization of parts and contrasted it with *construction* in the sense that we use the term today. The verb *design* as we now know it would correspond to *projeter*, a term that implies planning as well, and that he uses less often than *composer*. Because of this, one easily misses the apparently slight but in fact momentous shift in viewpoint that Mary himself did not verbalize.

The difference between composing (in my sense) and organizing is subtle. Mary changed the thrust of Durand's method from the formal arrangement of elements to the configuration of function and structure, and thus from the *act* of composing a form to the *action* of organizing a process. By directing his students' competence toward technology and industry and away from a preoccupation with form and 'taste', he established the organizing as one of the principal aspects of design. In contrast to Durand's original four drawings, or his later vague *assez grand nombre de dessins* (quite a few drawings), Gourlier's and then Mary's students had forty precisely circumscribed analytical drawings to complete as part of the course, and

> Moreover, the students are required to draw the various parts of buildings in the field, like floors and trusses in timber as well as in iron, vaults, stairways, etc., according to the processes and at the scales required by practice. (Lavallée/Dumas, 1831, p. 33)[122]

Mary's colleagues implemented this approach in their classes, too, especially Perdonnet, who also dealt peripherally with building structures. The whole long chapter (chapter IX, 'De la disposition des gares') at the beginning of the second volume of *Traité élémentaire des chemins de fer* in 1856 discussed the organizational aspects of station design (pp. 1–118). Perdonnet presented around fifty examples in detail with only a brief sub-section on 'Décoration architectonique des gares' (The architectural decoration of stations) at the end (pp. 112–118). Because Cornet's *Album* also included six detailed plans of railway stations in the 1849 edition (pls. 28–33), and fifteen in the 1853 edition (pls. 33–39, incl. 36 bis and 39 bis and ter), we know that Perdonnet cultivated the organizational aspect of design in his course too.

---

122. De plus, les élèves sont exercés sur le terrain à tracer les épures, grandeur d'exécution et d'après les procédés qu'on doit suivre dans la pratique, de différentes parties de construction, telles que: planchers et combles, tant en bois qu'en fer; voûtes; escaliers, etc.

Thinking about design in this way enabled Perdonnet to employ Mary's new design method at all scale levels and for all purposes—station halls, carriages, infrastructure, and locomotives. His students could use it to organize buildings where form and space are important criteria, and to create machinery where the focus lies only on function and economy qualified by material characteristics.

The distinction between composition and organization is therefore significant: Mary expanded the usefulness of the method from architecture to a general design process, and thus from what had originally involved almost exclusively art and 'taste' and was therefore restricted to architecture or bridges, to anything that architects or engineers might be called upon to design that involved materials, structure, functional behavior, and fabrication. Engineers who designed railways, bridges, or industrial plant could also design the machinery to build them using the same creative process. Mary created a tool that enabled *centraliens* to cross borders seamlessly between structural, machine, or industrial design, which is why it is so difficult to categorize the graduates of the school's early period professionally. Mary set a crucial educational marker in the pedagogical process that impacted all the disciplines involving building and manufacture. Even those who primarily worked as architects, like Trélat, Muller, Demimuid, Denfer, or Lavezzari differed in their breadth of knowledge and skill from the typical Beaux-Arts-trained architect.[123] Trélat introduced Mary's method in his curriculum at the École Centrale d'Architecture from 1865, and the novel ECP concept also influenced, as Levine pointed out (Bélier et al. 2012, p. 172), the difference between the design of the iron structures of Labrouste's early Sainte-Geneviève reading room (1843–50) and his study hall at the Bibliothèque nationale over a decade later 1862–68).

Durand had taken a first step toward this goal in proposing the *logic* of geometry to replace the rigid geometrical *rules* of proportion. This was a first move to define designing as a system. The lines in his 1813 diagrams no longer marked off the proportional relationships between parts of a column or a building.[124] He replaced their use in the organization of his plans with axes of symmetry and linear organization.[125] This new geometry liberated the plan from the confinement of the

123. This agrees with César Daly's opinion rather than with Picon, who understands the two professions as following the same goals (see Picon 2000, p. 44).

124. *Précis* 1813, pt. 1, pls. 5–8/10 and pt. 2, pl. 4.

125. *Précis* 1813, pt. 2, pls. 19 and 21, *marche à suivre*.

Académie's preoccupation with proportion and beauty. Mary took a second step by transforming Durand's geometrical *logic* into a *tool* to regulate and modularize function and structure. As mentioned before, he shifted his viewpoint from form to structure, and the lines to the spaces between them: the axes became simple lines in his organizational grids, and they defined structural spans.[126] Mary's grids could be used to create any object or organization in extrapolation of the concept behind Monge's descriptive geometry and Leblanc's drafting method. Geometrical grids were no longer restricted to building design and could be used to create machinery or other objects. The act of designing became a systematic, organizational process. Grids could even be used as an abstract principle, for example to organize critical paths in site management, as in the transition from traditional on-site *construction* to off-site component manufacture combined with on-site *assembly* that would soon characterize the industrialization of iron construction.

Mary took yet another step. As a consequence of his understanding design as a process, he considered building primarily as an act rather than as a 'fact'—as *constructing*, not as *construction*. He introduced a new twist to Vitruvianism, in which he stressed design and building as linked technological *processes* instead of interpreting architecture as a static *art object*. Through this typical nineteenth-century shift in orientation from space and symbol—that is, from visually based *meaning* (form, space, and beauty) to the dynamism of *making* (manufacturing and constructing)—he helped define a new approach to building.

All former design stratagems were transformed by Mary's method, which linked design to industry. Rondelet and Bruyère had both based their teaching on construction in the Vitruvian sense of *firmitatis* or stability, a static concept that is closely related to durability (*firmitas*), rather than the dynamic sense of manufacture.

Rondelet was a mason and an architectural practitioner who was versed in academic classicism but also interested in technology—that is, in masonry construction techniques and stereotomy. He also addressed iron structure in his work, as a result of having studied and worked with Soufflot.[127] Rondelet also

126. Durand placed columns placed on grid lines too (*Précis* 1813, pt. 2, pls. 1–2/4/8–13/15–16), but that is not the way he used his axes in designs. This is demonstrated by the fact that columns are not placed on the axes in 1813 pt. 2, pls. 19 and 21, the *marche à suivre*.

127. Soufflot had designed the iron roof over the staircase in the Louvre in 1779 shortly before he died, and Rondelet finished his Sainte Geneviève church, now the Panthéon, using substantial iron reinforcement in the structure.

claimed to have rediscovered *pisé*, the ancient Roman technique of concrete construction.[128] His view was state of the art when he published it, but from our vantage point it was backward-looking because he had based it on traditional site construction whereby artisans make parts from raw materials on site.

Bruyère looked more to the future, especially in his iron bridge proposals. His view of iron construction implied the concept of *prefabrication* and on-site *assembly*, both of which are processes. This is what Mary picked up on and developed in his course. Today we consider this a truism: in our view it is the only conceivable approach if we choose to use iron in construction, but it was a novel idea at the time. Eck had inadvertently documented the transition from iron structure as a cottage industry in his first volume in 1836 to a few cases of a more industrial treatment of iron in the second in 1841. Mary supported the transition conceptually through his method.

By grounding the abstract Durand method in practice, and by refocusing it from an object orientation to a process orientation, Mary operated as a *vulgarisateur*, but one who critically modified the conceptual model on which he based his system. The results were immediately apparent in the rapid industrialization of iron production, in which *centraliens* like Ferdinand Mathieu (grad. 1838) participated as technical director of Schneider et Cie., or in the standardization of component manufacture and construction like Barrault did with the structure of the 1855 Exhibition building. Many French and Belgian iron industries—Cail, Fives-Lille, Goüin, Schneider, Zorès[129]—came to rely on the theoretical-technical expertise of ECP graduates. Like Flachat's consulting firm had done from the very beginning, they thrived on their employees' inventive ability to see the essence of problems, and to formulate them and find solutions. This 'ECP mindset' became dominant and spread even to those who had not attended the school. It manifested in the results that contactors like Baudet-Leturc, Eiffel, Joly, Moisant, and Roussel produced. Joly survived as Nouguier Kessler et Cie. until the beginning of the twentieth century, while the others continued successfully and eventually amalgamated in several steps to form today's Eiffage Métal.

---

128. Rondelet was the one to popularize the technique as architecture, but the honor of the supposed rediscovery of the technique that was still practiced in rural construction in France (and in fact all around the Mediterranean) belonged to François Cointereaux, whose four cahiers entitled *École d'architecture rurale* appeared in 1790–91, and who claimed to have invented or reinvented the technique.

129. Although the founding families of these firms had not studied at the ECP, the *centraliens* who reached senior positions in many of them contributed to the standardization of iron manufacture based on their ECP education.

'Connaissance des matériaux'

Mary followed the design and planning part of his course with a second section that deals with construction technology. 'Construction des édifices' includes a substantial sub-section on building materials 'Connaissance des matériaux' and how to use them (pp. 105–234). It includes a chapter on wooden trusses (pp. 213–224), followed by a brief ten-page treatment of iron trusses that remained unchanged from version to version (pp. 224–234), and a substantial 43-page-long foray into the composition and behavior of mortars, cements, and concrete, both historical and modern (pp. 131–174). Mary had used concrete during his years in practice, had participated in the development of the tremie for underwater casting (Mary 1831 and 1832), and was a close friend of the concrete pioneer Vicat.

The chapters on cements and concrete lie outside the scope of this study, but they were relevant to the pioneering theoretical and practical development of reinforced concrete in France. As with iron, Mary did not stress or develop concrete further himself, although his successors did, and toward the end of the century, ECP graduates devised systems, structures, and calculation methods in reinforced concrete too. A prominent example was the pioneering reinforced concrete frame that Paul Cottancin (grad. 1886) patented in 1889 and featured in Anatole de Baudot's church Saint Jean de Montmartre (1894–1904). He did this with the same self-assuredness that Barrault had displayed half a century earlier in his iron structure of the Exhibition hall of 1855. The only difference was that the architect also participated in the innovation process, thanks to the pioneering efforts of Baudot's teachers Labrouste and Viollet-le-Duc.[130]

Another example was Louis Gellusseau (grad. 1906), Auguste Perret's engineer whose expertise supported the Perret Brothers' development of their larger-scale reinforced concrete structures between 1911 and 1936, including the seminal 1911–13 Théâtre des Champs-Elysées and probably the Sacré-Cœur d'Oran Cathedral 1903–13.[131] Gellusseau contributed to the technical underpinning of Perret's more

130. Baudot was closely allied with the ECP as he was among the first teachers in Émile Trélat's École Spéciale d'Architecture in 1865, along with Viollet-le-Duc, Émile Muller, and Henri de Dion.

131. Gellusseau did not participate in the earlier, technically less demanding 25bis rue Franklin apartments (1903–04) or the Garage de la rue Ponthieu (1906–07). The first was detailed and erected by the Hennebique agent Latron et Vincent and was similar to Hennebique's own 1 rue Danton (1898–1900; Abram 1987, pp. 82/86) that had been designed by another ECP graduate, Édouard Arnaud (grad. 1888). The second was built by the Perret Brothers themselves and was also engineered

complex, long-span structures, rather than to the formal development of the concrete frame that had been François Hennebique's contribution.

The treatment of stone, wood, mortars, and cements in Mary's course classifies him as affiliated with the construction approach to design that Picon calls 'Vitruvianism' (Picon 2000, p. 23). However, his view builds on what Bruyère had taught and is even more tightly integrated than a traditional construction bias: he views their technological characteristics as major design determinants. In other words, Mary understands the choice of materials—along with function—as one of the two chief regulating mechanisms for construction and structure. Durand had discussed materials too, but he understood them as addenda to form and space.

Whole sections of Mary's transcript substantiate this interpretation. They deal with building machines, tools, and methods of manufacture pertaining to brick, mortars, and cements (pp. 115–193, pls. XXXVII–XLIII), with chapter six, 'Échafauds, étaiments' (Scaffolding and shoring) (pp. 260–265) and chapter seven, 'Mode d'évaluation des ouvrages' (How to calculate structures) dealing with cost (pp. 265–270). Mary built on Bruyère, Camus de Mézière, Rondelet, Mandar, and Eck's publications, and his approach was comprehensive.

Again, it was Perdonnet in railway construction and Ferry, Walter, Faure, and then Callon in machine construction who amplified Mary's treatment of materials in their courses. Perdonnet, for example, wrote on the advantages and drawbacks of using wrought iron, cast iron, or wood in the construction of carriage wheels (*Traité élémentaire* 1856, vol. 2, pp. 144–148). He discussed various materials for rails and crossties (1855, vol. 1, pp. 379–381/419), or he considered types of wood and how to prepare them under headings like 'Préparation des bois' for crossties (1855, vol. 1, pp. 399–402) and carriages; 'Nécessité d'employer des bois bien secs' (The need to use well-seasoned wood); 'Nature des bois, Caractères des bois secs' (Characteristics of seasoned wood); or 'Tôle employée pour les panneaux' (Using panels in sheet iron; 1856, vol. 2, pp. 181–183). Perdonnet treated materials in small- and medium-sized construction, and Mary discussed them for large-scale structure, rationalizing built form in terms of technology and function. Pfammatter confirmed Mary's focus on process (Pfammatter 1997, pp. 132–146) but without bringing Bruyère's or Mandar's contributions into the equation.

Mary taught his preference for materials and fabrication well before the Crystal Palace in London initiated a general architectural awareness of the new material

---

by the Hennebique company (Abram 1987, pp. 86/88).

iron, its off-site manufacture, and its on-site assembly as a constituent design concept in 1850–51. His examples, many of them taken from Bruyère, are both formally and structurally modular. Most had probably been largely constructed on site, but they contain repetitive, prefabricated components like iron columns or composite elements like trusses that imply prefabrication.

The fact that Mary chose these examples over others indicates that he evidently preferred Bruyère's trend to prefabrication and assembly over Rondelet's on-site construction. He exploited what he chose more comprehensively than his teacher had in the *Études* by including function, plan organization, and structure, often with dimensions, construction, and materials in great detail. He was also influenced by Mandar's fastidiously detailed plans of the *maison de campagne*, which he republished. Mary's discussions of public facilities (pp. 44–90) are as detailed as Mandar's example, with the signal difference that they appear to imply prefabrication. One example among many is his description of the structure of the lateral wing of Gauché's wine depot (p. 58).

The final section of the transcript (pp. 235–270) deals with structural elements like stairs. It also illustrates his interest in both construction and the act of constructing by including a chapter on the design of scaffolding. This was possibly also influenced by Bruyère's treatment of the same topic (*Études* 1838, vol. 2, Recueil VII).

The multidisciplinary character of the curriculum

In case the students still did not make the connections among design, materials, and industry on their own, it was brought home to them in Perdonnet's railway class and also by Ferry, who was a recognized expert in industrial organization (Pothier 1887, p. 61). The remarkable synergy that developed between the teachers and their courses supplemented Mary's brief chapters in his text and made the ECP's pedagogy a success. Aside from his own courses, Mary was responsible for the integration of the various other courses as far as the field of building was concerned (Pothier 1887, pp. 117–118, cited in Pfammatter 1995, p. 287). For example, Mary's practical construction approach to water supply in his course segment *Détails pratiques sur la distribution des eaux* supplemented Bélanger's theoretical treatment of the same topic as recorded in his *Notes sur le Cours d'hydraulique*, where Bélanger referred to Mary and his statistical tabulations (*Notes* 1849–50, pp. 50 and 63). Mary's role in this integration made sense, as the whole program revolved around the issue of design, with each course contributing material and specialized viewpoints to the whole.

The ECP's curriculum and the commitment and team spirit of its professors was pervasive. The faculty based its pedagogy on that of their own professors Monge, Durand, Prony, and Bruyère, who had in turn been influenced by Perronet's earlier educational system. However, the ECP program went far beyond what they had experienced themselves, and they enabled their students to formulate and tackle novel, hitherto undreamt-of problems. Their influence is particularly prominent in Eiffel's and Moisant's oeuvres, but it is also evident in the work of all the others.

The integrative effect of the program is documented in de Dion's 1861 book *Cathédrale de Bayeux. Reprise en sous-œuvre de la tour centrale.* The report that he wrote as Flachat's employee is almost entirely devoted to problems of construction procedures and materials along with some information of archaeological interest.[132] The treatise makes fascinating reading as a blow-by-blow description of a complex and dangerous construction process: the underpinning of the central tower of the Gothic cathedral of Bayeux (1855–59), the consolidation of its foundations, and the replacement of the four piers that support it. The reader can follow the decision-making process, the reasoning behind it, the calculations, the support by means of scaffolding, the demolition, the reconstruction, and the careful decentering of the tower step by step and in dramatic detail. These details involved innovative procedural elements, including sinking five secondary foundations in open caissons to support the timber scaffolding around each of the tower's supporting piers, post-tensioning the tower with pre-heated iron rods, and winching 3000 tons of fissured and compromised masonry that had been solely supported on four points and buttressed by severely damaged structure onto the scaffolding. It illustrates the value of the ECP's innovative pedagogy involving the complex interaction between metallurgy, machine construction, foundation technology and geology, static and dynamic loading conditions, an understanding of humidity and its influence on material and structural behavior, the compressibility and strength of masonry and timber scaffolding, and the logistics of transportation and the building process. The twenty-seven-year-old de Dion had graduated in mechanical engineering a mere four years before he began this project, but he had already cut his teeth in practical construction on the erection of several iron bridges. He had Flachat's experience and knowledge supporting him in his work, but nevertheless this unprecedented and risky venture must have taxed his abilities to the limit. De Dion's report demonstrates the success of the ECP's multi-disciplinary approach in producing innovative graduates with unrivalled problem-solving capabilities.

The graduates were schooled in the newest knowledge and research, in the production of the new material, iron, and in the development of production facilities. Their education made them aware of the potential that iron represented for design, and Mary's system-oriented design method gave them the tools to innovate. This was a more comprehensive approach to design and the use of materials than any traditional architectural or engineering education offered. A British articled apprenticeship, the science-biased education at the ENPC, the exclusively analytical bias of the EP, or even the hybrid German system could simply not compete.

---

132. Henri de Dion's older brother Adolphe was an archaeologist and may have been unofficially involved in the project or otherwise influenced his brother's comments.

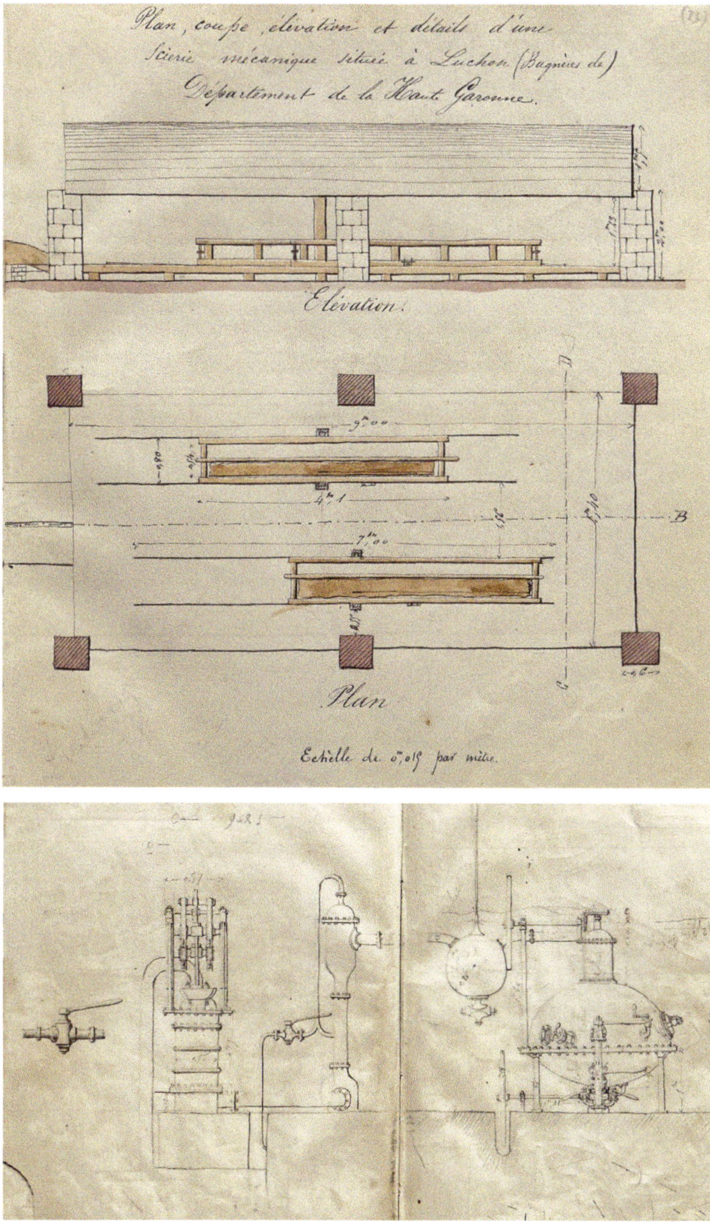

Fig. 103 - Three pages (pp. 23 and pp. 42–43) from Antonio Montero's notebook in Thumeloup's course *croquis à la main* 1849–50 showing the careful observation of construction and material, as well as the detailing of iron machinery. Material, construction, and detailing concerns transcended the boundaries between the courses and were common to them all. (Source: Peters Collection)

In the Monge tradition of field trips, Mary took his students to building sites around Paris. This provided yet another opportunity to enhance class teaching. Among the sites he chose were many hydraulic projects that were his own specialty, and where the students inevitably encountered iron construction and machinery. Pothier (1887, p. 230) recalled a visit to Louis-Georges Mulot's new artesian well in Grenelle (fifteenth arrondissement) in 1841, with which Mary was involved as director of municipal engineering (Reynaud 1870 and Gayant et al., p. 5), albeit before the celebrated cast-iron tower was added in 1852. The well was drilled to supply fresh water to a large section of Paris that had suffered from the lack of clean water in the 1832 cholera epidemic that had almost killed Lavallée and closed the school.

Mary's construction course was further enhanced by the requirement to document structures in the field during the summer break, supplementing what the students had learned of engineering language, analysis, and recording from Thumeloup, Olivier, and Faure. The idea was based on Colladon's and Leblanc's haptic approaches to knowledge. The first-year students documented either a

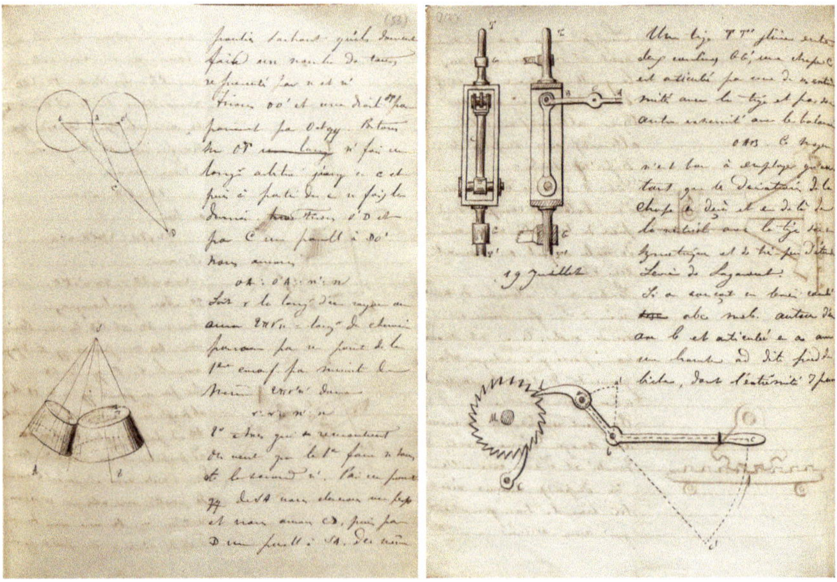

Fig. 104 - Two pages (pp. 53 and 112) from Ignacio Mendiola's notebook for Faure's course *construction des machines* (1849–50). They apply the skills that the student had learned in Thumeloup's drawing class and Olivier's descriptive geometry to the analysis and documentation of machinery. (Source: Peters Collection)

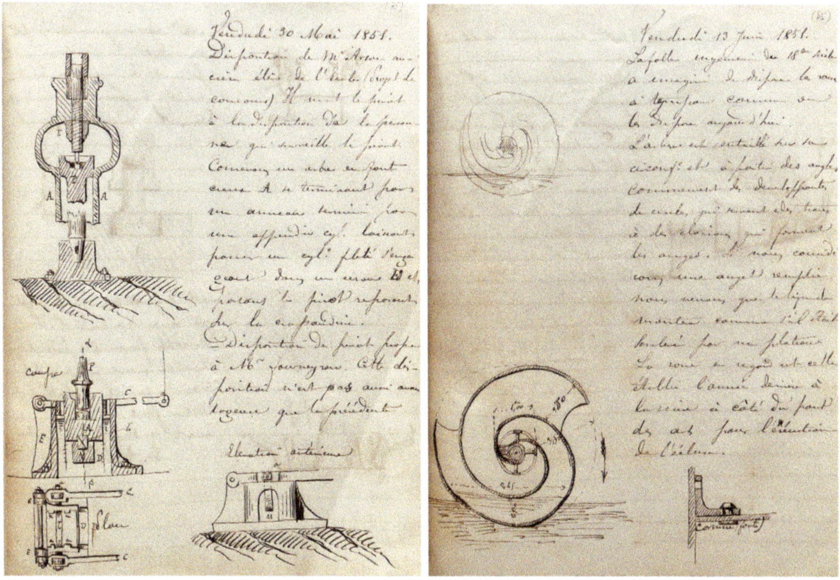

Fig. 105 - All the students, whether gifted in drawing or not, learned to use the visual medium to observe and document problems and solutions and to manipulate what they documented in design. These two pages (pp. 75 and 85) from Antonio Montero's notebook from Faure's course *turbines* (1850–51), display a keen observation of the assembly, connection, and functioning of machine parts. (Source: Peters Collection)

machine or a building in as much detail as possible, and the second-year students wrote a comprehensive report on an industrial plant. The border-crossing interests of all the current and former teachers and the integrative principle of the curriculum functioned well, and the students were very much aware of the latest developments in industry and in iron. This is well documented in student notebooks from various courses between 1849 and 1852.

The industrial, mechanical, civil, and structural engineers, the architects, planners, chemists, and all the others who were formed by this pedagogy between 1830 and the 1860s were eminently employable in the burgeoning construction-related and manufacturing industries. In their teaching, Mary and his colleagues had combined science and technology into a new industrial culture. The structural engineers among them created a distinctive French school of iron construction that was characterized by industrial prefabrication based on repetitive, modular design. They carried their integrative approach to design and industry forward toward what

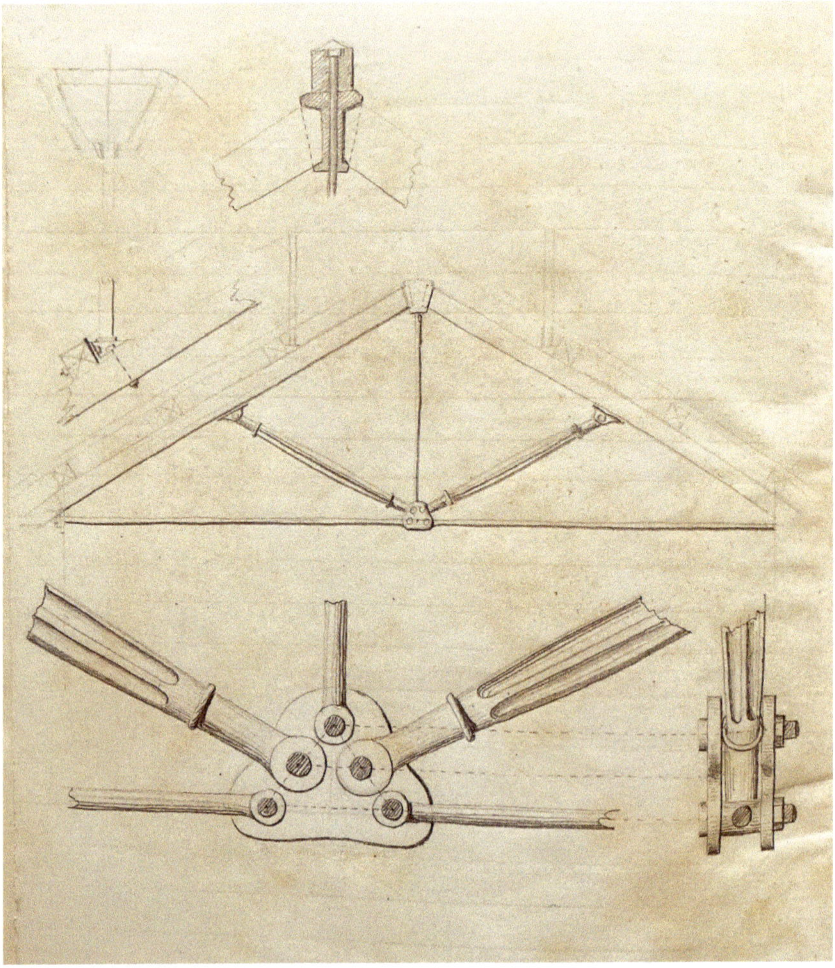

Fig. 106 - A border-crossing freedom between disciplines was encouraged by all the teachers. This is illustrated by Antonio Montero's sketch of a wrought- and cast-iron detail (p. 262) with indication of a connection in wood for a Polonceau truss in his notebook in Ferry's metallurgy course *métallurgie du fer* (1850–51). (Source: Peters Collection)

would become the aesthetics of the Modern Movement a century later, into a new culture of visual interpretation based on structure and materiality.

The results of this shift in aesthetic viewpoint had begun to manifest publicly between 1850 and 1860 in Flachat's railway stations and bridges, Barrault's Exhibition building, and Labrouste's second library far more consciously than it had in the earlier cast-iron structures like the warehouses, factories, greenhouses, and even the celebrated Crystal Palace that was—the name says it—far more renowned in the public eye for its light-flooded interior and its extensive glazing than its structure.[133] While the Crystal Palace was an influential early example of serialized but not industrialized manufacture on a grand scale that served as an inspiration for further structures—like the Empress Eugénie's Halles de Paris—cast-iron construction on the whole turned out to be a dead end, and it did not lead directly to industrialized iron construction systems.

The new aesthetic was based on structures that were built from industrially manufactured and standardized wrought-iron elements like angle irons, channel members, sheet metal, or the iconic I-beam, and it quickly spread to all forms of public structure, culminating demonstratively in Eiffel's Tower, which was presented to the public in 1889. Cultural arbiters were shocked and protested the tower, but the public embraced it, as it did the brash skyscrapers of the First Chicago School of architecture that Jenney was instrumental in promoting.

Thanks to the soberly functional path that Mary had blazed toward modern construction in his course, Barrault, Baudet, Contamin, de Dion, Eiffel, Jenney, Moisant, Polonceau, and their cohorts, many of whom have only been briefly mentioned in passing, were not structural formalists or historicists: rather, they pragmatically developed new structural types and gave them form. As builders they were not 'contaminated' by the rising tide of historicism that was beginning to seep through architectural design around mid-century. Nor were any of them engineering formalists in the sense that they were not limited by the classical models of structural theory—arch, beam, cantilever, post, suspension system, etc. They were able to experiment and to develop new conceptual models like the truss, two- or three-dimensional frames, or membranes.

In some of the most prominent examples of their work, however, architects were called upon to decorate their structures. The masked Vierendeel-type *entretoise B* truss in the 1855 Palais de l'Industrie was an example, in which the details were iconographic (decoration) rather than the enhancement of structural reading (ornament).[134] The arch on Eiffel's tower by Stephen Sauvestre, a graduate of the

133. The idea of the London Crystal Palace as the origin of a modern architecture in iron and glass was propagated by historians and theoreticians of the Modern Movement like Sigfried Giedion.

134. The cases in which architects and engineers collaborated to create structural form, like Labrouste and Baudet's Bibliothèque nationale or Baudot and Cottancin's Saint Jean de Montmartre, are noteworthy exceptions.

first class of Émile Trélat's school who had studied under both Viollet-le-Duc and de Dion, was a non-structural decorative device meant to create a symbolic entrance to the Exhibition of 1889, and the nave of Moisant's Grand Palais was embellished with Art Nouveau forms by the Beaux-Arts graduate Henri Deglane.

Despite this, these builders were confident in treading new ground. Writing in 1857, Barrault explained how they trusted their own judgment in the development of theory, in structural type, and in formal matters—and this was their strength. Their confidence helped them become the most prominent among the many builders that the ECP sent out into the professional arena. Many others after them participated in defining what proliferated as modern iron construction. The examples they set spread to engineers all over Europe and the Americas and from there it spread worldwide.

It was fortunate for the development of engineering education that the pedagogues Lavallée had hired were so diverse in their backgrounds and approaches and so willing to collaborate. Their interests dovetailed, each filling out lacunae in the others' abilities. Bélanger, Péclet, and Olivier were the theoretician-*vulgarisateurs*. Bélanger was also the consummate scientist, deeply involved in his research, and Olivier branched out from Monge and Hachette's geometry and Leblanc's drafting to explore the industrial implications of his field. Both Bélanger and Olivier were fully cognizant of the need to communicate their concerns to the practitioners who constituted their target group. Perdonnet, whose legacy consists of the many books that expanded and deepened what he taught, was not only an enthusiastic advocate of the new transportation system, but also a politically active pedagogue, equally involved in the politics of education and in pedagogical thought. He published about his profession as a railway pioneer and about his views on education. Mary was the methodologist and the integrator of all the disparate elements of education. All of them, alongside the chemists, physicists, metallurgists, economists, and mathematicians worked well together (thanks to Lavallée's managerial ability to keep their egos in line) and successfully educated generations of constructors. To name just one example of this cross-fertilizing collaboration, the *conseil d'études* delegated the metallurgist and industrial organizer Ferry in 1838 to choose a *répétiteur* for machine construction and the hydraulic engineer and designer Mary to choose a *répétiteur* for the railway course (Pothier 1887, p. 165).

I have used only a few of the most eminent students to demonstrate the effects that the ECP's pedagogy exerted on practice. Among the many other builders who would merit closer study are Tommaso Agudio who was prominent in railway construction in northern Italy, or Ernest Deligny (grad. 1842), who was only briefly mentioned in conjunction with the development of the I-beam. He was a railway and mining engineer. Then there was Flachat's relative Ivan Flachat (grad. 1850), railway engineer and entrepreneur. Other prominent graduates were the Swiss railway engineer and inventor of several compound systems for steam locomotives Anatole Mallet (grad. 1858); the railway and bridge builders Félix Mathias (grad. 1840) and Ferdinand Mathieu (grad. 1838), who was chef des ateliers de construction of Schneider et Cie. and the inventor of the cantilever construction method in bridge construction; Ferdinand's brother Henri Mathieu (grad. 1841); Léon Molinos (grad. 1851); Émile Muller (grad. 1844), who became Mary's successor for the course *constructions civiles*; and certainly Jules Petiet (grad. 1832), the eminent railway pioneer, Eugène Flachat's partner and director of the school after Perdonnet's death; or the ironworker turned engineer Charles Pronnier (grad. 1851). The German-French Théophile Seyrig (grad. 1865), who collaborated with both de Dion and Eiffel for several years, has only briefly appeared in this account. He too would merit a full discussion, as would the industrialist, politician, and architect Émile Trélat (grad. 1840), who taught construction at two schools from the 1850s, one of which he founded; Jules Govin (grad. 1869); and Paul Joseph Bodin (grad. 1871); and all those who headed departments in the great French contracting firms of the time—Fives-Lille, Cail, Schneider (like Ferdinand Mathieu), and Goüin (later Batignolles). Then there were the architect-engineers or design engineers who crossed between the disciplines. Contamin and Moisant were among these, but there were others like Émile Lavezzari (grad. 1853), Jenney, René Demimuid and Sylvain Périssé (both grad. 1858), Édouard Arnaud (grad. 1888), and Louis Gellusseau (grad. 1906), whose hybrid work and contributions it would be well worth examining more closely. Barjot and Dureuil (2008, pp. 60–63) and *Les anciens élèves de l'École Centrale 1832–1888* name many more who chose to work innovatively with iron in those early years and whose contributions also deserve to be illuminated in detail.

However, the six—Polonceau, de Dion, Eiffel, Jenney, Moisant, and Contamin, each of them from a different social background—will serve here to illustrate the effect that the school exercised on the industrialization of iron construction in more detail.

The result in iron: a brief characterization of six careers

What united these most celebrated graduates (as well as Barrault, Baudet, and all the others) despite their differences was their common educational culture. They had learned to draw conclusions from what they observed and how to use these conclusions to see their way to solutions for the novel problems they defined. They had learned how to employ scientific analysis and mathematical methods to turn seeing into innovation in technological creativity, and they had learned to design. The diversity in their careers illuminates the wide applicability and success of the ECP's pedagogical experiment.

Polonceau

(Jean-Barthélémy) Camille Polonceau (1813–1859) was raised in the profession as the son of Antoine-Rémi Polonceau, an eminent road, bridge, and hydraulic engineer under Napoleon. He graduated from the ECP in 1836 at the top of his class—*hors ligne* (off the chart) as Perdonnet wrote (Perdonnet in Comberousse 1879.1, p. A118). As Perdonnet's protégé, he made a rapid career at the very beginning of railway construction in France, and he was especially sought after for his ability to design economical locomotives and other equipment, as well as being celebrated for the equally economical roof truss that bears his name. Even before he graduated, Perdonnet employed him to help survey the two railway lines from Paris to Versailles, and then design the roundhouses and sheds for his and Flachat's startup company, the Chemin de fer de Versailles (left-bank line) in 1837. After graduation, Polonceau first continued his work on that line, and after it opened, he advanced to director of the Chemins de fer d'Alsace in 1842, where he developed the Strasbourg-Basel Line in collaboration with Flachat's office. That is apparently where he perfected his successful Polonceau truss.

Fig. 107 - Camille Polonceau was a graduate of the first years of the École Centrale. He was Perdonnet's protégé, friend, and colleague, and was known for his locomotive constructions and the economical composite truss that bears his name and that enjoyed widespread use in the nineteenth century. (Source: Guillet, *Cent ans de la vie de l'École centrale,* 1929, pl. 63)

Although some sources claim that Polonceau patented the truss, there is no record of it in the French Patent Office. However, he exhibited it in 1837 (Perdonnet in Comberousse 1879.1, p. A119) and published it in the first issue of César Daly's *Revue générale d'architecture* (no.1, Jan. 1840, cols. 27–32 and pl. 2). It caused a sensation and was republished in translation the same year in the *Allgemeine Bauzeitung* (vol. 5, pp. 273–280). The Polonceau truss is an extremely efficient arrangement of thin, hinged, wooden, wrought-, and cast-iron members. His analytical design approach and the way he used different materials to achieve the economic sparseness of the design harks back to Bruyère's influence, but it also owes a great deal to Mary's interest in materials.

Polonceau's truss was one among several that were developed in those years, and that may have been the reason that he did not, or could not, patent it. We previously saw Whipple's analogous 1841 structure. Rohault de Fleury's far earlier roof trusses in the conservatory in the Jardin des Plantes (1833) had a configuration that presaged Polonceau's (Pfammatter 1997, p. 85). Matthew Clark also built trusses that closely resembled Polonceau's in the Winter Palace in St. Petersburg after the 1837 fire, and again in 1841–42 (Fedorov 1992, pp. 79/83–84). Furthermore, Ardant published a composite truss with iron members in 1840 (plate XXV) that he had derived from Palladio, while Rudolf Wiegmann published another, also in 1839. Neither Ardant's nor Wiegemann's were as simple or as clearly defined as Clark's and Polonceau's. However, there were even earlier similar trusses that had been common in Manchester workshops between 1815 and 1824 (Fedorov, Heres, and Lorenz 2022, vol. 1, pp. 274–278) and probably also elsewhere in England, and it was unlikely that the observant Polonceau could have avoided seeing those on his visit with Perdonnet.

As iron construction was developing rapidly in France in conjunction with the spread of the railway at the time, it was Polonceau's truss that became the model for future developments. Its first large-scale use was in the shed of the Gare Saint Lazare, the terminus of the right-bank Paris-Versailles line (1842–53) and in the sheds at Paris-Auteuil on the inner belt line connecting to Saint Lazare (1853–54).

Polonceau later took over the traction side of the Chemin de fer d'Orléans, of which Lavallée was one of the directors and Jullien the engineer-in-chief, and he designed several notable locomotives and other railway infrastructure that he did patent. He was closely involved with the Saint-Simonian railway developers and was a founding member of the Société des ingénieurs civils de France, of which he became president in 1856. He also taught the economics of railways at the ECP (1851–52), and together with his mentor Perdonnet he published the multi-volume *Portefeuille de l'ingénieur des Chemins de fer* (1843–46) with addenda up until the year of his death (1859). He died young at the age of forty-six, but his career had been prolific, and his truss made a major impact on the subsequent development of iron construction.

De Dion

Count Joseph-Louis Henri de Dion (1828–1878) was born into an ancient aristocratic military family. He was first educated in Fribourg, Switzerland, because the family had sought refuge there during the anti-Bourbon revolt of 1830. His father appears to have been a military officer, but the son opted for a professional career and joined the ECP from 1848 to 1851.

De Dion belonged to one of the most prolific graduating classes in the history of iron construction. He graduated together with Léon Molinos and Charles Pronnier, as well as Louis Lasvignes, who assisted him in the reconstruction of the tower of Bayeux, and Louis Taviel de Mastaing, who later taught at both the ECP and Trélat's École Centrale d'Architecture. Their education lasted a year

Fig. 108 - Henri de Dion graduated from the mechanical engineering division, but worked as a theoretician, a structural designer, and a teacher. (Source: Wikimedia Commons, open access)

longer than most because of the Revolution of 1848 and the ensuing confusion. Like their classmates Antonio Montero from Montevideo (who never graduated) and Ignacio Mendiola from Puerto Principe, now Camagüey in Cuba (who graduated in 1852),[135] de Dion's first year at the ECP must have been split between the academic year 1848–49 (fall 1848–spring 1849) and the academic year 1849–50 (fall 1849–spring 1850). This was certainly an effect of the February Revolt of 22–24 February 1848 in Paris that put an end to the reign of King Louis Philippe and led to the formation of the short-lived Second Republic. During this time the school was closed, and professors and students participated in the revolt and subsequently in an experimental right-to-work industrial program in national workshops organized by the new government. It is unknown whether de Dion—as a former refugee aristocrat– participated or held back. However, like many others at the time, including the founders of the ECP and the railway financiers and entrepreneurs, he was a convinced Saint-Simonian in the

135. Neither Montero nor Mendiola became eminent enough to be mentioned in Guillet, but their eleven notebooks in the Peters Collection document the courses that de Dion also followed—nine courses at the ECP that represent roughly a third of the curriculum.

economic sense and thus a believer in technological progress as a motivating socio-economic force.

The normal school year only resumed after the workshops closed on 26 May 1848 (Pfammatter 1997, p. 121)—that is, presumably with the following semester that began in November. The school may have remained closed, or perhaps closed to incoming students until then. We may presume that the political disruption and the cholera epidemic raging in Paris at the time (Pothier 1887, p. 211) affected the curriculum for at least another year due to the many students who had to catch up. De Dion, who graduated in the machine construction and mechanics division, is said—like Polonceau before him—to have ranked first in his graduating class.

His professional life developed into a perfect example of the advantages of an ECP education. He combined theoretical studies, teaching, and practical construction with industrial management throughout his career. He began in Eugène Flachat's office following the path that many young ECP graduates took in the early years,[136] joining the firm right after graduating in 1851. He quickly worked his way up from the design office for the Saint-Germain railway, gaining experience in iron construction that he applied to several of Flachat's other railway projects as they gradually merged into the Chemin de fer de l'Ouest. He participated in the design and construction of several bridges in the Paris region, among them Flachat's 1851–52 Clichy-Asnières Bridge, the first large wrought-iron plate girder in France. This key project was erected by Alexandre Lavalley as Goüin's technical director, and de Dion's former classmate Léon Molinos as his *chef du bureau d'études*. Ernest Deligny participated, too, as a member of Flachat's team.

De Dion then gained further experience working on roofs for several railway stations and sheds. He proved himself so successful at this work that as early as 1854–55 he was put in charge of Flachat's consulting team on a major railway bridge at Langon, an even longer plate girder, over the Garonne on the Bordeaux-Sète Railway.[137] It had been designed by Stanislas de La Roche-Tolay (grad. ENPC 1841)[138] under chief engineer of the line Paul Régnauld

136. Most of this information comes from Molinos and Seyrig 1879.

137. Many of the participants on this project were Saint-Simonians. They included the contractor Ernest Goüin, Lavalley as his technical director, and Nepveu.

138. Reynaud 1883, vol. 2, p. 64. Seitz erroneously states that Régnault designed the bridge (p. 21).

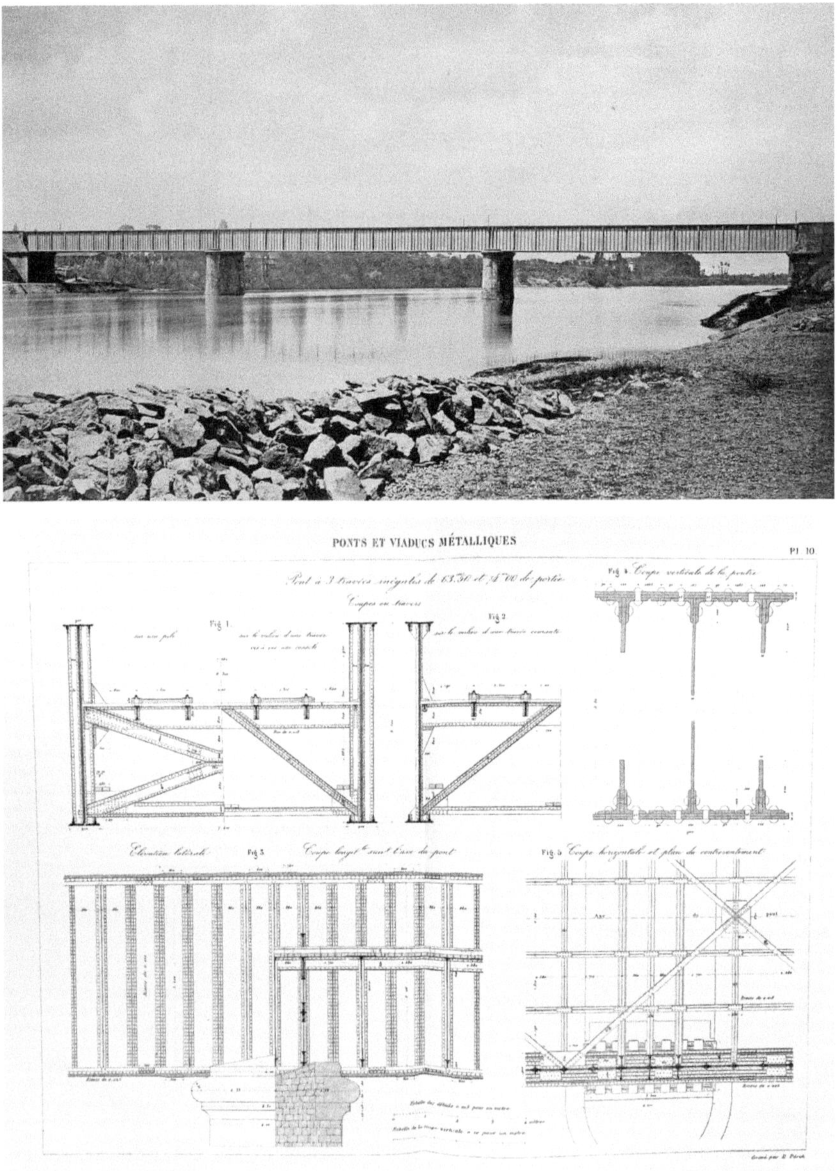

Fig. 109 - The Langon Railway Bridge over the Garonne (1855) was one of the early wrought-iron plate girders in France. It was designed by Stanislas de La Roche-Tolay and built by the Goüin Company with Alexandre Lavalley as responsible contractor. Both Ernest Goüin and Lavalley had previously worked in Britain and had imported structural techniques and construction methods to France. (Source: Reynaud, ed., 1883, vol. 2, *Les Chemins de Fer*, pl. 42) Details of the bridge. (Source: Régnauld 1870, pl. 10)

(grad. ENPC 1849).[139] The unusual case of a younger engineer supervising an older one is understandable when we know that Régnauld was the nephew of the Pereire brothers, who owned the railway line. There, too, de Dion worked with the contractor Lavalley, who had spent many years in England working on and studying iron bridge construction. He is one of the forgotten pioneers of the establishment of iron bridge construction in France from whose knowledge and experience many young engineers profited.

It is possible that the systematization of iron bridge design was at least in part influenced by Mary, whom Régnauld had experienced during his studies.[140] However, the primary expert who influenced the construction system—the use of rolled iron sections and riveting—in both bridges was Lavalley, who right after graduating from the EP, had spent four years in the iron industry in England (1842–46) working his way up from laborer to engineer before joining Goüin. It was his practical expertise that had aided the first French rolled-iron bridge constructions from Asnières (1851–52) on. Lavalley had made an extra study trip in 1852 for the specific purpose of importing the most up-to-date knowledge of iron construction and calculation techniques to France for the Langon Bridge (*Mémoires et compte rendu* 1892.2, p. 1003). Léon Molinos worked under Lavalley both at Asnières and at Langon, so that the latest knowledge of wrought-iron construction technology and detailing was being simultaneously transmitted to the young ECP graduates. The experience served as a kind of post-graduate education for both the mechanical engineer de Dion and the constructor Molinos.

Then Flachat put the 27-year-old, whose interests encompassed both structural analysis and practice, in charge of the complex and ultimately successful stabilization effort of the main tower of Bayeux Cathedral between 1855 and 1859.[141] As we saw, this unusual and difficult project perfectly suited both his original focus on machine technology as well as his experience in iron construction, and it documented the advantages of the ECP's border-crossing curriculum. The successful completion of the dangerous task earned him the knight's cross of the Légion d'honneur.

---

139. Régnauld had taken temporary leave of absence from the Ponts et Chaussées from 1852 to 1869 to serve as chief engineer of his uncles'—the Pereire brothers—Chemin de fer du Midi. However, Léonce Reynaud writes (1883, vol. 2, p. 64) that Jules Saige was the responsible chief engineer for this structure. He was certainly more experienced than Régnauld.

140. At ENPC Mary was professor for *navigation intérieure* from 1842.

141. De Dion's former classmate and assistant Louis Lasvignes (grad. 1851) took over the final part of the work in 1858 (De Dion 1861, p. 5, footnote).

While working on the Bayeux problem, he might have met Labrouste, who may have been the government's inspector or the architect assigned to the project, and possibly Viollet-le-Duc, too, who was one of the diocese inspectors (although not the one responsible for Bayeux).[142] After completing the restoration, de Dion left for Spain, working for the Belgian firm founded by François Pauwels,[143] where he designed and built several bridges and then a sugar factory in 1862 on Guadeloupe in the Caribbean, which he also ran for two years.

De Dion returned to France for health reasons, and in 1865 he developed a pioneering course called *stabilité des constructions* that he taught briefly to architecture students in fellow-*centralien* Émile Trélat's École Centrale d'Architecture. It was the first engineering technology course taught as part of an architectural curriculum in France. He based it on a geometrical approach to structural design that derived from what he had experienced in Olivier's classes, which in turn were based on the tradition of a graphical approach to structural knowledge that French engineers from Monge to Prony had developed.[144]

De Dion's promising teaching career was brief because further consultations and projects called him back to Spain and then to Turkey, where he designed a streetcar system for Istanbul. Then he went to Chile, where he worked in the guano industry, then back to Spain again where he developed the port of Santander on the northern coast (Guillet 1929, p. 312), and finally to Bolivia. During the Franco-Prussian War (1870–71), he returned home again to serve in the Paris region, building defense works under enemy fire for which de Dion was promoted to officer of the Légion d'honneur.

---

142. De Dion's interaction with Viollet-le-Duc may, however, only have dated from 1865, when they both taught at Trélat's school.

143. François Pauwels's construction company, founded in 1852, merged into the Compagnie générale de matériels de chemins de fer when it was chartered by Pauwels and others in Brussels in 1856. Charles Nepveu's bankrupt construction company joined them in 1857 (Compagnie générale 1856; Lewis, 1989). Evidently the company was still active under the name of Pauwels when de Dion worked for it in Spain.

144. It is unlikely that his course could have been inspired by personally communicated knowledge of Carl Culmann's contemporary teaching at the Eidgenössische Polytechnikum. It could not have come from Maurice Koechlin, Eiffel's later collaborator, who only returned to France from Zurich in 1877, or from Culmann's book *Die Graphische Statik,* which had appeared in 1866 but would only become available in a partial French edition in 1880 after Maurice Lévy had published *La Statique Graphique et ses Applications à L'Art des Constructions* in 1874.

During this hectic professional period, he continued working on graphical methods of calculation for designing and dimensioning non-orthogonal trusses. Together with Krantz and his former classmate Molinos he evaluated and recommended Eiffel and Seyrig's project for the Maria-Pia Bridge in Porto, and together with Seyrig he calculated and dimensioned it in 1876 using the method he had developed. The year before he had suggested doing the same for several restored cast-iron bridges that had been destroyed during the Franco-Prussian war[145] (*Mémoires et compte rendu*, 1875, pp. 90–93/546–587 + pl. 74), and he presented his general approach to the calculation of trussed iron arches to the Société in 1875 (*Mémoires et compte rendu* 1875 pp. 523–526). One aspect of de Dion's theoretical work that has not been studied so far is how much Théophile Seyrig contributed to his system of calculation and construction (Molinos and Seyrig in *Mémoires et compte rendu* 1879.1, p. 277), especially as Seyrig detailed it in the appendix to his and Molinos's biographical study (*Mémoires et compte rendu* 1879.1, pp. 434–474). Whether or not de Dion also worked on developing Eiffel and Seyrig's component construction system that later evolved into the 'kit of parts' as well is not recorded, but possible.

In 1877, the year before he died, he was elected president of the Société des ingénieurs civils, where he had been an active participant for years, taking part in discussions on the role of state versus private engineers (*Mémoires et compte rendu* 1872, pp. 123–139). In his inaugural presidential address he presented a broad and scholarly overview of the history and development of statics, beginning with Galileo, Hooke, and Mariotte. He admired Poncelet (whom he had probably known) for his ability in *vulgarisation* (*Mémoires et compte rendu* 1877.1, pp. 75–76). De Dion's lively account was peppered with telling personal anecdotes of the pioneers he had studied under and collaborated with.

Backed by his extensive international experience in iron construction and his work in structural analysis, his fame as a constructor in iron rests on his final structures for the Paris Exhibition of 1878 that were only finished after his early death at forty-nine. In his eulogy, his successor as president of the Société Henri Tresca said that he had been *vraiement l'homme des constructions en métal* (truly the master of iron construction; *Mémoires et compte rendu* 1878, p. 301).

145. These bridges had been built by Georges Martin, Émile Martin's second son, and manufactured by the family firm.

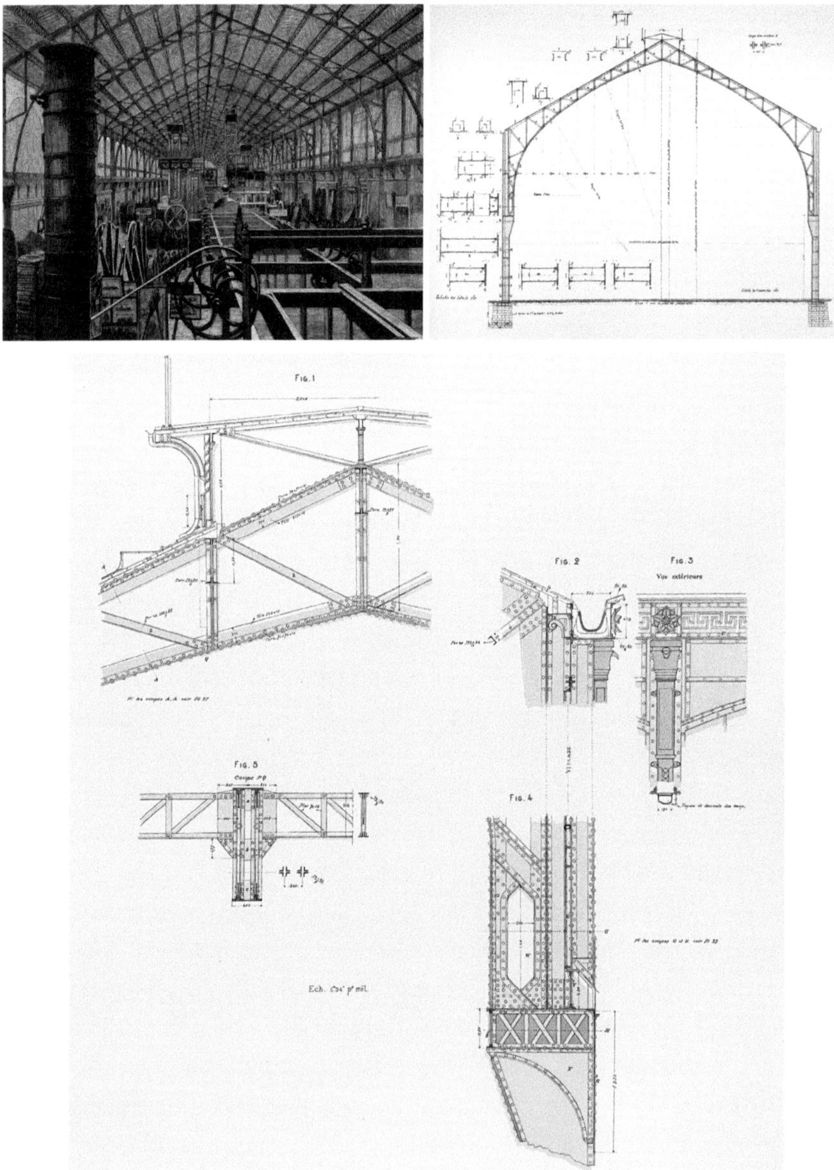

Fig. 110 - De Dion's chef d'oeuvre was the Galerie des Machines at the 1878 Paris Exhibition. Although he did not invent the wrought-iron fixed portal frame, he developed a form that was simpler than any previous version of that span size. The elegance of his design testified to his border-crossing ability in structural theory, construction, and manufacture. View into the hall in 1878. (Source: Simon de Vandières, *L'Exposition universelle de 1878 illustrée*, 1879, pl. 45, p. 89). Cross-section and detail. (Source: Vierendeel, *La Construction Architecturale en fonte, fer et acier*, 1897–1902, pls. 37, 38)

The 35.6-meter span machine hall for the 1878 Exhibition was a masterpiece of simplification. He designed it in wrought iron using riveted, trussed-girder frames that he calculated according to the methods he had previously developed for arches.

De Dion's goal was to create an open hall free of tension members, avoiding the complications that Krantz's flat arched roof in 1867 had experienced because of the thrust it generated, and the residual horizontal force at the simply supported footings (see further on under 'Eiffel').

De Dion avoided the problem by conceiving the structure of his hall as a rigid, fixed frame in which the verticals were designed to resist bending.[146] Conceptually similar to Moisant's far smaller portal frame in the Menier Mill in 1872, this represented a shift in viewpoint. All previous large-span halls had relied on arches to minimize or completely avoid lateral thrust. Barrault's 48-meter arches in the 1855 Exhibition building had been one example, and most long-span halls since then had been built either as tied arches with tension members below grade, or in the form of two- or three-hinged arches.[147] De Dion's solution used fixed, monolithic frames that he calculated graphically using his new, simple method. The frames were manufactured by Fives-Lille, Eiffel, and Schneider, and they as well as Baudet, Moisant, Cail, François Rigolet (the

146. Eiffel had used the term *ferme* (frame) once (Eiffel 1867, p. 3), but had calculated it as an arch. The difficulty that engineers experienced in calculating frames before de Dion developed his simple graphic method is evident from Eiffel's lengthy, 167 + 2-page *Mémoire sur les épreuves des arcs métalliques* (1867) that he undertook for Krantz's structure. He could only base his calculations on vol. 1 of Charles Bresse's 1859 *Cours de mécanique appliquée* and Auguste Albaret's applications in the *Annales des Ponts et Chaussées* (1861 and 1862).

147. Predecessor long-span iron structures included William Henry Barlow and Rowland Mason Ordish's world-record 75-meter tied arch with its tension member below the tracks in St. Pancras Station in 1866–68 and the three-hinged arch that was first proposed as an idea in a brief paragraph by François Mantion, Perdonnet's successor in 1864 for the railway course at the ECP, when discussing his railway bridge at Saint-Denis in the *Annales des ponts et chaussées* in 1860 (2e sem., pp. 220–221). Johann Wilhelm Schwedler popularized the two-hinged variant in the vault of a gasworks in Berlin in 1863. This was followed by the internationally influential two-hinged arches of the Unterspreebrücke in Berlin in 1865 and the Schlesischer Bahnhof in 1866–67 (Lang 1890, pp. 25–26 & footnote 22). Three further early three-hinged, cast-iron arch bridges are also known: the Hradecky Bridge of 1867 in Ljubliana by the Viennese engineer Johann Herrmann, followed by the now demolished Steiger Bridge (1869), and the Viennese Teggethoff Bridge (1872) in wrought iron by Anton Battig and August Köstlin—apparently on an idea suggested by the railway engineer Karl von Hornbostel. This last still exists, in part re-erected as the Kleine Ungarbrücke in Vienna.

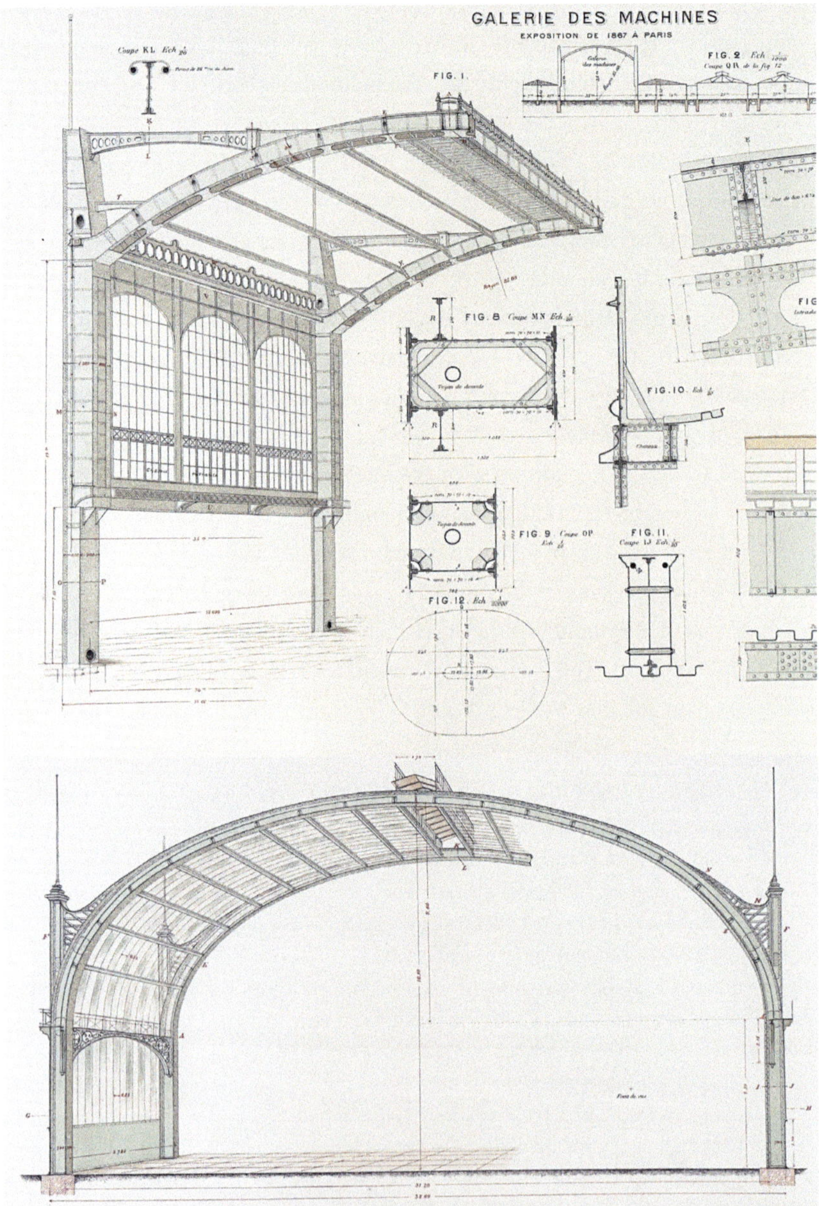

Fig. 111 - Two predecessors of de Dion's machine hall. Upper image: Jean-Baptiste Krantz's complicated fixed-arched structure of the 32-meter machine hall for the previous 1867 Paris Exhibition. (Source: Vierendeel, *La Construction Architecturale en fonte, fer et acier* 1897–1902, pl. 24) Lower image: The garden designer Heinrich Siesmayer copied and simplified Krantz's structure for the 31-meter fixed arches of the Palmengarten in Frankfurt in 1869. (Source: Vierendeel, pl. 25)

contractor on Reynaud's Amédée Lighthouse in 1865), and Henri Roussel erected them as well as the other iron structures of the Exhibition (*Mémoires et compte rendu* 1878.1 pp. 191–213 / 225–229 + pls. 115–118). The main contractors seem to have been Moisant, who built the external galleries, and Baudet and Moisant for the main hall. The following year Fives-Lille manufactured identical de Dion frames for the Las Delicias station hall in Madrid. They were erected by an engineer named Émile Cachelièvre.

Far simpler and more economical to manufacture than Schwedler's three-hinged arches, de Dion's elegant frames helped establish the monolithic concept in iron structure over the previously more common hinged arches or walls topped with simply supported Polonceau trusses.

De Dion's frame was not a novel concept, as Daly had claimed in 1878 and the *Mémoires de compte rendu* in 1879 (Vierendeel 1897–1902, pp. 151–152). Smaller fixed frames were not unknown in Germany and had also occasionally appeared in France, too, for example Moisant's 11.73-meter portal in the cross-section of the 1872 Menier Mill (Fig. 131).

De Dion's frame was larger, and although it was neither unprecedented as a type nor a daring span,[148] it represented, thanks to his capabilities in theory, a simple and elegant large-span solution, certainly better than Krantz's clumsy hall structure of 1867 or even Siesmayer's better version in 1869. His contribution demonstrated the advantages of combining an expertise in practice and theory, and the result was an excellent example of the beauty of engineering taught through *vulgarisation.* In this, his last project, de Dion established his reputation as a personification of the advantages of an ECP education. Remnants of his 1878 machine hall are preserved, especially in the Y-Hangar of the dirigible museum in Meudon.

The effect of the changes of 1848

There was a definite hiatus in the development of the ECP that can be traced in the graduates' approach to design. Polonceau and de Dion (and Barrault too) belonged to the earlier generation, those that studied—or in de Dion's case, began—before and up to the Revolution of 1848. Eiffel, Jenney, Baudet, Moisant,

148. It did not come anywhere near the 108-meter dome of the Vienna Exhibition in 1873 by John Scott Russell.

and Contamin began after, and one can sense a shift in their understanding of construction. This shift appears to have been the result of a later phase in the development of the curriculum, not only a change in the balance between basic and specialized courses in 1839–40 and the *travail de vacances*, but also in the development of the course material itself around mid-century, especially when Mary's course *architecture* metamorphosed into *construction*. It was also just after Bélanger had completed his transformation of the mathematical theory of mechanics in his *Cours de Mécanique* that he published in 1847.

While the first group approached construction as the design of forms or elements—Polonceau in his truss and Barrault and de Dion in the iron frame—Eiffel, Jenney, Baudet, Moisant, and Contamin were more concerned with designing and organizing general construction *systems* that could be industrially manufactured in series and that were applicable to many situations. The separation between the graduates is not radical, however. Time constraints had previously forced Barrault into system thinking in 1855, almost certainly as a result of Mary's very early teaching:

> In any type of construction work, a well-prepared project is the one essential requirement for a rapid and rational erection process, and my first step was to design the project....
> All the components that form the assembly need to interlink with precision, and to avoid slowdowns and cost overruns, site adjustment is forbidden....
> This is one point that one must insist upon in metal construction. Reducing the number of different parts to the minimum is not only the way to work economically, but also the way to work quickly....Just to cite one example, I saw to it that the ground-floor columns were all based on six models. When I first studied the project there were thirteen different ground-floor columns. (Barrault and Bridel 1857, p. 13)[149]

149. Dans un travail quelconque, un projet bien élaboré est la condition essentielle d'une marche rapide et rationnelle, et nous dûmes tout d'abord nous occuper de la rédaction de ce projet....
Toutes les pièces qui forment l'ensemble ont besoin de se raccorder avec précision et, sous peine de lenteurs et de dépenses excessives, l'ajustage sur place est interdit....
Il est enfin un point sur lequel il faut insister en matière de constructions métalliques. Réduire au minimum le nombre de pièces différentes est non-seulement un moyen de travailler économiquement, mais encore de travailler vite.... Pour n'en citer qu'un exemple, nous avons vu que les colonnes du rez-de-chaussée étaient toutes fondues sur six modèles. Dans notre première étude il y avait treize colonnes différentes au rez-de-chaussée.

Barrault's very early ability to shift to a new, systematic way of thinking about structure was the result of the courses he had been obliged to follow in machine construction, Perdonnet's early railway lectures, and Mary's class. It was inherent in the school's pedagogy from the very beginning, even before Bélanger began teaching and Perdonnet developed his mature course. From the outset, the curriculum had enabled border-crossing reactions in solving design problems. Barrault had refined his approach in Flachat's office and in his work in iron-smelting and in the design, construction, and running of iron mills. The school's mature system approach to design and construction subsequently impacted Labrouste's architectural development through Baudet's influence in 1860, especially in the stacks of the Bibliothèque nationale. Eiffel would follow this system approach toward construction to its ultimate realization.

Eiffel

Alexandre Gustave Boenickhausen, dit Eiffel[150] (1832–1923) came from a bourgeois family of German extraction in Dijon. His biography has been studied in detail in many publications. After a year's preparatory study at the Collège Sainte-Barbe in Paris,[151] and an unsuccessful attempt to enter the EP where he was judged *admissible* but did not manage to secure one of the available slots, he entered the ECP in 1852.

Eiffel had majored in chemical technology, planning to take over his uncle's factory, but quickly switched directions when discord in the family closed that opportunity to him. His apprenticeship in construction began with the Bordeaux bridge in 1857 under his first employer, the contractor Nepveu. When Nepveu sold his firm to Pauwels, Eiffel continued working for the latter, taking charge of the construction site. Eiffel had graduated as a chemical engineer, and although he had participated in all the ECP's construction and metallurgical courses, he

150. He later officially shortened his name to 'Eiffel'.

151. Montero's Sainte-Barbe chemistry notebook, begun in 1847 (it is not known when the course ended, as the Revolution of 1848 intervened), provides insight into what Eiffel learned before applying to the two grandes écoles. His teacher there was the same Jean-Augustin Barral whose name Eiffel later inscribed on the Eiffel Tower. Although chemistry was not one of the courses required for entrance to the ECP, prospective students at that time were required to submit examples of their previous education. The notebook for Barral's course shows the academic level demanded of the entrants, and chemical engineering was the field that Eiffel had originally intended to pursue.

Fig. 112 - Gustave Eiffel, the most famous of the École Centrale's constructors, graduated in 1855. His fame is due to the tower that bears his name, but his importance to construction rests on his development of a modularized and industrialized structural system that was flexible and could adapt to any orthogonal or random structural form. (Source: original photograph, open source)

was no expert. The system of rolled members connected by gusset plates that Eiffel used in his detailing of the structure was not his invention but had been developed by de La Roche-Tolay and Régnauld as the responsible designers. It is not certain, but Eiffel may also have been previously peripherally involved under Nepveu in the final phase of the earlier Langon Bridge too.[152]

Although Goüin had only been involved as contractor in the Asnières and Langon bridges, the knowledge that Régnauld and de La Roche-Tolay had gained from Lavalley's expertise is reflected in the detailing of the Bordeaux bridge as well. Régnauld published his book *Traité pratique de la construction des ponts et viaducs métalliques* in 1870 showing how he and de La Roche-Tolay had incrementally developed the beginnings of a standardized system using industrially rolled and built-up sections in riveted configurations with gusset plates for various spans.

152. Although Nepveu's firm only merged into François Pauwels's Belgian company in 1857 (under which the Bordeaux Bridge was erected), Nepveu had apparently also worked on the earlier bridge. The Langon Bridge was only finished very late in 1855, as the load tests were only carried out in mid-November (Régnauld, p. 151), so Eiffel may possibly have been peripherally involved in the project's final stages.

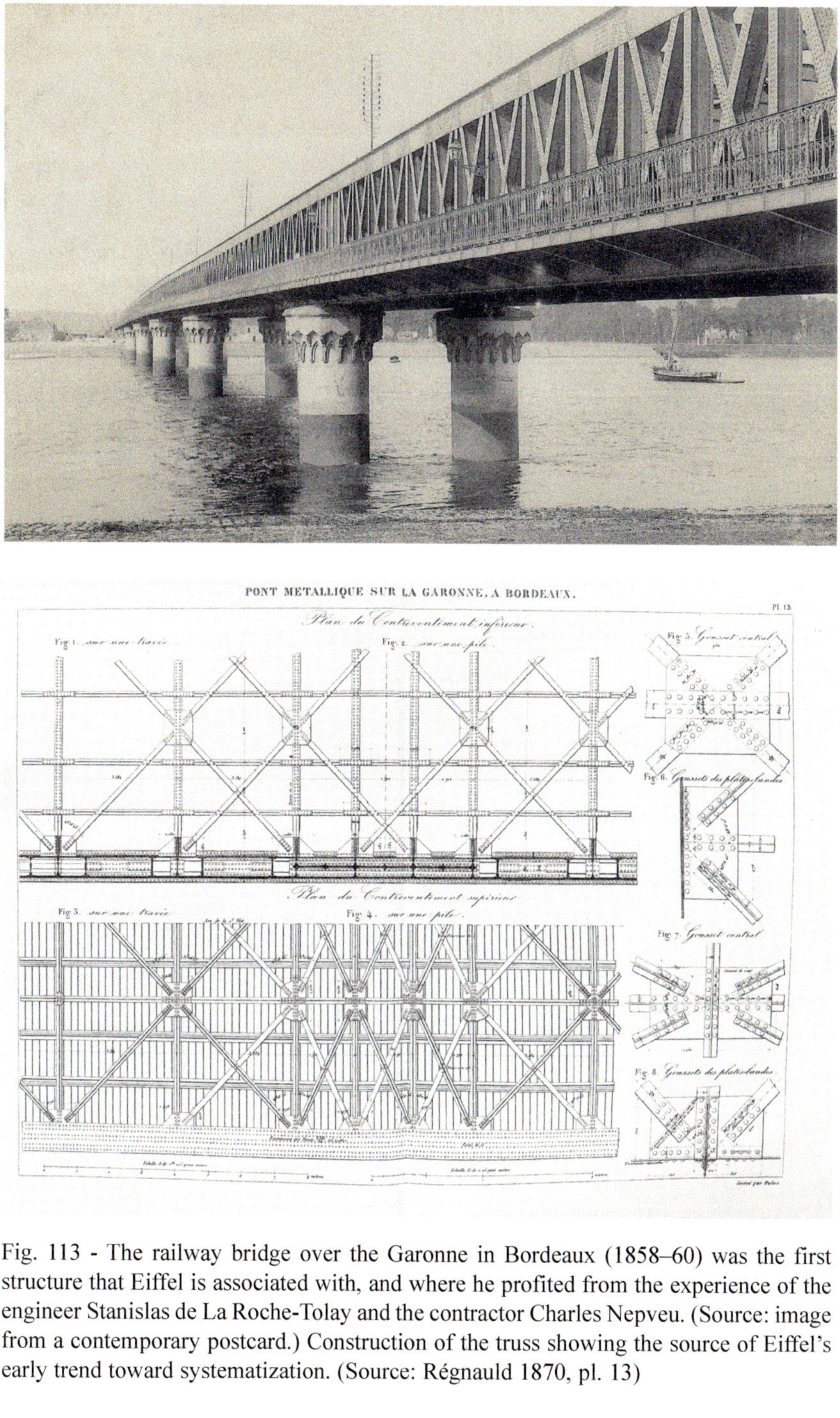

Fig. 113 - The railway bridge over the Garonne in Bordeaux (1858–60) was the first structure that Eiffel is associated with, and where he profited from the experience of the engineer Stanislas de La Roche-Tolay and the contractor Charles Nepveu. (Source: image from a contemporary postcard.) Construction of the truss showing the source of Eiffel's early trend toward systematization. (Source: Régnauld 1870, pl. 13)

When the Bordeaux bridge and the railway station St. Jean Bordeaux were finished in 1860, Eiffel continued as design engineer with Pauwels, working on at least three more bridges, one over the Lot at Capdenac and one over the Dordogne at Floriac, both in 1860–61, and then one over the Adour at Bayonne in 1864. The first two appear to have been built under Adolphe Jullien.[153] But who the responsible design engineers were for the Brive-Toulouse segment on which these two bridges were built, is not clear. The Bayonne Bridge was quite possibly a de La Roche-Tolay bridge, as he had been responsible for the rail line from Bayonne to the Spanish border. They were all box trusses like Bordeaux, but with different detailing. It is not clear whether Eiffel detailed all these himself, which would have been an unusual task for a young contractor early in his career. The different detailing in these bridges suggest that multiple design engineers were responsible for them, and Eiffel may only have supervised their manufacture and erection for Pauwels. Lavalley, Régnauld, de La Roche-Tolay, and perhaps other unnamed design engineers had all been involved in Eiffel's learning curve in this apprenticeship period.

Eiffel was promoted Pauwels's chef d'études des travaux d'usine (chief of design for factory prefabrication) for the French branch of the company in 1862, and this suggests that he may have undertaken a more responsible role in the design of the Bayonne Bridge. He stayed in that position for only two years before setting up as an independent consulting engineer in 1864.

Upon de La Roche-Tolay's recommendation he was immediately commissioned to calculate the arched girders of the structurally complicated roof of Jean-Baptiste Krantz's Pavilion du travail for the 1867 Paris Exhibition, for which he developed an experimental formula (Lewis 1989, p. 253 / Eiffel 1867, pp. 144ff). He derived it from Charles Bresse's theory and two articles that Auguste Albaret[154] had published in the *Annales des Ponts et Chaussées* in 1861 and 1862. Eiffel calculated the frame, if I understand correctly, virtually as an 'arc brisé' using Bresse's elastic theory of arches,[155] not as a stiff frame as de Dion would later do. He had trouble fulfilling Krantz's demand that the verticals should primarily resist axial forces and the footings not be considered as fixed.

153. Jullien was director of the Chemin de fer de Paris à Orléans, which had taken over the defunct Chemin de fer Grand Centrale, which owned the line on which these two bridges were built in 1857.
154. Albaret is otherwise only known for his work on Spanish and Portuguese railways and as inventor and manufacturer of construction and other machinery.
155. Eiffel 1867, pp. 1, 27–31, 55, 106, 112–118, 140–147.

Krantz had tried to counter the thrust of the arched roof by suspending it at the third points from an upward extension of the verticals. The idea was that the bending that this introduced in the verticals would in part compensate for the residual thrust that the arch exerted at the springing further down (Eiffel 1867, pp. 158–162), the rest being absorbed in the footings. The confusion that this unclear idea generated in the calculation and detailing of the structure contrasts with the straightforward solution that de Dion later developed for the same structural problem. Once again this testifies to the quality of an ECP education and the structural clarity that Bélanger was able to develop and communicate to his students.

Eiffel did his best to fulfill Krantz's requirements. He consulted with Tresca on the ductility of the material and conducted tests to confirm Bresse's theory with practical results. He carried these out on a trial arch in September and October 1865 together with the contractor Goüin's collaborator Ernest Fouquet (Eiffel

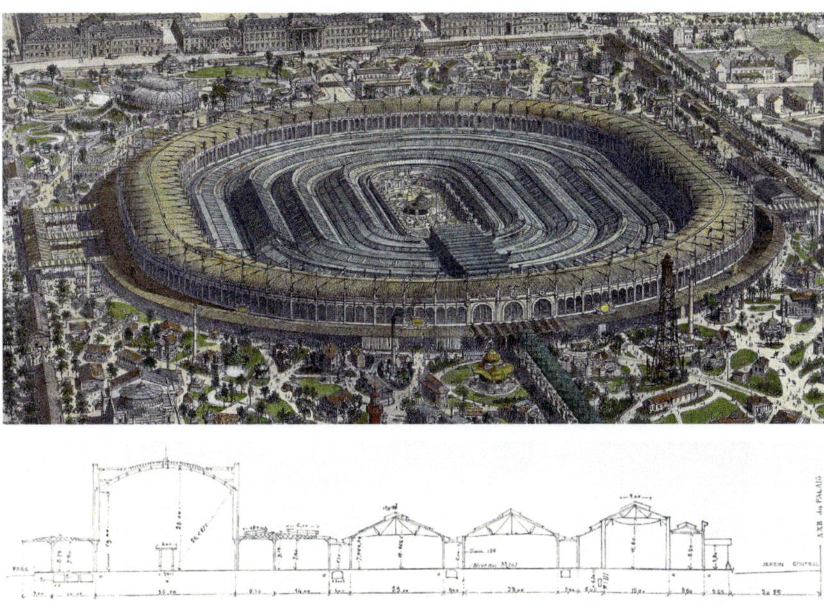

Fig. 114 - Bird's-eye view of the Palais de l'Exposition at the Paris Exhibition of 1867. (Source: *Vue officielle à vol d'oiseau de l'exposition universelle de 1867* contemporary lithograph by Eugène Cicéri and Philippe Benoist, detail) Cross-section through the bulding showing the Machine Hall and Eiffel's exterior canopy on the left. (Source: Ducuing, *L'Exposition Universelle de 1867 Illustrée*, vol. 1, p. 10)

1867, pp. 3–4), and he published the results in his 1867 *Mémoire sur les épreuves des arcs métalliques.*[156]

The cantilevered canopy that Eiffel was apparently then allowed to design on his own for the circumference of the building was far simpler and structurally clearer, visually contrasting Krantz's pragmatic, piecemeal approach to structure with the conceptual clarity that Eiffel had learned to appreciate at the ECP.

In 1866 Eiffel opened his own construction firm at Levallois-Perret, apparently taking over some of Pauwels's plant.[157] Two years later Théophile Seyrig joined him, investing a large sum in the firm but remaining an employee, while Eiffel became director of the company, which was named G. Eiffel et Cie. From 1868 to 1869 they built the Neuvial and Rouzat Viaducts on the Gannat to Commentry Railway for the Paris-Orléans Company under chief engineer Wilhelm Nordling. These were not yet Eiffel designs, as they had been developed by Pierre-Félix Moreau, chief designer at the consortium of Cail and Fives-Lille, who had previously designed and built the similar Bouble and Belon Viaducts.

Fig. 115 - Exterior view of the outermost ring of the Palais de l'Exposition 1867 showing the cantilevering canopy apparently designed by Eiffel. Comparing it to Krantz's frame, one can see the structural clarity that ECP engineers developed. (Source: Ducuing, *L'Exposition Universelle de 1867 Illustrée*, vol. 2, p. 153)

156. Eiffel mentions calculations that he had carried out earlier for dimensioning the arcs (Eiffel 1867, p. 139 footnote). De Dion had submitted a first version of his later frames of 1878 for that project.

157. The details of this period in Eiffel's career are complex. See Lemoine 1995.

The young firm's breakthrough came a few years later with the Douro Viaduct, for which Seyrig was the responsible designer, while de Dion and Molinos evaluated the design structurally and de Dion and Seyrig then carried out the calculations. Eiffel was already familiar with a modular approach to design and construction from his work on de La Roche-Tolay's bridges, but what was novel in Seyrig's design for the Douro Viaduct was that the system had to be flexibly adaptable to non-orthogonal structures. All the previous industrialized systems, like Flachat's at Asnières, de La Roche-Tolay's at Langon, Bordeaux, and Bayonne, or Pauwels's in Florient and Capdenac had resulted in orthogonal, linear structures with regular box-cross-sections. Eiffel and Seyrig adapted the system by distinguishing between variable and fixed criteria: member cross sections and the connector gussets were fixed, while their length and angles of connection were variable, creating a true 'kit of parts'.

Then, in 1879, during the early stages of the firm's involvement with the Garabit Viaduct, which was based on the Duoro Viaduct design, the partnership fell apart. Eiffel dismissed Seyrig, even though their contract of association still had two years left. Maurice Koechlin then joined the firm to replace Seyrig. Step-by-step, aided by Nouguier, who had previously worked for Goüin and then joined Eiffel in 1876, and Koechlin, and using de Dion's method of calculation (Eiffel, *Mémoire* 1889, vol. 1, p. 11), Eiffel began to perfect his and Seyrig's modular construction system in wrought iron.

Thanks to the fact that iron manufacture had been industrialized and the products standardized, he was able to pursue this more thoroughly than Bruyère had been able to do two generations before. The stages of the development progressed quickly from the frame for the Statue of Liberty begun in 1879 to its completion in 1883, and the system appeared even more clearly in the Garabit Viaduct, which had also begun in 1879 and was finished in 1884. Eiffel also refined the modern form of cantilever construction in this bridge, a method that he, Seyrig, and possibly also de Dion had pioneered in the Maria Pia Bridge in Porto 1877[158] (based on Ferdinand Mathieu's method for the Grandfey Viaduct), and again in the firm's work for the 1878 Paris Exhibition.

158. For the development of cantilever construction see Peters 2018.

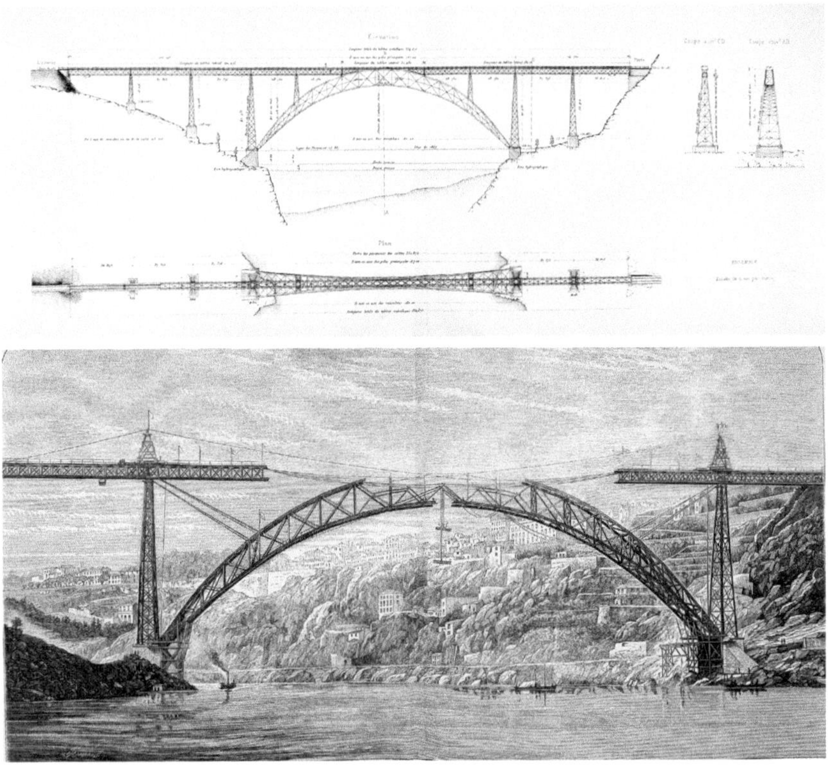

Fig. 116 - The railway bridge Maria Pia over the Douro in Porto was built in 1876–77 by the Eiffel-Seyrig firm. The project had been examined by Molinos and de Dion, and the structural calculations were made by de Dion and Seyrig. Seyrig is thus credited with the design of the structure that marks Eiffel's first attempt to standardize a non-orthogonal structural system in iron. (Source: Eiffel, *Notice sur le pont du Douro, à Porto*, 1879, pl. I) Wood engraving of the erection process by John Henry Rimbault. (Source: Eiffel, *Notice sur le pont du Douro, à Porto* 1879, pl. V, reproduced from *Engineering*, 21 June 1878)

Eiffel was an active self-promoter, using his name as a trademark for his business. He did credit his employees as long as they remained in the background, but in contrast to Flachat, who had always promoted his employees, or to his competitor Moisant, who invariably remained in the background himself, Eiffel attached his name prominently to the projects he built, the best-known being the 1000-foot tower. His partner Seyrig had apparently been fired for demanding a more prominent recognition of his contribution to the Garabit Viaduct, and he was never accorded the credit he deserved. Although he had been dismissed from the partnership while the bridge was in the early planning stages, Seyrig had undeniably

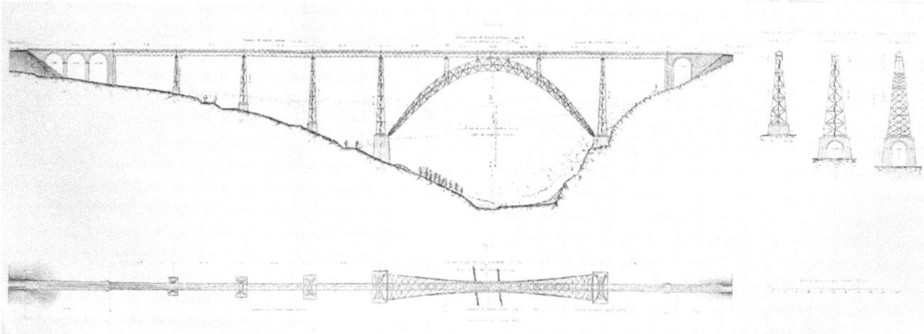

Fig. 117 - Théophile Seyrig (in the beginning), Maurice Koechlin, and Émile Nouguier worked on the Garabit Viaduct, which manifested the fully developed modular system. It had been originally proposed by Léon Boyer and detailed and in Eiffel's office. The salient characteristics are the standardization of connections and members and the varying length of members. Eiffel reused the Seyrig-de Dion calculation method for this project. It is unclear how much these engineers contributed to the development of the modular system. (Source: Eiffel, *Notice sur le Viaduc de Garabit*, 1888, pl. I) The cantilevering erection process of the Garabit Viaduct 1883. (Source: Eiffel, *Mémoire sur le Viaduc du Garabit*, 1889, atlas pl. 16)

played an important role in the Garabit Viaduct's concept. The design was his and the mature component system was possibly also partly his. The Garabit Viaduct was based on the Douro Viaduct and the success of that structure had secured the contract for the new one. Seyrig had contributed a great deal to Eiffel's success as

Fig. 118 - The Eiffel Tower, originally proposed and sketched by Koechlin and Nouguier, established Eiffel's fame worldwide and popularized his system, which influenced all subsequent iron and steel construction. It also entered public awareness

through the creation of the Meccano and Erector Set toys and impacted engineering education and design for over a century. The photographer Louis-Émile Durandelle photographed a series of images of the building process of the tower that was partly reproduced in Eiffel's monumental monograph *La Tour de Trois Cents Mètres* in 1900. (Source: photographs by Louis-Émile Durandelle, from Eiffel, *La Tour de Trois Cents Mètres*, 1900, vol. 2, pls. 3, 4, 6, 7, 8, 9, 11) he evidently had done in the development of de Dion's calculation system.[159] Molinos, whom Eiffel mentioned in several reports, may also have contributed to both the cantilever method and the standardized, flexible construction system. So did Nouguier and Koechlin. Both were repeatedly mentioned in association with all three structures: Liberty, Garabit, and the tower. Eiffel was a masterful coordinator of gifted collaborators and a visionary in the organization and development of industrialized iron construction.

The Eiffel-Seyrig system had the great advantage that it could easily be applied to rectilinear, curved and, when expanded one step more, to random forms in linked hierarchical subsystems. In the skeleton of the Statue of Liberty (1879–83), which was perhaps Eiffel's most complex structure, the rigid, symmetrical, and riveted primary frame is linked to a secondary system characterized by riveted, standardized rods of variable length, and then to a tertiary system of rods with bolted connections. This third subsystem of rods can rotate at the ends that are attached to the secondary system, and they attach on their other ends to the straps that connect the bronze plates of Bartholdi's sculpture. This three-tier system liberated the sculptural skin step by step from the rigid geometry of the primary system.

Where Bruyère had experimented with unique structural modules for each of his bridge designs, Eiffel and his team designed a universal component system based on his and others' experience as industrialists. In contrast to Bruyère, Eiffel was an industrialist as well as an engineer, and he established a successful iron prefabrication and what we would now call a design-build industry.

Eiffel went further than Mary too, carrying the Mary-Durand design method to its logical conclusion by creating a catalog of standardized connections using gussets as connectors. The activity of *connecting* became the element *connector*. It evolved into a primary component of the system. Under de La Roche-Tolay, Eiffel had used riveted gusset plates systematically in their mature form (see Fig. 113). Régnauld shows several intermediate forms in 1870 that demonstrate how they developed gradually, including a single example of a gusset in the bracing of the Langon Bridge (see Fig. 109).

Joly also published a contemporary example (Jolly 1863, pl. VI), which may have been the work of Joly's early ECP employees like Baudet or Moisant, who were of course trained on iron railway infrastructure. We do not yet see gussets for example in

159. Seyrig's abilities are evident in his successful bid against Eiffel and other prominent constructors to build the Dom Luis Bridge, a third structure of the same type, erected in Porto (1881–86).

Fig. 119 - Eiffel and Koechlin developed the characteristics of the system in the three levels of structure of the Statue of Liberty. The first level was the primary, symmetrical iron tower. The only variations lay in the gradual shortening and thinning of the diagonals from bottom to top and in the secondary framing for the raised arm carrying the torch. The second layer was the light framing that approximated the irregular shape of the skin. The connections were standardized, but the lengths of the members were variable. The rotating bars that attached the skin to the secondary structure with single bolts formed the third and final layer. The hierarchical concept of the three layers of structure was based (at a historical remove) on Durand's hierarchical design system tempered by Bruyère's interest in materials and then Mary's redefinition of the system in terms of construction process. (Source: aerial photograph, Peters, 1979) Fernique photograph 1881–83. (Source: Library of Congress [LoC] and BnF, Wikimedia Commons). Interior of structure (Source: Creative Commons generic 2.0)

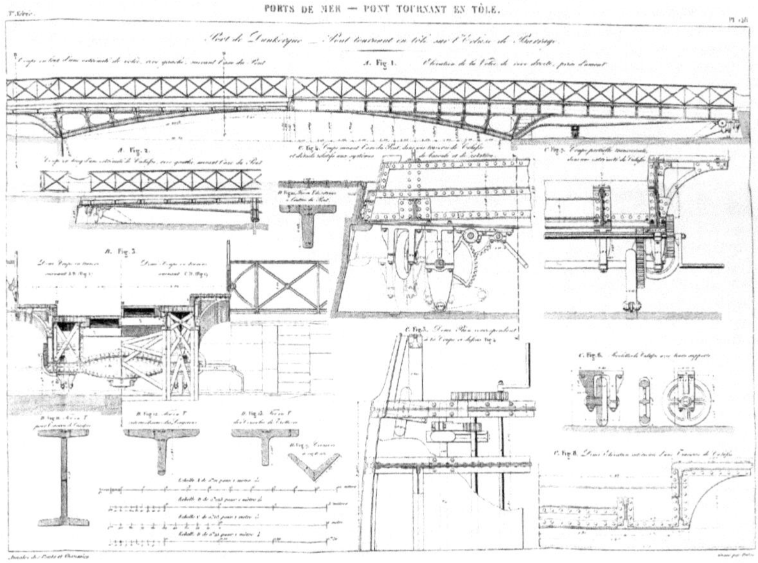

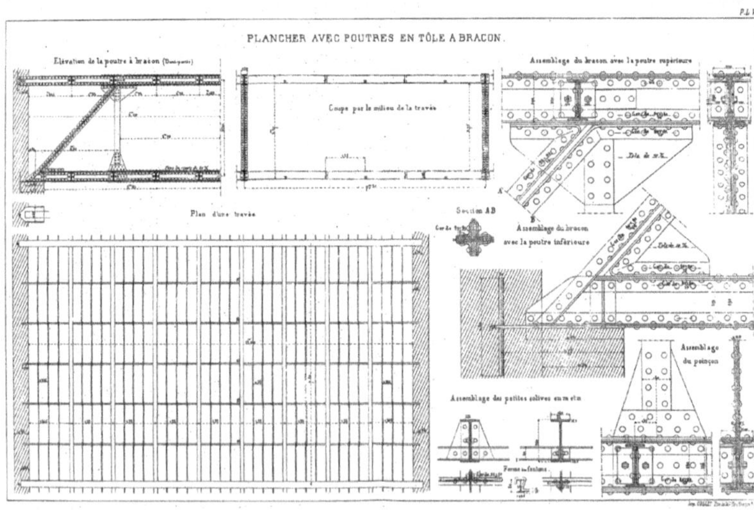

Fig. 120 - Upper image: The ENPC engineer Alexandre Plocq built a riveted, wrought-iron swivel bridge over the entrance to the wet basin in Dunkirk harbor in 1857. He did use the relatively new I-sections in his design, but connected them using brackets rather than gusset plates. (Source: Plocq 1859, plate 1, reprint of his article from the *Annales des Ponts et Chaussées* 1859). Lower image: A few years later, in 1863, Jolly shows the construction of a wrought-iron beam to support a floor. The *poutre à bracon* (braced beam) appears to have been among the first to use true riveted gussets to connect the members. Since the firm employed the ECP engineers Baudet and Moisant at the time, these novel details may have been a type developed by them. Moisant, like Eiffel, certainly used them later. (Source; Jolly 1863, album, plate VI from © gallica.BnF.fr)

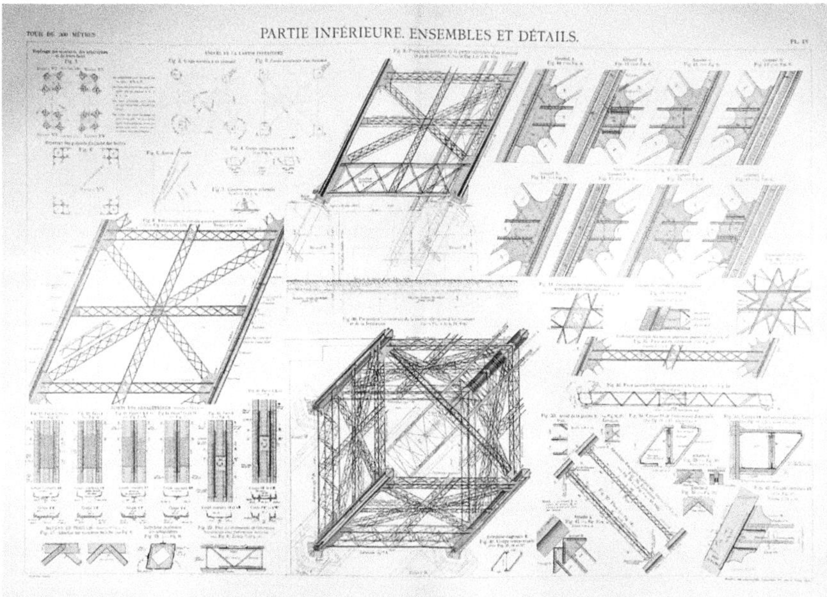

Fig. 121 - Details of the connections that characterized Eiffel's mature system. (Source: Eiffel, *La Tour de Trois Cents Mètres*, 1900, vol. 2, pl. IX)

the wrought-iron swivel bridge built by the ENPC engineer Alexandre Plocq just a few years before in 1857 (Plocq 1859, pls. 1 and 2).

Eiffel may have been aware of Neville's inconclusive attempts at developing gussets in the 1840s (James 1986), and he certainly must have known of the primitive connectors that North American bridge and lighthouse builders were using.

The critical difference between his mature kit of parts and de La Roche-Tolay's earlier system—or Krantz's hidden gusseted frame corners for the machine hall in 1867 (Eiffel 1867, p. 3)—was that Eiffel's connectors were conceptualized as standard system components that allowed varying angles of connection. The fully developed system as initiated in Douro and amplified in the later projects integrated standardized design with industrial manufacture and erection procedures. It completed what Bélanger, Perdonnet, Olivier, Walter, Ferry, Faure, and Mary had begun.

It is unclear to what extent Saint-Simonianism and Freemasonry influenced Eiffel's thinking. He certainly came into close contact with both ideologies early in his career. Virtually all the pioneering French railway engineers, industrialists, and financiers were Saint-Simonians or in sympathy with the doctrine. Paul

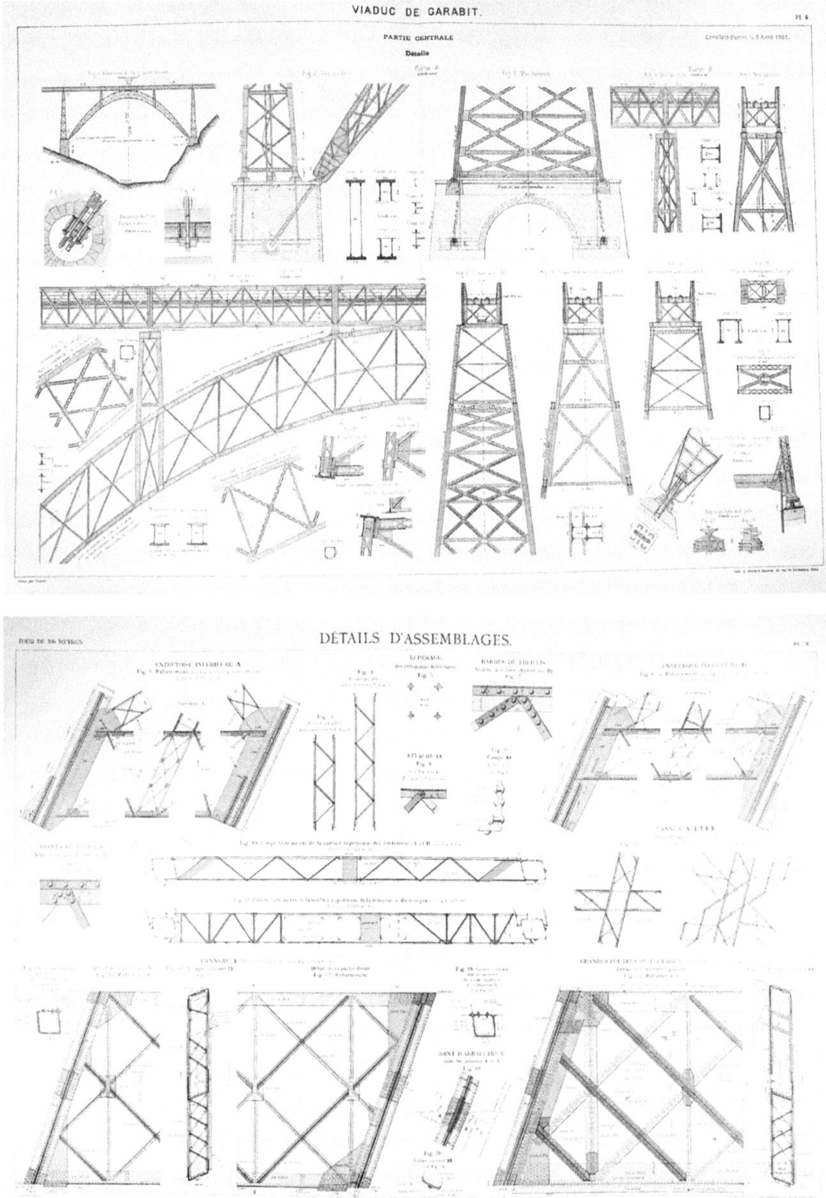

Fig. 122 - Comparing the two images, it is evident that the Eiffel 'kit of parts' system was more developed in the tower (1884–89) than in the Garabit Viaduct (1880–84), which incorporated more variants, and in which the connectors are less rigorously standardized. (Source: Eiffel: *Mémoire sur le viaduc de Garabit*, 1889, vol. 2 [atlas], pl. 4; Eiffel, *La Tour de Trois Cents Mètres*, 1900, pl. X).

Régnauld, the chief engineer of the bridge at Bordeaux on which Eiffel worked and Eiffel's employer Charles Nepveu, who was responsible for his introduction to contracting, were both Saint-Simonians and Freemasons. Many more Freemasons surrounded Eiffel's main works throughout his career: the sculptor of the Statue of Liberty, Frédéric Bartholdi; Maurice Koechlin, one of the original designers of both the Statue of Liberty and the tower; the principal governmental sponsors Prime Ministers Jules Ferry and René Goblet; and the Minister of Commerce and Industry Édouard Lockroy, who was also the Commissioner in charge of the World's Fair of 1889 (Pierrat and Kupferman 2009). Eiffel apparently repeatedly visited the Progrès et Vérité Lodge no. 8 à l'Orient in Bex, Switzerland, from 1881 with the sculptor Bartholdi, who often vacationed with him at his villa in Vevey and who it seems was a member (Bauer 2010). The Statue of Liberty was itself a monument to both Freemasonry and to Saint-Simonianism because it 'made the spirit flesh' by means of technology (see Tresch 2012, p. 211 citing Clot, 1858).[160] If one could demonstrate that Eiffel was indeed influenced by Freemasonry, one might be able to derive some philosophical insight into the hierarchical 'kit of parts' system.

Eiffel's industrial system was based on the simple hierarchical logic of the Durand-Mary rule-based progression from *component* to *element* to *subset* and finally to the whole *system*. Its practical success was guaranteed when he turned the theoretical system into his industrialized 'kit of parts'. Every other constructor had to reinvent a structural program and possibly retool a manufacturer for every new form and each new structure. But Eiffel could build any structure or form he encountered with his system, even non-orthogonal ones like the Garabit Viaduct and his tower, or completely random forms like the Statue of Liberty. Is there an intellectual link to the philosophy of Freemasonry here? This question is far more subtle and challenging than the usual discussion on the form and symbolism of the tower. Many open questions still await resolution.

Eiffel went beyond these ideologies and philosophies when he contacted Carl Culmann in Zurich to inquire whether he might know of someone who was fluent in French and was available to introduce graphic statics to his office.[161]

160. This is a transcendental idea that apparently exists in Freemasonic thought and contradicts traditional Christian theology.

161. This was shortly after Lévy had published his treatise in 1874 and began teaching. Perhaps it was Lévy who had recommended approaching Culmann?

His interest in graphic statics may have been kindled by de Dion when he had briefly taught construction to architecture students in the mid 1860s. Culmann apparently recommended his Franco-Swiss student Maurice Koechlin.[162]

Eiffel's 'kit of parts' view of construction entered popular culture. The relationship between his visual expression of system in structure and traditional French *imagerie populaire* has never been explored, but it is evident in the plates that he published in his monumental work *La Tour de Trois Cents Mètres* in 1900. As a result, Eiffel's thinking influenced popular cut-out paper models and it also became a broad-based didactical concept that was reflected in construction toys that influenced generations of future structural engineers right up to the mid-twentieth century. Frank Hornby's Meccano appeared in 1898 in England and was patented in 1901. It became known as 'mechanics made easy' and then as Meccano from 1908. In the United States Alfred Carlton Gilbert conceived an almost identical idea in 1911 and launched his Erector Set in 1913. Neither of these men was an engineer, and this shows how thoroughly Eiffel's concept had permeated general public awareness.

Eiffel is therefore justly considered to be the most original and influential of all the developers of the focused system approach to iron construction. He later placed the names of his teachers Bélanger and Perdonnet on the Eiffel Tower in 1889, possibly because he revered what they had contributed intellectually to his education. He also honored other former ECP professors under whom he had not studied: Coriolis, who had left early after the Revolution of 1830 but whose influence was still felt at the school, as well as the chemists Dumas and Jules Pelouze, neither of whom still taught when Eiffel studied, although their texts were still mandatory. He also honored engineers from his circle of acquaintances: Flachat, Polonceau, and Henri Tresca. More than a decade after his death, Flachat was still the shining model for all French civil engineers, Polonceau's iconic truss was ubiquitous, and Tresca was an older colleague to whom Eiffel may have felt that he owed his knowledge of the ductility of iron and who had assisted him in the first experiments on Krantz's frame in 1865 (Eiffel 1867, p. 8). He also honored Jean Augustin Barral, his chemistry teacher at the Collège Sainte-Barbe, but not Mary. It is unlikely Mary's absence was due to his not being a scientist, since neither were Perdonnet, Polonceau, and several others. Did he not agree

162. As Alsatian industrialists, the Koechlin family was well connected with the ECP and graduated four members under that surname and probably other relatives, too, between 1869 and 1890.

with his fellow *centraliens* that Mary—like Bélanger or Dumas whom he did honor—had been a key pedagogue in his education?

The reasons for Eiffel's eclectic choice of names have always remained obscure, and so any comment must remain speculative, but most of those he chose were *innovateurs* in science, industry, or construction rather than *vulgarisateurs*. We do know of the admiration that Eiffel felt for two of them, Flachat and Tresca, whom he evoked in glowing tributes on January 4, 1889, in his maiden address as president of the Société des ingénieurs civils (*Mémoires et compte rendu* 1889.1, p. 24). Eiffel also included de Dion's name on the tower, as well as those of the iron industrialists Ernest Goüin, Ernest Schneider, and Jean-François Cail, possibly in recognition of their contributions to his own thinking on monolithic system development in iron construction.

Jenney

William Le Baron Jenney (1832–1907) came from a wealthy, ship-owning family in Fairhaven, Massachusetts, and after a few years sailing, during which he later claimed to have made astute observations relating to construction practices, he began to study engineering at the Lawrence Scientific School at Harvard. He quickly became disenchanted by its abstract, scientific bent and transferred to the more practically oriented ECP in 1853.

Fig. 123 - William Le Baron Jenney graduated in 1856, worked as both architect and engineer, and eventually became the mentor of the developers of the steel-framed skyscraper in Chicago. (Source: Wikimedia Commons)

One might think that it would have been more natural for him to have chosen the United States Military Academy at West Point, or if he were unqualified or uninterested in a military career, perhaps Rensselaer Institute, which had graduated its first civil engineering class in 1835 and had just been reorganized as a three-year college in 1850 under the new name Rensselaer Polytechnic Institute. Another option would have been Union College, which had offered a fully ECP-influenced program under Gillespie from 1845 on. Both of these institutions are located in upstate New York near Albany which, like West Point, lay just 160 kilometers by road from home. All three institutions still oriented their curricula to French engineering education at the time and did in part depend on French texts or their translations as well. Jenney may simply have felt that all three schools were just too close to home. So perhaps he decided to go to the only source that accepted foreign students or those who did not have a previous French education and indulge his wanderlust at the same time.[163]

Jenney's subsequent development followed a less direct route than Eiffel's. Very different experiences influenced his professional path. Before even going to Paris to study, Jenney had taken advantage of his father's connection with the shipping industry to sail to the Philippines in 1849. In Manila, as he later reported, he was particularly intrigued by bamboo birdcages and probably also by southern Chinese bamboo construction, a system that was and still is common, especially for scaffolding and other temporary structures. After rounding Cape Horn, he stopped off in San Francisco, where a building boom was in progress to satisfy the needs of the 1848–49 Gold Rush. Entrepreneurs were importing prefabricated, ready-made housing in wood and iron, especially from Bellhouse out of Manchester in that year, and local contractors were beginning to erect houses using a novel, light-wood construction system to undercut these imports. Their system had originated in the midwestern United States (Cavanagh 2002) and was soon to develop into the popular construction type known as the balloon frame.

Structures that use bamboo and tied connections or nailed two-by-fours follow a specific logic. Their weak stick-like members and simple connections cannot individually withstand forces well, but in quantity they behave 'statistically' and act as though they were monolithic (Peters 1989). The quantity of members and

163. The ENPC had just opened to foreign, non-EP educated students as well in 1851–52, but this may not have been common knowledge at the time.

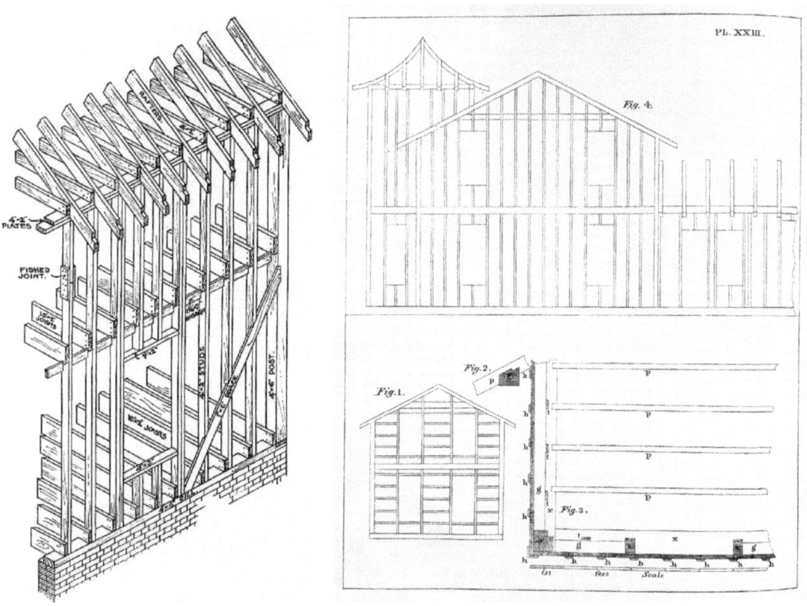

Fig. 124 - The light wood frame evolved in the Mississippi Valley in the mid-nineteenth century and eventually developed into the balloon frame and then the Western platform frame. The many too-thin members and weak nailed connections work together to distribute stresses throughout the structure and act together to resist forces as though the structure were monolithic. (Source: anonymous, public domain) The first published images of balloon framing appeared in Oliver Smith's *The Domestic Architect* in 1854. (Source: Smith, *The Domestic Architect*, 1854, pl. 23)

connections compensates for a lack of quality in the individual component. In fact, quantity created new qualities, like monolithic behavior—for example during earthquakes in the case of the balloon frame or typhoons in Chinese bamboo scaffolding.

Such structures also evoke new formal issues involving rhythm, geometry, and repetition. There is an analogy here to Durand's modular ideas and the repetitive forms of the Revolutionary Architects that Jenney would later encounter in Paris. The young Jenney could hardly have recognized the conceptual issues like the monolithic and structural potential that quantity and repetition offered in iron at the time, but he might have as he matured into the profession when it was revealed to him a few years later in Mary's method coupled with all his other courses. He did at least later claim that bamboo bird cage construction had drawn his attention in Manila. In his later writings they probably reminded him of his own work in iron.

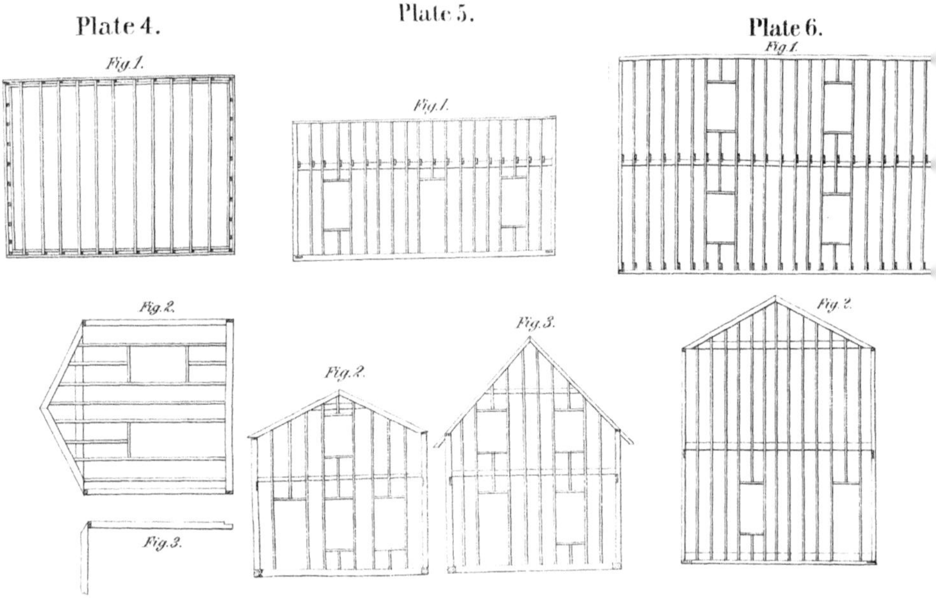

Fig. 125 - The system achieved immediate and widespread popularity through William Bell's *Carpentry Made Easy* in 1858. (Source: Bell 1858, pls. 4, 5, 6)

Fig. 126 - The light bamboo structural system of Southeast Asia is often used as scaffolding in building construction. It is another quasi-monolithic system in which the many weak members and connections act together to absorb energy through deformation, and are thus able to adapt to external forces. (Source: student photograph, Hong Kong, 2000)

Fig. 127 - Two photographs of a temporary bamboo opera house on Lantau Island, Hong Kong, 2000. (Source: Peters)

The sum of Jenney's professional training in the Bruyère tradition, together with Durand's and Mary's design strategies and taxonomic method, Perdonnet's railway construction, Bélanger's theory, Faure's machine construction and industrial organization, Ferry's metallurgy course, and Thumeloup's drawing lessons, his brief year in Mexico on the Tehuantepec Railway, and his return to France in 1858, where he worked on Hiram Berdan's[164] project for a mechanical bakery for the French army,[165] gave him an unusually broad background in design and construction. It certainly contributed to his ability to recognize new structural

164. Berdan is best known for his Sharpshooter Regiment in the United States' Civil War (1861–65), but he was a mechanical engineer and a prolific inventor of weapons and other machinery. He contributed another element to Jenney's path that helped to define the frame of mind necessary to mentor the development of the tall building.

165. Mechanized bakeries, mills, and bread-making generally appear to have been a major concern to military engineers in the earlier nineteenth century, ever since Bruyère devoted a plate to the organization of the 1819 Saint-Lazare prison bakery (Bruyère, *Études,* vol. 2, 1828, Recueil VIII, pl. 13), and Philippe Benoit, mechanical engineer and founding member of the ECP, published his book, *Manuel complet du boulanger, du négociant en grains, du meunier et du constructeur des moulins* in 1829.

possibilities, and in retrospect what he had previously observed on his voyages may have consciously or subconsciously influenced him when he had learned to see at the ECP.

Persistent sources, including Wikipedia, erroneously claim that Jenney worked for Moisant in Paris, but Moisant had not even graduated in 1858.[166] Moisant set up his construction firm in 1866 and began his work on the Menier Mill in 1869. Jenney, who had worked in Paris for Berdan (1858–61) and then had served in the Union army during the Civil War (1861–65), resigning in May 1866, subsequently worked for half a year in western Pennsylvania before establishing his office in Chicago in 1867, just a year after both Moisant and Eiffel had begun as contractors.

With ten years of international experience in machinery and structure and all of his experiential imagery behind him, Jenney began his career. In contrast to our five other graduates, and despite his incomplete education in architecture, he was not only intrigued by construction but also with the formal side of building, and with the expression of structure and material in form and detail.

We do not know whether Jenney knew about Moisant's pioneering structure in the Menier Mill (1869–72), or about Louis-Auguste Boileau's earlier 1854–55 churches Saint Eugéne and Saint Cécile 1854–56 and his 1862–64 Saint Marguerite in Le Vésinet with their early, but not very successful iron structures. He probably did know about Moisant's Bon Marché frame, which was contemporary with the earliest tall buildings in Chicago. James Fergusson had published his pioneering treatise on the acceptance of modern materials in architecture, *History of the Modern Styles of Architecture*, in 1862 and expanded it in 1874. In it he dealt with engineering structures and what he called ferro-vitreous art (Fergusson 1874, vol. 4, 'Book XI Civil and Military Engineering', subheading 'Ferro-Vitreous Art', pp. 556–559). They helped form Jenney's ideas about the relationship between iron structure and form. Fergusson's anti-Ruskinian comments on 'ornamental engineering' as an enhancement of structural and functional reading were novel, as were his views on the influence of the use of glass and iron in the design of the Crystal Palace of 1851. However, like Ruskin, he refused to accept it as architecture due to its lack of solidity. An even better-known influence on Jenney was Viollet-le-Duc's wrought-iron framed house in *Entretiens sur l'architecture* (1872, vol. 2,

166. Bernard Marrey, *Un capitalisme idéal* 1984 seems to have been the source for this misinformed statement, a statement that Fargues repeats (p. 18).

pp. 303–351 and atlas, pls. XXXIV–XXXVI). Jenney may well have been made aware of Moisant and Saulnier's Menier Mill (1869–72) if he had carefully read Viollet's footnote (1872, vol. 2, p. 334). Every building professional in Chicago knew Fergusson and Viollet. In fact, Jenney used the *Entretiens* in his teaching at the University of Michigan in 1877 (Steiner 1984, p. 203, endnote 45). He may have used the original as only the first volume that had no iron structures had appeared in English until that year (in two versions, 1875 and 1877).[167] It certainly influenced the development in Chicago more than Viollet's otherwise more popular *Dictionnaire raisonné*.

Jenney's development, perhaps like that of many others who carried the ECP's pedagogical characteristics abroad to other cultures, crossed borders and mediated between fields in a different way than de Dion or Contamin. His imagery was broader, but he was less philosophically determined than Eiffel and, unlike him and Moisant, not interested in contracting either. His conceptual models were more complex than the others', but his influence on iron structure was just as incisive. He was more successful as a catalyst and mentor than as a designer or builder himself. Jenney, like many others who did not become as well known, went through a complex transition, shifting cultural dimensions and navigating between the professions of architect and engineer, for which his education eminently suited him.

The other ECP graduates we have looked at influenced by example: Polonceau with his truss, de Dion with his impact on frame development, and Eiffel with his tower. These were the clearest examples. In contrast, Jenney in Chicago—like Eugène Flachat a generation before him in France—mentored a whole generation of the most prominent professionals. Among these, the best known are the structural engineers Lightner Henderson, Corydon Purdy, Edward Shankland, and Charles Strobel, as well as the architects Daniel Burnham, William Holabird, Martin Roche, and Louis Sullivan. Together they spearheaded a novel school of thought, the First Chicago School.

Just as Jenney and his Parisian fellow students had admired Perdonnet, Bélanger, Olivier, and Mary, Jenney's protégés clearly revered him. Perhaps for political and public-relations reasons, they elevated him to the role of the inventor of the skyscraper, using as their shaky alibi the fact that Strobel as

167. The second volume of Bucknell's 1875 translation is not dated, but it probably appeared later than 1877.

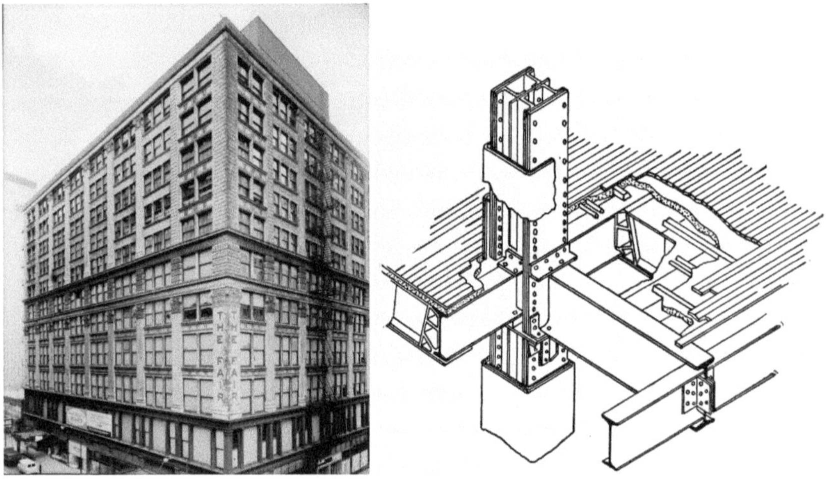

Fig. 128 - View of William Jenney's Fair Store 1891 in Chicago, the second stiff steel-framed building in which the walls and the skin are non-load-bearing. (Source: © HABS IL-1060) Detail of the stiff steel frame. (Source: Freitag, *Architectural Engineering*, 2nd ed., 1902, p. 143)

Carnegie Steel's Chicago representative had convinced him to use the firm's new rolled steel beams in lieu of wrought iron in the uppermost floors of the Home Insurance Building in 1885. However, this misattribution by no means diminishes the significance of Jenney's role and his influence in the development of the steel frame in Chicago, a role as influential as that of Eiffel as developer of the modern 'kit-of-parts' building system in Paris.

Jenney's firm built prolifically after the great fire of 1871 in Chicago, and he was responsible for one of the first truly modern, all-steel tall buildings, the Fair Store, in 1891 (his former employee Daniel Burnham and his partner John Root had opened the very first all steel-framed building, the Rand McNally Building, two years before). He may have no firsts to his name, but his true value to skyscraper construction was his role as the mentor of young professionals.

Moisant

Onésime Armand Moisant (1838–1906) is less known today simply because he never built an iconic structure that bore his name like Eiffel's tower. Nor did he use his name as his trademark as Eiffel did. Although he also built many bridges and part of the elevated section of the Parisian metro system, the bulk of his

Fig. 129 - Armand Moisant, a graduate of 1859, became Eiffel's greatest competitor and the builder of an even wider range of iron structures. The first self-stabilizing iron-framed building, the Menier Mill, can be attributed to him. (Source: Guillet, *Cent ans de la vie de l'École centrale,* 1929, pl. 64)

practice lay in buildings that he carried out in collaboration with architects. The architects were, of course, always named as the creators, despite the fact that it was often the structural systems that were noteworthy. So his name receded undeservedly into the background. He seems never to have published a notice, brochure, or book on his work, nor did he ever present his work or contribute to the discussions of the Société des ingénieurs civils as de Dion and Eiffel frequently did. He had become a member in 1866, the year he founded his firm (Eiffel had joined the Société in 1857, two years after graduating). However, Moisant was no less important among the professional engineer-contractors in France. He introduced the portal frame in his buildings, and many of the structures he was involved with as builder helped establish iron as an accepted architectural material.

He was the son of a prosperous freeholding farmer, and thus came from a very different social background than the others. He finished his secondary schooling in the neighboring town of Tours and then moved to Paris in 1855 to prepare for the entrance exam to the ECP the following year. He attended a respected private preparatory school, the Institution Jauffret (Fargues 2004, p. 11), and was thus not a *barbiste* from the usual preparatory Collège Sainte-Barbe, as Eiffel and many of the other aspirants to the *grandes écoles* had been.

Fig. 130 - Carreau du Temple (1863–64) designed by Jules de Mérindol, erected by Entreprise Joly with Moisant as site engineer. (Source: Wikimedia Commons) Interior (Source: Fernando Javier Urquillo Studio Milou architecture, Feb. 2014, Wikimedia Commons CC BY-SA 4.0)

Like his life, Moisant's professional career has not been studied in any detail. Upon graduation in 1859, twenty-eighth in a large class of 112,[168] he found employment first at the Compagnie des chemins de fer du Nord under Jules Petiet and then at Pierre-François Joly's Constructions métalliques d'Argenteuil. This was the same firm that had trained the future contractor Émile Baudet (grad. 1858), who later built the iron structures of Labrouste's Bibliothèque nationale. Moisant apparently stayed there for five years, where he may have been associated with the structures for the 1867 Exhibition.

Victor Baltard's Halles Centrales was another of Joly's projects that Moisant must have been associated with in some capacity. César Jolly and Théophile Joly's book documents the activities of the firm at the time that Moisant was employed there. In 1864 Moisant was site manager at Jules de Mérindol's Carreau du Temple market, a much smaller structure, but similar in concept to the Halles centrales, and the four market pavilions at the Place du Marché Saint Honoré.

In 1866, the same year Eiffel opened his firm, Moisant bought a small factory building and established himself in manufacturing and contracting. His firm later grew into the Ateliers Moisant-Laurent-Savey. He was the other design-build contractor in this group of graduates, whereby he was more often a collaborator of designers than a designer himself. He evidently preferred solving problems that were posed to him rather than the challenge of defining new problems like Eiffel.

168. The number in Moisant's graduating class has been quoted as 220. But *Les anciens élèves de l'École Centrale 1832–1888* (École Centrale 1889) lists 112.

Fig. 131 - The Moulin Menier, built in 1872, is probably the world's first self-stabilizing iron frame with non-load-bearing walls and portal frames. It is attributed to the architect Jules Saulnier, but the structure was designed, detailed, and carried out by Moisant, who later built many such structures. (Source: Myrabella, Wikimedia Commons, CC BY-SA 3.0) Elevation and section. (Source: *Nouvelles Annales de la Construction*, 1872, p. 13/14 - © from gallica.BnF.fr)

A mere three years after founding his enterprise, Moisant built the structure for which he is rightly best known, the Moulin Menier (1869–72) for the chocolate firm of the same name, designed by the architect Jules Saulnier.[169]

It was a pioneering, self-stabilizing, wrought-iron framed factory building, a major advance on Louis-Auguste Boileau's churches in Paris and Le Vésinet that had both been stabilized by masonry walls like almost all previous cast-iron frames. Not only did the building have non-load-bearing walls, it also used portal frames as stiffening and thus did away with interior diagonals that would have hampered the use of the interior space. Moisant's role in the development of this structure was clearly primary, as he subsequently built many more pioneering iron frames, whereas the architect of record, Saulnier, did not. The advance made by the Menier mill in 1872 is similar to that made by Bruyère's Pont sur le Crou in 1808, two generations before. Both changed the material from cast to wrought iron, which in turn changed the structural concept.

Although Jules Saulnier is said to have studied for a time at the ECP, too, albeit without graduating, this is surely a confusion with the Société's member Ernest Saulnier (*Mémoires et compte rendu* 1864, p. 22), *ingénieur mécanicien de la*

169. It is interesting to note that the industrialists Henri and Gaston Menier, sons of Émile Menier, who had built the mill, later employed the innovative ENPC engineer Armand Considère to design a pioneering concrete-frame factory building (1906–08) with the architect Stephen Sauvestre, the erstwhile architectural collaborator of Eiffel's and graduate of Émile Trélat's school (Hornstein-Rabinovitch 1990, pp. 215/218).

Fig. 132 - Four years after the success of the Moulin Menier, Moisant designed and erected Au Bon Marché. The architects were first Alexandre Laplanche, then Louis-Charles Boileau (see Gerald R. Larson, 'The Architecture Professor', online). The cornerstone had been laid in 1869, but the structure was only begun after the French-Prussian War of 1870–71. It was the largest iron-framed building of its time and had no internal diagonals. (Source: *L'Illustration: journal universel*, 23 March 1872, p. 205 lithograph by Charles Fichot)

Monnaie who had died in 1866 (*Mémoires et compte rendu* 1866, p. 56). At any rate his and Moisant's structure is considered the first true, self-stabilizing iron-skeleton building. It was not a one-off structure, but the beginning of the modern development. After all, it was immediately followed by Moisant's similarly novel iron structure for the world's first department store, Au Bon Marché (1870–87, to which Eiffel contributed a lesser part in 1879). This was the largest iron-framed building of its time and it was entirely stiffend by portal frames. We might therefore consider it to be the origin of modern iron building frames.

Louis-Charles Boileau, the chief architect of the Bon Marché, was active in the use of iron structure, but his designs were not system-oriented, so the systematization of the structure and detail solutions of the department store can be attributed to the engineer. The influence of this structure on Jenney and the Chicago development has not been explored.

Moisant was innovative in many ways and deserves to be better known. Aside from the portal frame, one of his specialties was the construction of suspended glass ceilings with roof structures above that had to solve the intricate problems of rain, condensation, and heat insulation as well as lighting. These were border-crossing questions for which an ECP education was ideally suited. Moisant installed the first of many ceilings of this type in the Bon Marché in the 1870s and the largest in the Grand Palais in 1900. Here, too, the relative contributions of architect and engineer to the design development have not been clarified. The innovative aspects of all these designs were material-physics and structure-determined, both of which were the special expertise of ECP graduates and Moisant's strength.

Unlike Eiffel, whose structures frequently determined the form, most of Moisant's most interesting structures were hidden and subordinate to the architectural form. They were collaborative efforts with architects, including the Menier mill with Saulnier and the Bon Marché with Boileau. He also collaborated with other engineers. These collaborations explain why Moisant's contribution to the systematization of iron construction has been comparatively neglected. Where Eiffel introduced the use of iron visibly to the public in his bridges and the tower, attaching his name to all he built, Moisant did not publish or exploit his name and more often explored the full technical complexity of its use inside buildings.

Fig. 133 - San Marcos Cathedral in Arica (1876; then Peru, now Chile) was one of two projects that Moisant and Eiffel collaborated on. It was a prefabricated iron structure that was manufactured in France and exported. (Source: Heretiq, Wikimedia Commons, CC BY-SA 2.5)

Fig. 134 - Street view of the Grand Palais. The building was largely detailed and built by Moisant for the Paris Exhibition of 1900 with a team of architects. It replaced Barrault's Palais de l'Exposition of 1855. The structure uses a combination of cast and wrought iron, or possibly steel, which has caused expansion and settlement problems. (Source: *Exposition universelle de 1900, album avec 50 photos*. Paris: 1900. Wikimedia Commons).

Moisant's firm became Eiffel's hitherto little-acknowledged design-build competitor in iron construction, but they also occasionally collaborated on projects such as St. Mark's Cathedral at Arica in Chile (Peru at the time) or the Crédit Lyonnais headquarters in Paris, both in 1876.

The Crédit Lyonnais glass ceilings were surely by Moisant. An exhaustive list of his many structures exists, most of which have since disappeared. They ranged widely in size and type: major railway stations;[170] parts of the elevated sections of the northern and southern circular routes of the Parisian Métro (1900–05); bridges,[171]

170. Among these were the Gare de Lyon (1895) and the Gare d'Orsay (1897–98) in Paris, Tours in 1898, and the last in Dijon in 1905.

171. Including the metalwork for what may have been a steel bridge, the three-span Pont du Midi over the Rhone in Lyon (1890–91, demolished 1962). The use of steel is however not certain. The first all-steel bridge (using Bessemer steel) had been the three-arch spandrel structure designed by Émile Cheysson and built by the contractor Henri Joret at the 1867 Exhibition in Paris and moved the following year to span the Vilaine near Sainte-Anne-sur-Vilaine, where it still stands. These were followed by Benjamin Baker and John Fowler's Firth of Forth Bridge in Scotland in 1890, Jean Résal

Fig. 135 - Images of the construction process of the Grand Palais by Moisant. (Source: Picard, *Exposition universelle internationale de 1900*, 1903, vol. 2, two pls. after p. 86, one facing p. 90)

market halls, churches, and theaters, several of them in South America; the iron framing for the prefabricated French Pavilion at the 1876 Philadelphia Exhibition; and three pavilions at the 1878 Exhibition, two designed by de Dion. The third, the Pavilion des Travaux Publics, was modified from his Philadelphia pavilion of 1876. He also built six of the pavilions at the Exhibition of 1889 including the Travaux Publics, the Chilean and the Maison Coloniale pavilions, the grand Central Dome and two of its side pavilions, the Passerelle de l'Alma, the lateral halls of Contamin's Galerie des Machines, and three structures for the Exhibition of 1900 including the entire iron structure for the Grand Palais, an unusually complex structure that suffered from subsidence and differential expansion.

The volume of his total industrial output was far larger than Eiffel's. Moisant concerned himself only with the manufacture and construction of his structures

---

and Amédée Alby's Pont Alexandre III (under Paul Rabel) in 1900, and Paul-Joseph Bodin's (grad. 1871) Viaduc de Viaur in 1902 (constr. Batignolles).

in contrast to Barrault, who had been proud to display the formal qualities of iron structure as valid form, or to Eiffel, who dominated all aspects of his designs. The élan of the earlier ECP engineers who were out to prove the aesthetic viability of exposed and undecorated iron construction was beginning to be suppressed by 1900.

Moisant had other sides to his interests that his ECP education had prepared him for. He established a provident and retirement fund for his employees and their spouses in 1896 which, like de Dion's inclusion of the laborer in his reconstruction of the Bayeux Cathedral tower in 1855–59, may have been influenced by Perdonnet's often-voiced concern for the workman. Moisant also had an abiding family-linked interest in agriculture that stemmed from his origins. Like all the others, he had taken courses in agricultural and industrial chemistry, and from 1878 on he developed an industrially organized model enterprise consisting of two farms in his vast Domaine de la Donneterie, twenty kilometers north of Tours.

Contamin

Victor Contamin (1840–93) was of an even more modest background than Moisant—so modest, in fact, that it was never specified beyond that single adjective, perhaps indicating that those who did mention it considered his origins shameful. His personal life, like that of most of the important ECP engineers, has not been adequately studied. Born in Paris, he apparently spent his first years in Vienna (Comberousse in *Mémoires et compte rendu* 1883.1, p. 742). Despite his presumably disadvantaged background, he managed to attend the Collège Chaptal in Paris, where he excelled as a student, and he studied at the ECP from 1857 to 1860, graduating second in his class of 116 as a mechanical engineer like de Dion. After a sojourn of three years in Barcelona developing gasworks, he briefly designed infrastructure for the public health authority in Paris and then joined the Chemin du fer du Nord in 1863 (where Moisant had briefly preceded him), and he rose through the ranks to become the company's chief engineer by 1890 (Guillet 1929, p. 312).

He was known as an expert in the metallurgy of iron and steel and in the development of railway structures, as well as for his inventiveness in analytical mechanics, and was mentored in that field by Bélanger. He served as *répétiteur*

Fig. 136 - Victor Contamin, a graduate of 1860, was the most socially disadvantaged of the eight students and one of the best in his studies. Although he worked mostly in railways and as a theoretician and teacher, his Galerie des Machines in 1889 was an influential world-record span. (Source: photograph by E. Appert, Wikimedia Commons)

for Louis de Mastaing in *mécanique appliquée* at the ECP from 1865 to 1872.[172] After de Mastaing's death that year, he took over the course himself and taught it for the rest of his life. He became recognized as an expert in material technology after publishing his course manuscript in 1878 as *Cours de résistance appliquée.* As a result, Alphonse Alphand, the general director of the 1889 Paris Exhibition and an engineer himself (grad. ENPC 1839), hired him on the basis of this publication alone to oversee the calculation of all the iron structures, including Eiffel's tower and the Galerie des Machines.

This iconic machine hall is the one project that is most closely associated with Contamin's name. The structural system, the detailing, and the calculation

172. The course was published two years after his death as *Cours de mécanique appliquée à la résistance des matériaux, leçons professées à l'École centrale des arts et manufactures de 1862 à 1872, par M. L. de Mastaing rédigées par M. G. Courtès-Lapeyrat.*

were his. His decisive influence on the structural design is proved by a letter that Contamin sent to an unnamed editor dated April 22, 1889:

> In two or three days I will gladly forward you the note on the Galerie des Machines that you request. I will summarize the considerations that led me to eliminate the tension members and to adopt hinges, and I will provide you with the statistical details that may interest your readers. (Contamin 1889 to unknown, Peters Collection)[173]

I have not found the popular account that Contamin expected from his unnamed correspondent, but he did explain the reasoning behind his decision in his maiden address as president of the Société on January 3, 1890 (*Mémoires et compte rendu* 1890.1, pp. 23–25). It was, he said, a question of economy and especially of safety. Multiple welds would have been needed to manufacture tension members for such long-span arches. Guaranteeing their quality was a cause for concern, and so was the quality of their connection to the arcs, the only bolted connections required in an otherwise riveted structure. Perhaps the memory of the failure of the far smaller marché Saint-Martin the previous decade still reverberated in his mind. The collapse in 1879 had been tentatively attributed to failed tension members (*Mémoires et compte rendu* 1880.1, p. 120).

A further argument that had led him to his three-hinged solution was the difficulty of assuring the stability of the hall's foundations. The subsoil in the Champ de Mars was heterogeneous because the area was peppered with uncharted areas of fill. Moreover, he stated, the longer the span, the better the economy of strengthening the foundations over the use of tension members. Last but not least, the three-hinged structure substantially simplified calculations and the distribution of material. The hinges moved and did not transfer bending moments. This made it easy and therefore inexpensive to accommodate unpredictable stresses caused by foundation subsidence or thermal expansion. Time was of the essence, and any simplification was an advantage. These were all compelling considerations that had everything to do with material technology, statics, manufacture, and construction method, and not at all with space or form.

---

173. …j'aurai le plaisir de vous adresser, d'ici deux ou trois jours, la note que vous me demandez sur la Galerie des machines. Je vous y résumerai les considérations qui m'ont amené à supprimer les tirants et adopter les articulations et vous communiquerai les données statistiques qui peuvent intéresser vos lecteurs.

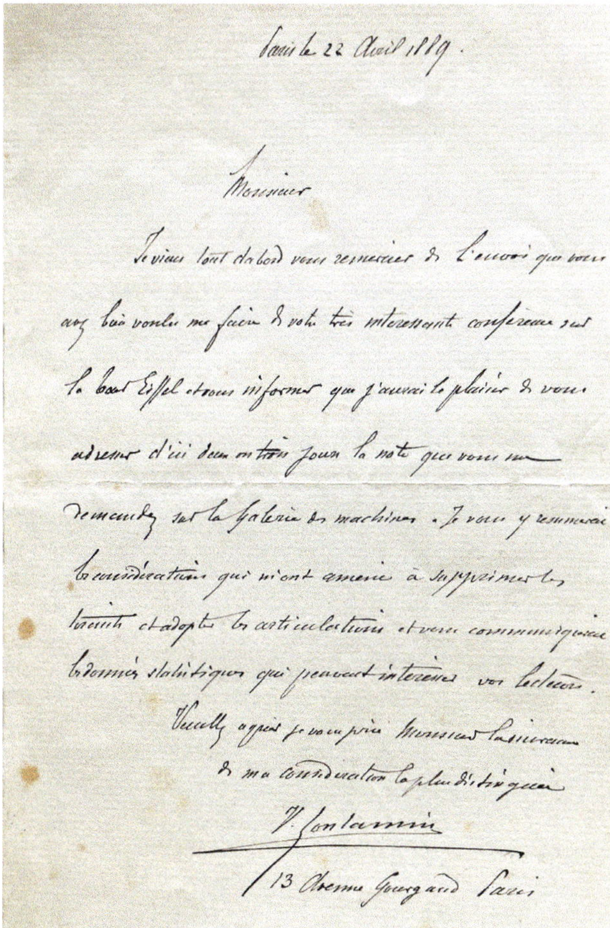
Paris le 22 Avril 1889.

Monsieur

Je viens tout d'abord vous remercier de l'envoi que vous avez bien voulu me faire de votre très intéressante conférence sur la tour Eiffel et vous informer que j'aurai le plaisir de vous adresser d'ici deux ou trois jours la note que vous me demandez sur la Galerie des machines. Je vous y résumerai les considérations qui m'ont amené à supprimer les tirants et adopter les articulations et vous communiquerai les données statistiques qui peuvent intéresser vos lecteurs.

Veuillez agréer je vous prie Monsieur l'assurance de ma considération la plus distinguée

V. Contamin

13 Avenue Gourgaud Paris

Fig. 137 - Contamin's letter dated April 22, 1889, documenting that the three-hinged structure of the Galerie des Machines was indeed his and not the architect Dutert's. (Source: Peters Collection)

Although the Beaux-Arts-trained Ferdinand Dutert had been appointed by Alphand to be responsible for the huge structure (*Mémoires et compte rendu* 1890.1, p. 19) and was thus officially the architect of record, neither the structure nor its detailing nor even the bold idea had originated with him. Like many of the other pioneering iron structures of the period, the idea for the huge, glazed hall had originated in a proposal by an ECP graduate.

The long-span space was first proposed in a meeting of the Société des ingénieurs civils in 1886 in a presentation by Émile Lavezzari on the forthcoming

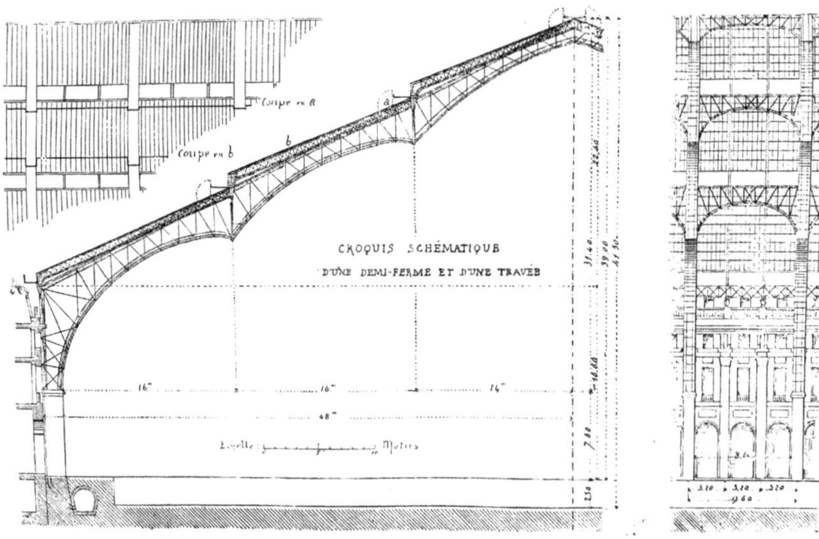

Fig. 138 - Émile Lavezzari's 1886 idea for a Galerie des Machines for the 1889 Paris Exhibition. The hall was to cover the Palais Royale garden with a glazed roof 96 x 230 meters in size. It would be a third larger than the then-record span of 75 meters in London. (Source: *Mémoires et compte rendu* 1886.1, p. 217)

Exhibition. Lavezzari was one of the earlier graduates of the ECP (grad. 1853) to work primarily as an architect. From 1881, like Émile Muller,[174] Viollet-le-Duc, and René Demimuid before him, he served as architectural member of the Société des ingénieurs civils' steering committee (he had joined the Société in 1874). In a meeting of the Société on March 5, 1886, Lavezzari presented and discussed his proposal to cover the 96 x 230.4-meter open space of the Palais Royal with a record-breaking glass-covered hall. The span would have been almost a third longer than the then-record span of 75 meters by William Henry Barlow and Rowland Mason Ordish in the St. Pancras Station in London, to which he compared it. The session is recorded in the *Mémoires et compte rendu* 1886.1, pp. 207–225. As a trained engineer, Lavezzari of course knew what the ambitious idea entailed, what technical difficulties it would pose, and which members of the Société might be involved in the project. He presented a preliminary calculation of materials and costs and credited de Dion's

174. Muller was president of the Société in 1872–73.

Fig. 139 - The clear span of The Galerie des Machines at 113.5 meters was as much a sensation as the Eiffel Tower at the same Exhibition and even more influential in construction. (Source: Exterior, photograph c. 1900 by Hippolyte Blancard [1843–1924]) Interior of the Galerie des Machines 1889. (Source: photograph 1889, Wikimedia Commons) The hinge at the base of the arcs that spanned 113.5 meters. (Source: photograph, 1889–1910)

influence on the design of his trusses, also acknowledging Trélat's expertise on ventilation and lighting (p. 219). He further referred to the innovative glass-maker Léon-Alfred Appert (grad. 1856), and to Charles Herscher, an expert in mechanical ventilation and member of the Société who was not an ECP graduate. He illustrated his proposal with a cross section (Fig. 138).

We do not know what Lavezzari's relationship was, if any, with Dutert, but he was Contamin's 'cher collègue', as the latter stated in his eulogy at Lavezzari's funeral in July 1887 (*Mémoires et compte rendu* 1887.2, pp. 251–253). How close the relationship really was personally and professionally is not recorded, but both were officers of the Société, Lavezzari a council member and Contamin vice-president, and Contamin had presided at the meeting during which Lavezzari had presented his idea. Lavezzari died on July 13, 1887, well before the designs for Contamin's even more ambitious 113.5 x 429-meter hall

Fig. 140 - Rotating an arc segment into position. (Source: Monod, *Exposition universelle de 1889*, p. 240). Joining the upper hinge. (Source: *L'Exposition de Paris 1889*, vol. 1, p. 4)

on the Champ de Mars were finalized. At 113.5 meters span[175] and based on Contamin's expertise in theory and the construction of railway structures, the arcs were to be only slightly greater than Lavezzari's, but the hall was almost twice as long. Had the two men discussed or perhaps even begun to develop the idea together? Contamin's final project was certainly different. Lavezzari had, for instance, not shown any proposal for how to deal with the lateral thrust of the great roof trusses, but his idea for an extremely long-span, glazed hall had borne fruit. Tenders for Contamin's great arcs, then still at 120 meters and not yet fixed at the definitive span of 113.5 meters, had only been invited in May, shortly before Lavezzari had died, and those for the foundations after that (*Mémoires et compte rendu* 1887.1, pp. 670/671). The first arc was only erected in April 1888 (*Mémoires et compte rendu* 1888.1, p. 426).

175. The span size is taken from Eiffel, Contamin, and Fouquet 1889, p. 11. The total length of 429 meters is taken from the article by P. Legrand in *La Science Illustrée,* 1889, 1er sem., p. 358. Legrand gave the arc span as 110.6 meters, and this was repeated in several of the popular accounts issued during the Exhibition.

Contamin mentioned but did not trumpet his contribution to the Exhibition building. He was an exceedingly humble man, and his border-crossing role as material theoretician, teacher, and practitioner in railway technology and building structures—aside from this one hall—was not conducive to public fame outside the field, like Eiffel's work, nor was it publicized by his students, like Jenney's. However, his service to both the railway and the construction industry was well recognized during his lifetime and highly valued in the profession. He supported the interconnection between the development of construction materials, structural systems, and detailing as no other.

Concluding thoughts

It is not a simple matter to separate out those aspects of the ECP's pedagogy that enabled the development of modern construction: the whole didactical concept contributed. The content of a curriculum is only part of an educational process. It must also be communicated by an appropriate method contingent on an intellectual concept and backed by a school of thought. The founders developed their concept based on Monge's *vulgarisation* pedagogy, and it took the form of the incorporation of science *into* technology, not as its application *to* technology. The faculty employed a tripartite experiential pedagogy to achieve this: an intellectual approach (using both spoken and written communication), as well as visual and haptic forms. This distinguished the ECP's curriculum from all other models of engineering education that relied solely on practice or on the intellectual communication of subject matter.

The fusion of theory and practice that the new didactical system of *vulgarisation* made possible restored the field of construction to what Perronet and Monge had begun and carried it forward into the industrial era. This enabled the development of iron construction and its corresponding industry, helping to move it from an experimental to an industrialized phase. Scientifically based technologists created our modern infrastructure in a process that was unique in the history of construction. The ECP changed the whole culture of engineering theory from an elitist effort to emulate science into a down-to-earth technological field that crosses the border between science and pragmatism. It institutionalized *vulgarisation* as a valid intellectual concept. Although scientists and theoreticians were slow to accept this hybrid form of thinking, and some never did, practicing engineers could be both theorists and pragmatists at the same time. The ECP proved that the two modes of thought could be compatible.

The basic characteristic of the pedagogy is that the school trained its graduates to see, as the chapter 'The conceptual issues' defined seeing. All the many and subtle aspects of the program instilled the ability to see. They were linked through the collaborative chemistry between the teachers that combined theory and practice in each of them in different ways. The intellectual dialogue that they conducted with one another and with their students supported their efforts. The dialogue was strongest in the earlier years, but it persisted until the last of the original faculty retired. Many of the critical steps in this process were details in the technological sense. They were small, but their impact was great. The net result was modern technological thinking (Peters 2000), a development that can be traced back to Monge's revolutionary educational convictions. The salient points of this process were:

1. Upgrading *vulgarisation* from a pedagogical method to a didactical principle. A hierarchical prioritization of the means to communicate information replaced the simplification of information, and practical examples replaced analogy through comparison. The first prevented the bowdlerization of information and the second rendered each theoretical postulate immediately tangible. This change epitomized the concept of the school, and it was implemented in part by teaching theory through objects. The process involved a close pedagogical collaboration between teachers like the organizer Perdonnet, the designer Mary, and the analyst Bélanger, and by practitioners outside the school who functioned as mentors, like Flachat or Petiet and other graduates who in turn hired the younger ones. The linkage of theory to practice intensified this process through apprenticeships and workshops during study and in a controlled post-graduate phase through practical work on the chance ambiguities that actual problem-solving presents, including what to theorists were secondary and thus negligible factors. As a result, the ECP graduates learned to see and enjoyed a common culture of border-crossing ability and the innovative capacity to integrate analytical and creative thought.
2. Bélanger's work successfully overcame the long-recognized conundrum that theory is not directly applicable to practice. He recognized that it must be translated into technological terms and incorporated into the requirements of practice without being falsified through simplification. It had to be made useful through a process of *vulgarisation*, and he accomplished this by stressing linguistic and analytical precision. But he qualified it, too, by his broad view of theory, his practical bent that even embraced approximation, and his realization that statics can be viewed as a special case in dynamics.

3. Colladon intensified this translation process by introducing the opposite process: that empiricism nourishes theory—in other words, that there is yet a further intellectual level of learning-through-doing in a workshop setting, and this introduced a significant haptic component into the pedadogy.
4. Perdonnet contributed a further aspect with his anti-elitist position that skilled and experienced craftsmen had a great deal to contribute to education from their manual experience. He accomplished this inclusion by introducing Pestalozzi's educational concepts to the school in many forms, including the stress on *Anschauung* (observation that leads to seeing), the importance of physical work, the involvement of heart, hand, and mind, as well as the mentoring principle that Monge had also practiced. He was supported in this by Eugène Flachat's unofficial post-graduate mentoring.
5. When Bélanger split mechanics into theory and machine construction in 1838, he rendered both relevant as design tools for what would later become iron construction at a larger scale in building. This pedagogical split enabled the teachers to actively re-link theory to practice and to make the connection between small-scale and large-scale design that led seamlessly from courses and exercises at the ECP to practice in the building world, from analytical mechanics through practical machine construction to design practice, both of machinery as a small-scale issue and structure at a large scale. This occurred initially in the railway industry as demonstrated by the two plate volumes that Cornet and Jacquin produced for Perdonnet's lecture course.
6. The guiding principle that all students were obliged to take general courses across all specialties, including from the other majors that the school offered. This followed Dumas's dictum *la science de l'ingénieur est une* and it fostered the incorporation of science and practice by teaching theory through objects. The distinction that the ENPC engineer Nordling made between theory and practice in his critique of Contamin in 1866 highlights the flexible advantages of the border-crossing education that the ECP students enjoyed. The new school of thought that they propagated dissolved the boundary between the theory that analyzes and the synthesis that is creative design.
7. The school attached great importance to the tripartite pedagogy: haptic-*manual* and impalpable-*visual* experience coupled with the traditional *intellectual* mode of understanding through spoken lectures and reading. These are the three main and inseparable aspects of a learning process. It was a system that Pestalozzi had originally introduced into juvenile education, and the ECP

implemented it by means of work on actual machinery in the school and outside. Colladon introduced this early on, possibly influenced by Leblanc or Beckmann. Olivier seconded it through his use of models in teaching descriptive geometry, influenced perhaps by Polhem's example. Thumeloup, supported by all the others, schooled visual and haptic understanding through observation and manual recording. This led to an intensive schooling of three-dimensional and visual literacy in all the courses through hand-eye-brain coordination.

8. Thumeloup's art-based interpretation of drawing as a visual, analytical language of the engineer overlapped with Oliver's scientific or analytical approach to the geometrical construction of a visual understanding of space as the handwriting of the engineer. It was a novel conflation of two ways of learning to see that implemented the ECP's pedagogical concept. Together, and supported by all the other teachers, they defined its applicability to a broad range of uses, from structural design and Fernique's photography to industrial espionage.

9. Mary's special form of *vulgarisation* adapted Durand's hierarchical method and modified it from formal composition to functional and structural organization. The fixation on *object* became a focus on *process*. The means he used were function, materials, and structure as they had been exemplified in Bruyère's teaching. By adapting and melding the Durand and Bruyère models, Mary created the field of design as we know it today and taught it as a distinct engineering activity. He shifted from the idea of on-site construction to industrial prefabrication and on-site assembly, whereby both Bruyère's and Eck's influence is evident in Mary's use of many of their examples. The detailed synergy between Mary's method and Bélanger, Perdonnet, Walter, Ferry, Faure, and Callon's instruction in railway and machine design and construction transformed the method from an abstract architectural theory into the technology of construction. Supported by the example of the railway industry, this transformation encouraged the development of system thinking in structural design and in industrial development. The system approach to design became especially clear in the invention of new cross sections like the I-beam that Flachat began to develop, the serial production of iron components and of standardized elements like Polonceau's truss, and standardized connection types in all the large manufacturing and contracting firms. Not only did Mary influence structural and mechanical engineering education, but his reinterpretation and expansion of Durand's method also entered architectural pedagogy through Trélat's pioneering École Centrale d'Architecture and many other fields of design as well.

10. The esprit de corps of the Société des ingénieurs civils de France (originally an ECP foundation) and the ECP's Association amicale des élèves de l'École Centrale engendered a common intellectual base for the profession. The two societies were both encouraged from the beginning by Lavallée to establish a mutually supportive network, and they generated a new, complex 'school of technological thought'.
11. Finally, the breadth of the program covered virtually all the then known aspects of engineering technology, from mining and metallurgy to public and private construction, contracting, transportation, urban infrastructure including sanitation, to industry (gas and later electricity, textiles, glass, ceramics), and chemical as well as agricultural engineering. This not only gave the ECP's graduates a solid grounding but also a flexibility in all aspects of nineteenth-century technology, and it allowed them to create and expand their chosen fields. It gave the students a well-rounded education, one that opened them up to the definition of new questions for the new problems that arose in industry and construction, rather than merely providing solutions for existing ones.

By mid-century the necessity of incorporating all these features in an engineering education had become self-evident and natural, and Comberousse would write about the tripartite teaching method in the year of the fiftieth anniversary of the school in 1879:

> ...today the whole world knows that lectures do not suffice in the applied sciences. They must be supplemented by frequent exams, repeated experiments and workshops, drafting exercises that naturally make one aware of the overall concept that one has previously understood intellectually, in each and every detail. (Comberousse 1879.1, p. 27)[176]

Much of this truth has long since been forgotten by modern engineering educators. Our current ignorance of the didactical concept of *vulgarisation* and lack of a linkage between historical awareness, manual skill, technical competence, creative playfulness, and daring vision has impeded the ability to see, and has led

176. ...tout le monde sait aujourd'hui que, dans l'étude des sciences appliquées, les leçons orales ne suffisent pas. Il faut qu'elles soient accompagnées de fréquents examens, d'expériences et de manipulations répétées, de travaux graphiques qui, par une réaction naturelle, fassent comprendre dans tous les détails la solution générale dont l'intelligence a d'abord embrassé l'ensemble.

to a noticeable slowing of innovation. There have certainly been innovations in engineering since the mid-nineteenth century, but advances in structural and material technology have mostly been incremental and hardly as massed to the degree that the mid-nineteenth century experienced. This is not destiny, as I wrote in the introduction, but a lack of flexibility and of an ability to 'translate' knowledge from fields outside of engineering.

These necessary conditions for innovation were proven both by the ECP's graduates in the mid-nineteenth century and by the analogous concept in industrial design adopted by the Bauhaus in the early twentieth. The Bauhaus's founder Walter Gropius emulated the ECP's border-crossing and integrative pedagogical model in 1919. We do not know at present whether he was even aware of the ECP's original pedagogical model, but it had by then become inherent in design teaching. Mary's method had long gained widespread acceptance in architecture programs. The Bauhaus only existed for fourteen years; the ECP's most influential phase lasted about double that. However, the architectural and industrial design prowess that it developed through the inventiveness of its many graduates throughout the twentieth century rivalled the novelty of the ECP engineers' iron structures. Together they profoundly affected and continue to affect the technology and the physical appearance of our world.

Barrault, Baudet, Contamin, de Dion, Eiffel, Jenney, Moisant, Polonceau, and their many colleagues were unfettered in their industrial prowess and technical inventiveness (see Guillet 1929, pp. 311–317). The limitations of historicism as well as engineering and architectural formalism that clouded the vision of many engineers and architects who preceded and followed them did not impede their ability to see, but at the same time their professional development was profoundly influenced by the history of French technical education. Of the five who lived long enough to develop mature careers, Eiffel dominated the structures he built as designer-contractor, Contamin became a mediator between teaching, theory, material technology, structure, and detailing, Baudet and Moisant acted as innovative collaborating contractors, and Jenney primarily worked as a mentor and a technically biased designer-architect. This is not to forget the other three—Polonceau and de Dion, who pioneered the new combinations of theory and practice and invented new model products, and Barrault, who used his knowledge of the iron industry and metallurgy to proclaim the aesthetic aspect of structure with an engineer's self-assuredness and with even more authority than the architect Labrouste.

Bruyère had designed and built the world's first true iron truss in 1808, and he established the advantages of malleable wrought iron over brittle cast iron in construction. However, the material was then still far too expensive for large-scale use, so it was another half century until Moisant and Eiffel began to make the modular three-dimensional frame in wrought iron a viable option in bridge or tall building construction. Jenney and the group surrounding him followed a similar pattern as soon as steel became an affordable building material around 1890. They developed the modern steel frame with a curtain wall step by step as a suitable option for tall office buildings that required flexible, open plans and natural lighting. They built on Eiffel and Moisant's system approach and recognized and exploited the advantages that monolithic connections in steel offered for tall building structures.

The professional lineage that ran from Bruyère and Durand via Polonceau and de Dion to Eiffel, Moisant, Contamin, and Jenney determined the development of system design and construction in iron and steel. Their educational lineage stretched from Perronet, Beckmann, Monge, Prony, Rondelet, Durand, and Bruyère to Leblanc, Perdonnet, Colladon, Bélanger, Walter, Ferry, Faure, Thumeloup, Olivier, and Mary. It established a professional perspective that we now take for granted.

The role played by cultural factors also contributed substantially to the development. These heterogeneous factors included industrial mentoring, private and government participation, educational or socio-economic philosophy, and religion—which I only hinted at, the democratization of education, the development of machine and railway construction, the new building materials, and the accumulated experience of all who stood in the background.

Lavallée, the promotor and catalyst of the ECP's unprecedented success, was so devoted to his creation that he remained its director for five years after he transferred the school to the state in 1857 to assure its sustainability. His gift wisely stipulated that the school was to be advised by a committee of former students (the *conseil de perfectionnement*) and that any profit made by the school was to be devoted solely to its further development (Adolphe Bitard in *La Science illustrée* 1888, vol. 1, 1er sem., p. 82). While the French government subsequently guaranteed the continued existence of the school as a successful institution beyond the founders' tenure and Lavallée's lifespan, it did little to inspire any further radical innovations in the scientific-technical-industrial curriculum. This curriculum continued to evolve organically, for instance in the field of reinforced concrete, but no longer revolutionarily—that is, it enabled no fresh quantum leaps.

The original, cutting-edge innovations that to a large part created our modern civil engineering thinking had been entirely due to the initiative of inspired and dedicated individuals. For the ENPC it was Perronet who first combined theory and practice in engineering education. For the EP it was the towering intellect of Monge and his colleagues whose original pedagogical mission was first diluted by Napoleon in 1804 and then destroyed in 1816, as Pothier wrote (1887, p. 26), for military and political purposes. For the ECP a decade later, it was the four founders: Lavallée, Dumas, Olivier, and Péclet and their cast of supporting educators, especially Bélanger, Colladon, Perdonnet, and Mary.

However, the French government did continue to foster the development of innovative iron construction beyond Lavallée's intention of integrating education and private enterprise. It did this with a continuing International Exhibition program from decade to decade (1855, 1867, 1878, 1889, and 1900) to which many ECP graduates contributed, culminating, as far as iron is concerned, in Contamin's Galerie des Machines and the Eiffel Tower in 1889, and Moisant's Grand Palais in 1900. This only partly intentional, and perhaps even serendipitous synergy between private and public interests formed a unique and robust collaborative environment that secured the international influence of French iron construction into the early twentieth century.

Bibliography

Archives

The archivists and librarians of the following institutions have been very helpful and forthcoming with information on documents in their collections: Archives CentraleSupélec, École Polytechnique, Bibliothèque centrale. They are thanked in the acknowledgements and in the relevant footnotes.

The following material was used in the text or illustrations and is credited in the appropriate captions, text, or footnotes:

Archives de la Bibliothèque Sainte-Geneviève

Figs. nos. 54 (bsg MS 4273_0100); 59 (bsg MS 4273_0020r, bsg MS 4273_0021r, bsg MS 4273_0022r, bsg MS 4273_0023r, bsg MS 4273_0024); 61 (bsg MS 4273_0019); 64 (bsg MS 4273_0017), all open source

Archives du Collège Sainte-Barbe at the Archives de Paris, Direction des Affaires Culturelles

D50Z_275 and D50Z_280

Archives de l'École des mines / Mines ParisTech

Minutes of the meeting of the Conseil de l'École Royale des Mines: 09.11.1824 (p. 373); 7–9.10.1824 (p. 374). 16.11.1824 (p. 377); 02.05.1826 (p. 12); 09.05.1826 (p. 15); 08.05.1827 (p. 119) and 15.05.1827 (pp. 121. 124–125); 15.05.1827 (p. 128)

Archives de l'École centrale des arts et manufactures de Paris (1820–2016) (Archives nationales de France)

Fig. 79: 20170270/7

Archives de l'École nationale des ponts et chaussées

https://www.ecoledesponts.fr/sites/ecoledesponts.fr/files/documents/eleves_corps_civils_libres_courspreparatoires_1744_1930_fusionnee_maj16.01.2020.pdf (mise à jour janvier 2020) / (Bruyère, Louis). 1799–1810. 'État des divers rapports et mémoires relative à l'art de l'ingénieur et recueillis par M. Bruyère. Ponts en fer.' *ENPC mss.233 Tome 24BIS* / Eisenmann, Joseph. 1801–02 'Cours de mécanique appliquée. Programme (A)'. ENPC ms. 2002–3 / Figs. nos. 9 and 85 with permission. Also Bruyère ENPC ms. 2436 and Becquey de Beaupré ENPC ms. 2134.

Archives nationales de France

'A Messieurs des Membres du Conseil municipal de la Ville de Paris', Louis Bruyère F/1. (accessed online 17.08.2019) author: Navier / 'Rapport à l'Empereur sur l'état actuel de l'enseignement professionnel et sur les moyens de le perfectionner', author: Rouher, in: *Bulletin administratif de l'instruction publique.* Vol. 14 no. 162, June 1863, pp. 111–115 online: https://www.persee.fr/doc/baip_1254-0714_1863_num_14_162_13836 / Centre Historique des Archives nationales, *Atlas des Inspecteurs généraux des Édifices diocésains 1853* F/19/*/1822-F/19/*/1825, F/19/*/1940-F/19/*/1941) / Francearchives Portail national des archives. Inventaire-F13/10/A-F/13/139 Bâtiments civils, consulted online 17.12.2020.

Bibliothèque de Genève, Geneva, Switzerland

Fig. 31 (inventory number 009), open source

Bibliothèque nationale de France

Perdonnet, Auguste. 1831. *Cours sur l'exploitation des mines et sur les Chemins de fer. 1831–1832.* BnF V-14297 (text) and V-4804 (atlas) / Perdonnet, Auguste. 1833. *Leçons sur les chemins de fer 1833–34.* BnF 14298 (text) / Fig. 70 (HD-1019 (3)-FT6)

Centre de documentation et de recherche Pestalozzi, Yverdon-les-Bains, Switzerland

Johann H. Pestalozzi: Sämtliche Werke und Briefe, vol. 10. *Briefe aus den Jahren 1816 und 1817.* 1968, letter numbers: 4276; 4667; 4671; 4293

Institut national de la propriété industrielle (INPI), Courbevoie, France

Fig. 75 (Halbou Pat no. 8040 / 1BB8040), with permission

Library of Congress, Washington, DC, USA

Fig. 128 (HABS IL-1060), open source

Musée d'art et d'histoire, Geneva, Switzerland

Fig. 96

Structurae.net

Figs. 84 b (Pont de la Concorde) and 113 a (Passerelle Eiffel)

Union College, Schenectady, NY, USA. Art Collections and Exhibitions

Figs. 47 a, b, c, with permission

United States Patent and Trademark Office, USA

Fig. 93 (US Pat. 2064), open source

Lavallée family archive

Fig. 12, portrait of Alphonse Lavallée, and Fig. 13, portrait of Pierre Laurens, information by Pierre and Jacques Lavallée, provided by Olivier Dibos, great-grandson of Alphonse Lavallée. Lavallée, Pierre, manuscript diary, details / Lavallée, Jacques. 1983 'Le coton de Louisiane et la fondation de l'École Centrale', in: *France Louisiane*, communicated by Olivier Dibos 31.12.2019 / Luquet, Gérard Christian, Agnès Jacotot and Olivier Dibos. 2015. 'Arbre généologique partiel et simplifié de la lignée des trois Alphonse Lavallée'. Folding inset of the article *Éléments de biographie sur la vie et l'œuvre d'Alphonse Francis Marie Martin Lavallée (1873–1954) dit Alphonse Lavallée, Alexanor; revue des lépidoptéristes français* 26 (6) 2014

Roussel family archive

Detailed information from Benoît and Christian Roussel on the development of the Entreprise Roussel and the portraits of their ancestors. Roussel, Christian, 2015. *Roussel. Entreprise de Serrurerie et Constructions Métalliques 1797–1880*. manuscript / Fig. 59 (Jean-François Roussel), Fig. 61 a, b (François Roussel and Théodore Roussel), Fig. 63 (Henry Roussel)

Tom F. Peters collection

Various ephemera, photographs, book covers, and the following twenty manuscripts:

ALS Charles Frénet, Lyon to Théodore Olivier, Sept. 8th. 1834 / ALS Victor Contamin to unknown editor, Apr. 22nd. 1889 / manuscript dedication by Auguste Perdonnet to Marquis Charles de La Valette, undated (1865) in Perdonnet, *Traité élémentaire des chemins de fer* 3rd. edition 1865, vol.1. / manuscript: John G. James 1982 / three PhD dissertation manuscripts: William Addis 1986, Ulrich Pfammatter 1995, and Aleksandra Kosykh 2021 / thirteen manuscript notebooks: Montero, Antonio. 1847. *Cours de Chimie de Mr. Barral. Sainte-Barbe. 14, 8bre 1847.* Collège Sainte-Barbe / Montero, Antonio. 1848–49. *Cours d'Hygiène, Professeur M. Doyère. 1ere Année.* ECP / Montero, Antonio. 1848–49. 1e année, Cours de Géom. Descriptive, Professeur M. [Théodore] Olivier, [course] ECP / Montero, Antonio. 1848–1849 *École Centrale des Arts et Manufactures. 1e Année d'Études. Cours de Géom[étrie] Descriptive. Professeur M. Olivier. Nom de l'Élève Montero. Salle 9.* [exercises] ECP / Montero, Antonio. 1848–49. *Cours de Chimie. Professeur Cahours. Salle 9. 1ère Année.* ECP / Mendiola, Ignacio. 1849–50. École Centrale des Arts et Manufactures. Année 1849–1850. 1ère Année d'Études. Cours de Construct[io]n de machines. *Prof. M. Faure. Nom de l'Elève Mendiola. Salle 9.* ECP / Montero, Antonio. 1849–1850 [École Centrale des Arts et Manufactures… 1e Année d'Etudes. *Cours*] *de Croquis à la main. Professeur M. Thumeloup. Nom de l'Élève Montero*. ECP / Montero, Antonio. 1850–1851. École Centrale des Arts et Manufactures. 1e Année d'Études. *Cours d'Architecture. Professeur M. Mary. Nom de l'Élève Montero. S[alle] 5. Constructeur*. ECP / Montero, Antonio. 1850–1851. École Centrale des Arts et Manufactures. 2e Année d'Études. Cours de Turbines. *Professeur M. Faure. Nom de l'Élève Montero. Constructeur*. ECP / Montero, Antonio. 1851–52. École Centrale des Arts et Manufactures. 3e *Année d'Études. Cours de Métallurgie (du fer). Professeur M. Ferry. Nom de l'Élève Montero. Constructeur*. ECP / Montero, Antonio. 1851–52. *Cours de Chimie analytique. Prof. Péligot. 3ème Année. Constructeur, Salle 13.* ECP / Carimantrand, Jules, *Cours de métallurgie, M. Ferry* [1856/57] / Terrier, Paul. 1861–62. École centrale des Arts et manufactures, Année scolaire 1860 1861. *2e année d'études. Album de croquis de 2e et 3e années. Nom de l'Élève: Paul Terrier–Salle 5 No. 11* Manuscript, subtitled: *Travail des Vacances de 1861*, but with multiple other sketches and notes for further years. ECP / Lafont, C. École Centrale des Arts et Manufactures, Année 1887 1888 2e *année d'études. Cours de Mécanique Professeur Monsieur Levy Nom de l'Élève C. Lafont, Salle 10 No. 8.* ECP.

Contemporary original publications

Albaret, Auguste. 1861. 'Calcul des arcs métalliques à section constante et à section variable soumis à des charges uniformément réparties suivant l'horizontale', in: *Annales des Ponts et Chaussées* 1861.2, pp. 90–124.

Albaret, Auguste. 1862. 'Mémoire sur le calcul des arcs métalliques dans le cas des grandes surcharges', in: *Annales des Ponts et Chaussées* 1862.2, pp. 205–363.

Allgemeine Bauzeitung mit Abbildungen. Redigiert und herausgegeben von Prof. Christ. Fried. Ludwig Förster... 119 vols. 1836–1918 Vienna: Verlag von L. Förster's artistischer Anstalt in Wien.

Anon. 1863. Obituary of James Bunning in *Gentleman's Magazine* 1863, pp. 799–800.

Arago, François. 1828–35. 'Notices scientifiques sur les machines à vapeur', in: *Annuaire du Bureau des longitudes*. Paris: Bachelier.

Ardant, Paul-Joseph. 1840. *Études théoriques et expérimentales sur l'établissement des charpentes à grande portée.* Metz: Lamort.

Babbage, Charles. 1833. *On the Economy of Machinery and Manufacture.* 3rd edition. London: Charles Knight. (The first and second editions are dated 1832).

Barlow, Peter. 1817. *An essay on the strength and stress of timber, founded upon experiments performed at the Royal Military Academy, on specimens selected from the Royal Arsenal, and His Majesty's Dock-Yard, Woolwich: preceded by an historical review of former theories and experiments, with numerous tables and plates. Also an appendix on the strength of iron, and other materials*. London: J. Taylor.

Barrault, Alexis. See also Flachat, Eugène.

Barrault, Alexis and Gustave Bridel. 1857. *Le Palais de l'Industrie et ses annexes. Description raisonnée du système de construction en fer et en fonte adopté dans ses bâtiments*. Paris/Liège: E. Noblet.

Beckmann, Johann. 1777. *Anleitung zur Technologie, oder zur Kentniss der Handwerke, Fabriken und Manufacturen, vornehmich derer, die mit der Landwirthschaft, Polizey und Cameralwissenschaft in nächster Verbindung stehn. Nebst Beyträgen zur Kunstgeschichte*. Göttingen: Wittwe Vandenhoeck.

Beckmann, Johann. 1780–1805. *Beyträge zur Geschichte der Erfindungen.* 5 vols. Leipzig: Kummer.

Beckmann, Johann. 1806. *Entwurf einer algemeinen Technologie*. Göttingen: Röwer.

Bélanger, Jean-Baptiste. 1840–42. *École centrale des arts et manufactures. Cours de mécanique générale et de théorie dynamique des machines. 1er (-3e) partie.* Paris: Lith. Clouet.

Bélanger, Jean-Baptiste. 1842. *Résumé de leçons de géométrie analytique et de calcul infinitésimal.* Paris: L. Mathias.

Bélanger, Jean-Baptiste. 1842–55. *Notes sur le Cours d'hydraulique.* Multiple ENPC/ECP transcripts, especially here the ECP editions 1849–50 and 1854–55.

Bélanger, Jean-Baptiste. 1847. *Cours de Mécanique ou Résumé de Leçons sur la Dynamique, la Statique et leurs Applications à l'Art de l'Ingénieur. Première partie. Dynamique et statique générales. – Hydrostatique.* Paris: Carilian-Goeury & V. Dalmont / L. Mathias.

Bélanger, Jean-Baptiste. 1864. *Traité de Cinématique*, and on the half-title: *Série de Traités sur la Mécanique et sur les Théories de ses Principales Applications à l'Art de l'Ingénieur I. Cinématique.* Paris: Dunod / Gauthiers-Villars / Eugène Lacroix.

Bélanger, Jean-Baptiste. 1864. *Traité de la Dynamique d'un Point Matériel*, and on the half-title: *Série de Traités sur la Mécanique et sur les Théories de ses Principales Applications à l'Art de l'Ingénieur II. Dynamique d'un Point Matériel.* Paris: Dunod / Gauthiers-Villars / Eugène Lacroix.

Bélanger, Jean-Baptiste. 1866. *Traité de la Dynamique des Systèmes Matériels,* and on the half-title: *Série de Traités sur la Mécanique et sur les Théories et ses Principales Applications à l'Art de l'Ingénieur III. Dynamique des Systèmes Matériels.* Paris: Dunod / Gauthiers-Villars / Eugène Lacroix.

Bell, William E. 1858. *Carpentry Made Easy; or, the Science and Art of Framing, on a New and Improved System. With specific Instructions for building Balloon Frames, Barn Frames, Mill Frames, Warehouses, Church Spires, etc. Comprising also A System of Bridge Building; with Bills, Estimates of Cost, and valuable Tables. Illustrated by thirty-eight Plates and near Two Hundred Figures.* Philadelphia: Howard Challen.

Benoit, Philippe and Jean-Sébastien Julia de Fontanelle. 1829. *Manuel complet du boulanger, du négociant en grains, du meunier et du constructeur des moulins.* Paris: Roret.

Bernoulli, Christoph. 1829. *Vademecum des Mechanikers oder praktisches Handbuch für Mechaniker, Maschinen- und Mühlenbauer und Techniker überhaupt nach Robert Brunton.* Stuttgart/Tübingen: Cotta.

Bernoulli, Christoph. 1834–35. *Elementares Handbuch der industriellen Physik, Mechanik und Hydraulik.* 2 vols. Stuttgart: Cotta.

Birlë, Alcide Cauvet, Gustave Eiffel, Émile Trélat, Léon Appert and Charles Talansier. 1889. "Nécrologie de Émile Muller, extrait du bulletin de l'Association des anciens élèves de l'École centrale des arts et manufactures, année, 1889–1890," pp. 59–74.

Bonaparte, Charles-Louis Napoléon. 1844. *Extinction du paupérisme.* Paris: Pagnerre.

Borgnis, Giuseppe Antonio. 1819. *Traité complet de mécanique appliquée aux arts, contenant l'exposition méthodique des théories et des expériences les plus utiles pour diriger le choix, l'invention, la construction et l'emploi de toutes les espèces de machines; 1818–20*. Paris: Bachelier.

Borgnis, Giuseppe Antonio. 1821. *Théorie de la mécanique usuelle, ou, Introduction à l'étude de la mécanique appliquée aux arts: contenant les principes de statique, de dynamique, d'hydrostatique et d'hydrodynamique, applicables aux arts industriels, la théorie des moteurs, des effets utiles des machines, des organes mécaniques intermédiaires, et l'équilibre des supports.* Paris: Bachelier.

Bresse, Charles. 1859–60. *Cours de mécanique appliquée professé à l'École Impériale des Ponts et Chaussées.* 2 vols. Paris: Mallet-Bachelier.

Brunet, François. 1809. *Dimensions des fers qui doivent former la coupole de la halle aux grains, calculées pour l'exécution du projet de M. Bélanger, architecte des monuments publics, on y trouve une méthode nouvelle pour rendre les caissons réguliers dans un voûte sphérique extradossé, d'une autre méthode générale, pour faire la coupe des pierres par une même opération.* Paris: Firmin Didot.

(Bruyère, Louis). 1799–1810. 'État des divers rapports et mémoires relatifs à l'art de l'ingénieur et recueillis par M. Bruyère. Ponts en fer.' *ENPC mss. 233 Tome 24BIS* notebook.

Bruyère, Louis. 1823/1828. *Études relatives à l'art des constructions recueillies par L. Bruyère.* 2 vols. Paris: Bance.

Cachin, Joseph Marie François. 1820. *Mémoire sur la Digue de Cherbourg, comparée au Breakwater ou Jetée de Plymouth.* Paris: Firmin Didot.

Castor, Antoine. 1869. *Recueil d'appareils à vapeur employés aux travaux de navigation et de chemins de fer. Fondations à l'air comprimé, ponts, quais, batardeaux etc.* 2 vols. Paris: Eugène Lacroix. This is the final and complete edition of all Castor's treatises and editions with slightly varying titles as published in 1856/60/61/64/67.

Chevallier, Temple, and James F. W. Johnston. 1838. 'Education of Students in Civil Engineering and Mining in the University of Durham', in: *The London and Edinburgh Philosophical Magazine and Journal of Science,* 3rd series, July 1838, pp. 1–8.

Chevallier, Victorin. 1871. 'Notice nécrologique sur M. Minard, inspecteur général des Ponts et Chaussées en retraite', in: *Annales des Ponts et Chaussées*. 1871, 2e sem. pp. 1–22.

Christian, Gérard Joseph. 1819. *Vues sur le système général des opérations industrielles, ou plan de technonomie*. Paris: Huzard.

Christian, Gérard Joseph. 1822/23/25/25. *Traité de mécanique industrielle, ou, exposé de la science de la mécanique déduite de l'expérience et de l'observation: principalement à l'usage des manufacturiers et des artistes.* 3 vols. + atlas. Paris: Bachelier.

Clot, Louis. 1858. *La vérité bibliographique: M. Émile Pereire.* Paris: H. &. C. Noblet.

Cointeraux, François. 1790–91. *École d'architecture rurale...* [varying titles] Paris, sometimes Lyon: author and others (4 'cahiers').

Colladon, Jean-Daniel. 1893. *Souvenirs et mémoires. Autobiographie de J.-Daniel Colladon.* Geneva: Aubert-Schuchardt.

Comberousse, Charles de. 1879.1. *Histoire de l'École Centrale des Arts et Manufactures depuis sa fondation jusqu'à ce jour.* Paris: Gautier-Villars.

Comberousse, Charles de. 1879.2. *Cinquantième anniversaire de la fondation de l'école centrale des arts et manufactures: compte rendu de la fête des 20 et 21 juin 1879.* Paris: Gauthier-Villars.

Comolli, Antoine. 1879. *Les Ponts de l'Amérique du Nord. Étude, calcul, description de ces ponts. Comparaison des systèmes Américain et Européen.* 2 vols. Paris: Levévre.

Compagnie Générale de Matériels de Chemins de Fer. 1856. *Statuts.* Bruxelles: Imprimerie de Deltombe.

Contamin, Victor. 1874. *Cours de Résistance appliquée. Année scolaire 1873–74.* [Paris]. 2nd ed. 1878. Paris: J. Dejey.

Coriolis, Gaspard-Gustave. 1829. *Calcul de l'effet des machines, ou Considérations sur l'emploi des moteurs et sur leur évaluation: pour servir d'introduction à l'étude spéciale des machines.* Paris: Carilian-Goeury.

Coriolis, Gaspard-Gustave. 1835. *Théorie mathématique des effets du jeu de billard.* Paris: Carilian-Goeury.

Cornet, Germain (with Auguste Perdonnet). 1849. *Album des chemins de Fer. Résumé graphique du Cours professé par M. Auguste Perdonnet, à l'École centrale des Arts et Manufactures.* Paris: Librairie Scientifique-industrielle de L. Mathias (Augustin). 60 pls. A second edition followed in 1851, and a third in 1853.

Cornet, Germain (with Auguste Perdonnet). 1853. *Album des chemins de Fer. Résumé graphique du Cours professé par M. Auguste Perdonnet, à l'École centrale des Arts et Manufactures. Troisième édition revue et augmentée.* Paris: F. L. Mathias. 78 pls.

Coste, Léon, and Auguste Perdonnet. 1830.1. *Mémoire sur les chemins à ornières.* Paris: Bachelier.

Coste, Léon, and Auguste Perdonnet. 1830.2. *Mémoires métallurgiques sur le traitement des minérais de fer, d'étain et de plomb en Angleterre; faisant suite au voyage métallurgique de MM. Dufrénoy et Élie de Beaumont, ingénieurs des mines.* 2 vols. (text and plates) Paris: Bachelier. Second edition 1837–39 included the names of Dufrénoy and Élie de Beaumont as authors.

Coulomb. 1773. 'Sur une application des règles, de maximis et minimis à quelques problèmes de statique, relatifs à l'architecture', in: *Mémoires de mathématique et de physique présentés à l'Académie Royale des Siences, par divers savans, & lûs dans ses assemblées.* Vol. VII, 1773, Paris, 1776, pp. 343–382.

Crelle, August Leopold. 1846. *Mémoire sur les différentes manières de se servir de l'élasticité comme force motrice sur les chemins de fer.* Berlin: G. Reimer.

Cremona, Luigi. 1872. *Le figure reciproche nella statica grafica.* Milan: G. Barnardoni. Printed together with, and mostly catalogued under: Felice Casorati, *La proprietá cardinali degli strumeni ottici anche non centrati.* Further editions were published later that year under his own name (without Casorati's paper) and in 1879.

Cremona, Luigi. 1874. *Elementi di calcolo grafico: ad uso degli Istituti Tecnici.* Turin: G. B. Paravia e Comp.

Cresy, Edward. 1847. *An Encyclopædia of Civil Engineering, historical, theoretical, and practical.* 2 vols. London: Longman, Brown, Green, and Longmans.

Culmann, Carl. 1851. 'Der Bau der eisernen Brücken in den Vereinigten Staaten von Nordamerika. Ergebnisse einer im Auftrage der Königl. Bayerischen Regierung in den Jahren 1849 und 1850 unternommenen Reise durch die vereinigten Staaten'. *Allgemeine Bauzeitung.* Vienna, vol. 16 , pp. 69–129 and pls. 387–397.

Culmann, Carl. 1866. *Die Graphische Statik. Mit 235 in den Text gedruckten Holzschnitten und 36 Tafeln.* Zurich: Meyer & Zeller.

Daly, César. See *Revue générale d'architecture.*

Dartein, Fernand de. 1888. *Léonce Reynaud: Sa vie et ses œuvres.* Paris: Dunod.

Delaware Bridge Company's Illustrated Album of bridges, roofs and viaducts, with general description. 52 Wall Street New York U.S.A., The. c. 1880. New York: Evening Post Steam Presses. The company was a subsidiary of the Cooper Hewett Company and existed 1875–1883. This was its first catalogue, published around 1880. The internal title was: 'The Delaware Bridge Company, 52 Wall Street, New York, U.S.A., Engineers and Contractors for the Construction of Bridges, Roofs, Viaducts, and similar structures. C. Macdonald, President and Engineer. Wm. M. Fincke, Secretary and Treasurer'.

Dhombres, Jean. See Fourcy.

Dion, Henri de, and Louis Lasvignes. 1861. *Cathédrale de Bayeux. Reprise en sous-œuvre de la Tour Centrale par M. E. Flachat. Description des travaux par MM. H. de Dion et L. Lasvignes, Ingénieurs, Anciens élèves de l'École Centrale.* Paris: A. Morel et Cie.

Du Camp, Maxime. 1875. *Paris ses organes, ses fonctions et sa vie dans la seconde moitié du XIXe siècle*. 5th ed. 6 vols. Paris: Hachette.

Dufour, Guillaume-Henri. 1834. 'Description d'un pont construit à Genève d'après un nouveau mode de suspension', in: *Bibliothèque Universelle des Sciences, Belles-Lettres et Arts. Science et Arts*. Vol. 57. Genève: Bibliothèque Universelle, pp. 229–268.

Dufrénoy, Ours-Pierre, and Jean-Baptiste Élie de Beaumont. 1827. *Voyage Métallurgique en Angleterre, ou Recueil de mémoire sur le gisement, l'exploitation et le traitement des minerais d'étain, de cuivre, de plomb, de zinc et de fer, dans la Grande-Bretagne*. Paris: Huzard.

Duleau (= Alphonse-Jean Claude Bourguignon). 1820. *Essai théorique et expérimental sur la résistance du fer forgé, contenant Des Expériences sur des barres chargées parallèlement et perpendiculairement à leur longueur; sur un arc placé entre deux appuis fixes, et chargé en un ou plusieurs de ses pointes; sur des barres qu'on essaie de tordre; avec des applications des résultats de ces expériences à l'aide des constructions*. Paris: Ve. Courcier.

Dumas, Jean-Baptiste. 1828–46. *Traité de chimie appliquée aux arts*. 8 vols. Paris: Béchet.

Dumont, Gabriel Pierre Martin. 1774. *Parallèle de plans des plus belles salles de spectacles d'Italie et de France, avec des détails de machines Théâtrales, mis au jour par le sieur Dumont*. Paris. This formed part of Dumont's *Recueil de plusieurs parties de l'architecture sacrée et profane* (1767) and is the most comprehensive version of parts that were issued previously.

Dupin, Charles. 1813 and 1815. *Expériences sur la flexibilité, la force et l'élasticité des bois: avec des applications aux constructions en général, et spécialement à la construction des vaisseaux: faites dans l'arsenal de la marine, à Corcyre, en 1811... 1er mémoire présenté à la première classe de l'Institut de France, le 12 avril 1813*. Nimes: Imprimerie Blachier-Belle. This version was issued as a preprint. The finalized, much longer form of the paper included all the statistics, the mathematical formulation of his theory and two plates in the expanded reprint of the report in the *Journal de l'École Polytechnique*, vol. 10, 1815, pp. 137–211.

Dupin, Charles. 1818. *Mémoires sur la Marine et les Ponts et Chaussées de France et d'Angleterre, contenant deux Relations de Voyages faites par l'auteur dans les ports d'Angleterre, d'Écosse et d'Irlande, dans les années 1816, 1817 et 1818; la description de la jetée de Plymouth, du canal Calédonien, etc.* Paris: Bachelier. Translated into English the same year.

Dupin, Charles. 1819. *Essai historique sur les services et les travaux scientifiques de Gaspard Monge*. Paris: Bachelier.

Dupin, Charles. 1820–24. *Voyages dans la Grande-Bretagne, entrepris relativement aux services publics de la guerre, de la marine, et des ponts et chaussées, au commerce et à l'industrie, en 1816, 1817, 1818 et 1819* [et 1820]. Paris: Bachelier. 6 vols. (in 3), 3 atlases (often in 1). Republished twice in 1825–1826 (Bachelier) and 1827 (Bruxelles: Palmier et Jobard). Partially translated into English 1822–25. The second edition (1: 1825, 2: 1825, and the first edition of 3: 1824) is the one consulted here.

Dupin, Charles. 1825–26. *Géométrie et Mécanique des Arts et Métiers et des Beaux-Arts. Cours normal à l'usage des Artistes et des ouvriers, des Sous-Chefs et des Chefs d'ateliers et de manufactures; Professé au Conservatoire royal des arts et métiers.* 3 vols. Paris: Bachelier / Ve. Courcier.

Duplaix, Marcelin. 1897. *Constructions métalliques*. Paris: ECP. Various editions 1893–97/1902–12/1912–24/1924, the last edition with '*et en béton armé'* added.

Durand, Jean-Nicolas Louis. 1802/1805, *Précis des Leçons d'Architecture données à l'École Royale Polytechnique*. 2 vols. Paris: Firmin Didot. Further editions in: 1809, 1813, 1817, and 1819/1821.

Durand, Jean-Nicolas Louis, and François-Tranquille Gauché. 1816. *Choix des projets d'édifices publics et particuliers, composés par les élèves de l'École Royale Polytechnique dans les concours d'architecture qui ont lieu chaque année* Paris: Durand.

Durand, Jean-Nicolas Louis. 1821. *Partie Graphique des Cours d'Architecture faits à l'École Royale Polytechnique depuis sa réorganisation*. Paris: École Polytechnique.

Durand, Jean-Nicolas Louis. 1833. *Recueil et Parallèle des Édifices de Tout Genre Anciennes et Modernes, remarquables par leur beauté, par leur grandeur, ou par leur singularité. Augmenté de plus de 300 autres bâtimens et de l'histoire générale de l'architecture de J. G. LeGrand, ouvrage publié sous la direction de Mr. les Professeurs de l'Académie I. et R. des Beaux-Arts.* Venice: Giuseppe Antonelli. The first edition appeared 1799/1801. This edition was the last issued in Durand's lifetime.

Dutens, Joseph. 1819. *Mémoire sur les travaux publics de l'Angleterre, suivis d'un mémoire sur l'esprit d'association et sur les différens modes de concession et de quinze planches avec une carte générale de la navigation intérieure, indiquant les deux systèmes des grands et des petits canaux de ce pays.* Paris: Imprimerie royale.

Eck, Charles Louis Gustave. 1836/1841. Vol. 1. *Traité de Construction en Poteries et Fer, à l'usage des bâtimens civils, industriels et militaires, suivi d'un recueil de machines appropriées à l'art de bâtir. Dédié à MM. les architectes, officiers du génie, et entrepreneurs de maçonnerie et de serrurerie... Avec 66 planches gravées par Hibon.* Vol. 2: *Traité de l'Application générale du fer, de la fonte, de la tôle et des poteries dans les constructions civiles, industrielles et militaires, dans celles des ponts fixes ou suspendus, des chemins de fer, des écluses et des digues à la mer, etc., etc.; suivi d'un aperçu sur l'art d'ériger les tuyaux de cheminées en briques d'après un nouveau système dédié à tous les constructeurs... Mémoire sur la construction de nouveaux planchers destinés à rendre les bâtiments incombustibles; par feu P.-D. Bazaine... Avec 80 planches gravées par Hibon et Ad. Leblanc.* And on the half-title: *Deuxième volume. Traité de l'application du fer, de la fonte et de la tôle dans toutes les constructions.* Paris: J. C. Blosse/Carilian-Goeury et V. Dalmont, éditeurs.

École Centrale des Arts et Manufactures. 1889. *Les Anciens Élèves de l'École Centrale 1832–1888.* [Paris: ECP].

Eiffel, Gustave. 1867. *Exposition Universelle de 1867. Mémoire sur les épreuves des arcs métalliques de la Galerie des Machines, faites par ordre de la Commission Impériale.* Paris: Lith. du Sénat. Alternative title on one of the only three copies known, and the one used here: *Mémoire sur les Épreuves des arcs Métalliques de la Galerie des Machines du palais de l'Exposition Universelle de 1867. suivant décision de la Commission Impériale. par G. Eiffel, Ingénieur Civil. Constructeur à Levallois Perret près Paris.*

Eiffel, Gustave. 1889. *Mémoire présenté à l'appui du projet définitif du Viaduc de Garabit...Extrait des Mémoires de la Société des ingénieurs civils Juillet 1888.* Paris: Baudry. 2 vols. (text and plates – with variant title: *Mémoire sur le Viaduc de Garabit. Description, Calculs de résistance, Montage, Épreuves et Renseignements divers*).

Eiffel, Gustave, Victor Contamin, and Ernest Fouquet. 1889. *Note sur les constructions métalliques. Congrès international des Procédés de Construction.* Paris: Imprimerie et Librairie Centrale des Chemins de Fer / Chaix.

Eiffel, Gustave. 1900. *La Tour de Trois Cents Mètres.* 2 vols. (text and plates) Paris: Lemercier.

Fairbairn, William. 1856. *Useful information for engineers; being a series of Lectures delivered to the working engineers of Yorkshire and Lancashire; together with A series of Appendices, containing the results of experimental inquiries into the strength of materials, the causes of boiler explosions, etc. Second edition.* London: Longman, Brown, Green, Longmans & Roberts. The foreword to the first edition is reprinted in the second, that appeared the same year.

Fairbairn, William. 1857–58. *On the Application of Cast and Wrought Iron to Building Purposes. Second edition, greatly enlarged, with corrections and additions. To which is added a short treatise on wrought iron bridges.* London: John Weale. The first edition had appeared in 1854. It was expanded again from the second edition in 1864 and reprinted in 1870.

Fairbairn, William. 1864–65. *Treatise on Mills and Millwork.* 2 vols. I: *On the Principles of Mechanism and Prime Movers.* II: *On Machinery and*

Transmission and the Arrangement of Mills. Comprising treatises on wheels, shafts, and couplings; engaging and disengaging gears; and mill architecture; and on corn, cotton, flax, silk, and woollen [sic] mills; to which is added a description of oil, paper, and powder mills, including a short account of the manufacture of iron. 2nd ed. London: I: Longman, Green, Longman, Roberts & Green; II: Longmans Green & Co.

Favaro, Antonio (tr. Paul Terrier). 1879/85. *Leçons de statique graphique.* 2 vols. 1: *Première partie. Géométrie de position* (1879), 2: *Deuxième partie. Calcul graphique avec appendices et notes du traducteur* (1885). Paris: Gauthier-Villars.

Fergusson, James. 1855. *The Illustrated Handbook of Architecture: Being a concise and popular account of the different styles of architecture prevailing in all ages and countries.* 2 vols. London: John Murray.

Fergusson, James. 1862. *History of the Modern Styles of Architecture.* London: John Murray. Vol. 4 of *A History of Architecture in all countries, from the earliest time to the present day.* 2nd expanded edition 1874.

Figuier, Louis. 1857–1914. *L'Année scientifique et industrielle, ou Exposé annuel des travaux scientifiques, des inventions et des principales applications de la science à l'industrie et aux arts, qui ont attiré l'attention publique en France et à l'étranger.* From 1868 on: *accompagné d'une nécrologie scientifique.* Paris: Hachette. vol. 10 (1866 for 1865) & vol. 12 (1868 for 1867), vol. 17 (1874 for 1873), from a series of 58 that ran from 1857 to 1914.

Figuier, Louis. 1868–70. *Les Merveilles de la Science, ou Description populaire des inventions modernes.* 4 vols. Paris: Furne, Jouvet et Cie. Many undated subsequent editions.

Figuier, Louis. (tr. Sir Edward Burnett Tyler). 1870. *Primitive Man.* London: Chapman & Hall. Translation of *L'Homme Primitif.* 1870. Paris: Hachette.

Figuier, Louis, (editor). 1887–1905. *La Science illustrée, Journal Hebdomadaire publié sous la Direction de Louis Figuier.* 36 semi-annual volumes. Paris. Figuier contributed until his death in 1894. The volume used here: 1893, 2e semestre.

Flachat, Eugène. 1836. *Établissemens commerciaux: Docks de Londres, entrepôts de Paris, projets de docks à Marseille.* Paris: Levrault.

Flachat, Eugène, Alexis Barrault, and Jules Petiet. 1842–46. *Traité de la fabrication du fer et de la fonte envisagée sous les rapports chimique, mécanique et commercial.* 3 vols.+ 86 pls. (vols. 2 & 3 *...sous les trois rapports*). Paris: Librairie Scientifique-Industrielle L. Mathias. 2nd ed. 1851/52, 2 vols. + pl. vol. (92 pls.) Liège: Avanzo.

Flachat, Eugène. 1866. 'M. Alexis Barrault', obituary in: *Mémoire et compte rendu des Travaux de la Société des ingénieurs civils.* Vol. 19. 1866, pp. 534–539.

Förster, Ludwig. See *Allgemeine Bauzeitung.*

Foulques, L. 1829. 'École centrale des manufactures et des arts, à Paris', in: *Bulletin de la Société Royale d'Agriculture, Sciences et Arts de Limoges,* n° 3 tome 8, (Mai 1829) pp. 82–85.

Fraisse, William. 1838. *Communication à établir entre les lacs Léman et d'Yverdon. Au moyen d'un canal ou d'un chemin de fer. Projet d'association, pour couvrir les frais présumés des études à faire sur le terrain.* Lausanne: Imprimerie et Librairie de Marc Ducloux.

Galloway, Elijah. 1827. *History of the Steam engine from its earliest invention to the present time.* London: B. Steill.

Gayant, Paul, Léonce Reynaud, and Achille Lefort. 1870. *Discours prononcés sur la tombe de M. Mary, inspecteur général des ponts et chaussées en retraite, membre du Conseil général de la Somme, ancien professeur à l'École impériale des ponts et chaussées, et à l'École impériale centrale des arts et manufactures. 12 janvier 1870.* Paris: Dunod.

Gillespie, William Mitchell. 1870. *Notes on the strength of materials and the stability of structures...printed for the civil engineering classes in Union College.* Schenectady, NY: Wiseman, Russell & Co.

Girard, Antoine-Gervais. 1815. *Préceptes de Rhétorique, tirés des meilleurs auteurs anciens et modernes.* 5th ed. Lyon: A. Rodez. The first edition appeared in 1787.

Girard, Pierre-Simon. 1798. *Traité analytique de la résistance des solides, et des solides d'égale résistance. Auquel on a joint une suite de nouvelles expériences sur la force, et l'élasticité spécifiques des bois de chêne et de sapin.* Paris: Firmin Didot / Dupont.

Gourlier, Charles, Léon-Marie Dieudonné Biet, Edme Jean-Louis Grillon, and Eugène Tardieu. 1825–50. *Choix d'édifices publics projetés et construits en France depuis le commencement du XIXe siècle; publié avec l'autorisation du Ministre de l'intérieur.* 3 vols. Vol. 1, 1825–36; vol. 2 1837–44; vol. 3, 1845–50. Paris: Louis Colas.

Gourlier, Charles. 1848 *Notice historique sur le service des travaux des bâtiments civils à Paris et dans les départements depuis la création de ce service en l'an IV (1795). Première partie de 1795 à 1848.* Paris: Imprimerie nationale. 3rd edition 1895.

Gourlier, Charles. 1855. *Rapport sur les matériaux de construction et de décoration en substances minérales, naturelles et artificielles: Fait à la Commission française du jury internationale de l'Exposition universelle de Londres.* Paris: Imprimerie Impériale.

Guillet, Léon. 1929. *Cent ans de la vie de l'École centrale des Arts et Manufactures 1829–1929.* Paris: ECP.

Guyot de Fère, François-Fortuné, *Annuaire des artistes français. 1re année 1832.* Paris: Guyot de Fère.

Hachette, Jean-Nicolas Pierre. See Monge 1795 and 1835.

Halbou, Alphonse. 1849. See under archives: Institut nationale de la propriété industrielle INPI, Pat. no. 8040 / 1BB8040.

Haupt, Herman. 1851. *General Theory of Bridge Construction: Containing Demonstrations of the Principles of the Art and Their Application to Practice; Furnishing the Means of Calculating the Strains Upon the Chords, Ties, Braces, Counter-Braces, and Other Parts of a Bridge or Frame of Any Description. With practical illustrations.* New York: D. Appleton & Co.

Hodgkinson, Eaton. 1824. 'On the transverse strain and strength of materials', in: *Memoirs of the Literary and Philosophical Society of Manchester*. Second series, vol. IV. London.

Humber, William. 1861. *A Complete Treatise of Cast and Wrought Iron Bridge Construction, Including Iron Foundations. In three parts, theoretical, practical, and descriptive. Illustrated by numerous examples drawn to a large scale.* 2 vols. London: Spon.

Jacquin, A[ndré-Auguste?] (with Auguste Perdonnet). Undated (1864). *Cours professé par M. A. Perdonnet, Nouvel Album des Chemins de Fer*. 53 pls. Paris: Eugène Lacroix. Further editions with slightly altered text and published by Lacroix et Baudry.

Jolly, César, and Théophile Joly. 1863. *Études pratiques sur la construction des planchers et poutres en fer avec notice sur les colonnes en fer et en fonte*. 2 vols. (text and album with 9 plates). Paris: Dunod.

Journal des séances des écoles normales, recueillies par les sténographes et revues par les professeurs. 1795/96. Vol. 2. See Monge.

Jullien, Adolphe, (1832). 'France – Bruyère (Louis)', obituary in: *Revue encyclopédique, ou analyse raisonnée des productions les plus remarquables dans la politique, les sciences, l'industrie et les beaux-arts; recueil mensuel, publié sous la direction de M. Hyppolyte Carnot,* vol. 52. 1831. This volume was published early in 1832, as Bruyère died on 31.12.1831.

Kepler, Johannes. 1634. *Somnium seu opus posthumum De astronomia lunari*. Frankfurt: Sagani Silesiorum.

Kircher, Athansius. 1656. *Itinerarium extaticum quo mundi opficium.* Rome: Vitalis Mascardi.

Kircher, Athansius. 1657. *Iter extaticum II qui et mundi subterranei prodromus dicitur.* Rome: Vitalis Mascardi.

Kircher, Athansius. 1664/1665. *Mundus subterraneus, in XII libros digestus.* Amsterdam: apud Joannem Janssonium & Elizeum Weyerstraten.

Lang, Gustav. 1890. *Zur Entwicklungs-Geschichte der Spannwerke des Bauwesens. Ein Anhang zu den Lehrbüchern über allgemeine Baukunde und Brückenbau.* Riga: N. Kymmel.

Langlois, Charles-Victor, and Charles Seignobos. 1898. *Introduction aux études historiques.* Paris: Hachette.

Lanz, José Maria, Agustín de Betancourt y Molina, and Jean-Nicolas Pierre Hachette. 1808. *Programme du Cours élémentaire des machines pour l'année 1808.* Paris: Imprimerie Impériale.

Lanz, José Maria, Agustín de Betancourt y Molina, and Jean-Nicolas Pierre Hachette. 1808. *Essai sur la Composition des Machines.* Paris: Imprimerie Impériale.

La Science illustrée. See Figuier, Louis.

(Lavallée, Alphonse with Jean-Baptiste Dumas?). 1830. *Plan der Central-Schule der Künste und der Manufacturen, bestimmt, um Civil-Ingenieurs, Fabrik-Directoren, Manufactur-Chefs, &c. zu bilden Gegründet zu Paris durch den Herrn Director Lavallée, und die Herrn Professoren Dumas, Olivier, Péctet* [*sic*] *und Benoit.* Augsburg: C. Kollmann & Himmer.

(Lavallée, Alphonse with Jean-Baptiste Dumas?). 1831. *École Centrale des Arts et Manufactures. Juin 1831. – deuxième année d'existence. – 171 élèves. Les élèves qui ont reçu le diplôme en quittant l'École, restent en relation régulière avec elle; ils en reçoivent tous les renseignemens qui ont de l'intérêt pour eux, sur les industries qu'ils exploitent et sur les positions industrielles qui pourraient leur convenir.* Paris: ECP.

(Lavallée, Alphonse with Jean-Baptiste Dumas?). 1832.1. *École Centrale des Arts et Manufactures. Cette école est dirigée par M. LAVALLÉE qui est chargé spécialement de la correspondance. Les études et la discipline sont placées sous la surveillance immédiate du conseil des Professeurs.* Études. [Paris: ECP] no date but with the remark under 'Considérations générales': 'Les Cours de l'École centrale des Arts et Manufactures vont s'ouvrir pour la quatrième fois, après une expérience de trois années' (p. 1).

(Lavallée, Alphonse with Jean-Baptiste Dumas?). 1832.2. *Année Scolaire 1832–1833. L'École centrale des arts et manufactures, destinée à former des ingénieurs civils, des directeurs d'usines, des chefs de manufactures, des ingénieurs-constructeurs, etc., est établie à Paris, Hôtel de Juigné, rue de Thorigny (au Marais)* 'Notice sur l'École centrale des arts et manufactures, adressée, par une commission représentant les professeurs de cet établissement et par le directeur, au Conseil Général des Manufactures, le 27 février 1833'; 'Rapport fait à la société d'encouragement, sur l'École centrale des arts et manufactures, le 12 juin 1833, par MM. le Baron Thénard, le Comte de Lasteyrie, le Baron Ch. Dupin, Francoeur, Mérimée, Payen, Rapporteur'; 'Notice sur l'École centrale des arts et manufactures, adressée par une commission représentant les professeurs de cet établissement et par le directeur, au Conseil Générale des Manufactures, le 27 février 1833'; 'Rapport fait au Conseil Général des Manufactures, le 30 mai 1833, Par une Commission composée de MM. Bérard, Roman, Desrousseaux, Saulnier, et D'Arcet, rapporteur'. Paris: ECP.

(Lavallée, Alphonse, with Jean-Baptiste Dumas?). 1834. *École centrale des Arts et Manufactures, destinée à former des ingénieurs civils, des directeurs d'usines, des chefs de manufactures, des professeurs de sciences appliquées, etc...Personnel de l'école - Année 1834–1835.* Paris: ECP.

(Lavallée, Alphonse, with Jean-Baptiste Dumas?). 1836. 'École Centrale des Arts et Manufactures', seven-page pamphlet attached to *Annuaire pour l'an 1837, présenté au roi par le Bureau des Longitudes.* 1837. Paris: Bachelier.

(Lavallée, Alphonse, with Jean-Baptiste Dumas?). 1847. *École centrale des Arts et Manufactures, destinée à former des ingénieurs civils, des directeurs d'usines, des chefs de manufactures, des professeurs de sciences appliques, etc. Année 1847–1848.* Paris: ECP.

Lavollée, Charles-Hubert. 1872. 'L'École centrale des Arts et Manufactures', *Revue des Deux Mondes.* 2e période, vol. 99. 1872, pp. 415–438.

Leblanc, César-Nicolas. See also Walter de Saint-Ange.

Leblanc, César-Nicolas. 1830. *Choix de Modèles appliqués à l'enseignement du Dessin des Machines, avec un texte descriptif, ouvrage adopté par le Conservatoire royal des Arts et Métiers, par l'École centrale des Arts et Manufactures, etc.* 2 vols. Paris: author / Malher. Edition used here: second ed., 1838. Paris: Veuve Le Blanc.

Le Camus de Mézières, Nicolas. 1781/82. *Le guide de ceux qui veulent bâtir; ouvrage dans lequel on donne les renseignemens nécessaires pour réussir dans cet art, & prévenir les fraudes qui pourroient s'y glisser. Dédié au roi.* 2 vols. Yverdon. A second edition appeared in 1786 with a slightly variant title: *Le guide de ceux qui veulent bâtir; ouvrage dans lequel on donne les renseignemens nécessaires pour se conduire lors de la construction, & prévenir les fraudes qui peuvent s'y glisser. Dédié au roi.* Paris: B. Morin/A. Jombert.

Lévy, Maurice. *La Statique Graphique et ses Applications àux Constructions.* 2 vols. Paris: Gauthier-Villars. The second edition (1886–88) ran to 4 vols. + 4 plate vols.

Lonchampt, Adolphe-Paul. 1865. *Recueil de Problèmes posés dans les examens d'admission à l'École impériale polytechnique et à l'École impériale centrale des arts et manufactures ainsi que dans les conférences des principales écoles préparatoires.* Paris: Gautier-Villars.

Love, Georges-Henry. 1852. *Mémoire sur la résistance du fer et de la fonte; et emploi de ces métaux dans les constructions, basé principalement sur les recherches expérimentales les plus récentes faites en Angleterre.* Paris: Carilian-Gœury et V[ict]or Dalmont.

Malézieux, Émile. 1873. *Travaux Publics des États-Unis D'Amérique en 1870, Rapport de mission par M.M. ingénieur en chef professeur à l'École nationale des Ponts et Chaussées. Publié par ordre de M. le Ministre des travaux publics.* 2 vols. Paris: Dunod.

Malo, Léon. 1873. *Notice sur Eugène Flachat.* Paris: Société des ingénieurs civils.

Mandar, Charles-François. 1801. *De l'architecture des forteresses, ou de l'art de fortifier les places, et de disposer les établissements de tout genre, qui ont rapport à la guerre. Première Partie. Essai sur la fortification.* Paris: Magimel.

Mandar, Charles-François. 1818. *Détails de construction d'une maison donnés pour l'instruction de MM. les élèves de L'École Royale des Ponts et Chaussées par M. Mandar Ingénieur en chef, professeur d'architecture. Année 1818 Lithographie des Ponts.* Paris: ENPC.

Mandar, Charles-François. 1826. *Études d'Architecture Civile, ou Plans, élévations coupes et détails nécessaires pour élever, distribuer et décorer une maison et ses dépendances, publiées pour l'instruction des élèves de l'École royale des ponts et chaussées.* Paris: Carilian-Goeury.

Martelet, Émile. 1867. *Obsèques de Aug. Perdonnet Président de l'Association Polytechnique le vendredi 4 octobre 1867. Discours de M. E. Martelet au nom de l'Association Polytechnique.* Paris: Typographie Alcan-Lévy.

Mary, Louis-Charles. 1831. 'Canal de la Somme, Notice sur la fondation par caisson de l'écluse de Froissy', in: *Annales des ponts et chaussées.* 1er sem. 1831, pp. 318–334, pl. V.

Mary, Louis-Charles. 1832. 'De l'emploi du béton dans la fondation des écluses', *Annales des ponts et chaussées.* 1er semestre 1832, pp. 66–105.

Mary, Louis-Charles. 1833. 'Description d'un des ponts tournants construits sur le canal de la Somme', *Annales des ponts et chaussées.* 1er semestre 1833, pp. 243–259, pl. XL.

Mary, Louis-Charles. 1850. *Rapport de M. l'Ingénieur Mary, sur le projet de conduite et de distribution des Eaux d'Arcier, suivi de plusieurs documents Sur les moyens d'exécution de ce Projet et sur les résultats obtenus dans d'autres Villes par des travaux analogues.* Besançon: Imprimerie d'Outhenin Chalandre Fils.

Mary, Louis-Charles. 1854. *Détails pratiques sur la distribution des eaux.* Paris: ECP (course transcript). 2nd updated edition 1865.

Mary, Louis-Charles. 1862–63. *Cours de Construction professé à l'École Impériale Centrale des Arts et Manufactures 1862–1863.* Course transcript. [Paris: ECP]. The final variant of Mary's course.

Mary, Louis-Charles. 1862. 'Notice sur la vie et les travaux de M. Vicat, inspecteur général des ponts et chaussées', in: *Annales des ponts et chaussées*, 2e semestre 1862, pp. 180–216.

Mary, Louis-Charles. 1865. *École Impériale Centrale des Arts et Manufactures. Détails pratiques sur la distribution des Eaux.* Course transcript. [Paris: ECP].

Mary, Louis-Charles. 1866. *École Impériale des Ponts et Chaussées. Notes prises par les Élèves au Cours de Navigation intérieure. Mr. Mary, Professeur.* Course transcript. [Paris: ENPC].

Mary, Louis-Charles. 1868. *École Impériale des Ponts et Chaussées. Appendice aux Cours de Navigation Intérieure. Distribution des Eaux. M. Mary, Professeur.* Course transcript. [Paris: ENPC].

Mary, Louis-Charles. 1873. *Cours de routes et ponts professé à l'École centrale des Arts et Manufactures. Année scolaire 1861–1862.* Course transcript. Paris: J. Dejey et. Cie / J. Baudry. The final variant of Mary's course. The 1873 reprint has more material on bridges up until 1870 and 15 additional plates.

Mathias, Ferdinand, and Charles Callon. 1846. *Études sur la navigation fluviale par la vapeur.* Paris: K. Mathias.

Maw, William Henry, and James Dredge Jr. 1872. *Modern Examples of Road and Railway Bridges; illustrating the most recent practice of leading engineers in Europe and America.* London: Offices of 'Engineering'.

Mehrtens, Georg. 1887. *Eisen und Eisenkonstruktionen in geschichtlicher, hüttentechnischer und technologischer Beziehung.* Berlin: Ernst Toeche.

Mehrtens, Georg. 1908/1920 or 1821/1923. *Vorlesungen über Ingenieur-Wissenschaften. II. Teil, Eisenbrückenbau.* 3 vols. Leipzig: Wilhelm Engelmann.

Mémoires et compte rendu des travaux de la Société des Ingénieurs Civils fondée le 4 mars 1848. From 1860: + *reconnue d'utilité publique par décret impérial du 22 décembre 1860.* From 1894: + *Ingénieurs Civils de France…* 1 vol./year until 1874, thereafter 2 vols./year, from 1880 with separate pagination. 1898 3 vols. bound in 2. 1848ff Paris: Lacroix (later Siège de la Société).

Minard, Charles-Joseph. 1834. *Leçons faites sur les chemins de fer à l'École des ponts et chaussées en 1833–1834.* Paris: Carillan-Goeury. 2nd ed. included in his general treatise on railways. 1840.

Mitis, Ferdinand von. 1835. *Geschichte des Wiener Donau-Canales und Darstellung der Ursachen seines unvollkommen schiffbaren Zustandes.* Vienna: Fr. Beck'schen Universitäts-Buchhandlung.

Molinos, Léon, and Charles Pronnier. 1857. *Traité théorique et pratique de la Construction des Ponts Métalliques.* Paris: Lacroix et Baudry.

Molinos, Léon, and Théophile Seyrig. 1879. 'Notice sur Henri de Dion Ancien Président de la Société des Ingénieurs Civils', in: *Mémoires et compte rendu des travaux de la Société des Ingénieurs Civils de France*, Séance du 18 Avril 1879, vol. 1879.1, pp. 422–474.

Monge, Gaspard. 1773. 'Réflexions sur un tour de cartes. Par M. Monge, Professeur royal de Mathématique et de Physique à l'école du Génie', pp. 390–412. in: *Mémoires de mathématique et de physique, Présentés à l'Académie Royale des Sciences, par divers Savans, & lûs dans ses Assemblées.* Vol. VII, 1773, Paris, 1776. Paris: Imprimerie royale.

Monge, Gaspard, Alexis-Théophile Vandermonde, and Claude-Louis Berthollet. 1793. *Avis aux ouvriers en fer sur la fabrication de l'acier.* Paris: Département de la Guerre.

Monge, Gaspard. 1794. *Description de l'art de fabriquer les canons: faite en exécution de l'arrêté du Comité de salut public, du 18 pluviôse de l'an 2 de la République française, une et indivisible; par Gaspard Monge. Imprimée par ordre du Comité de salut public.* Paris: Imprimerie du Comité du Salut Public.

Monge, Gaspard (and Jean-Nicolas Pierre Hachette). 1795. 'Géométrie descriptive', in*: Journal des séances des écoles normales, recueillies par les sténographes et revues par les professeurs.* Vol. 2, pp. 338–368.

Monge, Gaspard. 1798–99 (An VII). *Géométrie descriptive: Leçons données aux Écoles normales; l'an 3 de la République, par Gaspard Monge, de l'Institut national.* Paris: Baudoin. This was the first publishing in book form. The complete course edited and completed by Jean-Nicolas Pierre Hachette was published as the second edition in 1811, with the title addition *nouvelle édition avec un Supplément par M. Hachette.* There is some confusion with another 1811 edition that has another Hachette addendum dated 1809 and a further 1812 printing.

Monge, Gaspard. 1835. *Traité élémentaire de statique, à l'usage des écoles de la marine... septième édition, conforme à la précédente. Revue par M. Hachette, Membre de l'Institut.* 7th edition. Paris: Bachelier. The first edition had appeared in 1788 and the third and last augmented one that he edited himself in 1798–99 (An VII).

Muller, Émile, and Jules Denfer. 1872. *Album de serrurerie conforme au cours de constructions civiles professé à l'École centrale des Arts et Manufactures par Émile Muller contenant l'emploi du fer dans la maçonnerie et dans la charpente en bois etc.* Paris.

Muller, Émile. 1873. *Cours de constructions civiles professé à L'École Centrale des Arts et Manufactures: extrait de la leçon sur la régulation des pressions dans un réseau de conduites à gaz.* Paris: J. Dejey & Cie.

Muller, Émile. 1875/80. *Notes du cours de constructions civiles: Détails pratiques.* See Mary 1865.

Muller, Émile. 1882. *Album du cours de constructions civiles professé à l'École Centrale des Arts et Manufactures.* Paris: Baudry / Lyon: Beaud.

Nasmyth, James (and Samuel Smiles, ed.). 1883. *James Nasmyth Engineer. An Autobiography edited by Samuel Smiles.* London: John Murray.

Navier, Claude-Louis Marie Henri. 1823. *Rapport à Monsieur Becquey, conseiller d'état, directeur général des ponts et chaussées et des mines; et*

Mémoire sur les ponts suspendus. Paris: Imprimerie royale. Republished in an expanded edition in 1830 (*deuxième édition augmentée d'une notice sur le pont des Invalides*) and translated into several languages.

Navier, Claude-Louis Marie Henri. 1826. *Résumé des leçons données à l'École des ponts et chaussées, sur l'application de la mécanique à l'établissement des constructions et des machines*. Paris: Carilian-Goeury. 2nd ed. 1833/38, 2 vols.

Navier, Claude-Louis Marie Henri. 1832. 'A Messieurs des Membres du Conseil municipal de la Ville de Paris', obituary speech for Louis Bruyère. MS. Archives nationales, Louis Bruyère F/14. Accessed online 17.08.2019.

Navier, Claude-Louis Marie Henri. 1835. *Note sur le mouvement uniforme des waggons* [*sic*] *dans les parties des chemins de fer qui sont tracées en ligne courbe. (Extrait des Annales des ponts et chaussées)*. Paris: Carilian-Goeury.

Nordling, Wilhelm. 1864. *Mémoire sur les Piles en Charpente Métallique des Grands Viaducs... Extrait des Annales des Ponts et Chaussées, tome VIII. 1864, suivi d'un appendice*. Paris: Dunod.

Olivier, Théodore. 1831. *École centrale des arts et manufactures. Cours de 1831–1832. Engrenages*. Course transcript.

Olivier, Théodore. 1836. *Mémoire sur le système des courbes à petits rayons des chemins de fer, de M. Laignel*. Paris: Huzard.

Olivier, Théodore. 1842. *Théorie géométrique des engrenages: Destiné à transmettre le mouvement de rotation entre deux axes situés ou non situés dans un même plan*. Paris: Bachelier.

Olivier, Théodore. 1843/1844. *Cours de Géométrie Descriptive*. 2 vols. 1: *Première partie. Du point, de la droite et du plan*. (1843) / Atlas (1844); 2: *Deuxième partie. Des courbes et des surfaces courbes et en particulier des sections coniques et des surfaces du second ordre*. (1844) / Atlas (1844). Paris: Carilian-Goeury et V[ict]or Dalmont. This work had been preceded by *Développements de géométrie descriptive* in 1843 and followed by four further works until 1852.

Olivier, Théodore. 1846. *De la Cause du déraillement des wagons sur les courbes des chemins de fer*. Paris: L. Mathias.

Olivier, Théodore. 1851/1852. *Mémoires de Géométrie Descriptive, théorique et appliquée*. 2 vols. Paris: Carilian-Goœury / V[ict]or Dalmont.

Pappworth, Wyatt. 1872. 'On the Fall of the Iron Dome of the Anthæum at Brighton', in: *Papers read at the Royal Institute of British Architects, Session 1871–72* [*Transactions RIBA*]. London: RIBA, pp. 138–139.

Partington, Charles Frederick. 1822. *An Historical and Descriptive Account of the Steam Engine comprising a general view of the various modes of employing elastic vapour as a prime mover in mechanics with an appendix of patents and parliamentary papers connected with the subject*. London: J. Taylor.

Péclet, Eugène. 1830. *Traité élémentaire de Physique*. 2nd ed. Paris: Malher. 2 vols. + atlas. The first edition of 1828 is not recorded, and the second edition appears to have been reissued in 1832 by Hachette.

Perdonnet, Auguste. See also Germain Cornet, André Jacquin, and Évariste Thévenin ed. 1864–67.

Perdonnet, Auguste. 1831. *Cours sur l'exploitation des mines et sur les Chemins de fer. 1831–1832*. BnF V-14297 (text) and V-4804 (atlas).

Perdonnet, Auguste. 1833. *Leçons sur les chemins de fer 1833–34*. BnF 14298 (text).

Perdonnet, Auguste, and Camille Polonceau. 1843–46. *Portefeuille de l'ingénieur des Chemins de fer*. 3 vols., atlas, 3 addenda: 1845, 1847–48, and paper 1855. 2nd ed. 1861. Paris: L. Mathias.

Perdonnet, Auguste. 1855–56. *Traité élémentaire des chemins de fer.* 2 vols. Paris: Langlois et Leclercq. 2nd ed., 2 vols., 1858–60; 3rd ed., 4 vols., 1865.

Perdonnet, Auguste, and Camille Polonceau. 1857–59. *Nouveau portefeuille de l'ingénieur des Chemins de fer*. 4 vols. (1: text; 2: documents; 3: captions; 4: 168 pls.) Paris: Lacroix-Comon/Lacroix et Baudry. 2nd ed. 1866–67.

Perdonnet, Auguste. 1859. *Notions générales sur les chemins de fer.* Paris: Lacroix et Baudry.

Perdonnet, Auguste. 1862. 'Des Grandes Inventions au point de vue des services qu'elles ont rendus ou qu'elles peuvent rendre', in: *Association Polytechnique. Entretiens populaires publiés par Évariste Thévenin 2e série (1861).* Paris: Hachette, pp. 205–283.

Perdonnet, Auguste. 1864–67, in: Évariste Thévenin, ed.

Perdonnet, Auguste. 1864. *Association polytechnique. Conférences publiques et gratuites faites dans le grand amphithéatre* [*sic*] *de l'école de médicine les dimanches 10 et 17 janvier, par Auguste Perdonnet, Président de l'Association polytechnique.* (Paris)

Perdonnet, Auguste. 1866. *Les Chemins de Fer. Conférences Populaires faites à l'Asile Impérial de Vincennes sous le Patronage de S. M. l'Impératrice.* Paris: Hachette.

Perdonnet, Auguste. 1867. *De l'Utilité de l'Instruction pour le Peuple. Conférences Populaires faites à l'Asile Impérial de Vincennes sous le Patronage de S. M. l'Impératrice.* Paris: Hachette.

Perronet, Jean-Rodolphe. 1820. *Description des projets et de la construction des ponts de Neuilly, de Mantes, d'Orléans. etc., du projet du canal de Bourgogne, pour la communication des deux mers par Dijon; et de celui de la conduite des eaux de l'Yvette et de la Bièvre a* [*sic*] *Paris. Avec soixante-seize planches. Approuvée par l'Académie royale des Sciences; dédiée au Roi Louis XVI par M. Perronet, chevalier de l'ordre du Roi, son*

Architecte, et premier Ingénieur pour les ponts et chaussées; des Académies royales des Sciences de Paris, Stockholm; de la Société royale de Londres, etc. 3rd ed. 2 vols. (text and plates) Paris: Firmin Didot Père et Fils. The first and second editions appeared with the text in a larger format in 3 vols. 1782–1783, 1786 + supplément, 1789, and 2 vols. 1788–1789.

Picard, Alfred. 1902–03. *Exposition universelle internationale de 1900. Rapport Général administratif et Technique.* 1902–03 Paris: Imprimerie nationale.

Playfair, Sir Lyon. 1852. *Industrial Instruction on the Continent. (Being the introductory lecture of the session 1853–1853).* London: Government School of Mines and of Science Applied to the Arts.

Playfair, Sir Lyon. 1872. *On teaching universities and examining boards. (Address to the Philosophical Institution of Edinburgh, January 31. 1872).* Edinburgh: Edmonston and Douglas.

Plocq, Alexandre-Alfred. 1859. *Mémoire sur l'exécution d'un pont tournant à deux volées en tôle construit, en 1857, sur une écluse de navigation maritime large de 21 mètres, à l'entrée des bassins à flot du port de Dunkerque par M. A. Plocq, ingénieur des ponts et chaussées. Extrait des Annales des ponts et chaussées, 2e cahier 1859.* Paris: Dalmont et Dunod.

Pole, William, ed. 1877. *The Life of Sir William Fairbairn, Bart. Partly written by himself. Edited and completed by William Pole.* London: Longmans Green. Reprint 1970 Newton Abbot: David & Charles.

Polonceau, Antoine-Rémi. 1829. *Observations sur les routes, suivies de propositions sur leur amélioration et sur leur entretien.* Paris: Carilian-Goeury.

Polonceau, Antoine-Rémi. 1839. *Notice sur le nouveau système de Ponts en Fonte suivi dans la construction du Pont de Carrousel.* Paris: Carilian-Goeury et V. Dalmont.

Polonceau, Camille. 1840. 'Notice sur un Nouveau Système de Charpente en bois et en fer', in: *Revue générale d'architecture et des travaux publics.* no. 1, janvier 1840, cols. 27–32 and pl. 2. Paris: Paulin & Hetzel.

Poncelet, Jean-Victor. 1826. *Cours de mécanique appliquée aux machines.* Published in 1832. Paris: École d'application de l'Artillerie et du Génie, and further edition in 2 vols. by François Xavier Kretz and Charles de Comberousse (1874/76). Paris: Gauthier-Villars.

Poncelet, Jean-Victor. 1828. *Résumé des leçons du Cours de mécanique industrielle, professé par M. Poncelet 1827–1828.* Paris: Clouet.

Poncelet, Jean-Victor. 1839/1841 (1829). *Introduction à la mécanique industrielle physique ou expérimentale.* Written in 1829 but only published in 1839 and 1841. Paris/Metz: L. Mathias/Mme. Thiel, and published again by François Xavier Kretz in 1870. Paris: Gauthier-Villars.

Pothier, Francis. 1887. *Histoire de l'École Centrale des Arts et Manufactures d'après des documents authentiques et en partie inédits*. Paris: Delamotte Fils et Cie.

Pozzo, Andrea. 1693/1700. *Perspectivae Pictorum atque Architectorum… I. Pars/II. Pars*. 2 vols. Rome: Komarek.

Prony, Gaspard-Clair François Marie Riche de. 1799 (Ventôse An VII). *Plan d'instruction des élèves de l'École nationale des ponts et chaussées, pour l'an 7 ; imprimé par ordre du Ministre de l'intérieur*. Paris: Imprimerie de la République.

Prony, Gaspard-Clair François Marie Riche de. 1800 (An VIII). *Mécanique philosophique, ou analyse raisonnée des diverses parties de la science de l'équilibre et du mouvement*. Paris: Imprimerie de la République.

Prony, Gaspard-Clair François Marie Riche de. 1802 (Germinal an X). *Instruction-pratique sur une méthode pour déterminer les dimensions des murs de revêtement, en se servant de la formule graphique*. Paris: H. L. Perronneau.

Prony, Gaspard-Clair François Marie Riche de. 1802 (Floréal an X). *Recherches sur la poussée des terres, et sur la forme et les dimensions à donner aux murs de revêtement; Suivies d'une méthode pratique, à la portée des Ouvriers qui ont quelque habitude de se servir de la règle et du compas, pour résoudre très-facilement les principaux Problèmes relatifs à la forme et aux dimensions des Murs de revêtement… imprimé Pour l'usage de l'École polytechnique et de celle des Ponts et chaussées*. Paris: Imprimerie de la République.

Rankine, William John Macquorn. 1856. *Introductory Lecture on the Harmony of Theory and Practice in Mechanics, delivered to the Class of Civil Engineering and Mechanics in the University of Glasgow, on Thursday, January 3, 1856*. London/Glasgow: Richard Griffin.

Rankine, William John Macquorn. 1858. *A Manual of Applied Mechanics*. London/Glasgow: Richard Griffin and Company.

Reibell, Félix (Joseph-Matthieu Sganzin). 1839–41. *Programme ou Résumé des leçons d'un cours de constructions: avec des applications tirées spécialement de l'art de l'ingénieur des ponts et chaussées. Ouvrage de feu M.-J. Sganzin*. 3 vols. Paris: Carilian-Goeury et V. Dalmont. Several expansions to 5 vols.

Régnauld, Paul. 1870. *Traité pratique de la construction des ponts et viaducs métalliques*. 2 vols. Paris: Dunod.

Revue générale d'architecture et des travaux publics ed: César Daly. 1840–1888. Paris: Paulin & Hetzel.

Reynaud, Léonce. 1850/1858. *Traité d'Architecture contenant des notions générales sur les principes de la construction et sur l'histoire de l'art. Partie 1: Art de bâtir. Études sur les matériaux de construction et les éléments des édifices*. (1850) / *Partie 2: Composition des édifices. Études sur l'esthétique, l'histoire et les conditions actuelles des édifices*

(1858). 2 vols + 2 plate vols. Paris: Carilian-Goeury et Victor Dalmont / Dalmont et Dunod.

Reynaud, Léonce. 1870. 'Discours' in: *Discours prononcés sur la tombe de M. Mary, Inspecteur général des Ponts et Chaussées en retraite, membre du Conseil général de la Somme, ancien professeur à l'École impériale des ponts et chaussée et, à l'École impériale centrale des arts et manufactures. 12 janvier 1870.* Paris: Dunod. See Gayant.

Reynaud, Léonce, ed. 1883. *Les Travaux Publics de la France. Routes et ponts – chemins de fer – rivières et canaux – ports de mer – phares et balises par MM. F. Lucas et V. Fournié – Ed. Collignon – H. de Lagrené – Voisin Bey – E. Allard. Ouvrage publié sous les auspices du Ministère des travaux publics et sous la direction de M. Léonce Reynaud.* The volume used here: *tome deuxième: Chemins de fer par Édouard Collignon avec 50 planches phototypées, 57 gravures et une carte en chromolithographie.* 5 vols. Paris: J. Rothschild.

Ritter, Wilhelm. *Anwendungen der graphischen Statik. Nach Prof. C. Culmann, bearbeitet von W. R.* 4 vols. 1888–1906 Zurich: Meyer & Zeller / Albert Raustein.

Rondelet, Jean-Baptiste. 1802–17. *Traité théorique et pratique de l'art de bâtir.* 8 parts in 7 vols. Paris: for the Author. Multiple reedited and supplemented subsequent editions.

Rouher, Eugène. 1862. 'Rapport à l'Empereur sur l'état actuel de l'enseignement professionnel et sur les moyens de le perfectionner', in: *Bulletin administratif de l'instruction publique.* Vol. 14 no. 162, June 1863, pp. 111–115 online: https://www.persee.fr/doc/baip_1254-0714_1863_num_14_162_13836.

Rousselet, Louis. 1888. *Nos grandes écoles militaires et civiles, illustré de 169 gravures sur bois.* Paris: Librairie Hachette et Cie.

Sganzin. See also Reibell.

Sganzin, Joseph-Mathieu. 1809. 2nd ed. École Impériale Polytechnique. *Programmes ou Résumés des Leçons d'un Cours de Construction, Avec des Applications tirées principalement de l'Art de l'Ingénieur des Ponts et Chaussées, conformément au système d'enseignement adopté par le Conseil de Perfectionnement de l'an 1806.* Paris: Goeury/Ve. Bernard. The third edition in 1821 added a plate that had been previously used as a student exercise: *Troisième édition, revue, corrigée et augmentée.* Paris: Mme. Ve. Courcier. A fourth edition was amplified and modernized in 1839–41 by Félix Reibell after Sganzin's death.

Sheilds, Francis Webb. 1861. *The Strains on Structures of Ironwork; with practical remarks on Iron Construction.* London: John Weale.

Smiles, Samuel. 1857. *The Life of George Stephenson, Railway Engineer.* London: John Murray.

Smiles, Samuel. 1859. *Self-Help: With illustrations of character and conduct.*

London: John Murray. From the second edition the same year: *With illustrations of conduct and perseverance.*

Smiles, Samuel. 1861. *The Lives of the Engineers, with an account of their principal works; comprising also a history of inland communications in Britain with portraits and numerous illustrations.* 2 vols. London: John Murray. Supplemented in 1864 with a third volume on George and Robert Stephenson. Subsequent editions expanded up to 6 vols.

Smith, Oliver. 1854. *The Domestic Architect: Comprising a Series of Original Designs for Rural and Ornamental Cottages, with full and complete Explanations and Directions to the Builder, embracing the elementary Principles of the Grecian and Cottage Styles, with primary Rules for drawing and shading, and the Rudiments of linear Perspective. Illustrated by over Two Hundred Engravings.* Buffalo: Phinney.

Stuart (Meikleham), Robert. 1824. *A Descriptive History of the Steam Engine by Robert Stuart, Esq. Civil Engineer illustrated by engravings of forty-seven engines.* London: Knight and Lacey.

Stuart (Meikleham), Robert. 1827. *Histoire descriptive de la machine à vapeur, traduite de l'anglais de R. Stuart, Précédée d'une introduction exposant la théorie des vapeurs; suivi de la description des perfectionnements faits en France, et des considérations générales sur l'emploi de ces machines.* Paris: Librairie scientifique et industrielle, Malher et Compagnie.

Sturm, Leonhard Christoph. 1713. *Gründlicher Unterricht, vor der Allen So wohl denen, welche in Bau-Sachen dem ærario vorstehen, Als auch Baumeistern, Oeconomis und curieusen Reisenden zuwissen sehr nöthigen Wissenschafft, Von Häng – oder Sprengwercken.* Schwerin/Leipzig: Russworm.

Terrier, Paul. See Favaro.

Thévenin, Évariste, ed. 1864–70. *Almanach général des chemins de fer.* 7 vols. Paris: Gosselin, and later Furne, Jouvet et Cie.

Thiollet, François. 1832. *Serrurerie et fonte et de fe, récemment exécutées. Application aux planchers et combles, aux ponts, escaliers et machines diverses etc. Ornemens en fonte ajustés aux portes, devantures de boutique, grilles, rampes d'escalier, candélabres, cheminées, poêles etc.* Paris: Bance, later and undated edition with slightly variant title, Paris: Librarie Centrale d'Architecture.

Thumeloup, Nicolas-Auguste. 1841. *École centrale des arts et manufactures. Aperçu sur le dessin architectural, accompagné de … notions élémentaires sur les proportions des ordres, des portes et fenêtres … par M. Thumeloup … Année scolaire 1840–1841.* Paris: Clouet.

Thumeloup, Nicolas-Auguste. 1842. *Leçons élémentaires d'architecture, expliquant les douze planches adoptées par l'École Centrale des Arts et Manufactures: ou aperçu sur les proportions des ordres, des portes, des fenêtres et des arcades d'après les édifices antiques et modernes d'Athènes, de Rome, de Florence, de Paris, et les ouvrages d'architecture de Vitruve, Palladio, Vignole, Scamozzi, Serlio, Bullant, Philibert Delorme, Chambray, Stuart et Revett, Percier et Fontaine, Quatremère de Quincy, Donaldson, Letarouilly, etc.* Paris: Carilian-Goeury/V. Dalmont.

Todhunter, Isaac, and Karl Pearson. 1886/93. *A History of the Theory of Materials and of the Strength of Materials from Galilei to the present Time. By the late I. T., edited and completed for the Syndics of the University Press by K. P. Galilei to Saint-Venant 1639–1850.* 3 vols. Cambridge UK: University Press.

Tredgold, Thomas, and Peter Barlow, ed. 1853. *Elementary Principles of Carpentry; a treatise on the pressure and equilibrium of timber framing, the resistance of timber and the construction of floors, centres, bridges, roofs; uniting iron and stone with timber, etc. with practical rules and examples. To which is added an essay on the nature and properties of timber, including the methods of seasoning, and the causes and prevention of decay, with descriptions of the kinds of wood used in building. Also numerous tables of the scantlings of timber for different purposes, the specific gravities of materials, etc. Illustrated by fifty-three engravings, a portrait of the author, and several wood cuts. ...fourth edition, corrected and considerably enlarged. With an appendix, containing specimens of various ancient and modern roofs.* London: John Weale.

Valérius, Benoît. 1843. *Traité théorique et pratique de la fabrication du fer avec un exposé des améliorations dont elle est susceptible, principalement en Belgique.* Brussels: Librairie scientifique et industrielle.

Vauthier, Louis-Léger. 1898. 'Notice nécrologique sur M. Ernest Deligny', in: *Mémoires et compte rendu* Paris: Société des ingénieurs civils de France. Vol. 3, pp. 418–420.

Vicat, Louis-Joseph. 1833. 'Recherches expérimentales sur les phénomènes physiques qui accompagnent la rupture ou l'affaissement d'une certaine classe de solides', in: *Annales des Ponts et Chaussées*, 2e semestre, pp. 201–268.

Vierendeel, Arthur. 1897–1902. *La Construction Architecturale en fonte, fer et acier.* 2 vols. (text and plates) Louvin/Paris: A.Uystpruyst (Lyon-Claessen) / Veuve Ch. Dunod.

Viollet-le-Duc, Eugène-Emmanuel. 1858–1868. *Dictionnaire raisonné de l'architecture.* 10 vols. Paris: B. Bance / A. Morel.

Viollet-le-Duc, Eugène-Emmanuel. 1863–1872. *Entretiens sur l'architecture.* 2 vols. + atlas. Paris: A. Morel et Cie. / Vve. A. Morel et Cie.

Walter de Saint-Ange, Charles. 1835–1838. With César Leblanc (for the plates). *Métallurgie pratique du fer, ou description méthodique des procédés de fabrication de la fonte et du fer, accompagnée de documens relatifs à l'établissement des usines, à la conduite et aux résultats des opérations ... avec atlas des machines, appareils et outils actuellement employés, renferment tous les détails nécessaires pour exécuter les constructions; dessiné et gravé par Le Blanc.* Atlas: *Métallurgie pratique du fer, ou atlas des machines, appareils et outils actuellement employés à la fabrication de la fonte et du fer, avec un texte méthodique relatif à la conduite et aux résultats des opérations, renfermant tous les détails nécessaires pour exécuter les constructions*. Paris: L. Mathias.

Weale, John. 1843. *The Theory, Practice, and Architecture of Bridges of Stone, Iron, Timber, and Wire; with examples on the principle of suspension: Illustrated by One Hundred and Thirty-Eight engravings and ninety-two wood-cuts.* 2 vols. + plate vol. London: Architectural Library (Weale).

Weale, John (ed.). 1844–49. *Quarterly Papers on Engineering. Edited and Published by John Weale* (variant title for vol. 6: *Papers on Engineering. Edited and Published by John Weale*). 6 vols. London.

Whipple, Squire. 1847. *A Work on Bridge Building: Consisting of two essays, the one elementary and general, the other giving original plans and practical details for iron and wooden bridges*. Utica: NY, H. H. Curtiss. More common is the second edition: *An improved and Enlarged Edition of the Author's Original Work.* 1869. New York: D. Van Nostrand. Reprinted unchanged in 1873 and 1883.

Wiegmann, Rudolf. 1839. *Über die Construction von Kettenbrücken nach dem Dreieckssysteme, und deren Anwendung auf Dachverbindungen.* Düsseldorf: J. H. C. Schreiner.

Willis, Robert. 1835. *Remarks on the Architecture of the Middle Ages, especially of Italy*. Cambridge UK: J. & J. J. Deighton.

Wood, Nicolas. 1825. *A practical treatise on rail-roads, and interior communication in general; with original experiments, and tables of the comparative value of canals and rail-roads. Illustrated by engravings.* London: Knight and Lacey. The second edition of 1831 and then 1832 was expanded: *A practical treatise on rail-roads, and interior communication in general. Containing an account of the performance of the different locomotive engines at and subsequent to the Liverpool contest; upwards of two hundred and sixty experiments; with tables of the comparative value of canals and rail-roads; and the power of the present locomotive engines, Illustrated by numerous engravings. Second edition.* London: Hurst, Chance and Co. / Longman, Rees, Orme, Brown, and Green. A third edition 1838 was expanded again.

Secondary Sources

A Brief History of the Institution of Civil Engineers with an Account of the Charter Centenary. Celebration June 1928 (author Henry Homan Jeffcott?). n.p., n.d. [1928 London: Institution of Civil Engineers].

Abram, Joseph. 1987. 'An Unusual Organization of Production: The building firm of the Perret Brothers, 1897–1954'. *Construction History*, vol. 3, pp. 75–93. London: The Chartered Institute of Building.

Addis, William. 1986. *Theory and Design in Civil and Structural Engineering. A Study in the History and Philosophy of Engineering. A Dissertation presented to the Department of Construction Management in Candidacy for the Degree of Doctor of Philosophy*. Unpublished manuscript, University of Reading.

Armstrong, Christopher Drew. 2017. 'The Académie royale d'Architecture (1671–1793)', in: *The Companions to the History of Architecture, Volume II, Eighteenth-Century Architecture*. Eds. Caroline van Eck and Sigrid de Jong. New York: John Wiley & Sons, pp. 1–29.

Barbin, Évelyne. 2021. *La théorie et la pratique dans les travaux et la correspondance d'Émile Clapeyron et de Gabriel Lamé (1818–1835), Bulletin de la Société des amis de la Bibliothèque et de l'histoire de l'École polytechnique (Sabix)*, n° 67, 2021. Especially part II: *Le retour à Paris: Le temps des entreprises entre science, industrie et politique (1832)*, pp. 47–70.

Barjot, Dominique, and Jacques Dureuil. 2008. *150 ans de génie civil: Une histoire de centraliens*. Paris: PUBS.

Barker, Michael. 1992. 'An Appraisal of Viollet-le-Duc (1814–1879) and his Influence', in: *The Journal of the Decorative Art Society, 1850–the Present* no. 16, pp. 3–13.

Bastian, Herbert. 2017. 'Die geheime Leidenschaft des Gaspard Monge', *Schach* 10/2017 pp. 32–43. Berlin: Exzelsior.

Bauer, Alain. 2010. *Dictionnaire amoureux de la Franc-maçonnerie*. Paris: Plon.

Belhoste, Bruno. 1989. 'Les origines de l'École polytechnique. Des anciennes écoles d'ingénieurs à l'École centrale des Travaux publics', in: *Les enfants de la Patrie. Éducation et Enseignement sous la Révolution française*. pp. 13–53. Paris.

Belhoste, Bruno, Antoine Picon, and Joël Sakarovitch. 1990. 'Les exercices dans les écoles d'ingénieurs sous l'Ancien Régime et la Révolution', in: *Histoire de l'éducation*, no. 46, pp. 53–109.

Belhoste, Bruno. 1994. 'De l'École des ponts et chaussées à l'École centrale des travaux publics. Nouveaux documents sur la fondation de l'École Polytechnique' in: *Bulletin de la Société des amis de la bibliothèque et de l'histoire de l'École polytechnique, (Sabix)* no.11, February 1994, pp. 1–69.

Belhoste, Jean-François. 'Fer, fonte et cuivre en construction. Le rôle pionnier de François-Joseph Bélanger (1744–1818)', in: Porrino, Matteo 2021.2.

Belhoste, Jean-François. 1996. 'Fabrication et mise en oeuvre du fer dans la construction: Grandes étapes d'évolution (XIII-XIXe siècles)', in: *Monumental*, no. 13, pp. 8–17.

Belhoste, Jean-François. 2004. 'L'invention du profilé riveté', in: *Le Paris des Centraliens, bâtisseurs et entrepreneurs*. Paris: Action Artistique de la Ville de Paris, pp. 67–77.

Belhoste, Jean-François, 2008. 'Les centraliens et la construction métallique de 1830 à 1914', in: Barjot and Dureuil, pp. 59–83.

Bélier, Corinne, Barry Bergdoll, and Marc Le Coeur. 2012. *Henri Labrouste. Structure Brought to Light. With essays by Martin Bressani, Marc Grinon, Marie-Hélène de La Mure, Neil Levine, Bertrand Lemoine, Sigrid de Jong, David Van Zanten, and Gérard Uniack*. New York: The Museum of Modern Art.

Berry, Sue. 2012. 'The Georgian provincial builder-architect and architect Amon and Amon Henry Wilds of Lewes and Brighton c. 1790–1850', in: *Sussex Archaeological Collection 150*, pp. 163–183.

Billaud, Baptiste, Perrine Henninger, and Mélina Martinez. 2018. *École centrale Paris: archives papier (1820–2016). Répertoire numérique détaillé.* Pierrefitte-sur-Seine: Archives nationales.

Bocquet, Denis. 2007. 'Engineers and the Nation in Italy (1750–1922): Local Traditions and Different Conceptions of Unity and Modernity', in: *History and Technology*, vol. 23, no. 3, pp. 227–240.

Bordes, Jean-Louis. 2010. 'Ernest Deligny (1842), Une vie d'ingénieur au XIXe siècle. 1820–1898', in: *Centralien* no. 604, pp. 60–63.

Bridel, Georges. 2001. 'Gustave Bridel (1827–1884)', in: *Alpenbahn-Ingenieure. Schweizer Pioniere der Wirtschaft und Technik*, vol. 69. Meilen: Verein für wirtschaftshistorische Studien. See also Bridel's obituary in *Mémoires et compte rendu* 1885.1, pp. 253–257.

Brucculeri, Antonio. 2009 'Building Construction and Architectural Practice: The Teaching of François Marie Jaÿ at the École des Beaux-Arts of Paris (1824–1863)', in: *Proceedings of the International Congress on Construction History, Cottbus*, vol.1, pp. 245–255.

Bruter, Annie. 2008. 'Le cours magistral comme objet d'histoire', in: *Histoire de l'éducation*, no. 120, 2008, pp. 5–32.

Bruter, Annie. 2011. 'Les cadres de la parole enseignante', in: *Histoire de l'éducation*, no. 130, 2011, pp. 5–29.

Carvais, Robert. 2021. 'The social status of Parisian building contractors in the 18th century. A hierarchical and ambitious professional 'body': between nobility and destitution', in: *The History of Building Trades and Professionalism. Proceedings of the Eighth Conference of the Construction History Society*, 2021. Cambridge, UK.

Cavanagh, Ted (Edwin), 2002. *Who Designed Your House? A Technological and Cultural History of Conventional Wood Construction.* Bethlehem, PA/Ann Arbor, MI: Lehigh University Press/UMI Dissertation Services.

Chalvatzi, Katerina Maria. 2020. 'Early Iron in Theatre Construction: The Théâtre du Palais-Royal in Paris', in: *Iron, Steel and Building. Studies in the History of Construction. Proceedings of the Seventh Conference of the Construction History Society.* Cambridge UK: Construction History Society, pp. 31–44.

Channell, David F. 1982. 'The Harmony of Theory and Practice. The Engineering Science of M. J. M. Rankine', *Technology and Culture*, vol. 23 (1982), pp. 39–52.

Château, Emmanuel. 2008. 'L'édition du métré et des séries de prix du Conseil des bâtiments civils', in: Jean Philippe Garric, Valérie Nègre, and Alice Thomine-Berrada, *La construction savante. Les avatars de la littérature technique. Institut national d'histoire d'art.* pp. 103–112. Paris: P. Picard.

Chatzis, Konstantinos, and Claudine Fontanon. 2011. 'L'expérimentation "en plein air" ou "grandeur nature"', in: *Documents pour l'histoire des techniques*, no. 20, 2011.

Cirio, Fabienne. 1992. 'Werner Szambien, Les projets de l'an II, concours de l'architecture de la période révolutionnaire', in: *Bulletin Monumental*, vol. 150, no. 1. 1992.

Collins, Peter. 1962. 'The Origins of Graph Paper as an Influence on Architectural Design', *Journal of the Society of Architectural Historians* no. 4, pp. 159–162.

Dale, Antony. 1947. *Fashionable Brighton 1820–1860.* London: Country Life.

Debré, Patrice. 1998. *Louis Pasteur.* Baltimore: Johns Hopkins. Original ed. 1995 Paris: Flammarion.

Denslow, William R. 1957. *10,000 Famous Freemasons*. Trenton, MO: Missouri Lodge of Research.

(Dibos, Olivier). 'Alphonse Lavallée, le fondateur de l'École Centrale'. Wikisource attached to the Wikipedia article on Lavallée. Accessed 05.11.2019.

d'Orgeix, Émilie. 2011. 'Le défilement comme stratégie éducative: Le cas de l'Enseignement militaire', in: Guy Lambert and Estelle Thibault, eds. *L'atelier et l'amphithéâtre, les écoles de l'architecture, entre théorie et pratique.* Wavre: Mardaga.

Dorn, Harold I. 1970. 'The Art of Building and the Science of Mechanics: A study in the union of theory and practice in the early history of structural analysis in England', dissertation, Princeton University. Ann Arbor MI: University Microfilms.

Dugas, René. 1988. *A History of Mechanics.* Mineola, NY: Dover. Republication of 1955 Neuchâtel: Éditions du Griffon, and in turn, a translation of *Histoire de la mécanique*. 1950 Neuchâtel: Éditions du Griffon.

Dupont, Jean-Yves. 2000. 'Le cours de Machines de l'École Polytechnique, de sa création jusqu'en 1850. Première partie – La genèse d'un enseignement: 1794–1806', in: *Bulletin de la Société des amis de la Bibliothèque et de l'histoire de l'École polytechnique (Sabix),* 25/2000, pp. 1–11.

Dupont, Jean-Yves. 2009. 'Les créateurs de l'École centrale des Arts et Manufactures. 1828–1839', in: *Centraliens 590,* (Dec. 2008).

Fargues, Madeleine. 2004. *Armand Moisant. De l'architecture métallique aux fermes modèles tourangelles*. Saint-Cyr-sur-Loire: Alan Sutton.

Fedorov, Sergey G. 1992. 'Matthew Clark and the origins of Russian structural engineering, 1810–1840s: An introductory biography', in: *Construction History* vol. 9. London: The Chartered Institute of Building, pp. 69–88.

Fedorov, Sergey G. 2000. *Wilhelm von Traitteur. Ein badischer Baumeister als Neuerer in der russischen Architektur 1814–1832. Zur Entwicklung der deutsch-russischen Beziehungen im Bauwesen der ersten Hälfte des 19. Jahrhunderts.* Berlin: Ernst and Sohn.

Fedorov, Sergey G., Bernard Heres, and Werner Lorenz. 2022. *Eiserne Eremitage. Bauen mit Eisen im Russland der ersten Hälfte des 19. Jahrhunderts.* 2 vols. Berlin: Ernst and Sohn.

Foce, Federico. 2009. '*Leçons* a confronto. Progressi della meccanica applicata alle costruzioni dal *Résumé* di Navier (1826, 1833) al *Cours lithographié* (1837) di Saint-Venant', in: *Actas del Sexto Congreso Nacional de Historia de la Construccíon*, Valencia, pp. 479–490. (Thanks to Antonio Becchi for this source.)

Fourcy, Ambroise. 1828. *Histoire de l'École Polytechnique* (1987 ed.) *Annexes par Jean Dhombres.*

Garric, Jean-Philippe. 2011. 'L'Académie Royale d'Architecture aux origines de l'art de la composition', in: Guy Lambert & Estelle Thibault, eds. *L'atelier et l'amphithéâtre, les écoles de l'architecture, entre théorie et pratique.* Wavre: Mardaga.

Gasparini, Dario, and Stephen Buonopane. 2019. 'Navier's 1823 Mémoire: Analyses of displacements, stiffness, and vibration of suspension bridges'. *ASCE Journal of Bridge Engineering* vol. 24, no. 7, July 2019.

Gillmor, Charles Stewart. 1971. *Coulomb and the Evolution of Physics and Engineering in Eighteenth-century France.* Princeton, NJ: Princeton University Press.

Gourisse, Daniel, n.d. 'II. L'Histoire de l'École Centrale Paris', pp. 49–73. Accessed online 03.11.2019.

Grattan-Guinness, Ivor. 1990. *Convolution in French Mathematics 1800–*

1840. Vol. 2: *The Turns*, Series: *Science networks historical studies*, vol. 3. Basel: Birkhäuser.

Grelon, André. 1994. 'Comberousse, Charles (de) (1826–1897). Professeur de Travaux agricoles et de Génie rural (1882–1897)', in: *Les Professeurs du Conservatoire national des arts et métiers. Dictionnaire biographique 1794–1955,* vol. 1 A-K. Paris, pp. 340–342.

Grelon, André. 2000. 'Du bon usage du modèle étranger: la mise en place de l'École centrale des arts et manufactures', in: *Bulletin de la Société des amis de la Bibliothèque et de l'histoire de l'École polytechnique (Sabix-278)*, Bulletin n° 26, 2000, pp. 47–52. Reprint numbered pp. 1–7.

Guedes, Pedro Paulo d'Alpoin. 2010. *Iron in building 1750–1855: Innovation and cultural resistance*. Unpublished PhD Thesis, University of Queensland, Australia, accessed online 20.10.2020.

Guillerme, Jacques. 2008. *L'Art du projet. Histoire, technique, architecture.* Wavre: Mardaga. Individual articles cited in the footnotes.

Hamon, Françoise. 1997. *Construire pour l'industrie. L'enseignement de l'architecture industrielle à l'École centrale des Arts et Manufactures 1832–1914.* Unpublished PhD thesis, Paris 4 1997. Cited in Nègre, Valérie 2011.

Herbert, Gilbert. 1978. *Pioneers of Prefabrication. The British Contribution in the Nineteenth Century*. Baltimore: Johns Hopkins.

Hornstein-Rabinovitch, Shelley. 1990. 'Architecture on the Edge: Stephen Sauvestre, the École centrale d'architecture, and marginalist practice', *The Journal of Architectural and Planning Research.* Chicago: Locke Science Publ. Co., 7:3. 1990, pp. 209–221.

Huerta, Sanitago. 2015. 'Designing by 'Expérience': Lecreulx Model Tests for the Design of the Abutments of the Bridge of Fouchard', in: D. Aita et al., eds. *Masonry Structures: Between Mechanics and Architecture.* Switzerland: Springer, pp. 21–47.

Jacomy, Bruno. 1984. 'À la recherche de sa mission. La Société des Ingénieurs Civils.' pp. 209–219, in: *Culture technique. Les ingénieurs.* N° 12, mars 1984. CRCT: Neuilly-sur-Seine.

James, John G. 1986. 'The Origins and Worldwide Spread of Warren Truss Bridges in the Mid-Nineteenth Century. Part I. Origins and Early Examples in the UK', in: *History of Technology*, ed. Norman Smith, vol. 11, London: Bloomsbury.

Jelliss, George Peter. 2019. *Chronology and Bibliography of Tours. Knight's Tour Notes*. Vol. 12.

Jewett, Robert A. 1967. 'Structural Antecedents of the I-beam 1800–1850', *Technology and Culture* vol. 8, no. 3, pp. 346–362. Baltimore: Johns Hopkins.

Johnson, H. R., and Alec W. Skempton. 1956. 'William Strutt's Cotton Mills. 1793–1812', in: *Transactions of the Newcomen Society* vol. 30, pp. 179–205.

Kaufmann, Emil. 1952. 'Three Revolutionary Architects, Boullée, Ledoux, Lequeu', in: *Transactions of the American Philosophical Society* vol. 53, no. 3. 1952, pp. 433–560.

Kosykh, Aleksandra, Werner Lorenz, and Konrad Frommelt. 2018. 'The Roof of the Marble Palace in Saint-Petersburg: A structural iron ensemble from the 1770s', in: *6ICHH Building Knowledge, Constructing Histories*. Brussels. Vol. 2, pp. 809–817.

Kosykh, Aleksandra. 2021. *Pioneers of Modern Construction. Building with Iron in 18th-Century Russia*. Unpublished manuscript of dissertation, Fakultät für Architektur, Bauingenieurwesen und Stadtplanung der Brandenburgischen Technischen Universität Cottbus-Senftenberg.

Krafczyk, Christina. 2016. *Constantin Uhde. Bauen in Braunschweig*. Braunschweig: Braunschweigische Geschichtsverein (*Quellen und Forschungen zur Braunschweigischen Landesgeschichte, Band 50*).

Kurrer, Karl-Eugen. 2002. *Geschichte der Baustatik*. Berlin: Ernst & Sohn.

Kurrer, Karl-Eugen. 2018. *The History of the Theory of Structures*. Berlin: Ernst & Sohn, 2nd English edition (translation and expansion of *Geschichte der Baustatik* 2002).

Lacourt, Anne. 2016. 'Liste générale des élèves du corps et des élèves civils de l'École des ponts et chaussées 1744–1930 et des élèves des cours préparatoires 1875–1920'. École nationale des ponts et chaussées, Direction de la documentation, des archives et du patrimoine, Mission archives, consulted online and at https://www.ecoledesponts.fr/sites/ecoledesponts.fr/files/documents/eleves_corps_civils_libres_courspreparatoires_1744_1930_fusionnee_maj16.01.2020.pdf (mise à jour janvier 2020).

Lavallée, Jacques. 1983. 'Le coton de Louisiane et la fondation de l'École Centrale', in: *France Louisiane*. Communicated by Olivier Dibos 31.12.2019.

Launay, Louis de. 1933. *Un Grand Français. Monge, Fondateur de l'École polytechnique*. Paris: Pierre Roger.

Lemoine, Bertrand. 1995. 'L'entreprise Eiffel'. *Histoire, Économie et Société*. 14 (1995), pp. 273–285, reproduced in: Robert Thorne, ed., *Structural Iron and Steel. 1850–1900*, vol. 10. *Studies in the History of Civil Engineering*. 2000 Aldershot: Ashgate, pp. 247–259.

Lewis, G. M. 1989. 'Gustave Eiffel—the Career of an Entrepreneur', *The Structural Engineer* vol. 67, No. 13/4, July 1989, pp. 252–260.

Lorenat, Jemma. 2019. 'Je n'ai point ambitionnée *[sic]* d'être neuf: Modern Geometry in Early Nineteenth-Century French Textbooks', pp. 69–122, in: Gert Schubring, *Interfaces between mathematical practices and mathematical education*. Springer. Series: International Studies in the History of Mathematics and its Teaching.

Luquet, Gérard-Christian, Agnès Jacotot, and Olivier Dibos. 2015. 'Arbre généologique partiel et simplifié de la lignée des trois Alphonse Lavallée'. Folding inset of the article *Éléments de biographie sur la vie et l'œuvre d'Alphonse Francis Marie Martin Lavallée (1873–1954) dit Alphonse Lavallée, Alexanor; Revue des lépidoptéristes français* 26 (6) 2014.

Madrazo, Leandro. 1994. 'Durand and the Science of Architecture', in: *Journal of Architectural Education* vol. 48, no. 1 (Sept. 1994), pp. 12–24.

Marrey, Bernard. 1984. *Un capitalisme idéal*. Paris: Clancier-Guénaud.

Mattatia, Ronald. 2018. 'L'abeille et le piston. De 1830 à nos jours', in: *Centraliens*, n° 658, mars-avril 2018. Accessed online 30.04.2020.

Maurer, Bertram. 1998. *Karl Culmann und die graphische Statik. Anhang mit umfangreichen Culmann-Texten. Bericht Nr. 26 (1998) Institut für Baustatik der Universität Stuttgart. Professor Dr.-Ing. E. Ramm*. Berlin, Diepholz, Stuttgart: Verlag für Geschichte der Naturwissenschaft und Technik.

Mende, Michael. 2006. 'The Crucial Impact of Improvements in both Steelmaking and Rolling on 19th and Early 20th-Century Building Construction', in: *Proceedings of the Second International Congress on Construction History*. Cambridge UK: Construction History Society, vol. 2. pp. 2159–2170.

Michel, Jean. 1981. 'The genealogy of the 'Grandes Écoles': Origins and development of the French system for the training of engineers', in: *European Journal of Engineering Education*, 5. 1981, pp. 189–214.

Middleton, Robin. 1999. 'The Iron Structure of the Bibliothèque Sainte-Geneviève as the Basis of a Civic Décor', p. 36 in: AA files, no. 40 (Winter 1999) pp. 33–52.

Montel, Nathalie. 1995. 'Alexandre Lavalley, Profession: ingénieur civil', in: *Histoire, Économie et Société,* vol. 14, no. 2, pp. 215–228. See also obit Lavalley in *Mémoires et compte rendu* 1892.2, pp. 13/320–321/1002–1013.

Nègre, Valérie. 2009. 'Some Considerations on Traité de l'Art de Bâtir by Rondelet and the Technical Literature of his Time', in: *Proceedings of the Third International Congress on Construction History*, May 20–24, 2009, Cottbus, Germany, pp. 1089–1096.

Nègre, Valérie. 2011. 'Architecture et construction dans les cours de l'École centrale des Arts et Manufactures (1833–1864) et du Conservatoire national des arts et métiers (1854–1894)', in: Jean Philippe Garric, *Bibliothèques d'atelier. Édition et enseignement de l'architecture, Paris 1785–1871*. Paris: Institut national d'histoire de l'art.

O'Connor, John Joseph, and Edmund Frederick Robertson. 2002. 'Théodore

Olivier' *MT Mac Tutor*, mathhistory.st-andrews.ac.uk online. Accessed 23.12.2020.

Peters, Tom F. 1987. *Transitions in Engineering, Guillaume Henri Dufour and the Early 19th Century Cable Suspension Bridges*. Basel: Birkhäuser.

Peters, Tom F. 1989. 'An American Culture of Construction', in: *Perspecta 25*. New York: Rizzoli.

Peters, Tom F. 1996. *Building the Nineteenth Century*. Cambridge MA: MIT.

Peters, Tom F. 2000. 'Technological Thought is Design's Operative Method', in: *Perspecta 31*, pp. 118–129. Cambridge, MA: MIT.

Peters, Tom F. 2002. *Illustrated Bibliographical List of the Works of Louis Figuier (1819–1894) with an illustrated account of science and technology from Figuier's viewpoint. Compiled and annotated by Tom F. Peters, edited by Philip A. Metzger*. Bethlehem, PA: Special Collections, Lehigh University Library and Technology Services.

Peters, Tom F. 2003. 'Technological Thought and theory, a culture of construction', in: *Proceedings of the First International Congress on Construction History* vol. 3, pp. 1629–1637. Madrid: Insitituto Juan de Herrera, Escuela Tecnica Superior de Arquitectura de Madrid.

Peters, Tom F. 2010. 'How the Introduction of Iron in Construction Changed and Developed Thought Patterns in Design,' in: *Before Steel: The introduction of structural iron and its consequences*, pp. 127–152. Sulgen/Zurich: Niggli.

Peters, Tom F. 2015. 'Religious Affiliation and Wooden Truss Construction in the German-Speaking World', in: *Proceedings of the Fifth International Congress on Construction History, Chicago, July 2015*, vol. 3, pp. 117–125. n.p.: Construction History Society of America.

Peters, Tom F. 2018, 'The El-Kantara bridge in Algeria: An early development of cantilevering construction and what we can learn from its documentation in text and image (1862–63)', in: *Construction History* vol. 33, no. 1, pp. 69–82. Cambridge: The Construction History Society.

Peters, Tom F. 2020. 'Louis Bruyère's innovative iron bridges and their contribution to the development of modern iron construction' in: *Iron, Steel and Buildings. Studies in the History of Construction. Proceedings of the Seventh Conference of the Construction History Society*. Cambridge UK: CHS.

Pfammatter, Ulrich. 1995. *Ursprung, Entwicklung und Bedeutung des polytechnischen und industriellen Unterrichtsmodells. Elemente zur modernen Architekten- und Ingenieurausbildung*. Manuscript. ETH Zurich 1995.

Pfammatter, Ulrich. 1997. *Die Erfindung des modernen Architekten*. Basel: Birkhäuser.

Picon, Antoine. 1988. *Architectes et Ingénieurs au Siècle des Lumières.* Paris: Éditions Parenthèses.

Picon, Antoine, and Michel Yvon. 1989. *L'Ingénieur Artiste.* Paris: Presse des Ponts et Chaussées.

Picon, Antoine. 1994. 'Christian Gérard Joseph (1778–1832) Directeur du Conservatoire (1816–1831)', in: *Les professeurs du Conservatoire national des arts et métiers. Dictionnaire biographique 1794–1955*, tome I, A-K, pp. 332–336. Paris: Institut national de recherche pédagogique.

Picon, Antoine. 1995. 'Charles-François Mandar (1757–1844) ou l'architecture dans tous ses détails', in: *Revue de l'Art* 1995, vol. 109, pp. 26–39.

Picon, Antoine. 2000. 'From "Poetry of Art" to Method: The Theory of Jean-Nicolas-Louis Durand', introduction to the edition *Précis of the Lectures on Architecture.* Los Angeles: Getty Research Institute.

Picon, Antoine. 2009. 'The engineer as judge: Engineering analysis and political economy in eighteenth century France', in: *Engineering Studies* 1, pp. 19–34.

Pierrat, Emmanuel, and Laurent Kupferman, et al. 2009. *Le Paris des Francs-Maçons.* Paris: Le Cherche midi.

Porrino, Matteo. 2021.1. 'Designing a ground-breaking structure: Notes on the cast-iron/wrought-iron dome of the former Halle aux Blés 1809–1813', in: *History of Construction Cultures,* Boca Raton: CRC Press, pp. 677–684.

Porrino, Matteo. 2021.2. *Les matériaux métalliques. Histoire d'une technique et sauvegarde du patrimoine du 19e siècle. Sous la direction de Matteo Porrino.* Paris: Infolio.

Portuondo, María M. 2009. *Secret Science. Spanish Cosmography and the New World.* Chicago: University of Chicago Press.

Rieber, Alfred J. 1990. 'The rise of engineers in Russia', in: *Cahiers du Monde russe et soviétique* vol. 31, no. 4, pp. 539–586.

Roegel, Denis, 2011. 'The great logarithmic and trigonometric tables of the French Cadastre: A preliminary investigation', http://locomat.loria.fr/cadastre/analysis.pdf. Accessed online 17.10.2024.

Ruggeri, Nicola. 2011. 'Il legno nel '700: Aspetti meccanici e d'anatomia', in: *Bollettino Ingegneri*, no. 6 pp. 3–16. (Thanks to Antonio Becchi for this source.)

Saddy, Pierre. *Henri Labrouste architecte 1801–1875.* Paris: Caisse nationale des Monuments historiques et des Sites.

Sakarovitch, Joël. 1988. *Épures d'architecture. De la coupe des pierres à la géométrie Descriptive XVIe - XIXe siècles.* Basel etc: Birkhäuser.

Sandras, Agnès. 2020. 'Auguste Perdonnet, soutien efficace des premières bibliothèques populaires? (Un ingénieur des chemins de fer passionné par l'instruction du peuple)' bai.hypotheses.org online (consulted 25.01.2021). Open source Bibliothèques des amis de l'instruction, bibliothèques au XIXe siècle, billets, personnalités.

Scheurer, Werner. 1991. 'Professor Hermann Josef (P. Adrian Opraem) Eisenmann (1758–1838), der Sohn des Haslacher "Apostelmalers" Bernhard Melchior Eisenmann', in: *Die Ortenau: Zeitschrift des Historischen Vereins für Mittelbaden* vol. 71, pp. 448–465.

Seitz, Frédéric. 2014. *Gustave Eiffel: Le triomphe de l'ingénieur*. Paris: A. Collin.

Skempton, Alec W. 1956. 'The Origin of Iron beams', in: *Actes du VIIIe Congrès Internationale d'Histoire des Sciences*. Florence.

Smith, Stanley. 1992. 'The Design of Structural Ironwork 1850–1890: Education, Theory and Practice', in: *Construction History* vol. 9. London: The Chartered Institute of Building, pp. 80–108.

Steiner, Frances H. 1984. *French Iron Architecture*. Ann Arbor: UMI Research Press.

Sutherland, R. James M. 1963–64. 'The introduction of structural wrought iron', in: *Transactions of the Newcomen Society* 36, London, pp. 67–84.

Szambien, Werner. 1984. *Jean-Nicolas-Louis Durand. 1760–1834. De l'imitation à la norme*. Paris: Picard.

Taite, James. 2022. 'The Fox & Barrett floors of Parliament Hill, Ottawa', in: *The Construction Historian* 11, Cambridge UK: The Construction History Society.

Tardini, Chiara. 2014. *Toward Structural Mechanics Through Wooden Bridges in France 1716–1841*. Cham etc: Springer.

Timoshenko, Stepan. 1953. *History of Strength of Materials. With a brief account of the history of theory of elasticity and theory of structures.* 1983 ed. New York: Dover.

Tresch, John. 2012. *The Romantic Machine: Science and Technology After Napoleon.* Chicago: University of Chicago Press.

Troitsch, Ulrich. 2007. 'Reiner Daelen (1813–1887) und die Walzwerktechnik. Ein Pionier der westdeutschen Eisenindustrie im 19. Jahrhundert', *Ferrum. Nachrichten aus der Eisenbibliothek* 79, pp. 26–45 Schlatt: Eisenbibliothek.

Tufte, Edward R. 1983. *The Visual Display of Quantitative Information.* Cheshire CT: Graphics Press.

Turak, Theodore. 1970. 'The École Centrale and Modern architecture: The Education of William Le Baron Jenney', in: *Journal of the Society of Architectural Historians* 29, no. 1, pp. 40–47.

Vacant, Claude. 2006. *Jean Rodolphe Perronet (1708–1794): Premier Ingénieur du Roi et directeur de l'École des Ponts et Chaussées*. Paris: Presse de l'ENPC.

Van Zanten, David. 1978. 'Félix Duban and the Buildings of the Ecole des Beaux-Arts 1832–1840', in: *Journal of the Society of Architectural Historians* vol. 37, no. 3, pp. 161–174.

Vérin, Hélène. 1984. 'Le mot ingénieur.', in: *Culture technique. Les ingénieurs.* N° 12, pp. 19–26, mars 1984. CRCT: Neuilly-sur-Seine.

Vérin, Hélène. 1993. *La Gloire des Ingénieurs: L'intelligence technique du XVI*[e] *au XVIII*[e] *Siècle.* Paris: Albin Michel.

Vierling-Claassen, Angela. 2017. 'Models of Surfaces and Abstract Art in the Early Twentieth Century', in: *Aesthetics of Interdisciplinarity: Art and Mathematics*. Cham: Birkhäuser, pp. 199–207. (Previously published as conference proceedings in 2010.)

Weiss, John Hubbel. 1982. *The Making of Technological Man.* Cambridge, MA: MIT.

Wetzg, Volker. 2004. 'Defizite der Ingenieurausbildung', in: *Beton- und Stahlbetonbau* 99 Heft 5. Berlin: Ernst & Sohn.

Wiebenson, Dora. 1973. 'The Two Domes of the Halle au Blé [*sic*] in Paris', in: *The Art Bulletin* vol. 55, no. 2, pp. 262–279.

Index of personal names

Page numbers in italics indicate figures

About the author

How creators think has preoccupied Tom F. Peters for over half a century, ever since he studied architecture at the ETH Zurich where he was drawn toward the engineering side of construction. As a result, he spent his career trying to 'coordinate' engineering and architectural thought, gravitating to history, where he was among the earliest researchers in construction history and one of the inadvertent founders of the field. Moreover, as a border-crosser between cultures—he was born in the United States, attended school in Mumbai and then Zurich, worked there and in Copenhagen and London, and taught at the ETH and various universities in the United States and Hong Kong—he is fascinated by the cultural variety of construction thought. This book, based as it is in nineteenth-century French education that functioned as a vehicle for the development of iron construction, is at the same time a reflection of the worldwide diversity of construction thinking.